Environment, Man and Economic Change

Stanley Henry Beaver, M.A., F.G.S., F.R.Met.S.

Photo: D. G. C. Judd

Essays presented to S. H. Beaver

Environment, Man and Economic Change

Edited by A. D. M. Phillips and B. J. Turton

Longman
LONDON and NEW YORK

LONGMAN GROUP LIMITED
London

Distributed in the United States of America
by Longman Inc., New York

Associated companies, branches and representatives throughout the world

First published 1975

ISBN 0 582 50114 8
Library of Congress Catalog Card Number: 74–81136

Set in IBM Baskerville 11 on 12 point

Printed in Great Britain by Whitstable Litho Ltd

Contents

Contents

Preface

This volume is published to commemorate the retirement of S. H. Beaver as professor and first head of the Department of Geography at the University of Keele, an event which coincided with the twenty-fifth anniversary of the establishment of this department. In an academic career covering more than forty years at the London School of Economics and at the University of Keele, S. H. Beaver has taught and been a colleague of innumerable geographers and it would have been an herculean task to have assembled a volume of essays by all those ever associated professionally with him. In this volume, therefore, the editors decided to restrict the list of contributors to those connected with him during his tenure of the chair at Keele. With one exception, the contributors have been colleagues and geography undergraduates at Keele who are now university geographers. The exception is M. J. Wise, who succeeded S. H. Beaver as Sir Ernest Cassel Reader in Economic Geography at the London School of Economics and who represents a link between Beaver's London and Keele appointments.

The topics of the essays have been chosen by the authors themselves: the editors have not sought to impose restrictions beyond those of space and some measure of standardisation of format. The wide scope of the contents is a pleasant and appropriate reflection of S. H. Beaver's broad involvement in geography. Although the essays present this great diversity of subject matter, they fall into three main sections – studies in the physical, the historical and the socio-economic environments.

We are indebted to our fellow contributors for their cooperation, enthusiasm and patience at all stages in the preparation of this volume. For information on S. H. Beaver's career before his appointment at Keele, we are grateful to Professors H. C. Darby and M. J. Wise. We would like to thank Mr G. Barber, Mrs M. Patrick and Miss J. Pass, members of the Cartographic Unit of the Department of Geography at Keele, for their advice on and preparation of the illustrations in this volume. All figures were drawn and photographed by the unit with the exception of Figs 5.1 to 5.7, which were prepared by Mrs K. Dalton and Misses A. Pritchard, R. Bate and P. C. Catt of Bedford College London. We are also greatly indebted to Mrs S. Beech for retyping the final version of the essays. All of us must thank the staff of Longman for their encouragement, advice and skill.

Keele
19 March 1975

A. D. M. Phillips
B. J. Turton

Acknowledgements

S. H. BEAVER: an appreciation	The author is indebted to Professor R. O. Buchanan and Mr D. J. Sinclair for reading and commenting on his manuscript.
Chapter 1.	The author acknowledges the assistance of Professor K. Cameron in checking the derivation of certain terms, and of Mr F. A. Barnes.
Chapter 2	The authors wish to thank Keele University first-year geography students from 1968 to 1973 for contributing as groups to the collection of data at St Mary's Bay. The assistance of Dr H. M. Tinsley and of Dr A. Polwart is also acknowledged.
Chapter 4.	The author is indebted to Dr G. Boulton and to Dr J. Shaw for critical reviews of the draft manuscript.
Chapter 5.	The author wishes to thank Messrs T. E. Beaumont, N. Bell and C. F. Page for the collection and analyses of data which formed the bases of Figs 5.4, 5.5, 5.6, 5.8 and 5.9, and Mr J. Challinor and Dr B. K. Holdsworth for information on which Fig. 5.3 was based.
Chapter 6.	The author is indebted to Mr G. Bostock for access to the pollen analysis sites. Dr S. R. Eyre kindly commented on the manuscript and the radiocarbon dating was financed by the Department of Geography, University of Leeds.
Chapter 9.	The author wishes to acknowledge permission granted by His Grace the Duke of Portland to consult archives in his possession at Welbeck Abbey. He thanks Mr B. E. Coates for obtaining a copy of Fig. 9.1.
Chapter 10.	The author wishes to acknowledge permission granted by His Grace, the Duke of Northumberland to consult archives housed in Alnwick Castle.

Chapter 11. The author is indebted to Major Robb for permission to consult archives housed at Great Tew relating to the Boulton estate. He is grateful to Mr F. V. Emery for commenting on a draft of this chapter.

Chapter 23. The author wishes to acknowledge permission granted by Dr D. J. Evans to base Figs 23.1 and 23.2 on his research.

For permission to reproduce photographs the editors are indebted to Aerofilms for Plate 3; to the Director, Institute of Hydrology for Plate 5; to the Chief Engineer, Trent River Authority for Plate 6. They wish to thank the Editor, *North Staffordshire Journal of Field Studies* and the Controller, HM Stationery Office for permission to publish Figs 8.2, 9.4 and 9.6 based on copyright material.

Abbreviations

BM	British Museum
BPP	British Parliamentary Papers
ME	Middle English
NRO	Northumberland Record Office
OD	Old Danish
OE	Old English
ON	Old Norse
ORO	Oxfordshire Record Office
OW	Old Welsh
OWS	Old West Scandinavian
PRO	Public Record Office
SHC	William Salt Collections for a History of Staffordshire
SRO	Staffordshire Record Office
Sw	Swedish

Abbreviations for titles of periodicals cited in the references are those sanctioned by the *World List of Scientific Periodicals* (4th edn.). In those essays in historical geography, the original areal and financial units have been maintained with no attempt to provide contemporary equivalents.

A. D. M. Phillips and B. J. Turton

S. H. Beaver: an introduction

Stanley Henry Beaver retired from the chair of geography in the University of Keele in September 1974. He was the foundation professor, being appointed in 1950, and his retirement coincides happily with twenty-five years of geography at Keele. During an academic career spanning more than forty years, first at the London School of Economics and then at Keele, he has established himself as one of the leading and most distinctive geographers of his generation. The purpose of this introduction is to review his career and his achievements.

Born in 1907 in Willesden, northwest London, Beaver was educated at Kilburn Grammar School where he came under the influence of C. B. Thurston, one of the outstanding geography teachers and textbook writers of the period after the First World War, and entered University College London in 1925. In a year of nine students, his abilities were recognised with the awards of the Morris Prize for Geology in 1926 and a University Scholarship in Geology in 1927–28. He emerged in 1928 with first-class honours in geography. At University College London, the staff then included Professors E. J. Garwood and L. W. Lyde, with Margaret Shackleton, H. A. Matthews and C. Daryll Forde. As an undergraduate Beaver not only read a range of general subjects but was also introduced to a wide spectrum of geographical interests: such a broad approach has remained one of the most distinctive characteristics of his published work and constitutes his opinion of an acceptable undergraduate course in geography. He was particularly influenced by the geological interests of Garwood and by Lyde's philosophy of geography as 'a body of knowledge coupled with an attitude of mind'.

After studying for a diploma in education, Beaver secured his first academic post in 1929 with his appointment as assistant in the geography department of the London School of Economics. He was promoted to lecturer in 1932 and was appointed Sir Ernest Cassel Reader in Economic Geography in 1946. During his first two years at the School, he developed his geological and economic interests with a paper on the Jurassic scarplands and an unpublished investigation of the east Midlands ironstone industry, 'The geography of the iron industry of Northamptonshire, Rutland and south Lincolnshire', which was awarded a master of arts degree in 1931. Perhaps the greatest achievement of these early years was the completion with L. Dudley Stamp of *The British Isles: a geographic and economic survey*, first published in 1933. Beaver was responsible for the chapters on iron and steel, mining other than coal and iron, non-ferrous metals, textiles and communications and, for their time, these were remarkable syntheses of the factors involved in the location and development of these industries. In the same year, Beaver married Elsie Barbara Rogers, who also had been a geography student at University College London, graduating in 1930.

In the 1930s he established his interests in field work and in transport geography. This period in the Joint School of Geography (King's College and the London School of Economics) marked the development of the theory and practice of geographical field work, and S. H. Beaver, L. D. Stamp and S. W. Wooldridge formed a powerful team that operated both in Britain and in Europe. Beaver's own interests in the Balkan Peninsula were fostered by his leadership of several Le Play Society excursions. A lifelong interest in transport geography (engendered by twenty-five years of living alongside one of London's 'girdle' railways) led to his taking over the organisation of the Stephenson Locomotive Society from 1932 to 1939. The two interests were integrated most effectively in a study of the railways in the Balkan Peninsula, which appeared in 1941. Indeed, with the outbreak of the Second World War the significance and value of Beaver's research into these aspects of human geography were recognised by the government; he served in the Naval Intelligence Division of the Admiralty at Cambridge from 1941 to 1943, and afterwards on an occasional basis. During this period he made a significant contribution to the handbooks that were produced by this division of the Admiralty for members of the armed forces. Although he provided a regional analysis of the islands of the East Indies in the handbook for the Netherlands East Indies (vol. 1, 1944), his major concern was with transport studies. Thus in the handbooks for France (vol. 4, 1942), Belgium (1944), Greece (vol. 2, 1944), Denmark (1944), Netherlands East Indies (vol. 2, 1944) and Yugoslavia (vol. 3, 1945), he provided chapters on railways, while for France (vol. 4, 1942) he also wrote an account of the road system.

After leaving the Naval Intelligence Division, Beaver's attention was directed towards the need to plan for postwar Britain and in 1943 he joined the newly formed Ministry of Town and Country Planning as research officer. His interests and skills as a geographer and geologist were employed profitably in his assignment to consider the land use problems associated with surface mineral extraction, an activity which he had investigated first in his research into the Northamptonshire ironstone industry. This acquaintance with the planning implications of extractive industries was carried through to the stage at which Beaver realised that the dereliction which often resulted from opencast activities constituted in itself a problem of immense proportions. His subsequent research into the distribution of derelict land in Britain, the possibilities for its reclamation and the significance of the latter in land use planning is probably Beaver's finest achievement in the field of applied geography, and the wealth of publications which appeared after 1946 is testimony to the enthusiasm and energy he devoted to this problem. Subsequently, his experience with the Ministry was placed at the disposal of one of the country's basic extractive industries with his appointment, together with S. W. Wooldridge, to the Advisory Committee on Sand and Gravel, which sat from 1946 to 1949.

Beaver's association with the west Midlands region began in the late war years with his survey of derelict land in the Black Country and in the north Staffordshire coalfield for the Ministry of Town and Country Planning. He contributed an appendix on the Black Country to the Staffordshire report in the *Land of Britain*, published in 1945. His work in these areas was acknowledged by Sir Patrick Abercrombie and Herbert Jackson, who asked him to act as consultant on mineral resources and dereliction in their preparation of the west Midlands and north Staffordshire plans. His continuing interest in planning and particularly his more recent concern with the rural environment have been reflected in his active membership of the Staffordshire branch of the Council for the Protection of Rural England since 1952 and his chairmanship of that branch since 1968 and in his representation of the Geographical Association on the Standing Committee on National Parks.

A more permanent association with this region with its distinctive industrial history and landscapes was established in 1950. In February of that year he became one of the thirteen foundation professors invited by Lord Lindsay to the new University College of North Staffordshire. During his tenure of the Keele chair he has combined successfully the three activities of developing his own department, the first new department of geography in the postwar period of university expansion, consolidating his professional career and furthering the interests of geography in general.

Many leading geographical bodies have accorded him the honour of their presidency, notably Section E of the British Association (1960–61), the Institute of British Geographers (1963), of which he is a founder member, and the Geographical Association (1966). The Council of the Royal Geographical Society made the Murchison Award to him in 1962 for services to economic geography and in 1968 he was created an honorary fellow of the Polish Geographical Society in recognition of his work in promoting Anglo–Polish academic cooperation, especially through the medium of the seminars held in the two countries. Since 1970 he has been the British Association (Section E) representative on the National Committee for Geography. He has been a frequent delegate at International Geographical Congresses since 1931 and has attended all meetings since the Washington congress in 1952.

In addition to his pioneering work at Keele, Beaver has played a major part in the promotion of geographical teaching through another new organisation, the Council for National Academic Awards (CNAA).

Degree studies in geography had been available, under the auspices of the University of London, in many colleges of technology before the creation of this body, but Beaver's work as a member of the Council's Arts and Social Studies Committee, and later as chairman of its geography board, ensured that the CNAA geography courses in the new polytechnics were firmly based in terms of staff and other resources. Reflecting his other academic interests, he also served as chairman of the Council's Town Planning Board from 1967 to 1972.

His efforts in establishing new teaching centres of advanced geography has been matched by his organisation of new texts. He has been editor of the Longman 'Geographies for Advanced Study' since its initiation in the late 1940s, and between 1950 and 1974 he has been responsible for the editing of twenty-seven volumes in this series dealing with various aspects of systematic and regional geography; four of these are his own translations from the French.

Throughout his career as professor of geography at Keele, Beaver has been identified closely with the local region, both in his unstinted support for geographical and allied studies of the Potteries coalfield and its environs, and in his promotion of climatological studies centred on the meteorological station established by him on the university campus. An active member of the North Staffordshire Field Club, he was elected president in 1961–62 and has been on the editorial board of the *North Staffordshire Journal of Field Studies* since its foundation in 1960. From 1962 to 1974 he was president of the local branch of the Geographical Association; he is a trustee of the Cheddleton Flint Mill Industrial Heritage Trust and a member of the Planning and Historical Committee of the North Staffordshire Pottery Industry Preservation Trust. His long association with the Institute of British Geographers and his deep-rooted interest in north Staffordshire were combined appropriately and effectively in January 1964 when, as president of the Institute and host to its annual conference at Keele, he delivered his memorable lecture on the evolution of the Potteries cultural landscape. His enthusiasm for this remarkable industrial region has been communicated in his lectures to Keele's freshmen, to school conferences, colleges and Geographical Association branches throughout the country and to the many parties of visiting geographers conducted around the Pottery towns.

The Keele meteorological data have formed the basis of numerous papers in journals and of regular monthly contributions to the north Staffordshire *Evening Sentinel* newspaper. A monograph on the climate of Keele, written jointly with Elizabeth Shaw, research assistant in the department from 1960 to 1962, summarises effectively much of these data. Many local north Staffordshire organisations, both public and private, have found Keele weather data invaluable for various purposes and have cause for gratitude to S. H. Beaver for the lucid manner in which the technicalities of weather statistics have been translated into appropriate terms.

At the new University of Keele, Beaver has been involved in much committee work, but his greatest endeavours have been in the planning of his most local of landscapes, the 'one square mile' of the university campus astride the 'keel of England'. Through his chairmanship of the Grounds Committee, a post occupied since its formation in 1952, he has ensured the maximum preservation and improvement of the parkland of the former Sneyd family estate compatible with the demands for physical expansion of the university. In this task he has displayed the same qualities of research and reasoning that characterise his more widely known works.

M. J. Wise

S. H. Beaver:
an appreciation

'Good evening', said Mr Beaver, 'and whatever you do, don't get dirty finger prints on my slides.' It was my first meeting with Stanley Beaver. As an undergraduate student of the University of Birmingham, I had been pressed into service one winter night in 1937 or 1938 to 'work the lantern' for a meeting of the local branch of the Geographical Association which was to be addressed by the distinguished visitor from London. For some reason, the university's lantern operator was incapacitated and I waited in the dark, vaulted corridor of the now, alas, demolished neo-Gothic Mason College building in Edmund Street for Mr Beaver's arrival. In he came, taut and eager, carrying his box of 3¼ inch square glass slides. 'This,' said Mr R. H. Kinvig, who was to preside over the meeting, 'is Mr Wise, a student who will operate the lantern for you.' And so our acquaintance began. Most probably Stanley Beaver has forgotten the occasion, though he will probably agree that the adjuration concerning his slides is one that he has frequently delivered. If I did not greatly impress him that night save as a moderately efficient lantern operator who answered the rap of his pointer smartly and kept his slides in order and at least reasonably clean, he certainly impressed me as a highly organised and most efficient lecturer.

In more recent years, freed from caring for antique lanterns, it has been easier to appreciate the qualities of S. H. Beaver's lectures. His delivery is clear. Their outstanding quality lies in the skill with which the material is organised: there is always a well marshalled presentation of factual material and the themes are well maintained. Few lecturers of his generation have been able to summarise so ably: read, for example, his presidential address of 1964 to the Institute of British Geographers on 'The Potteries'. In thirty pages or so of text the principal facts and the main themes in the evolution of the industrial landscape of that district since the early eighteenth century are clearly assembled. The illustrations to his lectures, and they are usually plentiful, are well chosen in relation to the content of the lecture: indeed, the lecture often flows out of the illustrations. Wherever possible the photographs employed are his own. There is always a joke or two, or a humorous aside: he is not a man who can resist an opportunity for a pun. Sometimes the humour is inadvertent. The story is still told in the London School of Economics of the pleasure given to a crowded audience in the Old Theatre by Beaver's explanation that one of his slides was not quite up to the usual quality due to the difficulty of photography 'from the top of a moving public convenience'.

His qualities as a lecturer were aptly referred to by the President of the Royal Geographical Society when thanking him for his lecture to the Society, on 20 November 1944, on the subject of 'Minerals and planning'. Said Sir George Clerk, 'It was informative, it was lucid, it was also spiced with a quietly pungent humour which I for one found most refreshing' (Beaver, 1944, p. 198).

Contributions to economic geography

In reviewing Stanley Beaver's geographical writing, first place must be given to his contributions to economic geography. At the International Geographical Congress held in Paris in 1931 he offered a paper on 'The iron industry of Northamptonshire, Rutland and south Lincolnshire' (Beaver, 1933). This was the first of a number of papers which have examined the special features of the iron-ore fields of the east Midlands. The 1931 paper gives an explanatory description of the Northampton Sands ore field and considers why until that time the industry had been mainly one of ore extraction and export rather than of smelting. The fortunes of an industrial district could, he thought, be related to the triangle of *place, period and people*. Place included the physical characteristics and resources, the capital invested, the transport facilities. Under the heading of period had to be considered the comparative advantage of a location at a given time. The heading people included those engaged in the industries of the district, and especially those responsible for decision-making on suitable techniques and locations. The decision to develop the steelworks at Corby had, he argued, brought for Northamptonshire the three elements of the triangle into conjunction.

At the succeeding International Geographical Congress in Warsaw, 1934, he enlarged his topic of discussion to cover 'The localisation of industry' (Beaver, 1935). At that time, Beaver was one of a very few geographers in Britain working in this field. True, G. G. Chisholm had given a strong lead and R. Ogilvie Buchanan's work was beginning to appear in print, but in the very different context of the agricultural industries of New Zealand. L. Dudley Stamp had started his Land Utilisation Survey of Great Britain, but Wilfred Smith's work was yet to appear. Beaver was concerned that economic geographers had tended to ignore the history of industrial development and the economic factors involved in the siting of an industry: they had placed too much faith in lame 'explanations' based on physical circumstances. Economists, like Alfred Weber, had ignored geographical circumstances. The geography of industry awaited investigation. Much of the paper was taken up by a straightforward review of the principal influences on the siting of manufacturing industries: the location of raw materials, sources of power, location of markets, labour supply and transport facilities. However, viewed in the context of its time and of the state of the field of study in 1934, a number of significant points were made, and Beaver can be seen to be opening up future fields of work for himself. The geographer's attention was directed to the examination of the inequalities in the distribution of raw materials. There was comment on the influence of changes in methods of obtaining power on the values of locations: the importance of momentum was alluded to. Principles such as Weber's, he seemed to say, needed to be studied in terms of actual industries and of a changing transport situation with road transport increasing in importance. Reference was made to the appearance of a new pattern of industrial location in Britain and especially to the attraction for industry of Greater London. 'On the outskirts of the Metropolis', he commented, 'a vast new industrial region has come into being since the war' (p. 194).

At this time he taught not only students reading for the BA and BSc honours degrees in the Joint School of Geography at the London School of Economics and King's College London but also BSc(Econ) students. For this latter group especially, with compulsory courses in economics and political science, the need lay to draw their attention to the geographical context of economic and political life, to the ways in which economic and political decisions had what would now be called 'spatial' consequences and led to the formation of great industrial regions with all their implications for the lives of their inhabitants and for the nation as a whole. *The British Isles: a geographic and economic survey* (Stamp and Beaver, 1933), written in collaboration with his more senior colleague L. Dudley Stamp, was an attempt to meet the need for an appropriate textbook and review which would reach an even wider audience. It is no disrespect to his great collaborator to suggest that a number of Beaver's chapters (for example, those on textiles, the iron and steel industries and the engineering industries) were outstanding, not least for the demonstrations of change in the location of industry and for the analyses of the relevant influences. Naturally, criticisms were made of a book which attempted so much but Beaver's studies of changes in the industrial areas, especially the coalfields, formed an example which was followed up by many of his students and by other workers in numerous papers and dissertations.

The importance attached by S. H. Beaver to the impact of technical change on the landscape was reflected in the selection of 'Technology and geography' as the theme of his presidential address to Section E of the British Association for the Advancement of Science in 1961 (Beaver, 1961). Three aspects of the environment were distinguished: physical, technical and economic. 'The interest which the technical environment has for the geographer', he said, 'is largely concerned with the changes in geographical values which are brought about by inventions and innovations' (p. 316). There was a careful study, first, of the working of minerals, illustrating the importance for location of demand and of transport costs. Technical changes had opened new possibilities for production, for example, of ironstone in Northamptonshire, and also for land reclamation and restoration. Geographical patterns of production and distribution of iron ores were contrasted with those of gravel. He returned to his favourite case of brick manufacture. The discussion continued with a resumé of geographical responses to technical change in the energy producing industries. His examples led to the conclusion that 'rather than emancipating man from the influence of the geographical environment, the increasing scale and complexity of industrial technique offer a progressively narrowing range of locational possibilities' (p. 324). Not a very different generalisation, in fact, from that reached in relation to agriculture and land use in Stamp's *The Land of Britain: its use and misuse* (1948) and it is at least possible that here lies one example of Beaver's influence on the work of the Land Utilisation Survey of Great Britain. Returning to 'Technology and geography', further examples were drawn from the oil refining, transport and motor manufacturing industries. His conclusion included a plea for closer attention by geographers to the techniques of industrial production. The understanding of geographical patterns could not be complete, he argued, unless it was informed by first hand knowledge and experience of how industry worked: this was especially important at a time when old cultural landscapes were being rejuvenated in an era of rapid technical advance.

In this address something of Stanley Beaver's interest in mechanisms is sensed. He likes to know how things work, whether they be steam locomotives, potteries, oil refineries or nuclear power stations. He wants to know also the mechanisms of organisation. It is his view that only in the light of such knowledge, weighed against careful study of terrain and of physical resources, does the examination of location and of patterns of industrial location become a meaningful exercise. His paper on 'Coke manufacture in Great Britain: a study in industrial geography' (Beaver, 1951) was such an exercise. It opened with a review of the history of coke production in which the reader may note the attention given to the technical properties of the coals employed. The hearth process was studied and its decline attributed mainly to the introduction of the hot blast process in iron smelting and to the development of the beehive coke oven. The growth of production of beehive coke was related to demand from railways for use in the early locomotives and to the rise of the Teesmouth iron industry. The Coppée oven was used mainly in South Wales and Beaver related this circumstance to the similarity in high carbon, low volatile coal types between South Wales and the Franco–Belgian coalfield, home of the narrow, sideheated ovens. Such discussions of the physical characteristics of materials, the changing techniques of processing and the resultant geographical patterns make Beaver's work distinctive among the economic geographers of his day. Others, perhaps, had deeper insights into history or firmer grasp of theory: none can compare with Beaver in his knowledge of material characteristics and the detail of technical change, and in his ability to interpret geographical change in the light of this knowledge.

He continued his study of coke production with a suggested explanation for the relatively slow introduction into Britain, compared with European districts, of byproduct recovery ovens. There was an account of modern developments in the coke industry, an analysis of the geographical distribution of coke ovens, an argument for the conservation – the use only for metallurgical coke production – of the limited reserves of Durham coking coal. The value of a resource was shown to vary according to demand and the techniques available for obtaining it, processing it and realising its utility. There are lessons in such studies for some at least of those who write today on resource problems.

Space prevents more than a passing reference to a third example of Beaver's interest in technical change. The reader of his presidential address to the Geographical Association in January 1967 on 'Ships and shipping: geographical consequences of technological progress' (Beaver, 1967) will find the author again concerned with 'the technological environment [which] acts as a filter between man and all the other aspects of the total environment, controlling his ability to adapt himself to them and to modify them' (p. 134). He gave examples of the direct geographical consequences of such changes as the substitution of oil for coal, the development of refrigeration: he discussed links between changes in international trade and changes in ship construction and methods of propulsion. 'Nothing', he concludes, 'ever stands still in economic geography . . . both subject matter and the methodology are constantly on the move' (p. 155). A main problem, of course, is to relate the time dimension to events occurring in geographical space. Beaver's employment of technical change in its bearing on geographical values is certainly one useful method for relating two dimensions.

Mention of the address on 'Ships and shipping' leads naturally to some further consideration of his interest in the geography of transport. He has been interested both in the detailed study of the layout of lines of communication, especially of canals and railways, and in the relationship between the construction of lines of transport and the location of industry and settlement. Those who have been fortunate enough to participate in one of his field excursions on railway geography, and I recall vividly one held in the South Wales coalfield during the meeting of the British Association at Cardiff in 1960, will remember not only his mastery of detail but also his ability to recreate the circumstances in which competing or complementary lines were constructed. These included the expectations of traffic held at the time, the availability or otherwise of capital, the evaluations made of the terrain and the degrees of engineering or commercial skill possessed by the proprietors and their agents. One of my students, now a well-known and respected author and lecturer on transport subjects, used to baffle me by asking me to identify photographs of railway stations, locomotives, signal boxes and junctions, selected from his collection. He would not easily have baffled S. H. Beaver. In addition to the mastery of such matters as the history of locomotive types – a hallmark of the railway enthusiast – Beaver possesses a feeling for place and geographical circumstance.

A paper on 'The railways of great cities' (Beaver, 1937) demonstrates this second quality. The reader is reminded of the function of a railway system in the organisation of the daily life of a great city. But the real purpose of the paper, I suspect, was to advance a simple model of the railway system of the city and to discuss how far the features of the model were present in the railway plan of the principal European cities. True, the author does not use the word 'model': he writes of 'an ideal railway plan', and perhaps it is for this reason that the paper has been overlooked by those who have recently been calling our attention to the value of 'models'. Another of my favourite papers is his 'Geography from a railway train' (Beaver, 1936), perhaps because I, too, have related the monthly articles on 'Locomotive practice and performance' in the *Railway Magazine* to mainline gradient profiles. The paper suggests a simple, but from my own knowledge effective, method of using railway journeys as a way of studying or revising the geology and scenery of a region. Geological information is transferred from a geological map to a gradient profile, which then accompanies the traveller on the journey. The method has lost some of its value now that diesel engines and electricity have 'evened out' the gradients of the railway system, and now that even the Lickey Incline has lost its thrill, but it is still useful. When I came to teach at the London School of Economics I was handed some hundreds of duplicated railway profiles which Mr Beaver, I was told, had prepared for students travelling to field excursions. Many have been distributed but the demand has fallen off in an age of motor transport.

It may seem that Beaver's interests have been largely in systematic rather than in regional geography. This conclusion would be premature. His contributions to *The British Isles* though dealing with topics systematically were nevertheless contributions to the understanding of a region of national scale. Some of Stanley Beaver's best writing has achieved for particular industrial districts an interweaving of related developments in mining, industrial growth, transport construction, population migration and town building. Few better examples of concise regional description exist than the first half of his *Report on derelict land in the Black Country* (Beaver, 1946); or his analysis of the main features of the cultural landscape of the north Staffordshire Potteries in the paper to which reference has already been made. Subjective, yes, and the better for it; reflecting one serious student's view of the origins and the qualities of the landscape and of the people who have made it and live in it, and containing also his appreciation of opportunities for conservation and for improvement. Such writing seems to me to provide Beaver's answer to those who argue that regional synthesis is an impossible task. 'I have done it', he seems implicitly to say. If one seeks his method, one finds it first in serious and purposeful field work and field recording, in the critical examination of the historical sources, in a thorough study of the physical setting: then through the working out in detail of changing patterns of mining, industrial development and urban growth, relating the patterns period by period. It can be done: alas, we may never see completed the book on South Wales which was in preparation when the outbreak of war assigned him to other work. I have been fortunate enough to see and to use the maps that were prepared for what promised to be outstanding writing about an industrial region.

A major preoccupation of the Department of Geography at the London School of Economics during the 1930s lay in the organisation and production of the Land Utilisation Survey of Great Britain. Dr (later Sir) Dudley Stamp was the director of the survey but, as he frequently explained, the survey owed much to his colleagues and students at the London School of Economics as well as to collaborators in other universities and in schools. Stanley Beaver was active in organising the field mapping and in checking the primary surveys. He wrote the county report on Northamptonshire, based in part on a manuscript by Dr D. M. Allen (Beaver, 1943), and he contributed to the Staffordshire report (Myers, 1945). However, his most impressive contribution lay in the writing of the report on the West Riding of Yorkshire (Beaver, 1941). On account of the large area to be covered and the complexity of the settlement and field patterns, the survey of the West Riding had given rise to many difficulties. The report was widely regarded as one of the most valuable in the series. As Stamp himself wrote in his introduction to the volume:

> The preparation of a report on such a complex area inevitably presented a herculean labour. My colleague, Mr Beaver, was persuaded to undertake the work and has carried it through with the thoroughness, precision and attention to detail which characterises all his work and has prepared and drawn himself a series of maps and diagrams which cannot but prove invaluable to all students of the area. He modestly claims that he has no special knowledge of the county, but has actually penetrated into all parts in connection with his well-known studies on the industrial development published in *The British Isles* and was

primarily concerned in the training of those who carried out so much of the field work and whose notes have been found so helpful (p. 82).

Contributions to planning

During the last three years of the Second World War geographers in Britain were discussing and contributing to practical work in preparation for the introduction of new policies which, it was hoped, would lead to a remaking of the national environment after the war. On my return from overseas service in 1946 an invitation to work for the West Midland Group on postwar reconstruction and planning caused me to become acutely aware of how much had been accomplished. One of the first volumes which came into my hands was Beaver's *Report on derelict land in the Black Country*. It seemed to me a model study. Care had been given to the difficult problem of defining 'derelict land'. There was a highly pointed analysis of the characteristics of the Black Country landscape, followed by a summary of the physical geography and geology and of the industrial history of the district which wasted no words and, as I still find, omitted no points of importance. The classification of derelict land was original. The survey itself was based on a system of numerical notation: Appendix A which describes the system used is still a valuable guide to the young field worker. At that time plans were afoot to extend the built-up area of the Black Country outwards on to valuable agricultural land while leaving unused the derelict land within its borders. Beaver broke down the distribution of derelict land by local authority areas, examining for each local government area the extent of the problem and ways in which it was being tackled. The 6 100 acres of derelict land, strictly defined, were classified in terms of the practical problems of reclamation and development.

Looking back on the progress that has been made in the study, and in the clearance, of derelict land in the Black Country and in other industrial districts also examined by Beaver, it has seemed to me that far too little weight has been given to the contribution made by his studies. Partly, no doubt, this arises from the circumstances in which they were undertaken. He was a civil servant at the time, as senior geologist in the Ministry of Town and Country Planning; his reports were circulated widely but never, in the formal sense, published. The analyses and reports are scientific in character, classifying characteristics, quantifying the problem, eschewing altogether the emotive approaches which, more recently, have become common. But the changed appearance of the Black Country and the use which has been made of the derelict land of 1946 tell their own stories of the value of Stanley Beaver's work. Characteristically, he ended his report with a touch of humour. After painting a dismal picture of the state of the environment, he put additionally another point of view:

> It must be realised, too, that here as elsewhere, the people *like* their derelict land because they have been brought up with it all around them – sentiments very well expressed by a small boy who accosted me one day during my field survey – 'Hey, mister, don't yow go an' build howsis on our bonk'! (p. 40).

Derelict land was only one aspect of S. H. Beaver's wartime responsibility in the Ministry of Town and Country Planning for the general field of minerals and planning. As always, his first concern was to obtain a full picture, at national level, of the facts of the distribution of mineral resources and of existing and likely future sites of extraction and primary processing. Policies for land use in the postwar period were under formulation and it was important to possess an accurate picture of areas that were affected by, or at a time of increasing demand were likely to be affected by, mineral working, and of the nature of the land use conflicts which were bound to arise. It seems fair to say that his task was in part educative: planners had to understand the geological and geomorphological problems involved in assessing and delimiting the distribution of minerals and to appreciate the economic and technical constraints which influenced the actual choice of sites. The firms engaged in the mineral industries also had to be helped to understand the geological picture and to see the other considerations involved in allocating land for urban growth or reserving land on amenity grounds. The objects of land use and of resource planning had to be clarified and explained to civil servants in his own and in other ministries, and to the public. Beaver's published work of the time, for example, his lecture at the Royal Geographical Society on 'Minerals and planning' (Beaver, 1944), or his disquisition to the Town and Country Planning Summer School at St Andrews on 'Surface mineral working in relation to planning' (Beaver, 1949), should be seen, at least in part, in this light.

In these and in other lectures and papers he argued consistently for the fuller availability of information concerning the mineral wealth of Britain. Decisions on where and when to mine or quarry should be taken only after full consideration of all the factors involved, geological, geographical, technical and economic, and in the context of an informed planning policy. Special attention was required to be given to problems of after-treatment and reclamation. Public attitudes needed to be aroused and informed. Beaver identified a number of areas of the country where specially severe problems were arising. First among these were the gravel working areas of the Thames valley. The ironstone areas in the east Midlands and the chalk quarrying area of the lower Thames were also singled out as requiring special attention. The severe problems of the coal mining districts, such as subsidence and the variety of forms of derelict land, also deserved special attention. His map of the areas where conflicts between the mineral extracting industries and town and country planning were arising showed a concentration of problem areas in the northwest to southeast belt across Britain described by Professor E. G. R. Taylor as the 'axial belt'. This distribution arose, he felt, from the incidence of interwar development in that belt.

Probably the most valuable single contribution made by Stanley Beaver to the understanding of the planning problems created by the extractive industries was as a member of the Advisory Committee on Sand and Gravel from 1946 to 1949. The importance of his work for this committee has, to some extent, been overshadowed by that of S. W. Wooldridge who was able to lend to its work his unmatched knowledge of the geomorphology of southeast England. Beaver could bring his detailed knowledge of the economic geography and the technical problems of the mineral extracting industries as well as his interest in transport problems. The importance of transport costs in determining the localisation of sand and gravel production was one of the most interesting of the findings of the first report: the origins and destinations of sand and gravel movements were traced. In a joint paper with Wooldridge, problems of land use competition were dealt with (Wooldridge and Beaver, 1950). The authors were of the view that it was necessary to define and preserve land for the maintenance of present supplies and for adequate reserves for the future: 'What is required', they argued, 'is that the industry should be given a measure of priority (over other claims on land) in chosen areas of wide extent' (p. 54). Such a policy ran counter to the view that the land market could most efficiently allocate land between competing uses. The two authors showed, in what remains a powerful argument for the principles of conservation, that variations in the nature of land could not be ignored: short-run market forces could not efficiently dispose of sand and gravel bearing land, a resource with long run demands. 'It would argue a lamentable ignorance of that awareness of environment for which the geographer strives, a peculiarly unecological blunder . . . if towns occluded their own building material' (p. 51). Planning for conservation was thus the main theme of the contribution: the theme was then applied to the 'service areas' of the industry. In the discussion which followed the joint paper R. S. Brewis, a member of the industry, expressed his appreciation to Wooldridge and Beaver for 'the outstanding help which they have given to our industry through the knowledge brought by them to the Advisory Committee's deliberations, and also for the unstinted sacrifice of their time for so long a period without reward of any kind'. 'Speaking for myself', replied Beaver, 'I have enjoyed every moment of it' (p. 57). We believe him and doubt not that he was at his happiest inspecting at first hand in the field the complex problems involved in the siting, operation and restoration of gravel workings.

Contributions to field work

No other man I know has Stanley Beaver's ability to enthuse over an Etruria Marl hole. It would be a mistake, however, to assume that his interest in field studies is limited to surveys of industrial areas, conspicuous in his work though these have been. Undoubtedly his interest had been stimulated in his undergraduate days as a student of geography and geology. As a university teacher he has always been an active leader of field excursions for students. As has been indicated, the gathering of evidence in the field, its mapping and the interpretation of the distributions revealed were integral parts of the method employed in his studies of applied geography.

Like many other university teachers of geography of the time, S. H. Beaver found opportunities through the Le Play Society both to undertake and to encourage field studies. Happily he has recorded an outline of the history of the Le Play Society which had sprung by stages from the body of the Sociological Society, a society founded in 1902 principally by Patrick Geddes (Beaver, 1962). The Le Play Society existed to extend, by field visits and research, ideas of Le Play and Geddes concerning the interrelationships of societies and environments which Geddes had expressed in his principles of community study based on the elements of place, work and folk. For Stanley Beaver and for many others like him the Le Play Society gave a focus of activity as a field study group and an opportunity for developing field studies in environments differing from that of industrial Britain. He was a member of the society for twenty-five years and chairman of its council in the late 1950s. It fell to him when the society was wound up in 1960 to make arrangements for the continuance of at least some of its interests. Much of the survey material was given a home at Keele where the records of the Institute of Sociology (Le Play House) are also kept. His role in the founding of the *Sociological Review* at Keele reflected his desire to continue the original interests of the Sociological Society in environmental problems.

At the height of the society's work, Beaver was a member of a number of its expeditions, notably to Bulgaria and Albania. Unfortunately no reports of these visits were published, though Beaver's own writings do embody material gained on the Bulgarian expedition. After the Second World War he led student groups of the society on excursions to Holland, Norway, Belgium and, on two occasions, to France. A paper written jointly with Dr Alice Coleman carries a report of the work of the

Norwegian party in Dale-i-Sunnfjord (Coleman and Beaver, 1955). That his association with the Le Play Society has left a lasting influence on S. H. Beaver's outlook there can be little doubt. The latest evidence for this statement comes in a short paper entitled 'Three tropical islands', in which, employing observations made on a recent journey, he studies comparatively the responses to the physical environments of societies in Viti Levu, Upolu and Oahu (Beaver, 1973).

Probably, his association with the Le Play Society reflected his interests in travel and in making and maintaining contacts with geographers in other lands. From the earliest days of his academic career he has been a regular contributor to International Geographical Congresses. When the Institute of British Geographers began a new venture in sponsoring jointly the Anglo–Polish Geographical Seminar of 1959, held at Nieborów, Beaver was an active supporter of the idea and a participant. A paper by him on 'The reclamation of industrial wasteland' is to be found in the published proceedings (Beaver, 1961). The second seminar of 1962 met at Keele: Professor Beaver led the British delegation and acted as a joint editor of the resulting volume (Beaver, 1964). Of course many others were also involved in arranging the seminars, but he was one of the active workers and without him some element of vigour and spark would have been lacking. The series of Anglo–Polish seminars continues and its success has led the Institute to organise a number of similar seminars with geographers from other lands.

He has also been assiduous in his concern for the maintenance of field records and has put precept into practice. Who else would have had the vision to convert unused buildings at a brickworks into a safe reference collection of air photographs, which but for his intervention would have been destroyed? Who else would have had the initiative to establish a meteorological station in the new University College at Keele and, often at cost to his own sleep and leisure, to ensure the collection of a continuous series of records? Those who come regularly to seek out air photographs and to study from them past patterns of land use and economic activity, and those who embody the records of the Keele station into analysis of the British climate have cause to be grateful to the foresight of the first geographer at Keele.

At the foundation of his studies lies, undoubtedly, his interest in maps as records of landscapes, as documents to be added to and, where necessary, corrected. In this field of interest, as in others, he has involved himself in the study of the techniques of production. For many years he has represented the British Association for the Advancement of Science, and sometimes also the Geographical Association, at the Ordnance Survey Map Users' Conference. His representation of their interests has been vigorous, as has been his reporting back. The British National Committee for Geography has a hard-working subcommittee on cartography.

S. H. Beaver has served on this, too, and has been found regularly in his place at the congresses of the International Cartographical Association.

Contributions to geographical organisations

This can only be an interim appreciation of Stanley Beaver's work for, to the pleasure and profit of his colleagues and friends, he continues to work actively and purposefully. Retirement from the chair of geography at Keele will bring a change in his daily routine but will provide new opportunities for activity. Among other ventures, no doubt, he will continue to translate and to edit for publication in Britain works by leading European geographers. Characteristically he has been, as I write in this week of August 1973, in his usual place at the meetings of Section E of the British Association for the Advancement of Science, held this year in Canterbury. Picture a slim figure, perhaps a little below medium height, sharp of feature, acute of eye; he moves, nimble of step, from meeting place to meeting place on the open, severe campus of the University of Kent at Canterbury in search of the most useful paper or discussion. Nothing seems to escape him in the lecture room; he sits attentive, ready with quick comment or question in discussion. At such meetings his comments are, and have been, discerning, sometimes sharp. There is a word of praise or encouragement for the well prepared and well delivered paper, for the new man making his first appearance on an academic platform: sometimes an acutely critical comment on the trite or superficial, or for the man whose contribution has proved for him an apparently pointless exercise in technique. A sniff or two of disagreement from Stanley Beaver has enlivened many an academic discussion – see, for a taste of his style on such an occasion, pp. 66 and 67 of volume 103 of the *Geographical Journal* for 1944.

Seeing him each morning at the meetings of the Section E Committee has reminded me also of his service to geographical organisations. Unlike many of his colleagues, he has never lost his belief in the value of the work of the British Association and especially in the opportunities for interdisciplinary discussions which its annual meetings provide. He has been, of course, president of the section and its representative on the council of the association. He has been a traditionalist in defending the role of the section, run on disciplinary lines, in the association's organisation. At the same time, he has never been unwilling to consider new ways of carrying on the association's work and of fulfilling the changed but still important role which it now plays in the world of science. However, it is, perhaps, in his work for the Geographical Association that his contribution to the work of geographical organisations is most clearly apparent. For many years he was the organiser of the annual conference held each Christmas vacation at the London School of Economics. As a member of its executive committee, he has been able to bring the experience derived from his contacts with the founders of the Association and with members who were active in its struggling days before 1939 to discussion on how best the Association should develop its activities in its more prosperous times since 1945. There is no firmer believer in the value of contacts between university teachers and school masters and mistresses and time spent in such activity is not for him, as it is, alas, for some, time wasted from research. It is a way in which his knowledge and experience can be put to uses profitable to teachers and through them to their pupils in schools and, in the long run, to the corpus of ideas and knowledge which forms a 'subject'. When he was president of the Geographical Association it was his aim to visit every branch of the association during his year of office and he travelled thousands of miles to speak to the smallest as well as to the largest and most prosperous branches. The branches of the Geographical Association in the Joint School of Geography, London, in the University of Keele and in north Staffordshire owe much to his energy and enthusiasm.

He was a member of the group of eleven which met in London in February 1931 and from whose discussions the Institute of British Geographers sprang. He played his part as a member of council in 1953 and 1954 and as president in 1964. There are frequent references in the *Geographical Journal* to his contributions to discussions in the Royal Geographical Society. He served on the council of the society from 1943 to 1947 and was awarded the Society's Murchison Grant in 1962. He has been a most active member of the British National Committee for Geography and through his regular attendance at the International Geographical Congress has carried forward his interest in the international organisation of geography. Beaver's belief, as one can derive it from his actions, has been that the benefits accruing from membership of a society or organisation are directly proportional to the contribution made through active participation.

He has never confined his membership of societies, as he has never confined his scientific interests, within narrow disciplinary walls. Mention has already been made of his work for the Le Play Society. His fellowships of the Geological Society of London and of the Royal Meteorological Society must be mentioned. The interdisciplinary nature of Beaver's work is, however, best illustrated in this context by his contributions to the work of one of the most distinguished of the local scientific societies of Britain, the North Staffordshire Field Club. It is typical of the man that one of the main activities of his presidential year was a field excursion under his direction to establish the margins of the Potteries industrial district (Beaver, 1963).

It is not possible to mention all the societies to whose work he has contributed. He was one of the founder members of the Geoids Amateur

Operatic Society in 1930 and took part in a number of their early performances on the stage of the Old Theatre in the London School of Economics. I would like to have seen Stanley Beaver as the counsel for the plaintiff in *Trial by Jury*, their first production. The Geoids were formed originally from students and staff of the Departments of Geography and Geology at the London School of Economics and King's College London, under the leadership of the late Professor S. W. Wooldridge. Today they are one of the leading amateur operatic societies in southeast England.

Then, too, there is the Stephenson Locomotive Society. The journal of the society carries many papers and notes from S. H. Beaver, some describing his experience of railway travel in other lands. There is a recent one on railway museums visited in the course of his journey round the world in 1972. The most recent comments on the changes since 1950 in railway services between Stoke-on-Trent and London, comparing experience under steam, diesel and electric traction. Meticulous as ever, there is a reference to the period when electrification work was in progress and when services from Stoke were routed over Midland rails into St Pancras. His contention has been that an understanding of the geography of transport must involve the student in a study of the technique and of the organisation of the industry itself. Being the man he is, that has also involved a contribution to the relevant societies and groups. He has a prodigious memory for the facts of railway history and organisation. Certainly in the days of steam – it may no longer be so – he knew by heart the railway timetables for his chosen lines. One could stand in the middle of nowhere with him watching a train pass to be told immediately its origin, destination and timing. It took him a long time to become reconciled to the idea of owning and using a motor car!

Conclusions

In many parts of the world, former students now occupying posts in universities, schools, departments of town and country planning and in other organisations may be found to testify to the vigour of the inspiration given to their early studies by S. H. Beaver. In recent years he has extended his influence on degree courses and on the studies of students through his service as chairman of the geography panel of the Council for National Academic Awards. The advice which he has given to polytechnics establishing their degrees under the authority of CNAA has been greatly appreciated, as I know from my acquaintance with one such institution.

In his own outlook on geography, Beaver is, perhaps, to be regarded as a traditionalist. He has worked broadly within a conceptual framework for geography which was widely accepted in Britain in his early academic days. It is one which stresses the importance in geography of the man – environmental relationship and which places emphasis on the examination of regional diversity.

He has not been greatly attracted by recent trends to highly specialist systematic studies and to the use of statistical techniques. Yet, as I have tried to show, within his chosen framework he has made highly individual and distinctive contributions to the study of the geography of industrial areas and to the illumination of resource problems. He has led a way forward for the geographers of his own and of the next generation in demonstrating ways in which geographical study could be successfully applied to problems of public concern and social relevance. The success of his applied studies has been founded on a very high level of technical competence: indeed, in this respect, he has set an example and a standard.

He is a man who has given himself fully to his subject and to the universities and organisations he has served. Societies of diverse aims and origins, from the Royal Society to the smallest branch of the Geographical Association, from the Stephenson Locomotive Society to the Boy Scouts, have commanded his loyalty. It has never been too much trouble for Stanley Beaver to travel often hundreds of miles to speak to a group of students, teachers, or planners, or to take part in a discussion. It is this willingness to give which marks him out, and it is this quality above all which, at the time of his retirement from his chair, earns our appreciation and our gratitude.

An appreciation which failed to mention Elsie Beaver would be incomplete. They were students together at University College London. Their children grew up in a home of warm affection, which is happily maintained. Stanley's portrait hangs in a central position in the house. He could hardly have involved himself in so many activities; nor could he have organised his travelling, lecturing, writing and administrative work on so tight a schedule had he not had a sound and secure base provided for him at home. While caring for him, Elsie has found the time to keep alive her own study of geography and, in recent years, to resume her teaching. The welcome that was always hers among geographers is now doubly earned. For the ventures and the achievements of Stanley Beaver's career to the present time she must share the credit.

References

Details of S.H.B.'s works are given in the Bibliography of his publications on p. 29.

COLEMAN, A. and BEAVER, S. H. (1955): *see* Beaver (1955).
MYERS, J. (1945): *see* Beaver (1945).

STAMP, L. DUDLEY (1948) *The Land of Britain: its use and misuse*, Longmans (3rd edn., 1962). Summarises and supplements the work of the Land Utilisation Survey.

STAMP, L. DUDLEY and BEAVER, S. H. (1933): *see* Beaver (1933).

WOOLDRIDGE, S. W. and BEAVER, S. H. (1950): *see* Beaver (1950).

A. D. M. Phillips and B. J. Turton

S. H. Beaver: a bibliography

This bibliography excludes contributions to newspapers, to encyclopaedias, to the handbooks of the Naval Intelligence Division of the Admiralty, sectional reports to the North Staffordshire Field Club, and reviews.

1931 'The Jurassic scarplands', *Geography*, 16, 298–307.

1932 *The Americas*, London, 254 pp.
'The railways of Czechoslovakia', *J. Stephenson Loco. Soc.*, 8, 79–84, 139–43, 193–6.

1933 (with L. D. Stamp) *The British Isles: a geographic and economic survey*, Longmans, 719 pp. (six editions).
'The iron industry of Northamptonshire, Rutland and south Lincolnshire', *Geography*, 18, 102–17.
'The localisation of the iron industry in Northamptonshire', *Comptes Rendues du Congrès Internationale de Géographie, Paris, 1931*, 3, 317–29.

1934 (with L. D. Stamp) *A Regional Geography: Africa*, Longmans, 271 pp. (seven editions).
'The Hellenic electric railway', *Rly. Gaz.*, 62, 674.

1935 'The localisation of industry', *Geography*, 20, 191–6.

1936 'The Bulgarian state railways', *Rly. Gaz.*, 64, 1204–7.
'Geography from a railway train', *Geography*, 21, 265–83.

1937 'Razpredelenieto na naselenieto v Balgaria [Distribution of population in Bulgaria]', *Izv. Bulg. Geogr. Druzh.*, 4, 110–17.
'The why and wherefore of water-softening', *LMS Mag.*, 14, 272–3.
'The railways of great cities', *Geography*, 22, 116–20.
'The localisation of industry with special reference to Britain', *Comptes Rendues du Congrès Internationale de Géographie, Warsaw, 1934*, 3, 287–93.
Map of the British Iron and Steel Industry, British Iron and Steel Federation.

1938 'The geological aspect of railway construction: gradients on the London–Birmingham line', *Mod. Transp.*, 39, 21–2.
'Geography and planning', *Nature, Lond.*, 141, 170–1.

1939 'Geographical aspects of current problems', *Nature, Lond.*, 143, 169.

1940 'Bulgaria: a summary', *Geography*, 25, 159–69.

1941 *Yorkshire, West Riding: The Land of Britain*, vol. 4, part 46, Land Utilisation Survey (LUS), pp. 81–186.
'Railways in the Balkan Peninsula', *Geogr. J.*, 97, 273–94.

1943 *Northamptonshire and the Soke of Peterborough: The Land of Britain*, vol. 5, parts 58 and 59, LUS, pp. 331–413.

1944 'Minerals and planning', *Geogr. J.*, **104**, 166–93.

1945 'The Black Country', Appendix 2, in J. Myers, *Staffordshire: The Land of Britain*, vol. 5, part 61, pp. 639–52.

1946 *Report on Derelict Land in the Black Country*, Ministry of Town and Country Planning, London, 45 pp.

1947 (with R. R. Rawson) 'Aerial photography and geographical studies', *Geography*, **32**, 131–4.

1948 'Minerals and land use', in L. D. Stamp, *The Land of Britain: its use and misuse*, Longmans, pp. 232–41.

'Physical background to natural resources of the region: geology and relief' and 'Mineral resources and the problems to which mineral working gives rise', in P. Abercrombie and H. Jackson, *West Midlands Plan*, Ministry of Town and Country Planning, London, pp. 39–49 and 51–83.

1949 'Surface mineral working in relation to planning', in *Report of Summer School, St Andrews*, Town Planning Institute, pp. 105–24. Reprinted in *Jl R. Instn chart. Surv.*, **29**, 325–39.

'Physical background and natural resources', Appendix 1, in P. Abercrombie and H. Jackson, *North Staffordshire Plan*, Ministry of Town and Country Planning, London, pp. 71–114.

1950 'Rehabilitation of derelict land', *Nature, Lond.*, **166**, 630.

(with S. W. Wooldridge) 'The working of sand and gravel in Britain: a problem in land use', *Geogr. J.*, **115**, 42–57.

1951 'The development of the Northamptonshire iron industry, 1851–1930', in L. D. Stamp and S. W. Wooldridge, eds, *London Essays in Geography*, Longmans, pp. 33–58.

'Evolution et localisation de l'industrie du coke en Grande Bretagne', *Revue Sci. econ.*, **86**, 93–116.

'Coke manufacture in Great Britain: a study in industrial geography', *Trans. Inst. Br. Geogr.*, **17**, 131–48.

1952 'A new meteorological station at Keele', *Geography*, **37**, 229–31.

1953 'Ten fronts in one week – and their graphical expression', *Weather, Lond.*, **8**, 37–9.

1955 'Land reclamation after surface mineral workings', *J. Tn. Plann. Inst., Lond.*, **41**, 146–54.

(with A. Coleman) 'Dale-i-Sunnfjord: a study in changing geographical values', *Geogr. J.*, **121**, 51–63.

1959 'Les industries textiles Britanniques', *Revue Nord*, **41**, 1–19.

1960 'Land reclamation', *Chart. Surv.*, **92**, 669–75.

1961 'Technology and geography', *Advmt Sci., Lond.*, **18**, 315–27.

'The reclamation of industrial waste-land for agricultural and other purposes', in *Problems of Applied Geography*, Polish

Academy of Sciences, Geographical Studies, 25, Warsaw, pp. 97–111.
'Studies in commercial geography: the location of industry; examples of industrial location in the United Kingdom; the British iron and steel industry; the geography of ocean shipping', *J. Inst. Bankrs*, 82, 53–62, 128–38, 324–34, 413–22.
1962 'The Le Play Society and field work', *Geography*, 47, 225–39.
1963 'A geographical agenda for north Staffordshire', *N. Staffs. J. Fld Stud.*, 3, 1–16.
'Beating the bounds of the Potteries coalfield', *N. Staffs. J. Fld Stud.*, 3, 81–7.
1964 'The Potteries: a study in the evolution of a cultural landscape', *Trans. Inst. Br. Geogr.*, 34, 1–31.
'Mineral resources and power', in J. W. Watson and J. B. Sissons, eds, *The British Isles: a systematic geography*, Nelson for Royal Society of London, pp. 280–96.
'The Potteries: a brief summary', in *Problems of Applied Geography II*, Geographica Polonica, 3, Warsaw, pp. 1–8.
1966 'Land reclamation after surface mineral working in Great Britain', *Comptes Rendues du XVIIIe Congrès Internationale de Géographie, Rio de Janeiro, 1956*, 4, 212–18.
(translator with E. D. Laborde) P. Gourou, *The Tropical World: its social and economic condition and its future status*, 4th edn., Longmans, 196 pp.
(translator) J. Beaujeu-Garnier, *Geography of Population*, Longmans, 386 pp.
1967 'Ships and shipping: the geographical consequences of technological progress', *Geography*, 52, 133–56.
(translator, with G. M. Yglesias) J. Beaujeu-Garnier and G. Chabot, *Urban Geography*, Longmans, 470 pp.
1968 *The Geology of Sand and Gravel*, Sand and Gravel Association of Great Britain, 65 pp.
'Changes in industrial land use, 1930–67', in *Land Use and Resources: studies in applied geography, Inst. Br. Geogr., Spec. Pubn.*, No. 1, pp. 101–9.
1970 'An appraisal of the problem', in W. G. Collins, ed., *Proceedings of the Derelict Land Symposium, 1969* (Leeds), pp. 3–9.
1971 (with E. M. Shaw) *The Climate of Keele*, University of Keele Library, 169 pp.
1972 'Le problème des friches d'origine industrielle en Grande Bretagne', *Hommes Terres Nord*, pp. 5–10.
'Meteorology and philately', *Weather, Lond.*, 27, 290–4.
'Geography through stamps', *Geography*, 57, 158–9.
1973 'Three tropical islands', *Horizon*, 22, 15–21.

1974 (translator, with E. Derbyshire) J. Tricart, *Structural Geomorphology*, Longman, 305 pp.

B. T. Bunting

1 The language of site: South Pennine dialect terms and their use to geomorphologists

'In youth', as C. E. Montague wrote, 'you easily fall in love with words, written and spoken. You come, like other lovers, to feel an unreasoned sensuous thrill of joy at a word because it is just what it is. . . . Apart from the interest of their use in any particular place . . . words become enamouring objects, with glowing hearts of their own, like red wine or rubies.' One doubts very much whether the most ardent geomorphologist would make such claims for the terms he currently uses to aid interpretation. So great is the uncritical use of unexplained foreign terms that geomorphological nomenclature, in using 'a Latin word, a Greek remark and one (or two) that's French' may well be in danger of terminological miscegenation.

The geomorphologist 'is characteristically involved in the interpretation of spatially distributed landform elements which may often be considered as the response features of an environmental model' (Krumbein and Sloss, 1963, p. 138). It is possible to assess the frequency, characterise the form and analyse the constituents of most discernible landforms if they are related to drainage texture (Cotton, 1963) or stream basin geometry (Melton, 1958). However there are few words and even fewer ideas available to describe the many non-generic, indistinct, quasiperceptible sites (Curtis *et al.*, 1965) which make up most of the land surfaces of the world – what Lewis (1959) has termed somewhat enigmatically 'specific features' as opposed to the 'landform types' for which, and to which, present jargon is best suited.

An acute and integrated approach to the characterisation of the relatively featureless landforms upslope from first-order streams has been achieved through studies of drainage basin morphometry (Morgan, 1972), though here a generalised model of the total slopes in relation to the most readily perceived agent of removal, a river or rill, has been merely converted into a mathematical form. In other geomorphic studies aerial photographs have been used to provide a complete coverage of an area. With these aids the vast extent of indescribable features (or 'featurelesses') becomes obvious, together with the scattered occurrence of the known and discernible landforms. However, through the process of photo-interpretation the latter are selected immediately and the former neglected, usually producing what could be called a pictogram. Words from foreign languages are used to describe such well-known features, ignoring the fact that English is one of the world's most versatile languages, possessing the largest vocabulary.

In the area to the east of Keele, in fact within the entire Peak District, there is an almost infinitely varied morphology. First, elevated areas of alternating permeable and impermeable rock types occur, of gritstone, sandstone, shale and of limestone and mudstone, alluvium and till. Second, dry crests and rocky screes, wet clay and shale surfaces, peaty valley heads and backslope seepage areas are encountered. A third contrasting element is provided by low-order streams, which may be either slightly or deeply incised. Various attempts have been made to express this wide variety of slopes and flats, both wet and dry, hard and soft. Bridges and Doornkamp (1963) related soil conditions to slope forms and dimension but confused the ordinal character of the one with the numeral depictions of the other. Johnson (1957) and many others frequently interpreted minor slope forms in terms of remnants of former land surfaces, while Straw (1962) was more successful in finding meaningful terms for his study of the site conditions of glacial remnants in Derbyshire. Earlier attempts (Wood, 1942) at producing an image of dynamism in an essentially static situation, by the selective treatment of uniform elements of landscape, also failed to render an adequate image of a confused reality. Only in the work of Starkel (1959) are a perception of the nature and significance of minor forms and a thorough attempt at naming them found, and, not for the first time, words from a Slavic language were the prime agent for depicting states of ground.

One of the problems is that attempts to describe landforms are constrained by terms, symbols and concepts which have always been in use within the science. Geomorphologists have failed to make use of the vocabulary of landform perception of the original settlement groups, the study of which traditionally has belonged to other researchers, namely historical geographers (Eyre, 1957), and to those toponymists or choronymists (Dorion, 1972) who may have closer knowledge of older, now defunct, topographical terms from sources other than the modern maps on which geographers rely. These settlement groups often had cause to be very cognisant of the properties, 'the very feel', of the sites where they lived. Hence their terms stemmed as much from their social psychology and practical needs as from their limited perception of the natural environment. Living close to the land and rarely travelling far, once settled, their original allocation of names was related often to properties of practical importance – the local relief features, the physical state of the site, its value for settlement location or for boundary definition. Some names apply to soil characteristics, some to rock types and some to distinct landforms (Poirier, 1965), and they become fixed and formalised because of their frequent usage and wide occurrence. More significant are the dialect forms, which may only be known in a limited locality and which may have been altered only slightly by phonetic change from their earliest form and still retain their original meaning (Boileau, 1971). These terms may reflect not only a keen appreciation of the natural state and potential value of the land, but also may contain some implications of the nature of the former vegetation cover.

Glossaries of soil science (Plaisance, 1958; Tavernier *et al.*, 1960) indicate that English and foreign dialect terms alike have enriched and enlivened the vocabulary of soil scientists, such terms being defined today either in accord with their etymology or at a discreet degree of variance from it. Indeed Plaisance expressly introduced some French dialect words in the hope that they would be useful to foreign pedologists. More recent trends in soil nomenclature (United States Department of Agriculture, 1960) have sought to avoid foreign language and folk elements, by using words and word elements derived from classical sources, but in so doing major semantic errors have been perpetrated. For example, the word clay in Latin is *argilla*, in Greek *argillos*, and this is quoted with modern French equivalents. However, the derivation is from Greek *argos*, meaning shiny and hence it is related to glazed or potters' clay. A better course would have been the direct use of 'clay', related to Indo-Germanic *klai* and Slavonic *gley*, in the development of a nomenclature.

It must be admitted that there are dangers attached to the use or adaptation of dialect terms by geomorphologists. Baulig (1956) maintains that they lack precision and generality, having different meanings in different areas, and quotes the conclusions of Burrill (1954) who found that in the USA some words have at least six widely different uses in different parts of the country. This is hardly the point, for a word can be chosen for a specific purpose, to describe a feature or site, provided that it be clearly defined and rigorously applied in accordance with its etymology. The great advantage of dialect terms is that they refer to the suitability of slope elements and states of ground for cultivation or other use, either directly or by implication. Apart from lending precision to geomorphological terms for small relief units, the adoption and wider use of dialect terms would go some way to meet Hartshorne's (1960) plea for a geographical geomorphology – that a basic requisite in studies of cause and effect relationships is the development of ways of describing landforms in terms of characteristics that are most significant in relationship with other factors. It follows that the terms would be meaningful to, and capable of use by, students of land use, forestry and soil, and others concerned with rural and marginal areas in upland regions.

The first part of the principle, of using new and strange terms or names for features which have previously been neglected or not described, is that the users of such names must be in a position to supply a set of identifying descriptions (Donnellan, 1972). Most of these descriptions have been gathered from the field and checked with etymological sources by the author. The second part of the principle is the need to test some of the names descriptive of features for which names are already commonly accepted or used; such names are often imprecise in so far as they are used variously or have no constant etymology.

Suggested dialect terms

Rock outcrops and upstanding rocks

The term 'tor' has gained wide usage among geomorphologists (Linton, 1955), where it is related specifically to crestal rock outcrops. However, the derivation is first from OE *torr*, a hill, a high rock or rocky peak, and may be further derived from Welsh *twr* or perhaps by OE from the Gaelic, *torr* (Charles, 1938), a possible Celtic loan word. Its most appropriate use is for a hillside with precipitous rock slopes: for example, Mam Tor, Higger and High Tors in Derbyshire. The word is not recorded as a common noun in Cornish and Breton (Kurath, 1954). The term tor is seldom used to describe upstanding stones or rock outcrops of small basal area and with joint-bound faces so typical of the southern Pennines; here they are called most frequently 'stones' or 'stands'. Features with the dimensions of Cornish tors are termed 'knotts' or 'pikes', derived from ON *knottr*, a rocky cairn or hillock (Thorson, 1936), and ME *pike* a pointed hill, while 'knoll' (OE *cnoll*) could be used to describe the larger features of resistant rock left isolated by slope retreat. Extensive plateau crest areas with low-angled slopes bordered by steep rock scarps may be termed 'nazes' from ON *nes*, a promontory derived through OE *naess*, a projecting ridge (Ekwall, 1960). This term implies a relatively broad flat summit. Narrower, rockier forms, resulting from closer-spaced dissection, could be termed 'nabs' from ON *nabbr*, in ME *knabbe*, a hilltop, or projecting knoll, ridge or peak. Frequently such features are terminated on the downslope side by spur-end rocks or 'knar(r)s', though the usual dialect term in west Derbyshire is 'gnarr'. Perception of soil conditions is usually rendered in terms of gravel or grit from the original gritstone, though the etymology of the term is difficult to trace. It possibly derives from OE *greot* meaning gravelly (Cameron, 1959) as in Greet Lands, Grislow and Gristly Field, dating from at least the mid-fourteenth century in OE *gristleland*, land difficult to work. The word is also known in the analogous Glamorgan areas, for example Greethill.

Dry, smooth crestal areas

The two most common terms for the crest of an escarpment are 'ridge' and 'edge', the latter being most frequently limited to a free face or steep hillside (Wright, 1898–1905). Ridge is derived from OE *hrycg* a long narrow, stony elevation of land and edge from OE *ecg*. Both are of such wide usage that no specific meaning can be assigned, though a possible variant of ridge is 'rigg' (ON *hrygar*), which, though etymologically similar, refers to a very narrow, rocky ridge between adjacent low-order streams. The term 'low' describes the moundlike areas of thinly bedded limestone or fine sandstone which may be found above wide spurs (nazes) or plateaus of the southern Peak District. The term is derived from OW (*hlaw*) and could be rendered lawe or lowe to overcome the apparent ambiguity to modern ears, or else loe (Ray, 1691).

A useful word to describe such low angle, crestal sites with rather thin soils and a broken sward is 'stang', possibly derived from OE *stanig*, meaning rocky land and related to the Nordic words ON *stong*, a pole or measure of land, and Sw *stång* (Thorson, 1936) which means an area of land or the boundary poles delimiting it (Ekwall, 1960). That transference of meaning can take place from a boundary feature to the area of land traversed or delimited by it is a device recommended to terrain analysts. The area of a stang is usually small – less than 4 hectares and with a height range of about 15 metres. The opportunity exists for an ascending scale of features with a definite nomenclature; forms larger than a stang could be termed 'mall', from Welsh *moel* meaning bald (Ekwall, 1960) or the crown of a hill (Cameron, 1959). A feature with smooth slopes of about 80 ha and a height range of 60 m is best described in this way. The Welsh term 'minn' (OW *mynydd*) is frequently found – Wincle Minn. Such features have steeper, smooth slopes. Rounded hills of small basal area could be termed 'copp' (Finn Copp and Pickup), the equivalent form with steep rocky slopes being pike.

Hillslopes and broad crestal areas

There is a wealth of terms available for differentiation of broad watershed slopes, among which are 'brecks' (OWS *brekka*, a slope), which is applied to relatively steep (12–20°), wide slopes, as in Norbreck (Ekwall, 1922) and often altered to brick (Sunbrick) on modern maps. 'Yeld', from OE *hielde*, a slope (Charles, 1938), could be applied to gentler (6–12°) slopes. For locally steeper, smooth slopes (20–30°), 'fall' (ME *falle*, a slope) could be used. This last provides an example of the ME use of a verb as a noun frequently noted in the north and east Midlands (Mustanaja, 1960). The terms rock scarp and free face therefore might well be refined, or at least freed, to describe slopes masked by loose rock or formed by joint planes respectively

The very marked notches or re-entrants in scarp faces are vividly invoked by the term 'nick'. There are many examples: Winyards Nick (SK 2581), Nick i'th' Hill (SJ 8759). Broader features could be termed 'gates', though 'scarth' (ON *skard*) survives in the north. Broader col-like forms with a flat base bordered by steep slopes are called 'wagues'. This rare term of doubtful origin possibly denotes a pole used to draw a rush cart (Wright, 1898–1905) and, by transfer, a defile wide enough to provide passage for so large a vehicle. A word of wider but varied usage is 'rake', OE *hrace*, throat or pass (Ekwall, 1960), which is usually a narrow stony path or cattle track. A most useful term for a dry grassy slope along the strike, above or below a scarp, is 'thwaite'; however, Smith (1956) found that, as applied to the shelving part of a hill, this was not the genuine appellative use, for the word refers to clearance of land.

Smooth dry slopes on backslopes

Various terms can be found to describe such low-angle slopes. 'Bench' is widely used and 'flat' is used indiscriminately by many workers. Subtle variants are available. 'Lench', from OE *hlenc*, hillside (Smith, 1961) or from *hlinc*, a slope or gentle ride (Cameron, 1959) further interpreted as a bank or rising ground (Charles, 1938), is applied in the field to rock-based ledges or to a broad level area with thin soils. Stone Lench and Lenche Moore are known (Cameron, 1959). The word flat in the southern Pennines is almost invariably linked to low angle or level slopes with deep, usually moist soils. For such uses the archaic, or at least sixteenth-century, rendering of flatt might be revived. Sites supplied with seepage moisture from springs on impermeable strata may be termed 'bents', defined as high shelving commons on the shallow declivities of a hill (Halliwell, 1904). Counterscarps formed by vigorous basal sapping could be termed 'sleets' (Cameron, 1959).

Low-order streams, seepage features and microforms on steep slopes

Valleyhead depressions on scarp slopes may best be characterised by the term 'slack', widely used in the Pennines. According to Cameron it is common in minor and field names in Derbyshire though rarely found in the plural. Derived from OWS *slakki*, it represents a shallow valley or hollow on a steep hillside (Thorson, 1936). The modern English form, and modern Danish dialect for this, is slank, a narrow hollow, though Addy (1888) interpreted it as ravine. Normally a slack has only seasonal water flow, though such names as Wet, Stony, Long, Top, Moss and Fearney Slack show the varied aspect of these sites even when restricted to depressions within one rock layer. Moist, wooded, deeply incised, features below scarps are frequently termed 'dumbles' (OE), variously corrupted to dell or dingle (Halliwell, 1904). 'Cown' may be used for complex, deeper hollows, incised into multilayered scarps. Cameron states that this is possibly of Celtic origin and is a variant of the well-known combe. A similar nordic word is 'brund' from ON *brunnr*, a spring or stream head or a brow above a spring.

Terms for first-order streams are numerous. Probably the most acceptable, because of brevity and a definite etymology, are 'sike' and its derivatives, 'syke', 'sick' and 'sitch', used both for natural straight, low-order streams and artificially-straightened streams and major ditches. Sike is derived from ME *sike*, while an earlier form is siche, from OE *sic* or ON *sik* (Cameron, 1959), all meaning a ditch or a small gully, alternatively a small stream in marshy ground (Charles, 1938). The word closely resembles the Danish dialect *sig*, a slow-flowing brooklet. The diminutive form 'strickle' or more correctly 'sickle' also exists. Sitch is interpreted as a modern form of the same OE *sicel* and means a small stream, especially in marshland, which can very often dry up (Cameron, 1959). The names of the features often express this seasonal drying-up of first-order streams and denote their origin in seepage areas rather than as perennial springs. The present Dryshed Bank (Aston, north Derbyshire) was, in 1674, Drysitch Bank (Cameron, 1959). An interesting related term is the verb 'sine', to drain (Wright, 1898–1905), and Siney Sitch exists!

A first-order stream rising from a strong spring, flowing fairly steeply for a short distance in a non-incised course, may be termed a 'gusset', possibly from OE *gyse*, a bursting forth of water; gutter is a more modern form. Low-order backslope streams incised into counterscarps are termed 'gills', OWS *gil*, usually interpreted as deep valleys narrowing at the upper end (Thorson, 1936). The broad dry-floored upper parts of gills may be termed 'hope', a blind valley (Kurath, 1954; Ekwall, 1922). The similarly derived Widdop is interpreted as wide valley (Smith, 1961). 'Clough' is often considered to have the same meaning as gill, though the latter is relatively steep and evenly graded, with the valleys gradually deepening downslope. Cloughs are typically very steep for a short distance downstream from their source, of gentler gradient downslope, maintaining this slope for considerable distances; consequently they have a greater length of deeply incised valley and are more typical of plateau areas, while gills are better represented on more steeply sloping backslopes. This generalisation fits in with their occurrence and with dialect usage, for cloughs invariably occupy ravines. Whether these terms are best used to describe a stream or its valley is a moot point, and

Kurath (1954) notes that by transference clough can be applied to the steep cliffs or bluffs of a gorge. There are many other names available from the southern Pennines to describe low-order streams: 'burn', 'brook', 'goyt' (or 'groit'). Virtually synonymous with clough is 'strind', a water channel (Smith, 1956), which may be reserved for less incised streams, less than 15 m in depth. Cameron notes a reference to clough or strynd for 1533, and to stryndes for 1285. On the floor of such strinds and cloughs a small wet, perhaps peaty area is found frequently separated from the main stream by a low ridge of storm debris. These are termed 'rin' or 'ryn', the original meaning of which was a small relatively stagnant area of water with thicket growth, while the ridge itself could be termed 'rinde', probably from OE, ridge or bank.

A frequently occurring term is 'grains' or 'granes', describing the usually wet land near the junction of two low-order streams. Derived from ON *grein*, the fork or a river, it can apply to the tributary of a river, as in Grain Water, or to the meadows or grassy plot at the fork of a river. It is similar to Swedish dialect *gren*, the angle which two branches of a tree make with one another (see groin, Elgee, 1912). Place names such as The Grainings and Grains-in-the-Water refer to similar features. Hence the word may be restricted to areas of wet land near the confluence of low-order streams and the word 'twizle' (OE *twisla*, confluence) may be used to describe the low ridge upslope from the junction of two such streams. Ekwall (1922) explains the name Entwistle in this way.

Terms descriptive of streams and valleys of higher order are numerous, though very inconsistent in their application. As the size of such features is such as to include many varied sites, it is inappropriate to list them, for many are non-specific and of widely varying common usage.

Moist sites, usually valley heads, seepage hollows and peatland

The wide, shallow, saucer-shaped forms at and above the head of many first-order streams, with deeply leached soil on sandstones, or gleys on shales overlying the sandstones, have been termed seepage hollows (Bunting, 1961). They have linear flow only in wet periods and are often flooded in late winter. Although the term hollow is encountered frequently in the Pennines, applied to many and varied sites, a useful specific term is that provided in, for example, Hollow 'Sponds' (SJ 9880), derived from OE *spon* or ON *spann*, meaning a shaving, though it can mean a bed or couch (Wright, 1898–1905). Although there is no linguistic relationship between the two words, spond recalls the effect of ponding of water, and thus the form and the state of such a site are adequately expressed. A term descriptive of a spond which has been eroded or degraded by headward gullying, and with an integrated linear flow, at least in winter, might well be 'leech' from OE *laecc*, meaning a stream or bog. Similar in form is 'lache' or 'laitch', interpreted (Ekwall, 1960) as a stream slowly flowing through bogland. Smith (1956) states that all these are north Midland dialect terms for a small stream flowing through boggy or muddy lands. Such names as Fenny, Wythen, Rushby and Black Lache are common, but leech might best be reserved for the site, and lache for the aspect of seasonal linear flow, though the two are variants of the same word.

Characterisation of wet, level or gently sloping areas sited away from first-order valley heads on watershed slopes presents few difficulties. The terms marsh and swamp are rarely used in upland areas and a clear distinction between 'moor' (OE *mōr*) and 'moss' (OE *mos*) suffices. Moss usually refers to a very wet, virtually flat interfluve or crest, which is covered with peat to a depth of 30 cm or more. Peaty valley heads, too, may be similarly described: Flat Moss or Hollow Moss. Moor should be restricted to drier interfluve sites of larger area and often with slopes in excess of 3°.

The wettest areas of moss are best referred to as 'mires' (ON *myrr*), which usage is confirmed in many parts of the central Pennines, though 'meas(e)' (OE *meos*) is preferred in the southern Pennines (Ekwall, 1960). A leech would be better drained than a mire, though the former may grade into the latter in an upslope direction. The close relation of the two terms is brought out by the older name for Leash Fen (SK 2973) recorded as Lechemeyre in 1503 (Cameron, 1959). Perhaps an even more extreme form of moss, permanently saturated, is 'quakeland', which may be defined as moving, trembling bog (Warrack, 1930), derived from the dialect Quaker or shaker grass (*Briza media*). Ray (1691) provides the adjective mizzy, meaning quagmire, which is perhaps derived from, or related to, mease. It is also a Lancashire dialect word, formed from the adjective (ME *misy*) meaning a swamp or bog.

The term 'haggs' (ON *hagg*) is used for a small area of eroded peat or broken ground within a peat moss; Warrack (1930) defines a hagg as a peat hole, a water hollow, dry in summer, wet in winter. Smith (1918) refers to peat broken up by haggs or channels, though the more common meaning of hag(g) is a mound or isolated hummock (Addy, 1888).

Perennially wet level sites on the lower slopes of crestal areas, often on a bevelled surface, may be termed 'flasks' from ME *flashe* (Thorson, 1936) meaning pool (Ekwall, 1960), or from OD *flask*, swampy ground, a pool. The term is not to be confused with flush, identified by Smith (1918) as the 'green gairs' of grassland around the outflow of springs or where two streams meet.

The backslopes of gritstone cuestas usually contain many smaller and more localised moist areas, most widespread of which are the narrow, elongated, moist areas trending downslope towards valley heads as percolines. Broader intervening areas of drier aspect have been termed intercols (Bunting, 1961). However, there are important variants of this general scheme. 'Swang' is often used to describe a wet area forming flat, grassy land at the foot of a declivity and liable to be flooded (Wright, 1898–1905); the term may thus be applied to sites on thin shale partings on gentle backslopes of up to 4°, or to those sites where accumulation of fines and slight gleying gives a distinctive soil, a wet surface and fresh growth of grass even in dry summers. It is apparently derived from ON *svangr*, a hollow place (Smith, 1918). Similar sites with moisture accumulation on gritstone backslopes, with no apparent change of rock type, result from seepage along bedding planes or from the collection of water at the junction of two or more percolines. On such sites relative overdeepening of the sandy soils is found. The areas are small and are marked by a change from *Calluna* to *Molinia*, *Erica* or *Vaccinium oxycoccus*. They may be termed 'scopes', which in Lancashire dialect means a kind of basin or scoop for lading water (Halliwell, 1904). The term also recalls the scooping effect of the upslope migration of this point of excessive corrosion, previously termed a seepage cusp (Bunting, 1961).

The obvious word 'hole' is used rather indiscriminately in much of northern England as a diminutive form of hollow rather than referring to a definite hole. Wet Head Hole, Turbury Hole and Foul Hole are some of the names attached to natural peaty depressions on backslopes. The idea is qualified where peat excavation has occurred, for such sites can be described as 'pitts', OE *pytt*, a hollow or pit (Ekwall, 1960), though the term is usually confined to artificial diggings – Peat Pitts or Raddle Pitts. Raddle (ochre) has various spellings (reddle, ruddle) and was used as a dye for marking sheep (Hunter, 1892). Finally the small pools of peaty water often found in transitional areas between moss and moor may be referred to as 'dub(b)s' or 'dubland', OE *dubb*, a pool or ME *dubbe*, a muddy pool or stagnant water (Cameron, 1959).

It is difficult to assign a definite meaning or usage to the expressive word 'wham'; features bearing this name range in size and form from small, wet, peaty sponds to large, fully developed ombrogenous bogs. Possibly the term is best kept to define an area of peatland sited on a broad, low-angled slope, usually a valleyside bench or broad ledge, for example Slade Wham (SE 1561). Such sites are often backed by a crestal sandstone which provides abundant seepage water, which may bisect the wham, as at Foul Sike Whams. The distribution of the term itself suggests that it may be derived from ON *hvammr*, a small valley or grassy morass of wide dimension and general flatness (Thorson, 1936). A variant is the peaty or grassy area elongated downslope along the line of an incipient stream or percoline, the 'mea', a term more often used in northeast Yorkshire than in the Pennines and derived from the dialect for cranberry (*Vaccinium oxycoccus*), the meaberry.

Sites with a subsoil of clay, till or shale often give rise to extremely wet areas, but only seasonally where the drainage gradient is away from the site. With the general impermeability of the subsoil, only the surface is wet in winter, the surface layers drying out and cracking open in the summer. Such areas may be termed 'slapes', probably from ON *sleipr* meaning a slippery place (Thorson, 1936). However, the derivation is obscure, for OE *sloep* (Cameron, 1959) means soft or smooth. Ekwall (1960) mentions that this word is only found in charters and Easther (1883) states that for the Huddersfield area, although familiar to some, it does not belong to the dialect. Examples of its use are Sleep Hill and Slape Gill, and it appears in field names, Slipe and The Slip. Slapes is common in the Midlands (*vox usitatissima*, Ray, 1691, p. 62). The frequent growth of *Juncus* on such sites is reflected in the word 'seeves' and Ray (p. 61) mentions 'Seavy ground . . . such as is overgrown with rushes'. It is perhaps dialect, seave from ON *sef*, sedge or rush, and it survives in Seavy Syke Hole (SD 8196). Similar to slapes is the term 'fleets' (Ward, 1939), which could be applied to the thin shale partings capping flattened spurs, forming wide ill-drained areas from which water flows on three sides.

The importance of the impermeability of shale partings and clayey areas is revealed where seepage and runoff from such sites moves downslope on to an underlying sandstone layer. Such sites have deep, moist soils underlain by a saturated pseudogley C horizon, even by deeply rotted rock. These deep soils have low resistance to erosion and are easily gullied. Such sites are often known as 'sypeland', from the dialect sipe, to percolate slowly (Warrack, 1930), or to soak through.

Conclusions

This essay has examined a selection of dialect words which could be used in future to categorise certain dry and moist sites and first-order streams for which no really definite terms exist. More distinct dialect terms could be found in the northern Pennines and in the Lake District for glacial forms, but here an attempt has been made to clarify those terms especially applicable to watershed slopes on sandstones and shales, with their widely varying moisture characteristics.

There is a richness of dialect terms relating to states of ground, with their implications of slopes of varying length and angle, and of soils of differing surface character, particularly of texture. Often similar terms are placed together for emphasis. Their qualification by words descriptive of vegetation, tenure, land use and other site characteristics are both colourful and varied – far too varied to have been dealt with here.

As the understanding of slope formation and landform development broadens, the need for objective, meaningful terms for describing slope elements and facets will increase and the necessity for characterising valley forms and stream types will be more urgent. It may well be that these terms from the Pennines and terms from other areas will have a role to play in both descriptive and empirical geomorphology.

References

ADDY, S. O. (1888) *A Glossary of Words used in the Neighbourhood of Sheffield*, vol. 22, part 57, English Dialect Society.

BAULIG, H. (1956) 'Vocabulaire franco-anglo-allemand de géomorphologie', *Publs. Fac. Lett. Univ. Strasbourg*, No. 130.

BOILEAU, A. (1971) 'Toponymie et contact des langues en Belgique', in H. Dorion, ed., *Place Names and Language Contact*, L'Université Laval, Québec, pp. 42–89.

BRIDGES, E. M. and DOORNKAMP, J. C. (1963) 'Morphological mapping and the study of soil patterns', *Geography*, **48**, 175–81.

BUNTING, B. T. (1961) 'The role of seepage moisture in soil formation, slope development and stream initiation', *Am. J. Sci.*, **259**, 503–18.

BURRILL, M. F. (1954) 'Toponymic generics in the United States', *Bull. Geol. Soc. Am.*, **65**, 1892.

CAMERON, K. (1959) *The Place Names of Derbyshire, parts I–III*, English Place-Name Society, Cambridge.

CHARLES, B. G. (1938) *Non-Celtic Place Names in Wales*, Monograph 1, London Medieval Studies.

COTTON, C. A. (1963) 'A new theory of the sculpture of middle latitude landscapes', *N.Z. Jl. Geol. Geophys.*, **6**, 769–74.

CURTIS, L. F., DOORNKAMP, J. C. and GREGORY, K. J. (1965) 'The description of relief in field studies of soils', *J. Soil Sci.*, **16**, 16–30.

DONNELLAN, R. S. (1972) 'Proper names and identifying descriptions', in D. Davision and G. Harman, eds, *Semantics of Natural Language*, Dordrecht, pp. 356–79.

DORION, H. (1972) *Les noms des lieux et le contact des langues*, L'Université Laval, Québec.

EASTHER, A. (1883) *A Glossary of the Dialect of Almondsbury and Huddersfield*, vol. 21, part 39, English Dialect Society.

EKWALL, E. (1922) *The Place Names of Lancashire*, Manchester.

EKWALL, E. (1960) *The Concise Oxford Dictionary of English Place Names*, 3rd edn, Oxford University Press.

ELGEE, F. (1912) *The Moorlands of North-eastern Yorkshire: their natural history and origin*, London.

EYRE, S. R. (1957) 'The upward limit of enclosure on the East Moor of north Derbyshire', *Trans. Inst. Br. Geogr.*, 23, 61–74.

HALLIWELL, J. O. (1904) *A Dictionary of Archaic and Provincial Words, Obsolete Phrases, Proverbs and Ancient Customs from the Fourteenth Century* 6th edn, London.

HARTSHORNE, R. (1959) *Perspective on the Nature of Geography*, Chicago, Rand-McNally; London, Murray, 1960.

HUNTER, J. (1892) *The Hallamshire Glossary*, London.

JOHNSON, R. H. (1957) 'An examination of the drainage pattern of the eastern part of the Peak District of north Derbyshire', *Geogr. Stud.*, **4**, 46–55.

KRUMBEIN, W. C. and SLOSS, L. L. (1963) *Stratigraphy and Sedimentation*, 2nd edn, San Francisco, W. H. Freeman.

KURATH, H., ed. (1954) *Middle English Dictionary*, University of Michigan Press.

LEWIS, G. M. (1959) 'Some recent American contributions in the field of landform geography', *Trans. Inst. Br. Geogr.*, **26**, 25–36.

LINTON, D. L. (1955) 'The problem of tors', *Geogr. J.*, **121**, 470–87.

MELTON, M. A. (1958) 'Geometric properties of mature drainage systems and their representation in an E4 phase space', *J. Geol.*, **66**, 35–56.

MONTAGUE, C. E. (1949) *A Writer's Notes on his Trade*, Penguin.

MORGAN, R. P. C. (1972) 'Observations on factors affecting the behaviour of a first-order stream', *Trans. Inst. Br. Geogr.*, **56**, 171–85.

MUSTANAJA, T. F. (1960) *A Middle English Syntax*, Memoires de la Societé Néophilologique de Helsinki.

PLAISANCE, G. (1958) *Léxique pédologique trilingue*, Paris.

POIRIER, J. (1965) *Toponymie, méthode d'enquête*, L'Université Laval, Québec.

RAY, J. (1691) *A Collection of English Words not generally used, with their significations and original, in two alphabetical catalogues, the one of such as are proper to the northern, the other to the southern counties*, 2nd edn, London.

SMITH, A. H. (1956) *English Place-name Elements*, 2 vols, English Place-Name Society, Cambridge.

SMITH, A. H. (1961) *The Place-names of the West Riding of Yorkshire, part II*, English Place-Name Society, Cambridge.

SMITH, W. B. (1918) 'The improvement of hill pasture', *Scott. J. Agric.*, **1**, 3.

STARKEL, L. (1959) 'Development of the relief of the Polish Carpathians in the Holocene', *Przegl. Geogr.*, **31**, 121–40.

STRAW, A. (1962) 'Glacial drift in the area around Bakewell, Derbyshire', *E. Midld. Geogr.*, **3**, 72–80.

TAVERNIER, R., JACK, G. V. and BOALCH, D. H., eds (1960) *Multilingual Vocabulary of Soil Science*, Rome.

THORSON, P. (1936) *Anglo–Norse Studies: an inquiry into the Scandinavian elements in modern English dialects*, Amsterdam.

UNITED STATES DEPARTMENT OF AGRICULTURE, SOIL SURVEY STAFF (1960) *Soil Classification: a comprehensive system, 7th approximation.*

WARD, G. H. B. (1939) 'Lost records and place names of Midhope Moors, Mickleden and Cutgate', *Sheffield Clarion Ramblers Annual Handbook*, pp. 69–81.

WARRACK, A., ed. (1930) *A Scots Dialect Dictionary*, 2nd edn, Chambers.

WOOD, A. (1942) 'The development of hillside slopes', *Proc. Geol. Ass.*, **43**, 128–40.

WRIGHT, J. (1898–1905) *The English Dialect Dictionary*, 6 vols, London.

E. Derbyshire, L. W. F. Page and Rosemary Burton

2 Integrated field mapping of a dynamic land surface: St Mary's Bay, Brixham

Of all extensive natural slopes in humid, mid-latitude terrains, the coastal cliff in weak rocks is perhaps the most vulnerable to rapid natural degradation. This vulnerability may be aggravated locally by the proximity of high-density human settlement. As public pressures on cliffed coastlines increase, Steers's (1961) plea for greater scientific endeavour in such areas will assume greater significance.

With such considerations in mind, several coastal sites were examined in 1968, with a view to regular, long-term monitoring of changes along a cliffed section of the coast, with the aim of assessing the relative importance of natural and man-induced processes. St Mary's Bay, 1.4 km east of Brixham, was ultimately selected for study (Fig. 2.1). The 62 m high cliff is developed mainly in the Eifelian Slates series of Middle Devonian age (Ussher, 1903), consisting of mechanically weak micaceous slates and shales in an anticlinal structure with much local contortion, across which a well developed slaty cleavage dips to the southwest. The more resistant Brixham limestones, with their abundant quartz veins, form the promontories to the north and south of the bay. Behind the cliff line is a landscape of smoothly convexo-concave valley and ridge forms running east to west with a relative relief of barely 15 m. The entire coastal section from Berry Head to Froward Point has been evaluated as category A by the Nature Conservancy (Countryside Commission, 1968 and 1969) and scheduled as a site of special scientific interest on biological and geological grounds.

The programme of study involved annual visits in late March from 1968 to 1973 by undergraduates being directed in basic topographical and thematic mapping. The distribution of phenomena was recorded each year as a first stage in the investigation of cause and effect relationships, the results of which allowed time to be considered as a factor in measurable landscape change (Fig. 2.2).

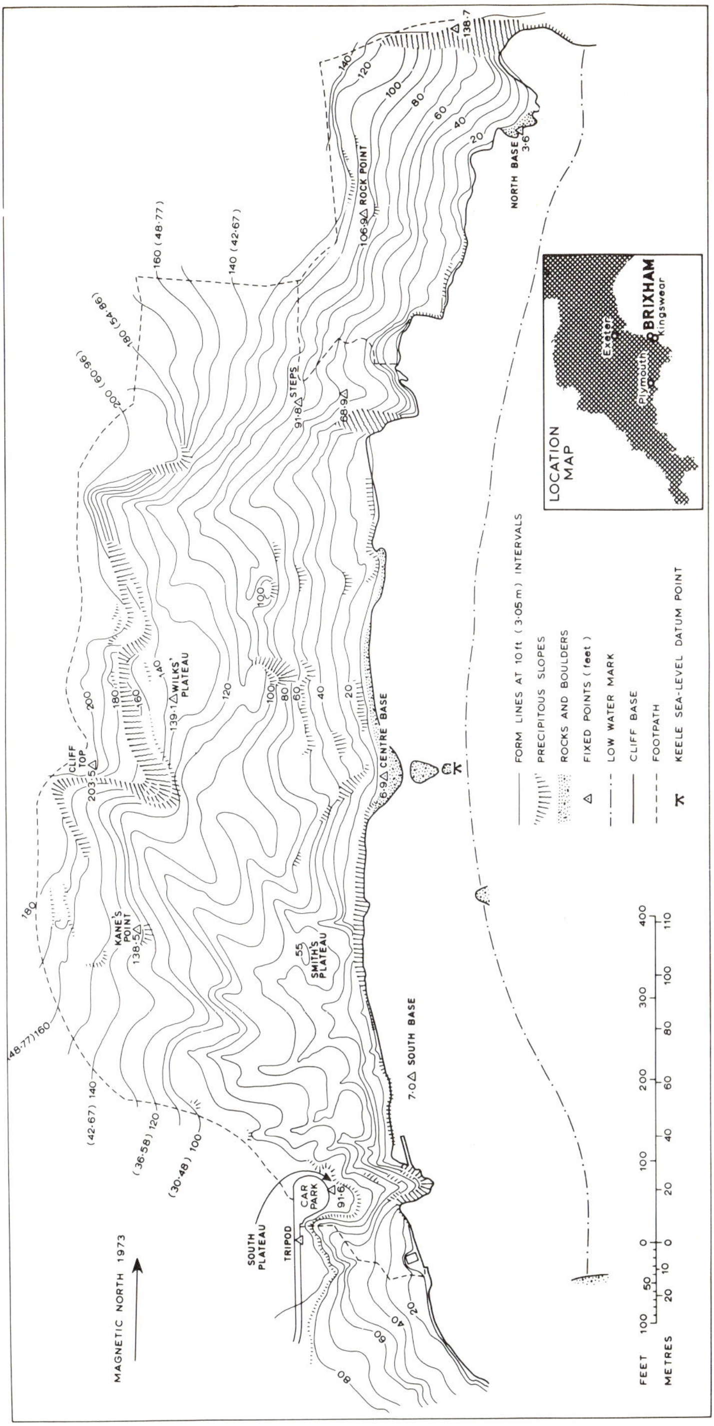

Fig. 2.1 Hypsometric map

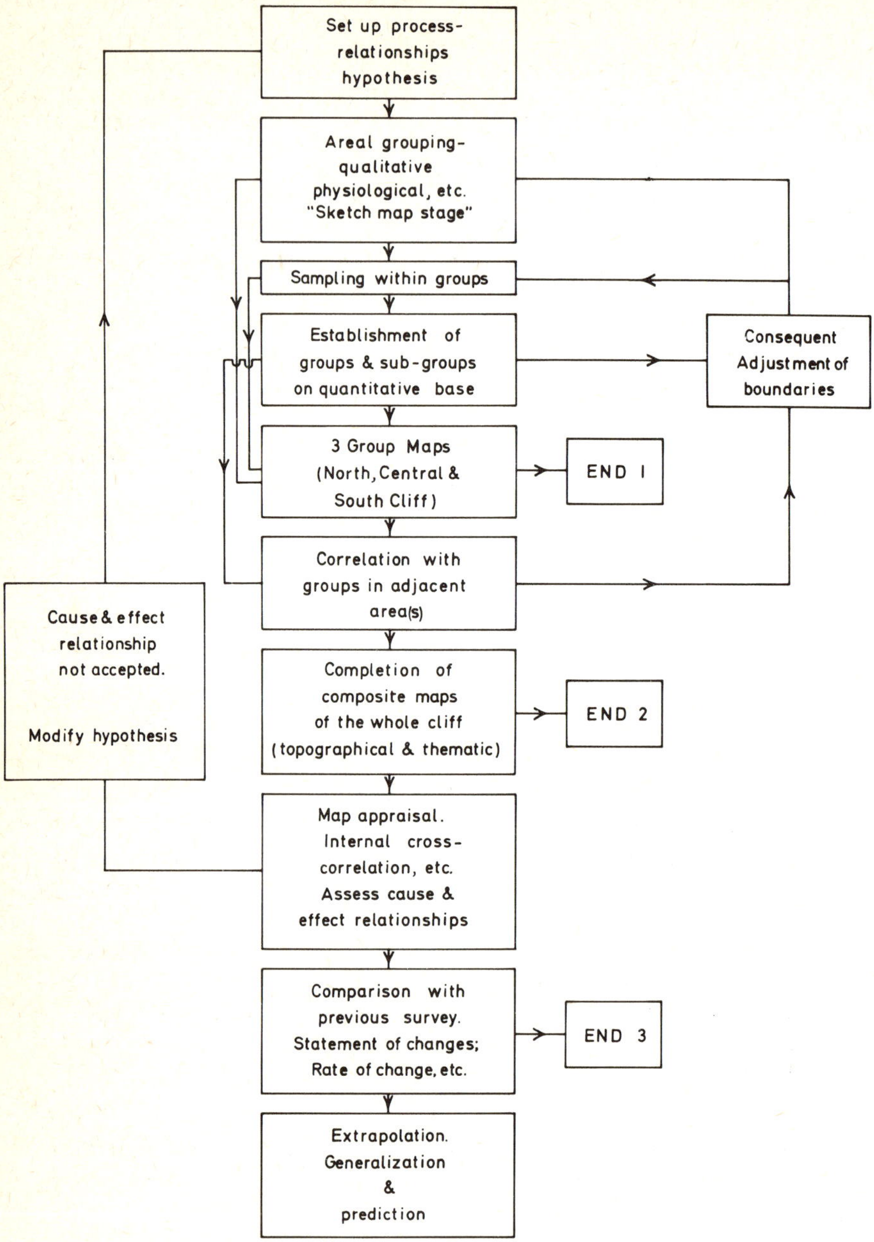

Fig. 2.2 Field survey: functional design

Mapping techniques and methods

A first requirement in work of this kind is a base map with sufficient detail to permit the adjustment of discrepancies contained in field plots executed by less accurate methods (Fig. 2.1). Between 1968 and 1970 control was fixed by plane table triangulation from a measured base on the beach. In March 1972 a theodolite triangulation was carried out to provide control points for three plane table parties, one working on each of the northern, central and southern sections of the cliff (Fig. 2.3). The density of the control was such that no point on the cliff lay more than 100 m from a control point. Cliff topography was mapped by standard plane table methods adapted to overcome the specific difficulties provided by dense vegetation cover and steep slopes. Form lines at 3 m (10 ft) intervals were inserted by interpolation from spot heights fixed by Indian clinometer and tied to the trigonometrical heights of the control points based on an arbitrary sea level datum point (Fig. 2.1).

Despite the inexperience of the student surveyors, only a slight adjustment of results has been necessary and the degree of internal consistency of the mapping is high. Independent traverses and point sampling confirmed this, and the results which form the basis of the conclusions have been checked in the field.

Results

The results of the field investigations are mapped to indicate the distribution and character of slopes, surface materials and vegetation for the 1970–73 period. Cliff features are shown in hypsometric and morphological maps (Figs. 2.1, 2.4 and 2.5). The cliff face comprises a steep, planar surface in the north, a complex area of intermediate gradients with a stepped form in the centre and, in the south, an area of spurs and re-entrants with gentle to intermediate declivity giving way in the extreme south to concave slope forms of steep over-all gradient. Absolute relief is greatest in the centre of the cliff (Cliff Top, 62 m KD[1]) with maximum relative relief (up to 12 m/225 m^2) southeast of this point. Some three-quarters of the surface slopes more steeply than 27°, but steep rock slopes (greater than 42°) account for only 8 per cent of the area mapped; only 2 per cent of the cliff slopes at less than 5°.

In view of these gradients it is not surprising that the soils and regolith show evidence of physical disturbance (Fig. 2.6). The *in situ* soils are of two main types, a rendzina-like profile associated particularly with the limestones and mudstones of the headlands, and a more extensive brown calcareous soil developed on the slates and shales underlying most of the cliff. The rendzina-type soils are best developed on the northwestern cliff top, being clay-rich with a good crumb structure and generally dark red-brown (2.5 YR 3/4) to reddish brown (2.5 YR 4/4 to 5 YR 4/4). Soil

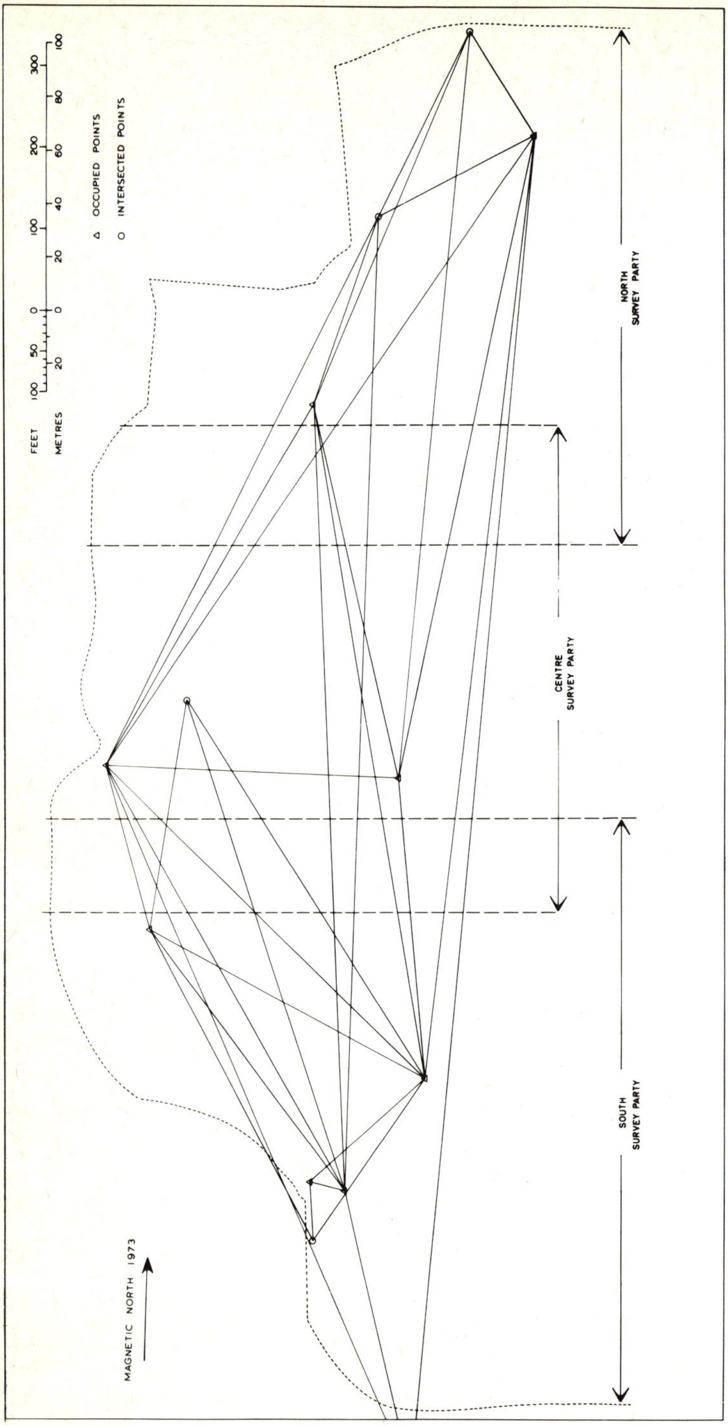

Fig. 2.3 Triangulation control diagram

Fig. 2.4 Breaks and changes of slope

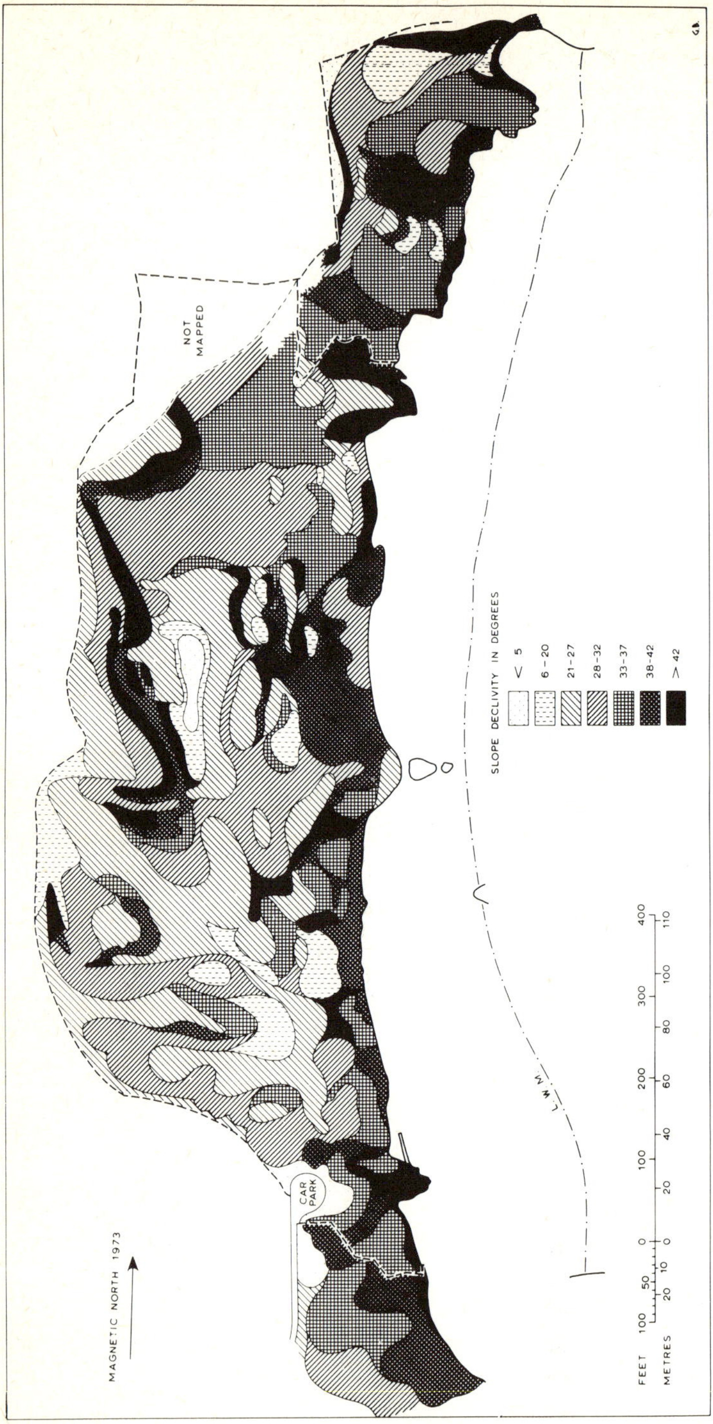

Fig. 2.5 Slope facets

reaction rises from the neutral range (6.8 to 7.5) in the mull layer to distinctly alkaline (8.6 to 8.9) immediately above the C horizon. Bulk densities range from 1 235 kg/m^3 in A horizons to 1 800 kg/m^3 in G horizons and 1 920 kg/m^3 in C horizons. These soils are generally thin (4 to 20 cm, averaging 8 to 10 cm) and free draining, except where rotational shear sliding is common, as in the vicinity of South Plateau. Here physical disturbance frequently produces multiple profiles, some of which may be affected by seasonal waterlogging. Where sliding replaces rotational slip, as on the north cliff, soils may be very shallow or in various stages of truncation.

The brown calcareous soils also display abundant soil fauna, roots and dispersed organic matter. Where best developed, however, these silty loams are thicker (A + (B) horizons 15 to 50 cm), more stony, generally drier and with a greater structural range (fine crumb to blocky with depth) than the rendzina-like profiles. Mulls range up to 12 cm in thickness above a yellowish red (5 YR 5/6) to dark brown (10 YR 3/3 to 4/3) or brown (7.5 YR 5/4), neutral (pH 6.8) A horizon overlying a yellowish brown (10 YR 5/4) to olive yellow (2.5 Y 6/4) transitional (B) horizon. Hydromorphic characteristics due to seasonal saturation are found in these soils, particularly along drainage lines, the only large area of gleys occurring beneath willow thicket in the south.

Degradation of the brown calcareous soils ranges from discrete masses of thick soils, complete with vegetation such as bramble and herbs, moving downslope following simple landslide rifting (about Kane's Point) to variously disturbed shallow phases on the generally steep slopes (Fig. 2.6). Small-scale slumping is important and terracettes are common, especially in areas with a bracken and herbs or disrupted blackthorn and bramble cover. Some areas are so steep, and granular creep, sliding and increments from rockfall are so frequent, that plant cover remains discontinuous. Such skeletal soils are transitional to scree.

Regolith on the cliff has been mapped and classified on the basis of mean diameter. This is controlled by three factors: the fissility of the bedrock, its history of surficial movement and its susceptibility to comminution in the course of weathering. Apart from the predominance of finer, silt-rich regolith in the extreme south and the very coarse debris making up the large cone in the extreme north, grain size varies widely over short distances.

It is evident from the overall form of the cliff that the major mass movements described over many years from several sections of the coast of southwest England (Woodward and Ussher, 1911; Osman, 1917; Aber, 1940, 1962) have also occurred here (Fig. 2.7). Many structures visible on the cliff are secondary, in that cleavages, joints and bedding planes have been widely exploited during superficial movements of the bedrock. Three mass movement provinces may be recognised on the basis of

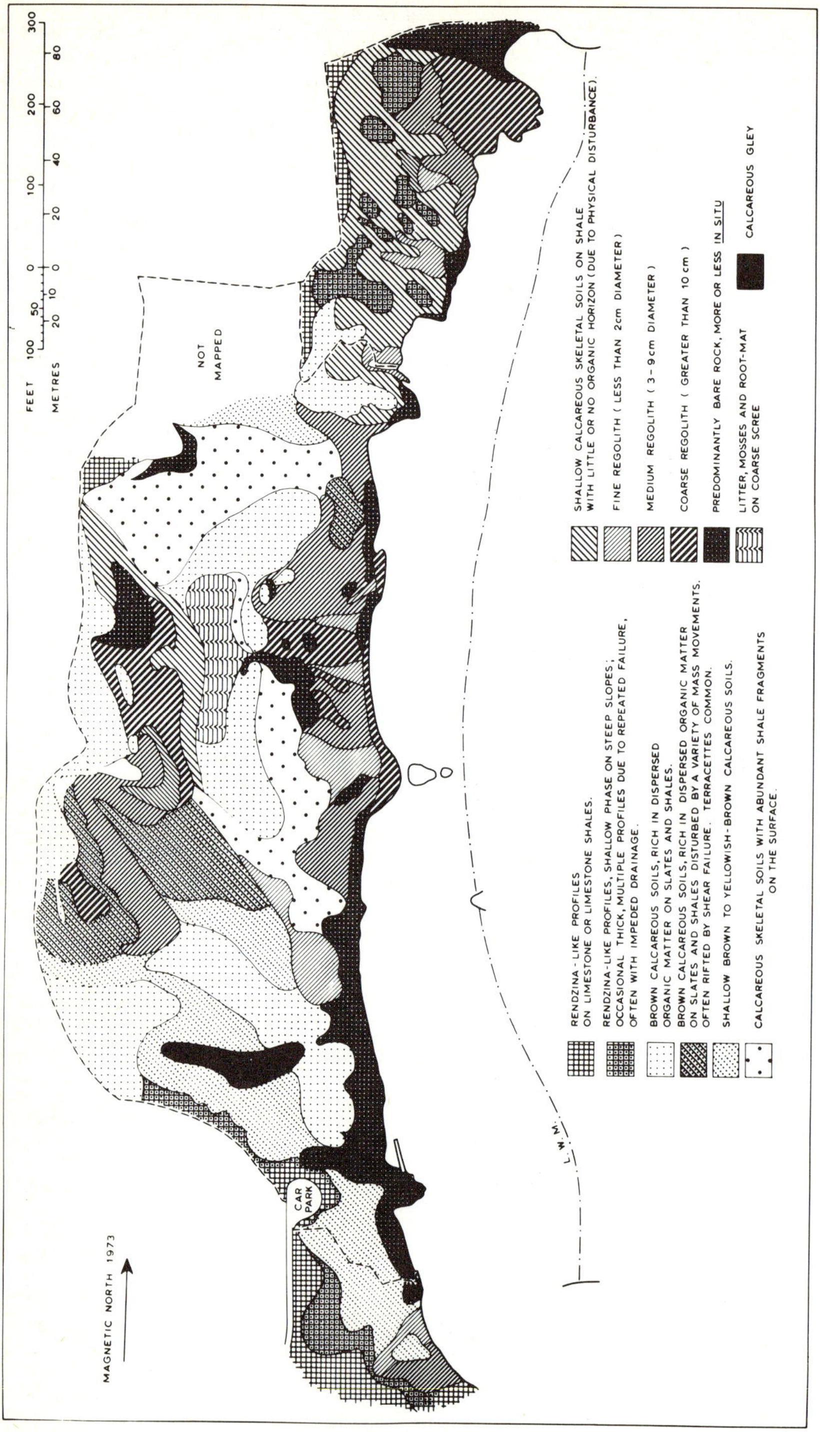

Fig. 2.6 Soils and regolith

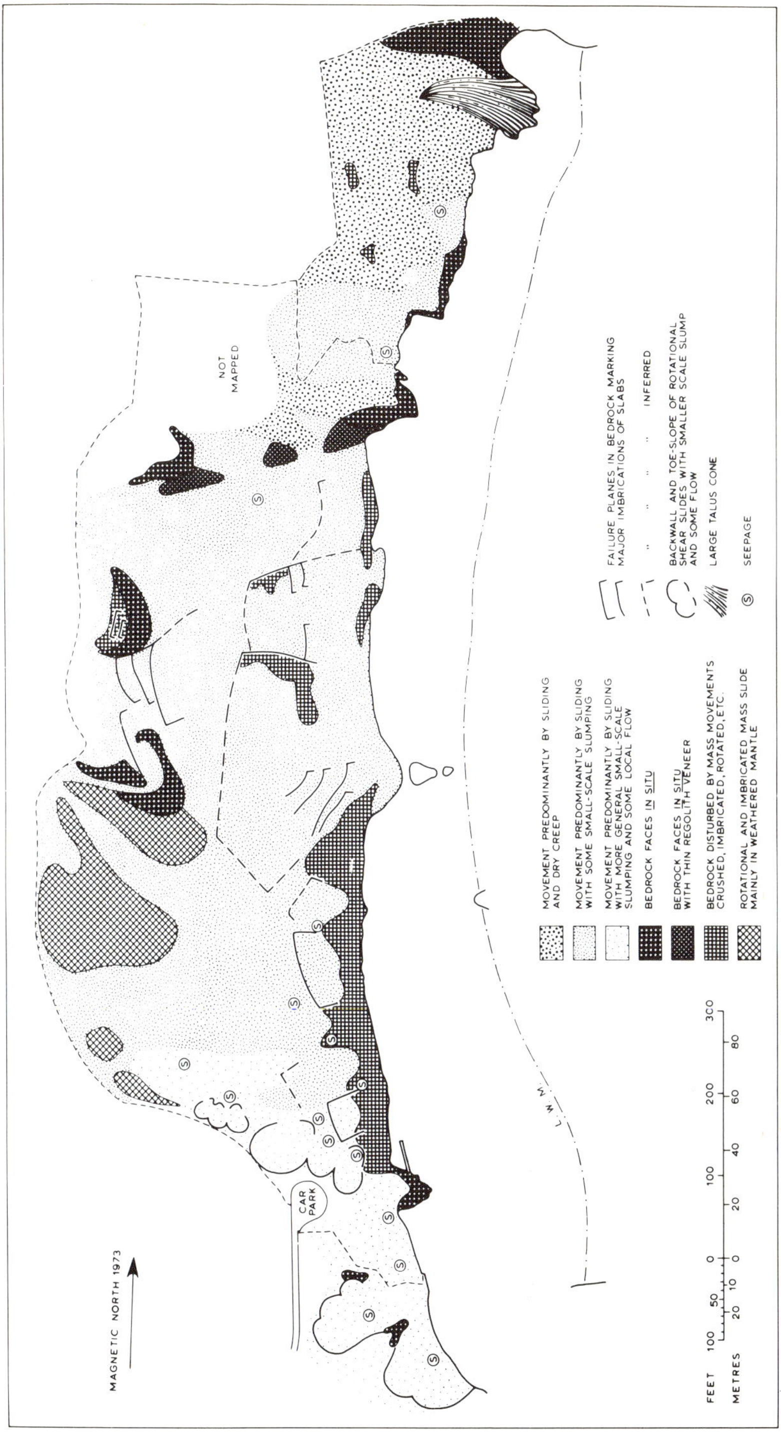

Fig. 2.7 Principal mass movement types

moisture levels, depth of the weathered mantle and the extent of bedrock disruption. The northern section of the cliff is a steep, rather dry environment, with few springs. The dominant controls on form here are the well developed slaty cleavage which dips steeply (about 40°) to the southeast and, to a lesser extent, the bedrock dips which are more variable. Disruption of bedrock is only superficial and the generally thin regolith is moving predominantly by the processes of rockslide, with seasonal talus and soil creep. A major morphological feature of this part of the cliff is the large debris cone made up of material ranging in size from silts to boulders measuring up to 4 by 3 by 2 metres. The large calibre of the material forming the bulk of this feature indicates that landsliding has been the major process involved, although parts of its surface still move by rockslide, talus creep and soil creep (Plate 1).

1. Talus creep on the debris cone at the north end of St Mary's Cliff. Coarse talus thrusting and overriding vegetation and mixed debris, August 1973

The large central section of the cliff (north steps to Kane's Point) is a moist area with many reversed slopes. The most important influence on form and drainage is the series of slab failures along slaty cleavage planes, which have produced the large promontory (120 by 80 m) seaward of Wilks' Plateau. This disrupted mass has become imbricated (Plate 2) and

2. Imbricated bedrock above centre cliff, St Mary's Bay, seen from the north, March 1973

has broken down further to produce the large talus slope to the north and south of Centre Base. In addition to large-scale disruption of the bedrock, rockslide and talus creep are important. The conditions of greater soil moisture and generally thicker regolith in comparison with those found on the north cliff result in a combination of rockslide, talus creep and small-scale slumping.

The southern part of the cliff face is distinctly moist with abundant seepages. Here, rotational shear sliding of a thick weathered mantle rich in silt and clay is the most notable type of mass movement, together with limited earthflow. Rock slide is less important than small-scale slumping, which forms terracettes on some grassy slopes.

A certain level of stability is required for the development of any plant cover, and a substantial and uninterrupted period of uniform conditions is necessary for the climax (in this case deciduous woodland) to become established. At present only shrub and herbaceous species are represented on the cliff (Fig. 2.8) and the vegetation succession appears to have been repeatedly disrupted by mass movements. However, the instability of the environment is not the only factor that has modified the development of the vegetation community. Soil type, drainage and exposure to wind also play an important role, and the distribution of different vegetation types on the cliff is a result of the interaction of these factors.

There are three main types of plant community: the scrub, generally dominated by blackthorn (*Prunus spinosa*), but with considerable variation in species composition; species-rich herbaceous communities in which grasses are dominant; and open pioneer communities in process of colonising bare ground.

One form of the pioneer vegetation is well represented on the skeletal soils of the cliff's northern section. On the steeper slopes bare rock accounts for at least 50 per cent of the total area and many of the species typical of the community are drought resistant. The pioneer species, colonising joints and crevices where soil begins to accumulate, include *Sedum forsteranum, Sedum album, Plantago* sp., *Bellis perennis, Festuca rubra, Centranthus ruber* and *Cardamine hirsuta*. Where a slightly deeper soil has been established, and a turf mat developed, a greater variety of species can be found, including *Galium mollugo, Galium aparine, Leodoton* spp., *Lotus corniculatum, Luzula campestris, Agrostis stolonifera, Trifolium* spp., *Poterium sanguisorba* and *Carlina vulgaris*. The shallow rendzina-type soils in the north can support a sparse scrub of *Rubus* sp., *Pteridium aquilinum* with *Rosa* sp., and a ground layer of *Hedera helix* and herbaceous species. *Prunus spinosa* is present only in a stunted form, barely reaching 0.6 m in height on steep and exposed slopes.

The blackthorn scrub adopts a more vigorous habit on the deeper

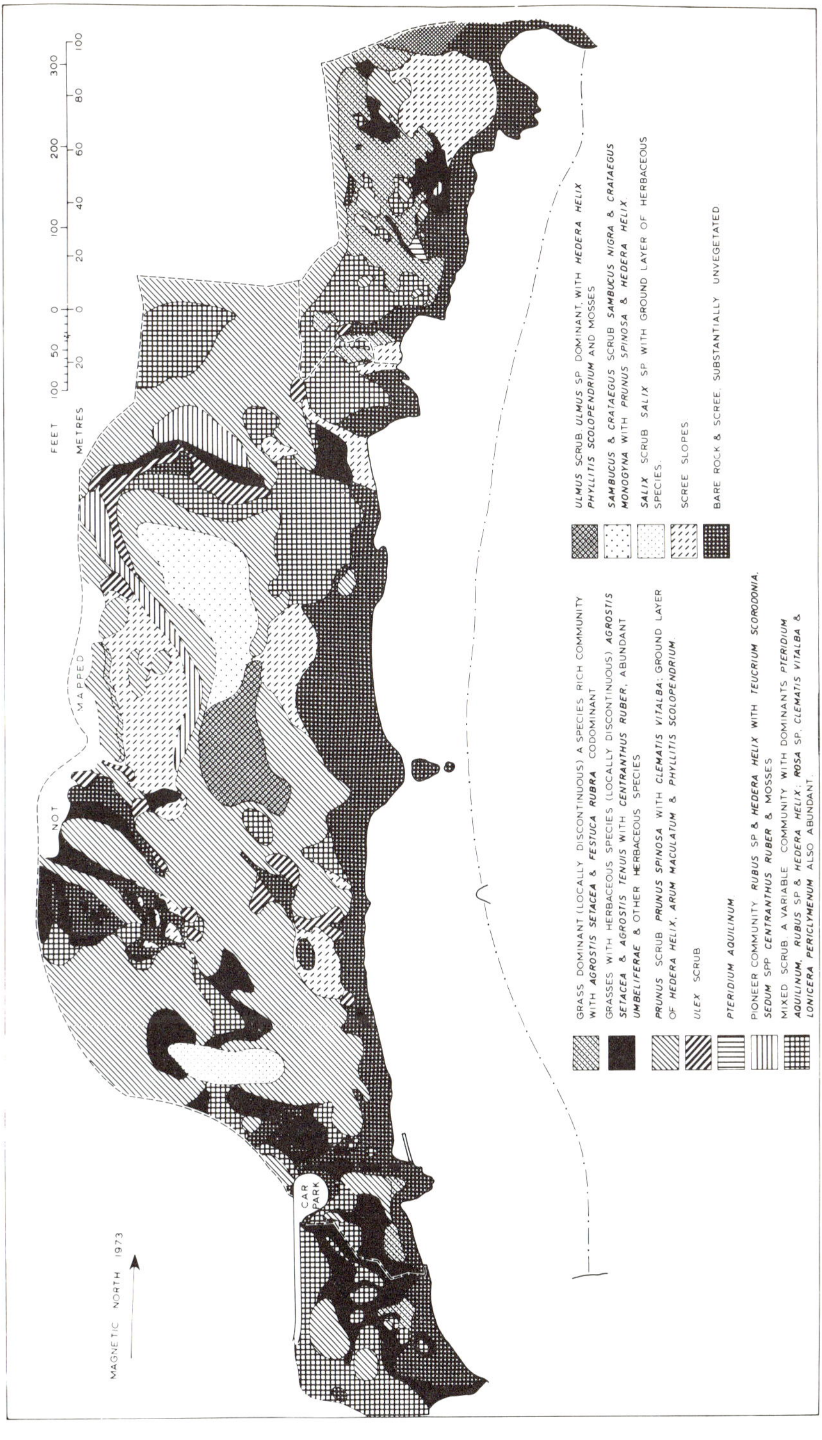

Fig. 2.8 Principal vegetation communities, March 1973

brown calcareous soils developed on bedrock near the northern cliff top, reaching in more sheltered locations a height of about 3 m. Few other species are present, but *Clematis vitalba, Rubus* sp., *Lonicera periclymenum* and *Hedera helix* make some contribution to the dense canopy layer. As little light can penetrate the canopy, the ground layer is sparse, with a high proportion of bare ground. *Hedera helix*, adopting a creeping habit, is dominant but *Phyllitis scolopendrium* and *Arum maculatum* are also present.

Scrub on the central section of the cliff is different in character. *Prunus spinosa* is replaced as dominant by *Ulmus* sp., or by *Sambucus nigra* with *Crataegus monogyna*. However, the composition of the ground layer is more or less uniform throughout the different types of scrub, with *Hedera helix* in both creeping and climbing forms, *Phyllitis scolopendrium*, mosses and *Arum maculatum* the most abundant species. At the western margin of Wilks' Plateau, the scrub rapidly gives way to a pioneer community that extends on to the lower slopes of an area of mobile scree. The general situation is one of transition from scrub first to a vegetation of *Rubus* sp., *Hedera helix, Teucrium scorodonia* and grasses on skeletal soils, then to an open community characterised by drought resistant pioneer species (*Sedum forsteranum, Sedum album, Plantago* sp., *Rumex* sp.), mosses and lichens, and finally to the bare regolith of the mobile scree in which fragments of vegetation are brought down with the scree from the cliff top (Fig. 2.9).

Scrub of very varied species composition is widespread throughout the moist area of finer regolith between Wilks' Plateau and South Plateau. Mass movements are likely to cause fundamental changes in the environmental factors affecting plant growth. For example, the aspect, angle of slope and exposure conditions are subject to change, which may increase the light reaching the ground layer of a community, leading to a more vigorous development of certain of its members. Soil structure and soil drainage are also likely to be affected by mass movement. These processes have the effect of initiating modifications of the community from within, and cause the differential growth of existing members of the plant association. The present scrub vegetation, where locally *Prunus spinosa, Rubus* sp., *Hedera helix, Rosa* sp., *Pteridium aquilinum* and even creepers (*Clematis vitalba* or *Lonicera peryclymenum*) assume dominance, has probably been modified by these processes, and the variation in species composition appears to be the result of such changes.

The abundant seepages in this section of the cliff have a profound influence on the composition of the scrub vegetation. *Salix* sp. replaces *Prunus* sp. as the dominant on the gley soils south of Kane's Point. The more open canopy allows a vigorous ground flora of herbaceous species to develop. *Oenanthe pimpinelloides, Scrophularia auriculata* and *Juncus* sp. are present under the *Salix* scrub in the early part of the season, and a

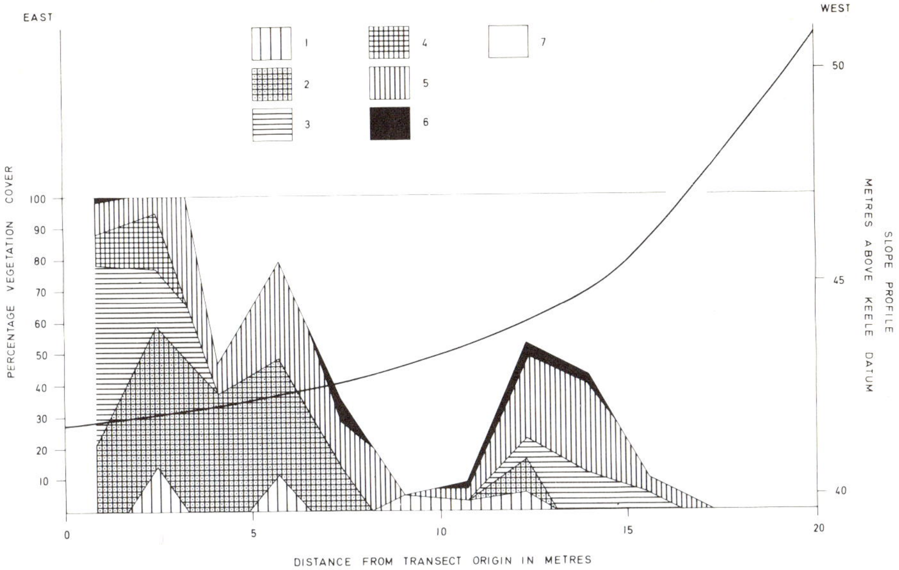

Fig. 2.9 Vegetation and slope transect along a bearing of 080° immediately south of Wilks' Plateau, showing transition from *Rubus–Hedera* association to bare scree, March 1972

1. **Herbs and ferns**
2. *Hedera helix*
3. *Rubus* **sp.**
4. *Teucrium scorodonia*
5. *Centranthus ruber*; *Umbilicus rupestris, Sedum* **sp.**
6. **Mosses and lichen**
7. **Rock**

variety of other herbaceous species including *Eupatorium cannabinum* occur in other moist sites on the cliff.

The south section of the cliff is also a moist environment, the shallow brown calcareous soils supporting a species-rich grass community with *Festuca* sp. and *Carex flacca* as codominants. The *Carex* becomes more abundant on the seepage lines. Many other species are present, but they cannot be described or quantified accurately because field surveys have been completed early in the season. However, the following species are present with some frequency: *Lotus corniculata, Gallium* spp., *Poterium sanguisorba, Centaurea* sp., *Betonica officinalis, Hypericum* sp. and *Dactylorhiza* sp. In contrast to the remainder of the cliff, the vegetation in the south extends into the spray zone. *Plantago maritima* is abundant near the shore but otherwise the vegetation appears little modified. The absence of species well adapted to a maritime environment is noteworthy. Apart from *Plantago maritima*, the only other specifically salt tolerant species that occur are *Beta vulgaris*, which has been found colonising the lower slopes of the large talus cone in the north, and *Crithmum maritimum* which occurs on bedrock surfaces within the spray zone.

Analysis

During the mapping, samples of fine and medium grades of debris were taken for estimation of the angle of internal friction following the method described by Carson (1971) for relatively coarse heterogeneous materials. The results are internally consistent, and accord with the general relationship suggested by Carson (1971; Carson and Petley, 1970) for threshold slopes, using shallow slide stability analysis on the model of Skempton and De Lory (1957):

$$\tan \theta = \tfrac{1}{2} \tan \phi' \tag{2.1}$$

where θ is the angle of surface slope and ϕ' the angle of internal friction of the debris. Most values of ½ tan ϕ' are close to the values for the slopes sampled (Table 2.1). Where slopes are maintained at angles far steeper than those suggested by the values obtained, they coincide with

Table 2.1 Representative values of surface slope (θ) and angles of internal friction of debris (ϕ') St Mary's Cliff, Brixham

Field sample number	Tan θ	θ (degrees)	½ tan ϕ' (degree equivalent)	½ tan ϕ'	ϕ' (degrees)		
						Fine debris	
S5*	0.5095	27	26	0.4829	44	Mean ϕ'	49.6°
S1	0.7002	35	37.5	0.7699	57		
S3	0.7002	35	33.5	0.6635	53	Mean difference between values of θ and ½ tan ϕ'	2.9°
C1	0.5774	30	24	0.4502	42		
C6	0.5543	29	26.5	0.5000	45		
N1	0.7813	38	32.5	0.6399	52	Extreme difference between values of θ and ½ tan ϕ'	6°
N5	0.6494	33	34.5	0.6882	54		
						Medium debris	
S7	0.6494	33	33.5	0.6635	53	Mean ϕ'	54.7°
S8	0.6494	33	33	0.6400	52		
C3	0.7002	35	34	0.6675	53	Mean difference between values of θ and ½ tan ϕ'	2.8°
C4	0.8391	40	34	0.6635	53		
C8	0.6249	32	26.5	0.5000	45		
C10	0.8391	40	40.5	0.8311	59	Extreme difference between values of θ and ½ tan ϕ'	6.5°
N2†	1.0000	45	44.5	0.9813	63		
N3†	0.7133	35.5	42	0.9020	61		
N4	0.7813	38	33.5	0.6635	53		

* Sample containing the greatest amount of fines.

† Samples containing the greatest amount of material with a mean diameter >6 cm.

Source: Field data.

imbricated masses of disturbed bedrock with landward dips of up to 45°, as between South Plateau and Centre Base. These results suggest that most of the unvegetated fine and medium calibre debris is in a state of conditional instability. Only the coarsest sample (N3) appears to be in a stable condition in the dry state. As the area sampled lacks even pioneer plant species, recent movement seems likely. This is best explained by reference to the presence of water, derived mainly from precipitation. On the basis of experimental results (Horn and Deere, 1962), it is probable that the coefficient of sliding friction of the platy scree may be reduced by between 24 and 50 per cent when a layer of water is present. This is likely to be of greatest significance on the steeper slopes covered by the coarser screes relatively poor in fines. It may also be a minor factor in slab failure along slaty cleavage planes. Temporary saturation, giving high pore water pressures and localised perched water tables, is the most probable means of effectively reducing the ϕ' values of the cliff debris, as shown for slopes underlain by slate on Exmoor (Carson and Petley, 1970)[2].

In addition to these movements of the unvegetated screes, many small-scale changes in the extent and character of certain vegetation communities have been recorded between 1969 and 1973. The most notable of these have been the extension of bare ground due to slipping in the extreme south during the 1970–71 winter, the disruption of the *Rubus–Pteridium* community adjacent to Kane's Point, and the shifting margins of the *Rubus–Hedera* and *Prunus–Clematis* communities below and north of Cliff Top. Other changes include the establishment of pioneer communities and their seral change on the large talus cone and in areas further south, notably between South Plateau and Smith's Plateau. Some of these changes could be the result of periodic movements due to instability of the vegetated areas of the cliff. Further circumstantial evidence suggests that the vegetated areas are not completely stable and have, in fact, moved in the recent past. Much can be deduced from the condition of the blackthorn scrub in the area between South Plateau and Smith's Plateau. Virtually all the scrub along the crest of the lower (seaward) cliff is dead; the remaining specimens show limited and stunted new growth and there are no signs of any regeneration. This is probably the result of the combined effects of exposure to salt and wind, the disruption of soil structure and water supply, and the exposure of roots at the cliff edge. The fact that a non-maritime plant species such as blackthorn is found in a dying condition at the lower cliff crest in the spray zone indicates that the community did not evolve in its present location. These observations suggest that not only the regolith, but also the vegetated slopes, are unstable and mobile in this section of the cliff.

The conspicuous absence of salt tolerant species in the plant communities in the spray zone and adjacent to the cliff edge has been noted.

Such species can only become established if the area of colonisation is reasonably stable. If the rate of erosion is greater than the rate of colonisation, it is unlikely that a salt tolerant spray zone community can ever become established. The absence of such communities could be taken as further evidence of the instability of the lower slopes of the cliff. Slope instability therefore is not confined to the areas of bare regolith. Skempton and De Lory (1957) have shown that the condition of limiting stability on an infinitely wide slope is expressed by the following equation:

$$(\tau=)\gamma z \sin\theta \cos\theta = c' + (\gamma - m\,\gamma_w)z \cos^2\theta \tan\phi' \; (=s) \qquad (2.2)$$

where γ is the bulk density of the soil, γ_w the unit weight of water, θ the angle of surface slope, c' the cohesion of the soil, z the depth below the surface of the failure plane, m the height of the water table above the slip surface, ϕ' the angle of internal friction of the debris, τ the shear stress and s the shear strength of the material. In the absence of certain data, notably pore-water pressures, rigorous analysis of the probable range of relationships between bulk densities, slope angles and angles of internal friction for individual sites is not practicable. But, comparisons employing generalised values of these entities may usefully be attempted if, following Carson and Petley (1970), the surface watertable case be assumed, that is m equals unity. This appears reasonable in the case of St Mary's Cliff where the shaly regolith is rarely deep and the transition to the less permeable bedrock generally rapid.

It is also reasonable to assume that the cohesion (c') of the regolith is zero in view of the low clay content of most of the debris. It accords with the results of more exacting tests by engineers on a variety of talus types and granular residual soils (Carson and Kirkby, 1972).

Insertion into equation (2.2) of the mean values for all material sampled on the cliff supports the field evidence provided by the vegetation studies that, in the saturated state ($m = 1$), shear stresses at the debris–bedrock interface exceed the shear strengths in the area colonised by shallow-rooting species on slopes greater than 32° with bulk densities of 1 920 kg/m^3, and on slopes greater than 30° with bulk densities of 1 800 kg/m^3. The fine scree would become unstable under the specified conditions at slopes between 28° and 30°, depending on the value of γ, while the value for medium scree (35°) is somewhat higher (Table 2.2). The data on slopes, regolith and surficial structures may be used to isolate those parts of the vegetated cliff most liable to sliding. Comparison of the results (Fig. 2.10) with the vegetation map (Fig. 2.8) makes clear the association of sliding with the *Prunus* and *Pteridium* communities and rotational slip and terracetting with the grass and herb communities.

The apparent instability of both vegetated and bare regolith on St

Table 2.2 Variation in the critical angle for typical cliff debris, St Mary's Cliff, Brixham

	Thickness of regolith (m)				
	1.0	*0.5*	*1.0*	*0.5*	
	$\gamma = 1\ 920\ kg/m^3$		$\gamma = 1\ 800\ kg/m^3$		
Critical angle	30°	41°	28°	40°	where $\phi' = 50°$ (fine regolith)
$(\tau > s)$	34°	44°	32°	44°	where $\phi' = 55°$ (medium regolith)

Source: Field data.

Mary's Cliff invites speculation as to changes occurring over a period longer than that of direct observation. In the absence of field records, aerial photography provides perhaps the best data on which to base an assessment of longer-term changes. Stereoscopic aerial photographs dating from 1946 were used to map cliff lines and vegetation types by normal photogrammetric methods. Plotting was on the basis of unchanging control points established precisely in the field in 1972. The vegetation was mapped as a series of physiognomic groups (Fig. 2.11). Due to the absence of ground control and the relatively small scale of the photography (1:5 000) it has been impossible to suggest with any certainty the species composition of the physiognomic groups, but many differences in the extent of bare ground, the distribution of the physiognomic groups as well as in the form of both cliff top and cliff foot are to be noted when comparisons are made with subsequent situations.

A high oblique photograph, dated 1954, exists (Plate 3), but as flight data are not available precise plotting of the outline form and vegetation of the cliff by normal photogrammetric methods is precluded. However, as control points on the base map could be identified on the photograph and ground heights were known over the whole area, it was possible to produce a map of the line of cliff top and cliff foot, together with plots of vegetation on a physiognomic basis. The vegetation groups, while physiognomically defined, have been described in terms of dominant component species by analogy with photographs of comparable quality and perspective taken throughout the 1969–73 period.

Outstanding changes emerge from a comparison of the 1946 and 1954 maps. There is a general colonisation of the north cliff by grass and herbs which form a widespread though discontinuous cover even on the talus cone (Figs. 2.11 and 2.12). Four discrete areas of scrub are also present: three of *Pteridium–Rubus* and one of *Prunus–Rubus*. The reduction in the area of bare regolith in the large re-entrant south of the north steps is

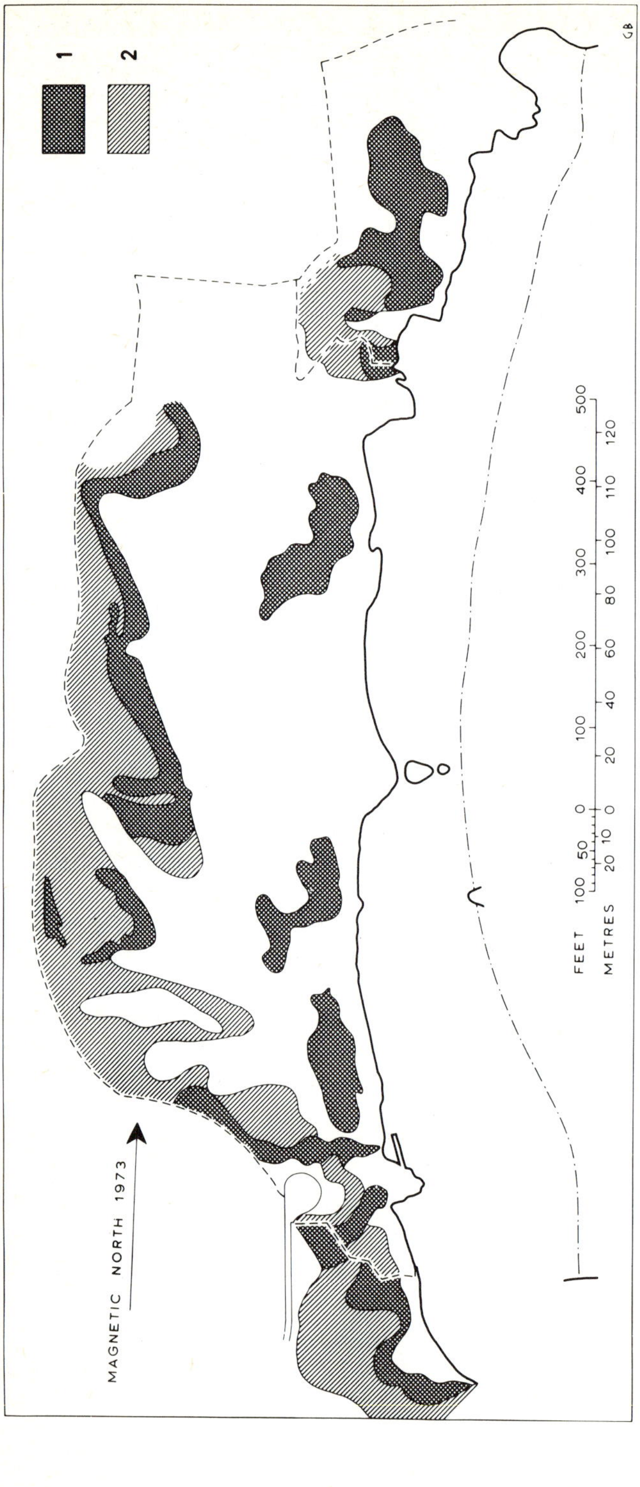

Fig. 2.10 Vegetated areas most susceptible to shallow sliding
1. Vegetated areas demonstrably unstable under present (1943–72) conditions
2. Vegetated areas potentially unstable under present conditions

3. Oblique aerial photograph (1954) showing St Mary's Bay from the south

of the right order and indicates development of the *Prunus* and *Pteridium–Rubus* communities in a period of relative stability. The most dramatic changes occurred between Cliff Top and South Plateau. These include the loss, largely by rotational shear slide, of some 6 m of the cliff top, together with over 100 m of hedgerow northwestwards from South Plateau; the off-setting of the footpath running southward from near Cliff Top, suggesting slab failure and surficial sliding; the increase in the extent of shallow sliding on all sides of Kane's Point, especially evident in the increased disruption of both footpaths and vegetation; and the advance of the cliff foot between Centre Base and South Plateau by an average of 5 m.

Notable changes also occurred between 1954 and 1973 (Figs. 2.8 and 2.12). In terms of the distribution of plant communities, there was some degradation of the pioneer communities on the northern talus cone and the scree slope above Wilks' Plateau, with the *Rubus–Teucrium–Hedera* community extending further up the scree in 1954. This suggests that

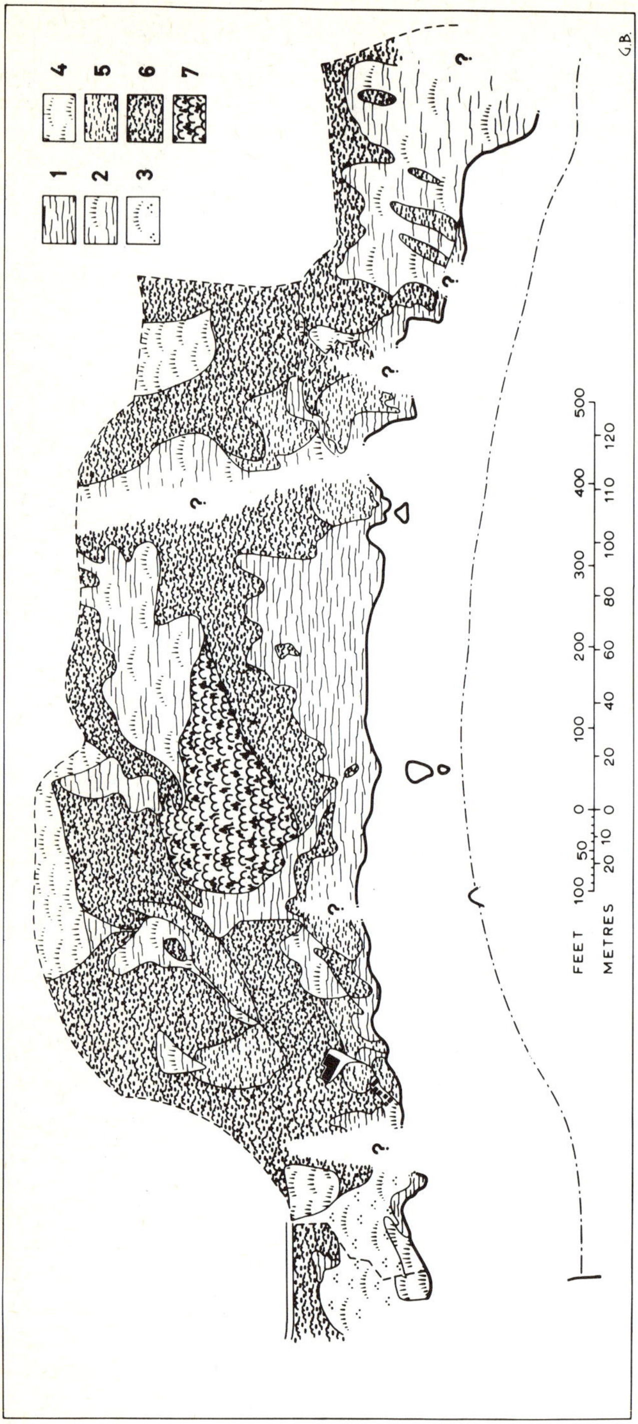

Fig. 2.11 Vegetation on a physiognomic basis in 1946

1. Bare ground
2. Grass cover
3. Scrub: of varying density, height up to 1.5 m
4. Scrub: very dense
5. Low scrub, height not greater than 0.75 m
6. Bush (low trees and shrubs up to 3 m)
7. High bush (tree cover over 3 m)
Based on air photographs

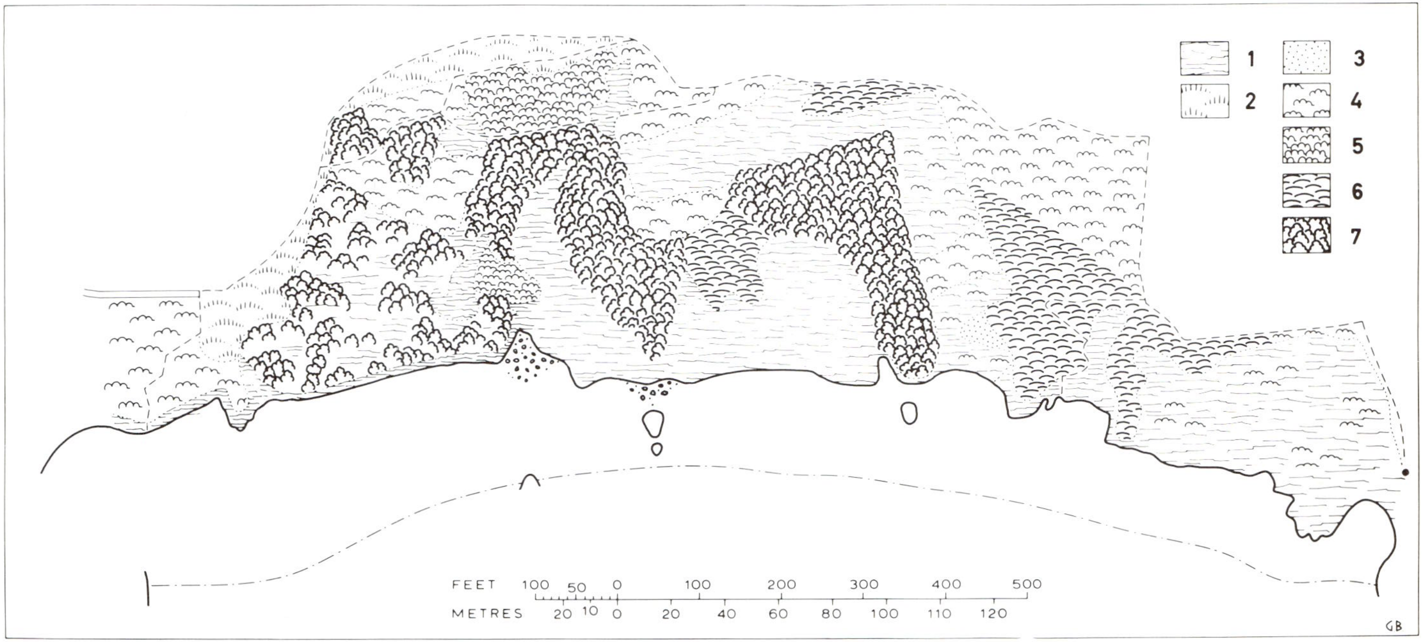

Fig. 2.12 Vegetation on a physiognomic basis in 1954

1. **Bare rock**
2. **Grass with areas of bare ground**
3. **Grass with herbaceous species**
4. **Grasses – undifferentiated**
5. *Prunus* **scrub;** *Pteridium aquilinum*
6. *Pteridium aquilinum* **with some scrub (probably** *Rubus* **sp. and** *Prunus spinosa***) and** *Clematis vitalba*
7. *Ulmus* **scrub with** *Clematis vitalba* **(and probably** *Hedera helix***)**

Based on air photographs

these two areas enjoyed a period of relative stability for several years before 1954, but have since become more mobile, a view reinforced by the disappearance by 1973 of the three discrete areas of *Pteridium–Rubus* scrub established between 1946 and 1954. Taking the centre and southern sections together, there appears to have been an increase in creepers (*Clematis* sp. and *Lonicera* sp.) in comparison with the 1954 situation, when they were limited to the *Ulmus* scrub area. This, again, is consistent with general mobility of the regolith. The tendency towards a decrease in the area of scrub (dominated by *Prunus* sp.) noted between 1946 and 1954 continued into the later period, a complementary increase occurring in the area of grass, pioneer vegetation and bare soil. Much of the *Prunus* scrub along the cliff foot was either dead or dying in 1973. The widespread shallow sliding and associated subsurface slab failure which this implies is confirmed by another major change, notably the appearance of a large area of willow scrub at some time between 1946 and 1973. This process has progressively altered the area west of South Base from a slope with continuous fall, as in 1946, to a series of tilt blocks with reversed slopes, impeded drainage and cliff foot seepages. Other tangible evidence of this sliding includes the abandonment of the refreshment hut existing in 1946 (Fig. 2.11) and the collapse and burial of the concrete steps and ramp which served the beach huts as late as 1954. By 1973 a retreat of 2 to 5 m in the southern part of the cliff front re-established the 1946 cliff foot line although there also occurred a local advance to Centre Base of 6 m due to undermining and rockfall.

Study of the changes in the form and vegetation of St Mary's Cliff between 1946 and 1973 indicates that the position of the cliff foot varies about a mean approximating the 1946 and 1973 positions by as much as 6 m in the south, but elsewhere, even below the actively mobile north cliff, variation is insignificant. Where the cliff foot position varies most, the strongest evidence of morphological, drainage and vegetation change is also found. The largest area of unvegetated cliff face lies immediately northwest of Centre Base. The apparent consistency of form and outline displayed by this cliff seems to be an expression of the balance it represents between rock and debris disruption, transportation and removal. Of the factors affecting the broad equilibrium of this cliff slope, the two of fundamental importance are soil and ground water, and wave action.

The frequency of the surface watertable condition is a critical consideration in assessing the likelihood of movement of the regolith, including whole segments of relatively shallow-rooting communities, especially those of *Prunus* and *Rubus* spp. This will fluctuate greatly with variation in such parameters as regolith thickness, cross-slope and downslope convexity and concavity as well as rainstorm incidence and severity. Taking a conservative estimate of void ratios of 0.5 and 0.3 for material of 1 800 and 1 920 kg/m^3 density respectively, a column of regolith 10 by 10 by 50 cm deep (a representative figure for regolith depth over large areas of the cliff face) would accommodate 20 and 12 cm of water respectively when saturated. Assuming that the mean condition of the soil approximates to field capacity where between 30 and 50 per cent of the voids are occupied by capillary and interstitial water, increments of some 10 to 15 cm of water for the lower density material and some 6 to 9 cm for the higher density regolith would be required to produce the surface watertable condition. Examination of daily rainfall records for Dartmouth and Kingswear reveals that between 1943 and 1973 falls of 6 cm in a single day occurred ten times. However, on the areas of denser regolith and well developed litter and root mat layers, it is probably acceptable to consider these critical precipitation amounts as accumulating over periods longer than twenty-four hours, perhaps three or four days. The total number of such events, excluding those occurring at the end of dry spells of more than three days' duration, was seventy-seven in the twenty-nine years examined (average 2.7 per annum). The total number of three to four day rainfall events giving rise to falls of at least 10 cm was five over the same period, one fall of this magnitude occurring in each of the years 1944, 1947, 1952, 1960 and 1961.

St Mary's Cliff faces due east and its vulnerability to oversteepening and collapse is likely to be high in prolonged or severe easterly winds especially at high water spring tides. Records of such wind conditions over the south and southwest coastal areas of England for the period 1943 to 1973 may be placed alongside the periods of surface watertable condition deduced on the basis of the above arguments. In the three years before the 1946 photography, easterly winds were not important in periods of probable high water tables. In 1947, however, all high rainfall rate events (two occasions with 6 cm or more in twenty-four hours; three occasions with 6 cm or more in three to four days; one occasion with 10 cm or more in three to four days) occurred in March, a month when one-third of all winds greater than 11 knots blew from the sector northeast, east, southeast. Moreover, two of the 6 cm events occurred only three days apart and coincided with high water springs. In the whole of the record examined, this period (4 to 12 March 1947) appears the most likely for the initiation of slab failure and acceleration of shallow sliding.

There followed a period of years which generally lacked this coincidence of critical conditions, an interpretation supported by the 1954 photographic evidence of colonisation, even of steep regolith slopes such as the north cliff and talus cone. Persistent and often strong easterlies occurred soon after, notably in January 1955 when 45 per cent of all winds blew from the easterly sector (about four times the average). The

4. St Mary's Cliff, view to the west–northwest, March 1972

easterly gale of 13 January coincided with a three-day wet spell in which over 6 cm of rain fell within one day of high water springs. The severe winter of 1961–62 is impossible to assess fully on the present basis because of the lack of appropriate daily rainfall records for all three local stations, but December 1961 received rather more than 50 per cent of all winds from the east, and the easterly gales of 24 December followed a 6 cm rainfall event and coincided with high water spring tides. It seems likely that this winter, with easterly gales as late as 7 March (within three days of high water springs), was one of high ground watertables in this region, perhaps exacerbated by a frozen surface crust from time to time. In general, however, frost does not appear to have been a major factor in the instability of cliff regolith. While the easterly gales of 17 and 18 March 1964 coincided with a 6 cm fall, this occurred at the neap so that conditions were not quite optimal for widespread mass movement.

The year 1965 contained three occasions when significant movement may have been stimulated. The coincidence of easterly gales, heavy rainfall and high tides occurred on 20 January, again on 30 and 31 January and on 22 and 23 December. Apart from the occurrence of heavy rain and easterly gales at the neap on 13 and 14 April 1966, the next coincidence of all three requisite conditions occurred between 15 and 27 February 1967. Three days of easterly gales coincided with over 6 cm of rainfall immediately following high water springs between 18 and 20

February 1969. This was the first such event to occur within the period of direct observation of the cliff. In this case it caused destruction of the lowest 3 m of the north steps and caused scree falls above the storm beach, especially in the central section. Although not individually associated with significantly heavy rainfall events, the easterly gales of February 1970 led to an increase in the area of bare ground in the far south of the cliff. Absence of easterly gales associated with heavy rainstorms in 1971 and 1972 accord with the minor amount of change recorded on the cliff at this time.

The recent history of St Mary's Cliff has been one of limited shallow sliding of regolith with creep and localised flow occurring several times a year, together with more rapid sliding of regolith, and sliding and rotation of discrete masses including bedrock in particularly favourable years under natural agencies. Examination of the cliff in successive springs and during the peak holiday season leaves no doubt that the impact of treading, being greatest when the cliff is dry, is negligible except on fine debris close to both sets of steps.

The consistent overall form of the cliff and the principal vegetation communities, especially the maintenance of the outlines of the areas of bare scree, belie the mass movements known to occur. The evidence above is convergent and suggests that, triggered principally by oversteepening of the lower cliff by wave action, the regolith, together with its covering vegetation and root mat, moves in large but relatively thin units by shallow sliding. The most impressive evidence of this process is provided by the communities dominated by *Prunus* sp. which can be seen to have occupied old areas of bare scree, not just by the normal process of colonisation but by progressive downhill sliding of large vegetated mats. There is some evidence that hedgerow blackthorn surrounding the upper cliff is the source of this cover. In addition to the photographic documentation of this process in the 1946–54 period, comparison of the 1954 and 1972 photographs (Plates 3 and 4) shows such a change on the slopes between Cliff Top and Smith's Plateau. Blackthorns situated about halfway between Smith's and South Plateaus at an altitude of about 15 m have proved, from study of tree rings, to be at least seventy-five years old. Given an origin in the cliff top zone due to cliff retreat by sliding in the fashion known to have occurred in the southern cliff top hedgerows between 1946 and 1954, a maximum annual average movement over the lifetime of the shrubs of about 50 cm appears probable. The area of willow thicket provides the clearest evidence of marked changes in the drainage of the south cliff due to movement of tilt blocks, and 1947 was the most likely year to have experienced notable movements of this kind. Tree ring counts from the most mature willow specimens in 1973 indicated a maximum age of twenty-six years, suggesting that growth began in the summer of 1947.

Conclusions

The mapping and analysis of a suite of related surface variables is a method well suited to the characterisation of modes and rates of degradation of active natural slopes. Areas of bare scree are at the critical stability angle throughout. Most of the vegetated parts of the cliff, contrary to first impressions, are also susceptible to movement, particularly by shallow sliding, so that vegetation migrates *en masse*. This process is triggered by marine undermining of the cliff base, leading to the rotation of tilt blocks and shallow sliding fed by cliff-top retreat by rotational shear slide. Thus the cliff consistently displays a vegetation cover of about 75 per cent, parts of which can be expected to slide on two or perhaps three occasions in most years. These debris and vegetation mats move intermittently at a long-term average rate of 0.5 m a year, and cliff top retreat has averaged 0.63 m a year in the south during the period under review, with an annual maximum of 1.03 m (northwest of South Plateau) and a minimum of 0.46 m per year (near Cliff Top). It is concluded that the cliff is a long-term equilibrium form, retreating by undermining at the foot and shear-sliding at the top to maintain an essentially uniform overall gradient. Most of the cliff debris and its vegetation is *in transit* under prevailing conditions of climate and tide, and regular monitoring will continue.

Notes

1. Keele Datum.
2. Great accuracy cannot be claimed due to the experimental deficiencies of the tilting box method. In material coarser than sands, this is subject to sample calibre errors. The high values for samples N2 and N3 probably reflect such errors. The importance of a water layer in reducing the coefficient of sliding friction may have been overestimated.

References

ARBER, M. A. (1940) 'The coastal landslips of south-east Devon', *Proc. Geol. Ass.*, 51, 257–71.

ARBER, M. A. (1962) 'Coastal cliffs: report of a symposium', *Geogr. J.*, 127, 303–20.

CARSON, M. A. (1971) 'An application of the concept of threshold slopes to the Laramie Mountains, Wyoming', in D. Brunsden, ed., *Slopes, form and process, Inst. Brit. Geogr. Spec. Pubn.*, No. 3, 31–48.

CARSON, M. A. and KIRKBY, M. J. (1972) *Hillslope Form and Process*, Cambridge.

CARSON, M. A. and PETLEY, D. J. (1970) 'The existence of threshold hillslopes in the denudation of the landscape', *Trans. Inst. Brit. Geogr.*, 49, 71–95.

COUNTRYSIDE COMMISSION (1968) *The Coasts of England and Wales: measurement of use, protection and development, study report No. 1*, London.

COUNTRYSIDE COMMISSION (1969) *Nature Conservation, special study report No. 2.*

HORN, H. M. and DEERE, D. U. (1962) 'Frictional characteristics of minerals', *Geotech.*, 12, 319–35.

OSMAN, C. W. (1917) 'The landslips of Folkestone Warren and thickness of the Lower Chalk and Gault near Dover', *Proc. Geol. Ass.*, 28, 59–83.

SKEMPTON, A. W. and DE LORY, F. A. (1957) 'Stability of natural slopes in London Clay', *Proceedings, Fourth International Conference on Soil Mechanics*, 2, 378–81.

STEERS, J. A. (1961) 'Coastal cliffs: report of a symposium', *Geogr. J.*, 127, 303–20.

USSHER, W. A. E. (1903) 'The geology of the country around Torquay', *Mem. Geol. Surv. England and Wales.*

WATERS, R. S. (1958) 'Morphological mapping', *Geography*, 43, 10–17.

WOODWARD, H. B. and USSHER, W. A. E. (1911) 'The geology of the country near Sidmouth and Lyme Regis', *Mem. Geol. Surv. England and Wales.*

A. M. D. Gemmell

3 Problems associated with the analysis of raised shorelines in glaciated areas

Although Jamieson recognised as early as 1865 that the traces of former shorelines in Scotland were not strictly parallel to present sea level, it was a considerable time before his views were generally accepted in Britain. Soon after 1865 Geological Survey reports began to classify raised beaches as the '100 foot', '50 foot' and '25 foot', according to their approximate altitude in the areas under investigation. This nomenclature was adopted widely and beaches were classified according to their approximate height above sea level. This practice led to the widespread misconception that raised shorelines were essentially horizontal, despite the fact that Jamieson had shown that this was not the case.

Donner (1959, 1963) was one of the first researchers to realise that Jamieson had been correct when he deduced that the warping of British raised shorelines possessed a definite and significant pattern. Donner produced maps showing isobases – lines joining points of equal uplift – on the '100 foot' shoreline which were subcircular and centred on a focus in the southwest Grampians. But he persisted in referring to the shoreline as the '100 foot' beach.

Sissons (1962) stated that, in theory at least, the older the shoreline the steeper should be its gradient. His work in the Firth of Forth indicated that the '100 foot' shoreline in this area was almost horizontal, but that lower shorelines were rather steeper. This led him to deduce that the '100 foot' shoreline was not in fact a synchronous feature, and should not be regarded as a true shoreline. Another problem to be considered was the accurate correlation of the various raised beach fragments. The most widely accepted method in Britain is that proposed by Sissons (1963), who suggested that precise levelling of the break of slope at the back of raised beach fragments would allow accurate height measurements to be used to separate and correlate shoreline remnants. By this method Sissons showed that rather than three or four raised shorelines in Scotland, in the area of the Firth of Forth there were some twelve to fifteen distinctive shorelines, all of which sloped eastwards along the Forth valley.

A similar multiplicity of shorelines has been elucidated from northeast Ireland and southwest Scotland (Stephens and Synge, 1966; Synge and Stephens, 1966; Stephens, 1968) using a different technique. Rather than survey along beach fragments, Synge and Stephens have determined the heights of fragments at a large number of sites. The individual fragments have been plotted on a graph of altitude against distance from an arbitrary origin, and, by linking fragments by lines drawn through groups of points, representations of former shorelines are produced. The present state of knowledge about raised shorelines in Scotland is such that it is recognised that a multiplicity of shorelines exists, but no system has been devised to link the shoreline assemblages from the various parts of the country. In this paper the objective is to discuss several of the difficulties associated with the collection and subsequent analysis of raised shoreline data in the hope that they may be allowed for in future research.

Problems and techniques of field study

The survey of the height of raised shoreline fragments is a form of sampling of the infinite population of altitudes possible on any such fragment. The greater the interval between readings, the greater the possibility that a significant feature will not be adequately sampled, and may even be totally unrepresented. Gerrard and Robinson (1971, p. 52) concluded that 'high levels of variability of slope angle . . . are introduced into slope profile data depending on the length of recording interval selected and the micro-relief of the slope'. If this is the case for slopes it must be applicable to raised beach profiles, which are really very low angle slope measurements, particularly where there has been isostatic adjustment.

Contemporary studies of raised shorelines have used two main sampling schemes. The first is based on a wide scatter of readings in areas where few fragments survive, only one or two altitudes being obtained for any one site. The chances of the points sampled in this way being representative of the total population of altitudes are slight. This wide scatter technique has been frequently used, notably by Donner (1959, 1963) and Synge and Stephens (1966). In all cases these were studies of large areas, often with discontinuous shorelines, and practical problems of time availability must have deterred a more detailed survey of the area.

McCann (1966), again working in a large area, the west of Scotland, regarded Donner's sampling scheme as inadequate, but used a technique only marginally more satisfactory. Like Donner, he selected a number of widely scattered shoreline fragments, but then found the altitude of several points on each fragment and selected one to be representative of the altitude of the shoreline in the area, a value used in all further calculations. This technique involves a considerable degree of subjectivity.

The second sampling scheme has been used mainly by Sissons and his associates in a prolonged and intensive study of raised shorelines in the Firths of Forth and Tay. Each feature was followed along its length, height readings being obtained at intervals of 50 to 80 m (Sissons, 1963, 1967*a*, 1967*b*, 1972; Sissons and Smith, 1966; Cullingford and Smith, 1966; Sissons, Smith and Cullingford, 1966; Smith, Sissons and Cullingford, 1969). Even at this intensity of survey, the sample may be of only moderate significance as it would still be possible to omit significant features, but it is a considerable improvement on other schemes. Sissons has criticised the use of widely spaced readings as a basis for the interpretation of raised shorelines: 'closely spaced accurate height measurements, however, readily enable the gently inclined raised beaches to be distinguished from the more steeply sloping outwash plains' (Sissons, 1967*a*, p. 168). Cullingford and Smith (1966) also point out that a single height reading from a terrace fragment can give no indication as to the direction and degree of slope of that fragment, information perhaps vital to the successful correlation of various fragments of a raised beach.

A considerable degree of variation exists in the accuracy of the different instruments used to determine the altitude of the shoreline fragments at the sampling points. Many of the earlier research projects were based on altitudes from an aneroid barometer, a notoriously fickle instrument, affected by prevailing weather conditions and not really suitable for accurate survey. It has been employed by Synge and Stephens (1966) and by Donner (1959, 1963), in both cases readings being taken from the local high water mark and later linked with Ordnance Survey Datum. Donner used the upper limit of barnacle growth to define the high water mark, Synge and Stephens the upper limit of seaweed. Both these limits are susceptible to local variations in marine conditions (Sissons, 1967*b*), although the problem may not be too serious as the deviations in local datum will be reflected probably in variations in the shoreline altitude. Synge and Stephens (1966) also used a hand level and staff for some of their survey work. While considerably more accurate than an aneroid barometer, concern has been expressed (Sissons, 1967*b*) that this technique may involve errors which may prove to be significant. Sissons, Cullingford and Smith, Gemmell (1971) and Gray (1974) have used precise levelling techniques linked directly with benchmarks to give altitudes accurate to 0.01 m. Andrews (1970*a*, 1970*b*), working in the Canadian Arctic, adopted the surface of water bodies of known elevation as local datum levels and surveyed accurately from these with a level or a theodolite. However, in more remote areas such as Greenland and the Antarctic, there may be no datum available apart from that for a particular locality. A mean sea level based on the midpoint between local high and low tides has been used in the south Shetlands (D. E. Sugden, pers. comm.), giving acceptable results.

The ideal sampling scheme should employ therefore accurate survey techniques, levelling being the most suitable. Such survey should be tied where possible to an accurate datum such as Ordnance Datum. The optimum frequency of sampling is hard to define, as a balance must be struck between the number of sampling points and the time available for data collection. Probably the best compromise is to survey the feature at intervals of 30 to 50 m whenever possible. In exposed situations like the west coast of Scotland individual remnants may exist only for short distances, being separated by breaks of several kilometres (Gray, 1972). It is advisable to survey these remnants at such intervals, as direct comparison between the overall slope and the variation in altitude along the slope of each remnant may be made, factors which may be helpful in their identification and correlation. A continuous trace of the variations in altitude along a given shoreline fragment would be even more valuable. This could best be obtained by a continuous recorder which would trace a profile of the ground it moved over. Data could then be abstracted from this profile at whatever sampling interval required. If the entire profile is under study, interpretation would be simpler if it were first digitised using a D-Mac digitiser.

Once the frequency of observations in the sample has been resolved, the questions of what part of the topographic expression of a shoreline is to be measured to determine its altitude, and of what relation the measured point bears to the level of the water plane at the time of its formation, must be resolved (Walton, 1966). At this point a distinction must be made between the 'marine limit' and 'raised beaches'. The marine limit is the uppermost point affected by the sea at any location, and is not necessarily coincident with a raised beach fragment. By its nature the marine limit is time-transgressive, at least in areas which have been glaciated. In contrast, the fragments of a fossil shoreline ought to be synchronous, as each shoreline should be the visible expression of a former sea level.

In general the marine limit is not measured as often as the shoreline fragments but has been used in some studies (Kirk, Rice and Synge, 1966; Stephens and Synge, 1966; Synge and Stephens, 1966; Andrews, 1970*a*, 1970*b*). It has been delimited in a number of ways: in many areas this limit has been regarded as being marked by the top of storm-produced shingle ridges but in others it has been deduced from deltas and ice-thrust boulder barricades. Washing limits also have been used extensively in this context but they are often obscured or altered in altitude by post-formational solifluction (Smith, 1966).

Raised shorelines can be either erosional or depositional in origin. The relatively rapid nature of isostatic uplift in recently deglaciated areas does not allow sufficient time for the formation of any particular shoreline. The most common expressions of late-glacial and postglacial shorelines are depositional beaches and notches eroded in unconsolidated material such as till. Rock platforms should be viewed carefully, as it is unlikely that they will have been cut in the short period of time since deglaciation (Gray, 1974)[1]. In many cases, postglacial deposits may be found on a preglacial or interglacial rock platform. Here the altitude of the postglacial beach should be determined by an examination of the upper limit of deposition on the platform.

Andrews (1970*a*) considers that delta lips are graded to the low tide level, and therefore give a deceptively low altitude. He has added 3 to 5 m to the altitude of the delta lip in order to make it comparable with neighbouring beach terraces. A highly subjective technique, it should only be used in areas such as the Canadian Arctic where sample points may be many kilometres apart, and where no other expression of the altitude of a former shoreline may exist than a delta. Smith (1966) doubts the value of deltas as indicators of former sea levels, because of their tendency to be eroded at the lip during the formation of lower shorelines, tending to give too high an altitude for the delta lip.

The part of the shoreline most commonly measured is the break of slope or the notch at the back of the beach. Andrews (1970*a*) separates the beach notch from the break of slope at the back of the beach: in the case of the latter he notes that the 'relationship of this point to the former water plane is not consistent and can be either above or below the water plane' (p. 121), while the former is assumed to reflect the maximum elevation of the water plane. In Britain, however, most researchers have measured the break of slope rather than the notch.

A problem for those who have used the break of slope at the inner margin of a shoreline is the inconsistency in the altitude of readings taken at short intervals. As Synge and Stephens (1966) note, the altitude of this point may vary markedly over short distances, especially on the west coast of Highland Britain where differences in fetch and exposure may have a pronounced effect. Sissons (1972), working in the Firth of Forth, has tried to overcome this difficulty and that of material slumping from the backslope of the beach, by taking measurements 10 to 20 m from the break of slope and by avoiding sampling areas where the true height of the break of slope is obscured by peat or alluvial fan deposits which have accumulated after the formation of the beach. In some locations the break has been located by boring and its altitude then recorded. This technique of sampling away from the break of slope is effective in sheltered areas such as the heads of long estuaries where beaches tend to be wide and flat, but in exposed areas with steep rocky coasts such as north and west Scotland beach deposits often display a marked seaward slope (Smith, 1966). Readings taken 20 m from the break of slope (where the beach is at least this width) may differ from it in elevation by 2 m. Gemmell (1971) took measurements at 5 m from the break of slope and, although not entirely satisfactory, the method proved practicable for sampling the narrow beach deposits of the west coast of Scotland. A better technique would be to use a particular point on the deposits related to sea level at the time of deposition, a point naturally varying in distance from the break of slope according to the seaward slope of the deposit.

The variation in altitude of the break of slope at the inner margin of the beach deposit has its parallels in presentday shorelines. Ritchie (1966, p. 79) working in South Uist in the Outer Hebrides, stated that 'on the west coast high shingle ridges are found up to 30 ft OD . . ., but these can be shown to be within the range of present coastal processes'. Variations in altitude of the break of slope may however be produced after the formation of the beach, to give false readings. The relevance of the readings will be adversely affected if peat is found on the beach, or if the sampling point is near a stream, as alluvial fans may not be readily visible to the naked eye.

A further difficulty arises when the slope at the back of the beach is cut in unconsolidated material. For example a section of raised beach measured just south of the village of Corrie on the east coast of Arran demonstrated some interesting variations in altitude. The measurements were taken 5 m from the break of slope at the back of the beach in every case and at 60 m intervals. When plotted they revealed that the southern part of the profile (A–B) was relatively smooth and low, while the northern part (C–D) was considerably higher and also displayed a far greater degree of variation in altitude (Fig. 3.1). The contrast was such as to suggest that here a feature other than a single raised shoreline exists (Gemmell, 1971).

The data obtained from this section of raised beach have been analysed statistically to indicate whether or not the two parts can be separated. This was done by applying a 't-test' to the null hypothesis that there is no significant difference between the means of the two sets of data (the northern and southern parts of the profile). For the purpose of the test the data were placed in one or other of the two groups on the basis of whether the cliff at the inner margin of the shoreline was cut in rock or till, as it had been observed that the higher section of the beach was roughly coincident with the extent of the till fossil cliff. The t-test gave a value of $t = 7.49$ (with 19 degrees of freedom), which is highly significant, and the null hypothesis can safely be rejected. Thus the

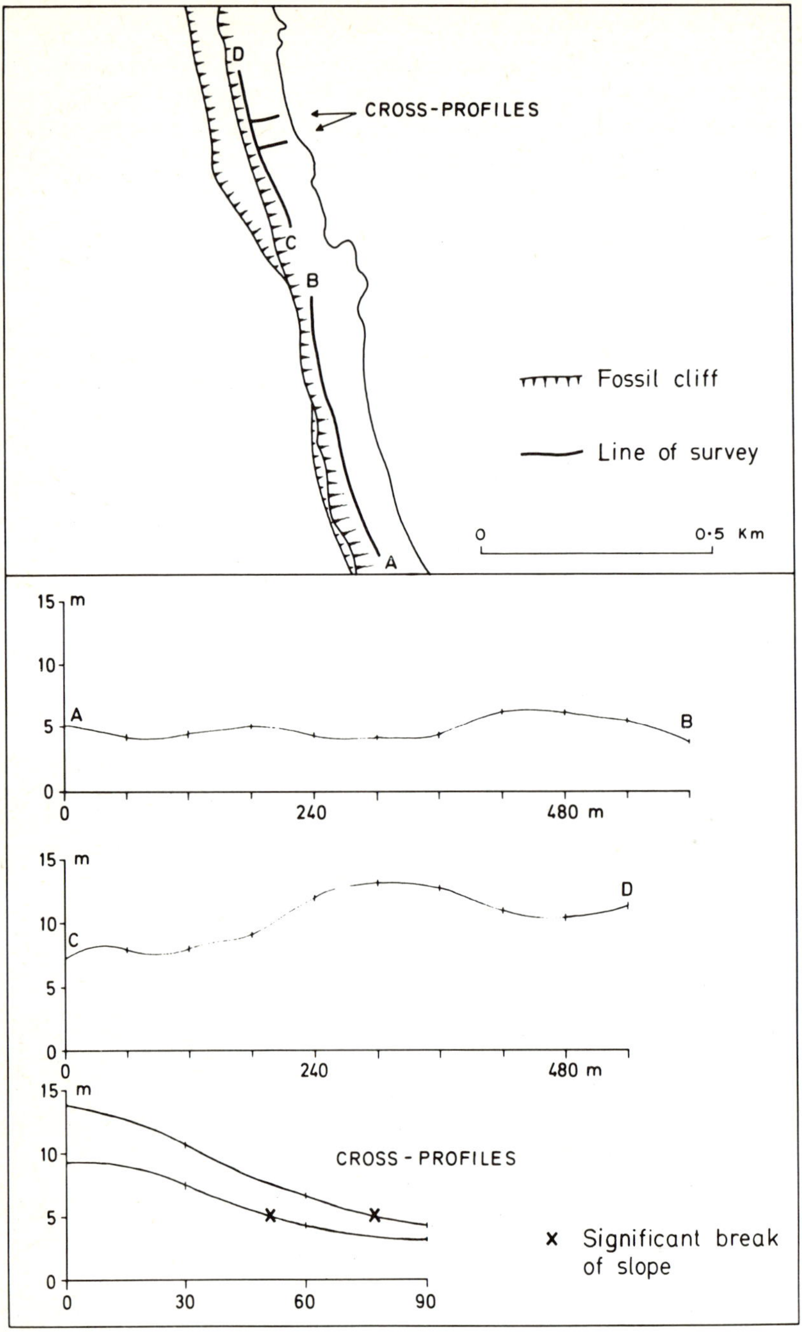

Fig. 3.1 Long- and cross-profiles of a raised shoreline on the east coast of the Island of Arran

samples from the foot of the rock cliff display a significant difference from those at the foot of the till cliff.

A closer examination of the area indicated that the *t*-test had revealed an important problem. The point measured at the foot of the rock cliff was the upper surface of raised beach sediments, but the samples taken at the foot of the till cliff were found on close inspection to represent the upper surface of a layer of till which presumably had soliflucted or crept down the cliff and spread across the beach sediments. This raised the apparent level of the surface and created a false break of slope at the back of the beach. In fact cross-sections from the cliff to the sea show a slight notch (X, Fig. 3.1) at about the same altitude as the beach below the rock cliff, suggesting that till slumped from the cliff either before or during the period at which this beach was deposited.

The number of cases of readings being taken of the altitude of a shoreline fragment with a till cliff along its inner margin is unknown; moreover no attempt has been made to determine whether or not there has been any slumping on to the beach to give erroneous altitude readings. In the field it must be ensured that the features being measured are indeed beach deposits. Sissons and Smith (1966) comment on the problems encountered by confusing outwash and raised beaches, and recommend accurate survey techniques to separate the two. Synge and Stephens (1967) contest Sissons's interpretations of the Firth of Forth shorelines, as they feel that many of his shorelines are sequences of ice-marginal deltas. Though this theory seems rather unlikely in the light of present knowledge, it emphasises the fact that disagreement about interpretations can always arise unless the nature of the feature being sampled and, whenever possible, its relationship to sea level at its formation are carefully determined.

Problems associated with data analysis

The aim of raised beach analysis is to produce as much information as possible on patterns of uplift and deglaciation in the study area. Altitudinal data must first be separated and then assigned to the most appropriate shorelines. It is no longer sufficient to group beach fragments which are of the same altitude, as has been done in the past. 'On a moderately exposed coast with a tidal range of 4 m the relationship of delta surfaces, storm beaches and wavecut terraces may be such that these synchronous deposits could vary in elevation with respect to mean high tide by 10 m' (Andrews, 1970*b*, p. 79). Having traced a shoreline fragment along its entire length, it is not enough to assume that it is synchronous in origin, for eustatic adjustments have resulted in some isostatically deformed beaches being buried partly by later eustatic deposition. As a result the presentday surface of the shoreline is not synchronous.

Ideally, datable material should be available at frequent intervals along each shoreline, so that fragments of the same age may be linked. This ideal is difficult to attain as a result of the paucity of datable material in many areas. In northern Canada (Andrews, 1970*b*) it has been possible to produce diagrams of approximate shoreline slopes by noting the altitude of contemporaneous beach deposits over a large area, and then linking them on this basis alone. This method appears to work well, but with the coarse grid of points that Andrews used much fine detail will have been missed.

The other widely used method involves the calculation of shoreline fragment gradients which are then projected to link up with one another. Fragments which lie on these projected gradients are considered to have a strong likelihood of belonging to the same shorelines. Problems arise with this method if a series of shorelines has only slightly different gradients and altitudes. However, strong evidence exists to suggest that the gradient of any given shoreline will be unique within any area. Andrews and Dugdale (1970) have adapted a formula produced from Scandinavian work by Norrman (1964) relating the age of any shoreline to its gradient in an area where the age and slope of two reference shorelines are known. The problem of this approach is the determination of the shoreline's gradient. As individual shorelines vary in altitude from place to place, the calculation of a sufficiently representative gradient poses considerable difficulties. The techniques used most often in the calculation of the shoreline gradient can be considered under two distinct categories, two- and three-dimensional representations.

Two-dimensional diagrams

There are two types of two-dimensional diagram. The first is the shoreline distance diagram, in which shoreline altitude and distance are compared (Fig. 3.2), allowing slope to be directly obtained. The second type, the shoreline relation diagram, plots the altitudinal differences between various shorelines and the reference shoreline (Fig. 3.3) for which a synchronous origin has been established. The value of this diagram is that it plots relative heights, and so should not be affected by local deformations such as might be caused by post-depositional tectonic movements. The technique has been used in Britain by Synge and Stephens (1966), and in North America Andrews (1970*b*) has extended the diagram to produce a nomogram enabling the age of a shoreline fragment to be determined from its altitude above a reference shoreline if the height of the latter at the point in question is known. This technique has been used by Gemmell (1973) in a study of the deglaciation of the Firth of Clyde.

Shoreline relation diagrams are useful in the analysis of the pattern of

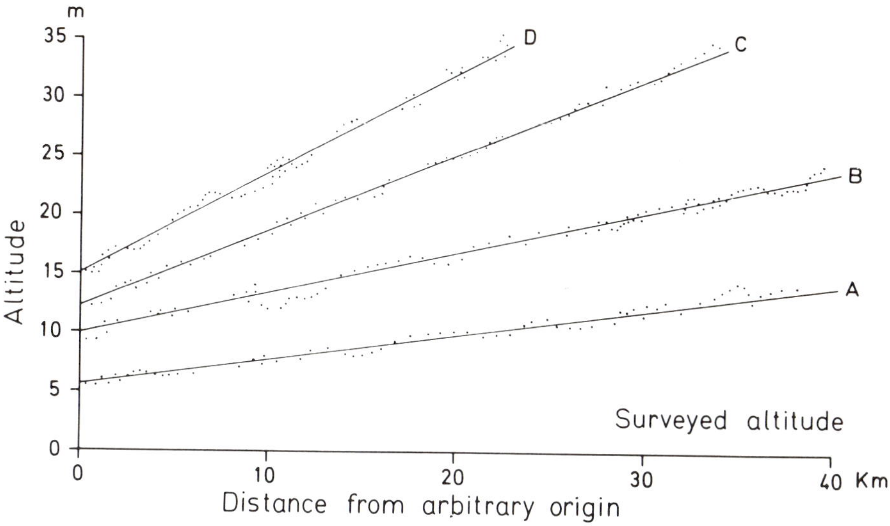

Fig. 3.2 Shoreline distance diagram constructed from hypothetical data. The X-axis represents the distance from an arbitrary origin, in practice usually taken as being outside the area being studied

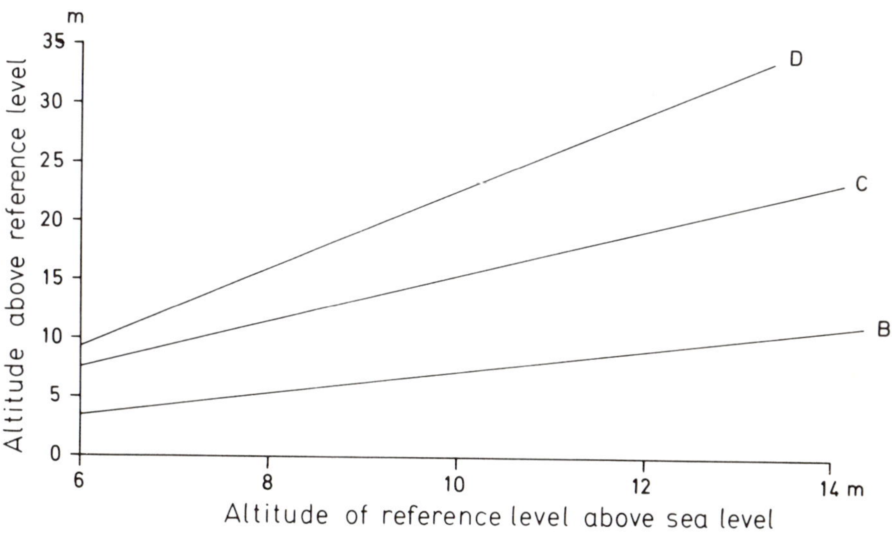

Fig. 3.3 Shoreline relation diagram constructed from the data in Fig. 3.2. Shoreline A (Fig. 3.2) is the reference level

raised beach fragments in an area, but their construction is subject to a number of constraints. To ensure their accuracy

> (1) the reference level has to be synchronous; (2) the elevation of this level and others must accurately reflect the position of a relative sea level; (3) older strandlines have also to be truly synchronous; and (4) post-glacial rebound must have been a smoothly decelerating function (Andrews, 1970*b*, p. 49).

The other two-dimensional diagram, the shoreline distance diagram, does not suffer from such constraints, although the shorelines drawn on it are truly synchronous. If they are not, the entire exercise is meaningless as the diagram cannot be expected to show the true variations of uplift which have occurred along the shoreline in question.

The problem with two-dimensional techniques is that they are essentially linear representations of an areal phenomenon, and must be regarded with caution. They are usually produced by projecting the altitudes of the shoreline fragments in question onto a baseline, and then by joining the points on this line. The selection of an orientation for the baseline is difficult. In theory, unless the initial depression was completely symmetrical about a central point, the uplift consequent on deglaciation would not have produced uniform rates of tilting within the isostatically readjusting area. Thus the possible projection planes, usually radiating from the centre of uplift, would fail to indicate a uniform tilt for the same strandline at different places. The convention therefore has been to use that projection plane which will give the *steepest* gradient for any given strandline, the gradient being designated as a maximum value. Andrews (1970*b*) has suggested that the orientation of the projection plane can be found by the 'triangle method' in which the direction of the local isobases is determined by a form of interpolation, and the projection plane is taken as being at right angles to the isobase.

Having decided the orientation of the plane, the next problem is the projection of the various points onto it. There are two possibilities; the point is projected to a place either normal to the plane (A, Fig. 3.4) or to where the appropriate isobase cuts the plane (B, Fig. 3.4). There may be quite significant differences in position between the two methods, the direct projection onto the plane tending to displace points towards the ice centre. Andrews (1970*b*) gives the error produced in this way as:

$$\text{Error} = R \cos \alpha - R$$

where R is the radius from the centre of uplift on which the site lies, and α the angle between the projection plane and this radius. The commonest technique of fitting a curve to the points on the projection plane is regression analysis, which produces a straightline representation of the shoreline tilt. As regression techniques produce 'best fit lines', there will

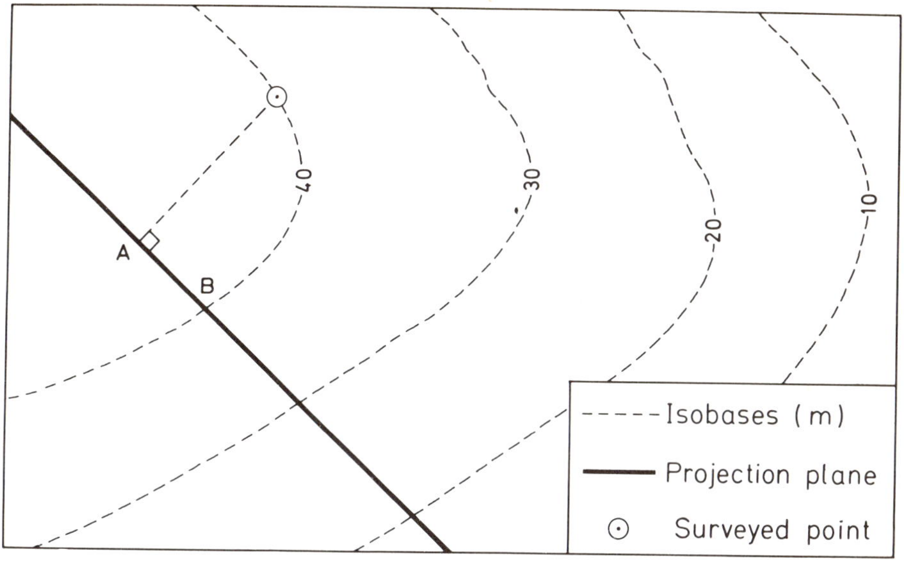

Fig. 3.4 Diagram illustrating the problems of projecting from a point on to a projection plane

be as many points above the line as there are below it, especially as the altitude of a given shoreline is likely to vary considerably. A more reasonable approach might be to represent a strandline not by the regression line itself but by a band representing the 95 or 99 per cent confidence limits. This would provide a more realistic appraisal of the range in altitude of any strandline without affecting its gradient. Although there is widespread use of regression lines for this analysis in Britain, difficulties are involved. Shorelines are represented as a thin line, giving the impression that the exact altitude of the shoreline can be predicted at any point on the diagram, whereas, as already shown, the true altitude may vary considerably from the expected height. The general use of regression analysis in the study of raised shorelines however has been to determine the overall *trend* in the slope of the shorelines, and not as a predictive technique. Another restriction is that the regression line will show the trend of the shoreline only within the area surveyed.

In the past regression lines have been fitted to shoreline data without regard to the critical assumptions for the validity of the technique (Poole and O'Farrell, 1971). Several of these assumptions are particularly applicable where the analysis is used on raised shoreline data. The first, that the values of the variables are observed *without measurement error*, is an extremely difficult condition to achieve in the field. It serves, however, to emphasise the importance of taking measurements consistently from the same facet of the raised shoreline, so that the data obtained are strictly comparable.

The second basic assumption, that the relationship between the variables is linear, is accepted tacitly in most British examples of raised shoreline studies in which regression techniques have been used. There is, however, no evidence that isostatic recovery shows a straightline relationship with distance from the centre of uplift, and Andrews (1970*b*) points out that curvilinear, polynomial and exponential curves may be more appropriate in many cases.

The third major problem is that of autocorrelation: that any given value is not independent of those on either side of it in the sequence. This does not prevent the derivation of an unbiased estimate of the regression coefficients, but there are other consequences particularly if the degree of autocorrelation is high. First, the estimates of the coefficients will have a large and inaccurately estimated variance, leading to inefficient predictions of the values of the dependent variable. Second, and even more important, the associated inferential tests will be invalid, so the degree of significance of the line cannot be calculated. This is crucial, because although a regression line can be produced from autocorrelated data, there is no way of obtaining estimates of the probability that this line is meaningful.

In considering this question, Doornkamp and King (1971) consider that readings along a slope profile are autocorrelated, but the problem with shorelines is different. On a presentday shoreline the altitude which the various beach features attain is a function of such factors as fetch, prevailing wind direction and tidal range. If the altitude was controlled purely by the tidal range, a degree of autocorrelation would be expected because the height of the tide at any one point is related directly to its height at adjacent locations. However, fetch and prevailing wind direction are factors in which autocorrelation will be low. For example, because the maximum fetch at one point is 5 km there is no reason for the fetch at an adjacent point not to be 25 km or 250 km, the only control being coastal configuration. Some of the factors controlling the altitude of the shoreline being studied would suggest a degree of autocorrelation in the data. However, their influence would be more than compensated by other factors and not enough to have a grave effect on the regression analysis.

With isostatically deformed shorelines the altitude of any feature is to a large extent controlled by the amount of isostatic uplift, a factor producing a steady rate of change in altitude throughout the shoreline system. This being the case, the individual altitude readings cannot be said to be independent of one another, and a high degree of autocorrelation is likely. In any study involving the calculation of regression lines from raised beach data, transformation of the data might be advisable before calculations are made, as in this way autocorrelation can sometimes be excluded.

Many studies of raised shorelines contain diagrams showing series of regression lines, each indicating the trend of a particular raised shoreline (Cullingford and Smith, 1966). Quite apart from the problem of the validity of such lines, these shorelines have all been projected onto the same projection plane, intimating that the isobases are parallel for each shoreline. According to Sissons (1967*b*), this is unlikely, a point over which he is in direct conflict with Synge and Stephens (1967). Andrews and Barnett (1972, p. 4) note that

> the propriety of projecting the direction of tilt of former water planes over considerable distances is not without question. The underlying assumption is that the direction of tilt is toward the locally dominant ice centre, whether this is the 'main' or a subsidiary one. In fact, as uplift will probably continue at each locus, even if at different rates, the possibility remains that the tilt direction is a composite one.

They add that migration of ice centres may take place in response to ice sheet dynamics. The combination of these two items of information suggests that uplift is sensitive to changes in location of ice centres, and that shorelines formed during different stages of the decay of an ice sheet should have individual isobase patterns. In turn this implies that for true comparison of the form of uplift of different shorelines, the projection plane should be calculated individually for each, a point of obvious concern particularly with regard to shoreline relation diagrams.

Three-dimensional diagrams

This type of diagram has an advantage over the two-dimensional representation in that it shows spatial variation in uplift over the entire area, thereby obviating the need to locate a projection plane from which to construct a diagram. The three-dimensional diagram frequently takes the form of a map over which isobases are superimposed. This is a technique which has been used for some considerable time in Scandinavia. Isobases may be plotted by simple interpolation using the network of shoreline altitudes as spot heights, a technique applied in Britain by Synge and Stephens (1966) and by Sissons (1967*a*) in a map of the general pattern of uplift in Scotland. This technique is often the only one that can be used because of the paucity of data in many areas. The results obtained should not be regarded as being of great precision, but as an overall guide to local patterns of uplift, a fact which may be in perfect accord with the aims of the study.

The problems of this interpolative procedure are essentially those of the data before processing. A major source of error is likely to be incorrect correlation of shoreline fragments over large distances, with the consequent production of distorted and false isobases. The most reliable method of correlating shoreline fragments is through the study of datable material found in the beach deposits.

A more sophisticated technique employed recently is trend surface analysis, which can best be described as a form of three-dimensional regression analysis, producing surfaces which show the general trend of individual shorelines, and particularly the spatial variations in their altitude and gradients. The technique has been used in Canada by Andrews (1970*b*), and in Britain by McCann and Chorley (1967) and by Smith, Sissons and Cullingford (1969) on shorelines in the northwest and the southest of Scotland respectively.

There are two main types of surface analysis, Fourier analysis and power series (Norcliffe, 1969*b*), the latter including the linear and polynomial models. Power series are employed in most of the shoreline trend surfaces published to date. Surfaces can be used to model a continuous distribution of a particular phenomenon over an area, or to predict values of discrete observations in an area not sampled at present. The use of the word 'discrete' is important in this context, for violations of this basic assumption have caused many of the problems encountered in trend surface analysis.

For a trend surface to be used at all, the number of data points must exceed the number of terms in the equation or the trend is invalid (Davis, 1973). Thus higher order polynomial trend surfaces should generally be avoided unless there is a very large amount of data. It is generally agreed that for raised beach work, linear and quadratic surfaces are the most useful (Tinkler, 1971; Andrews, 1970*b*). This is partly a function of the fact that raised beach data are often distributed around the periphery of land masses leaving large areas without information. As most computers cannot distinguish between an absence of data and zero values, these blank areas are treated as if they were areas at sea level. This is not particularly significant with linear and quadratic surfaces as they are relatively insensitive to even major anomalies, and instead show an overall slope, which is really all that should be asked of the technique (Doornkamp, 1972). However, with higher order surfaces, boundary effects begin to occur unless there is an even spread of data. Although the surface values are reasonably accurate in the areas for which there is data, elsewhere the values of the surface lack control, and cannot be used for prediction or interpretation (Unwin, 1968; Davis, 1973).

An associated problem of concern to raised beach researchers is the areal distribution of the surveyed points. On mountainous coasts, where shorelines tend to be discontinuous, beach data can be obtained only from areas where shoreline deposits are preserved. Thus the data assumes the pattern of a series of clusters of shoreline altitudes spread along the coast. This clustering of points affects the form of a polynomial trend surface. Robinson (1970, p. 32) noted that clusters might exert undue influence on the form of a mathematical surface: 'There are good geographical reasons for judging that clustered points, as long as the clustered sample points conform to the clustering of the total population, do have more importance in influencing the shape of the surface than does the isolated point.' Norcliffe (1969*b*) and Bassett (1972) are of the opinion that clustered data should be avoided, while Doornkamp (1972) adds that the data values should be normally distributed. There is a strong body of opinion against using trend surface analysis on precisely the kind of data which is obtained from studies of raised shorelines. However, Davis (1973, p. 352), while agreeing that clustering may be troublesome, admits that 'trend-surface methods may be more robust against the effects of clustering than is commonly supposed'.

Robinson (1970, 1972) and others have examined methods of avoiding or minimising the effects of clustering. One possibility involves some kind of transformation of the coordinates. Robinson (1972) tested a number of techniques on clustered cirque data from Tasmania, each of which involved selecting a representative value from each cluster. In one technique the selection was made by placing a grid over the area and choosing one point at random from each grid square. A second method was to select a point at random from each cluster, while a third produced a value by simply averaging the X, Y, and Z coordinates for each cluster. All these techniques yielded surfaces which correlated closely with that produced using the original data. This was taken to indicate that a reduction in the clustering of the data did not have an adverse effect on the form of the trend surface produced. While this may be true, it might imply equally that a reduction in clustering by some kind of selection technique does not lead to any reduction in the degree of distortion of the surface induced by the original clusters! Gray (1972), specifically concerned with raised shoreline data, advocated the use of a geographic mean within each sample, but similar problems will probably be encountered with this scheme. There are other statistical grounds for suspecting the validity of averages for overcoming problems of clustering; Doornkamp (1972) stated that the use of averages is not justified as it reduces the variability of the data, while Unwin (1973, p. 32) noted that 'although the less clustered distributions give better fits, the surfaces are also less probable'.

Polynomial trend surface models operate under many of the same restrictions as linear regression models, one of the most important being that the data should not be autocorrelated (Norcliffe, 1969*b*). As already noted with respect to the two-dimensional diagrams, raised shoreline data is autocorrelated, and so inferential tests with regard to its significance will be invalid. Gray (1972) proposed the use of a geographic mean value to overcome this problem. However, the clusters are autocorrelated if they are part of the same raised shoreline; and hence also the points representative of the clusters will be autocorrelated. A logical escape from this problem is by transformation of the data, though this will then result in a trend surface 'map' which is difficult to interpret.

Assuming that a valid trend surface model has been constructed, its mathematical and geographical significance must be tested. Unwin (1970) has illustrated the problems of assessing the mathematical significance of such models. With regard to geographical significance Unwin and Lewin (1971) noted that, if the altitudes used to generate the surface were all taken from the highest nickpoint at the back of the beach, the residuals should be negative everywhere, as all beach deposits should lie below that level. However trend surface analysis will give a best fit surface with both positive and negative residuals. Such a surface cannot be directly representative of any surface of erosion or deposition, but will indicate merely overall gradients.

The use of time series analysis techniques

Fourier techniques have been used mainly in the analysis of variations in time series, but there is no reason why they could not be employed in the study of spatial distributions, either in the linear or areal sense. Instead of the data being read at regular time intervals, it would be taken at regular intervals on the ground. As readings of the altitude of any raised shoreline fragments should be ideally at closely spaced regular intervals, such data could be theoretically suitable for Fourier analysis.

The great advantage of time series analysis in the study of former shorelines is that it is designed to deal with autocorrelated data. Fourier analysis is specifically concerned with analysing periodicity in data, and can break down a line or surface into a series of basic curves, estimating the contribution that each makes to the whole shape. Bassett (1972) considered that surfaces can be regarded either as a series of cyclical components with overlaid random noise, or as a two-dimensional autoregressive process. In either case time series analysis is a particularly suitable technique. Double Fourier analysis can be applied in much the same way as trend surface analysis to produce maps of mathematical surfaces, so the technique could be used to construct isobase maps of uplift in isostatically adjusted areas.

There are some underlying constraints in the use of Fourier analysis which have to be considered. The most important is that the technique can be used only with *stationary* series, series in which the moments (such as the mean) and the joint moments (such as autocorrelation) do not vary with time or distance (King, 1969). This constraint can be met for raised shoreline data by the subtraction of the linear trend surface, allowing the periodic components full rein (Davis, 1973). A more serious constraint is that some Fourier analysis programmes require the specification of the fundamental wavelength of a series and failure to do this correctly may lead to spurious results (Norcliffe, 1969*a*). A method of obtaining information about wavelengths in the series is by spectral analysis (Rayner, 1971, 1972), a technique which reveals the relative importance of the component wavelengths, and may provide a basis for subsequent Fourier analysis. At present little work has been done in these fields, although Thornes (1972) has experimented with spectral analysis in the study of debris slopes. In future increasing emphasis will probably be placed on the use of time series techniques in the study of slopes, and there seems little reason why this technique should not be extended to the analysis of raised shorelines.

Conclusions

In the past the collection and analysis of altitudinal data from raised shorelines have not always been based on the most appropriate techniques. These will vary with the aims of the study, but many techniques have been used incorrectly. The standards of data collection have improved recently with the emphasis by Sissons on regular and accurate surveying, though more attention could be paid to defining precisely what part of the spectrum of shoreline features is in fact being heighted. There are many weaknesses in the methods employed at present in the analysis of data. Not all results obtained to date are invalid but they should be viewed with some caution. The future application of time series analysis to the data may contribute substantially to the solution of the problems involved.

Note

1. An alternative view has been expressed recently by Sissons (1974).

References

ANDREWS, J. T. (1970*a*) 'Differential crustal recovery and glacial chronology (6,700 – OBP), west Baffin Island, NWT, Canada', *Arct. & Alp. Res.*, 2, 115–34.

ANDREWS, J. T. (1970*b*) *A geomorphological study of post-glacial uplift with particular reference to Arctic Canada, Inst. Br. Geogr. Spec. Pubn.*, No. 2.

ANDREWS, J. T. and BARNETT, D. M. (1972) 'Analysis of strandline tilt directions in relation to ice centres and post-glacial crustal deformation, Laurentide ice sheet', *Geogr. Annlr.*, **54A**, 1–11.

ANDREWS, J. T. and DUGDALE, R. E. (1970) 'Age prediction of glacio-isostatic strandlines based on their gradients', *Bull. Geol. Soc. Am.*, **81**, 3769–72.

BASSETT, K. (1972) 'Numerical methods for map analysis', *Prog. in Geogr.*, **4**, 217–54.

CULLINGFORD, R. A. and SMITH, D. E. (1966) 'Late-glacial shorelines in eastern Fife', *Trans. Inst. Br. Geogr.*, **39**, 31–51.

DAVIS, J. C. (1973) *Statistics and Data Analysis in Geology*, Wiley.

DONNER, J. J. (1959) 'The late- and post-glacial raised beaches in Scotland', *Suomal. Tiedeakat. Toim. Sarja III*, **53**, 1–25.

DONNER, J. J. (1963) 'The late- and post-glacial raised beaches in Scotland II', *Suomal. Tiedeakat. Toim. Sarja III*, **68**, 1–13.

DOORNKAMP, J. C. (1972) 'Trend-surface analysis of planation surfaces, with an East African case study', in R. J. Chorley, ed., *Spatial Analysis in Geomorphology*, Methuen, pp. 247–81.

DOORNKAMP, J. C. and KING, C. A. M. (1971) *Numerical Analysis in Geomorphology – an introduction*, Edward Arnold.

GEMMELL, A. M. D. (1971) 'The glaciation of the Island of Arran', unpublished PhD thesis, University of Glasgow.

GEMMELL, A. M. D. (1973) 'The deglaciation of the Island of Arran, Scotland', *Trans. Inst. Br. Geogr.*, **59**, 25–39.

GERRARD, A. J. W. and ROBINSON, D. A. (1971) 'Variability in slope measurements', *Trans. Inst. Br. Geogr.*, **54**, 45–54.

GRAY, J. M. (1972) 'Trends through clusters', *Area*, **4**, 275–9.

GRAY, J. M. (1974) 'The main rock platform of the Firth of Lorne, western Scotland', *Trans. Inst. Br. Geogr.*, **61**, 81–99.

JAMIESON, T. F. (1865) 'On the history of the last geological changes in Scotland', *Q. Jl. Geol. Soc. Lond.*, **21**, 161–203.

KING, L. J. (1969) *Statistical Analysis in Geography*, Prentice-Hall.

KIRK, W., RICE, R. J. and SYNGE, F. M. (1966) 'Deglaciation and vertical displacement of shorelines in Wester and Easter Ross', *Trans. Inst. Br. Geogr.*, **39**, 65–78.

McCANN, S. B. (1966) 'The main post-glacial raised shoreline of western Scotland from the Firth of Lorne to Loch Broom', *Trans. Inst. Br. Geogr.*, **39**, 87–99.

McCANN, S. B. and CHORLEY, R. J. (1967) 'Trend surface mapping of raised shorelines', *Nature, Lond.*, **215**, 611–12.

NORCLIFFE, G. B. (1969*a*) 'An alternative method of identifying the characteristic scale of a periodic spatial series', *Area*, **1**, 21–8.

NORCLIFFE, G. B. (1969*b*) 'On the use and limitations of trend surface models', *Can. Geogr.*, **13**, 338–48.

NORRMAN, J. O. (1964) 'Vätterbäckenets senkvartära strandlinjer', *Geol. För. Stockh. Förh.*, **85**, 391–413.

POOLE, M. A. and O'FARRELL, P. N. (1971) 'The assumptions of the linear regression model', *Trans. Inst. Br. Geogr.*, **52**, 145–58.

RAYNER, J. N. (1971) *An Introduction to Spectral Analysis*, Pion.

RAYNER, J. N. (1972) 'Applications of harmonic and spectral analysis to study of terrain', in Chorley, *Spatial Analysis in Geomorphology*, pp. 283–302.

RITCHIE, W. (1966) 'The post-glacial rise in sea-level and coastal changes in the Uists', *Trans. Inst. Br. Geogr.*, **39**, 79–86.

ROBINSON, G. (1970) 'Some comments on trend-surface analysis', *Area*, **2**, 31–6.

ROBINSON, G. (1972) 'Trials on trends through clusters of cirques', *Area*, **4**, 104–13.

SISSONS, J. B. (1962) 'A re-interpretation of the literature on late-glacial shorelines in Scotland with particular reference to the Forth area', *Trans. Edinb. Geol. Soc.*, **19**, 83–99.

SISSONS, J. B. (1963) 'Scottish raised shoreline heights with particular reference to the Forth valley', *Geogr. Annlr.*, **45**, 180–5.

SISSONS, J. B. (1967*a*) *The Evolution of Scotland's Scenery*, Oliver & Boyd.

SISSONS, J. B. (1967*b*) 'Comments on the paper by F. M. Synge and N. Stephens in Transactions No. 39', *Trans. Inst. Br. Geogr.*, **42**, 163–8.

SISSONS, J. B. (1972) 'Dislocation and non-uniform uplift of raised shorelines in the western part of the Forth valley', *Trans. Inst. Br. Geogr.*, **55**, 145–59.

SISSONS, J. B. (1974) 'Late-glacial marine erosion in Scotland', *Boreas*, **3**, 41–48.

SISSONS, J. B. and SMITH, D. E. (1966) 'Raised shorelines associated with the Perth Re-advance in the Forth valley and their relation to glacial isostasy', *Proc. R. Soc. Edinb.*, **66**, No. 8, 143–68.

SISSONS, J. B., SMITH, D. E. and CULLINGFORD, R. A. (1966) 'Late-glacial and post-glacial shorelines in southeast Scotland', *Trans. Inst. Br. Geogr.*, **39**, 9–18.

SMITH, D. E., SISSONS, J. B. and CULLINGFORD, R. A. (1969) 'Isobases for the main Perth raised shoreline in southeast Scotland as determined by trend-surface analysis', *Trans. Inst. Br. Geogr.*, **46**, 45–52.

SMITH, J. S. (1966) 'Morainic limits and their relationship to raised shorelines in the east Scotland Highlands', *Trans. Inst. Br. Geogr.*, **39**, 61–4.

STEPHENS, N. (1968) 'Late-glacial and postglacial shorelines in Ireland and south-west Scotland', in R. B. Morrison and H. E. Wright, eds, *Means of Correlation of Quaternary Successions, Proceedings of the Seventh Congress of the International Association of Quaternary Research*, **8**, 437–56.

STEPHENS, N. and SYNGE, F. M. (1966) 'Pleistocene shorelines', in G. H. Dury, ed., *Essays in Geomorphology*, Heinemann Educational, pp. 1–51.

SYNGE, F. M. and STEPHENS, N. (1966) 'Late- and post-glacial shorelines and ice limits in Argyll and north-east Ulster', *Trans. Inst. Br. Geogr.*, **39**, 101–24.

SYNGE, F. M. and STEPHENS, N. (1967) 'A reply to J. B. Sissons' comments', *Trans. Inst. Br. Geogr.*, **42**, 169–73.

THORNES, J. B. (1972) 'Debris slopes as series', *Arct. and Alp. Res.*, **4**, 337–42.

TINKLER, K. J. (1971) 'Trend surface analysis', *Area*, **3**, 270–2.

UNWIN, D. J. (1968) 'Some complications underlying trend surface analysis', in J. Doornkamp, ed., *The use of computers in geomorphological research, British Geomorphological Research Group Occasional Paper*, No. 6, pp. 9–18.

UNWIN, D. J. (1970) 'Percentage RSS in trend surface analysis', *Area*, **2**, 25–8.

UNWIN, D. J. (1973) 'Trials on trends', *Area*, **5**, 31–3.

UNWIN, D. J. and LEWIN, J. (1971) 'Some problems in the trend analysis of erosion surfaces', *Area*, **3**, 13–14.

WALTON, K. (1966) 'Vertical movements of shorelines in Highland Britain: an introduction', *Trans. Inst. Br. Geogr.*, **39**, 1–8.

P. Worsley

4 An appraisal of the glacial Lake Lapworth concept

A common theme in many recent papers on British glacial geomorphology has been the denial of many of the criteria for the recognition of glacial lakes as first established by Kendall (1902) in the Cleveland area. Following a re-examination of the field evidence in many areas, a common conclusion has been that the Kendall model of glacial lakes and subaerial overflow channels is inadequate to account for all the observed morphological relationships and an alternative hypothesis emphasising the role of stagnant ice and subglacial drainage has been proposed. This change in interpretation has followed, to a certain extent, the application by British geomorphologists of the work of the Swedish geographer Mannerfelt (1945). The latter studied deglaciation in mountainous areas and invoked mechanisms of subglacial drainage to explain anomalous meltwater channels. Two of the reasons for the ten-year delay in applying Mannerfelt's ideas effectively in Britain were that most of his work was published in Swedish and appeared during the Second World War. Thus the evidence, in the areas of south Scandinavia which Mannerfelt chose to study, that glacial lakes had existed during the decay of inland ice has been overlooked. Once the momentum of reinterpretation was generated, the possibility of widespread glacial lakes associated with deglaciation was ignored and hence a situation arose in which 'the baby was [literally] thrown out with the bath water'.

Experience of present glaciated regions reveals that glacial meltwater lakes are widespread and it is clear that there are inherent dangers in trying to deduce glaciological conditions from morphological data alone. This can be illustrated by the reaction of Stenborg (1969) to Clapperton's (1968) proposals concerning englacial drainage networks based on a study of channels assumed to have been formed by glacial meltwater. In this instance Stenborg, a glacial hydrologist with extensive knowledge of the Mikkaglaciären in Swedish Lapland, found Clapperton's inferences inconsistent with the type of networks developed within this particular glacier. A final note of caution may be sounded by reference to the suggestions made by the empirical process workers Schumm and Shepherd (1973). They have gone so far as to question the fundamental basis of the subglacial concept, namely the assumption that normal drainage systems are unable to erode 'humped' longitudinal profiles, although it is recognised that scale problems also are involved.

In the literature dealing with phenomena associated with the last glaciation in England, two examples of former glacial lakes are often quoted. First there is the Lake Eskdale–Pickering system, and, second, Lake Lapworth. Part of the former has been subject to reappraisal by Gregory (1965), who concluded that the evidence for a widespread lake in Eskdale was meagre and that a more satisfactory interpretative model would incorporate a stagnant ice mechanism similar to that applied by Sissons (1961). Indeed the latter mechanism is now invariably cited in analyses of relic drainage channels. In view of this general situation, it is perhaps remarkable that Lake Lapworth has not been comprehensively reappraised on the same basis, especially in view of its continuing popularity in many school textbooks (Robinson, 1969; Sawyer, 1970). Such books perpetuate a common misunderstanding of the Lapworth concept in that they relate the drainage diversion of the River Severn to the initial overflow of the Lapworth Lake. However Wills (1937) envisaged that not until the Ironbridge Gorge had become incised to about one-half its present depth, did the overflowing Lake Lapworth meltwaters contribute to its cutting. Shaw (1972*a*, p. 628), commenting on the Lapworth concept, stated that there was 'no evidence for a widespread lake of the Lake Lapworth type during the retreat of the Irish Sea ice'.

The time therefore is opportune for undertaking a re-examination of the accepted Lake Lapworth concept. The evidence for an extensive meltwater lake in parts of Staffordshire and Shropshire during the retreat of the last glaciation ice 15 000–18 000 years BP will be critically analysed: first, the historical development of the concept will be traced, leading to an examination of the theoretical possibility of a lake; and second, the available field evidence will be appraised as a basis for an assessment of the lake concept in terms of the current understanding and knowledge of the area. The origin of the Ironbridge Gorge will only be considered in so far as it has a direct bearing on Lake Lapworth.

Historical background

The idea of a proto-Lake Lapworth seems to have been derived from two independent suggestions made at the turn of the century. Lapworth postulated that the cutting of the gorge section of the River Severn was related to the diversion into it of the drainage of the upper Dee in glacial times (Watts, 1898). It is possible that Lapworth really meant the upper Severn of today when he referred to the upper Dee. Later, Harmer (1907) published a paper, previously presented to the Geological Society, in which he invoked a glacial lake overflow channel mechanism as being primarily responsible for the origin of the Ironbridge Gorge. Thus the general concept of a glacial lake impounded on the northwest side of the main watershed and extending from the southern Pennines across to the central Welsh uplands was established (Fig. 4.1). These were simply general speculations and it was left to Wills (1924) to test them against the detailed field evidence and, as might be expected, his conclusions involved a far more complex reconstruction of events than those implied in the basic proposals of Lapworth and Harmer.

Immediately west of the Ironbridge Gorge lies a relatively constricted valley referred to by Wills as the Buildwas Strath, containing at its eastern end an area of hummocky terrain underlain by sands, gravels and small patches of till. This association of lithologies was given the formational name, Buildwas Sands. The sequence was interpreted as the product of deposition in a deltaic marginal environment associated with a glacial lake named Lake Buildwas. The till masses were thought to be derived possibly from icebergs floating in the lake or from the renewed presence of glacial ice due to readvance. The general conformity in elevation of the uppermost Buildwas Sands outcrops and the upstream projection of the Main Terrace gradient of the Severn led Wills to think that the level of Lake Buildwas was the controlling factor in the relationship.

Concurrent with Wills's work, the Geological Survey was engaged in mapping to the north of Ironbridge and the results were published as survey memoirs (Pocock and Wray, 1925; Whitehead *et al.*, 1927, 1928). On the basis of shorelines, laminated sediments (varves), overflow channels and the nature of the fluvioglacial morphology, Dixon recognised the former presence of a glacial lake in the Newport area lying some 20 km northeast of Ironbridge, which he named Lake Newport. Collaboration between Dixon and Wills led to the realisation that their respective lakes Newport and Buildwas would have been separated only as long as the ice remained in contact with the northwestern flanks of the Wrekin. On the retreat of the ice, the two separate lakes could not have failed to amalgamate and, since Lake Newport had a slightly higher level than that of Lake Buildwas, Dixon and Wills argued that the threshold col at Ironbridge carrying the Lake Buildwas discharge would have gained dominance over that controlling Lake Newport at Gnosall, leading to the abandonment of the latter. Dixon suggested that the newly created larger lake should be called Lake Lapworth in recognition of Lapworth's original proposal (Fig. 4.2). Evidence to support the concept of a more extensive lake at the *c*. 90 m level was forthcoming from the Wem area (Pocock and Wray, 1925), where, it was claimed, lake

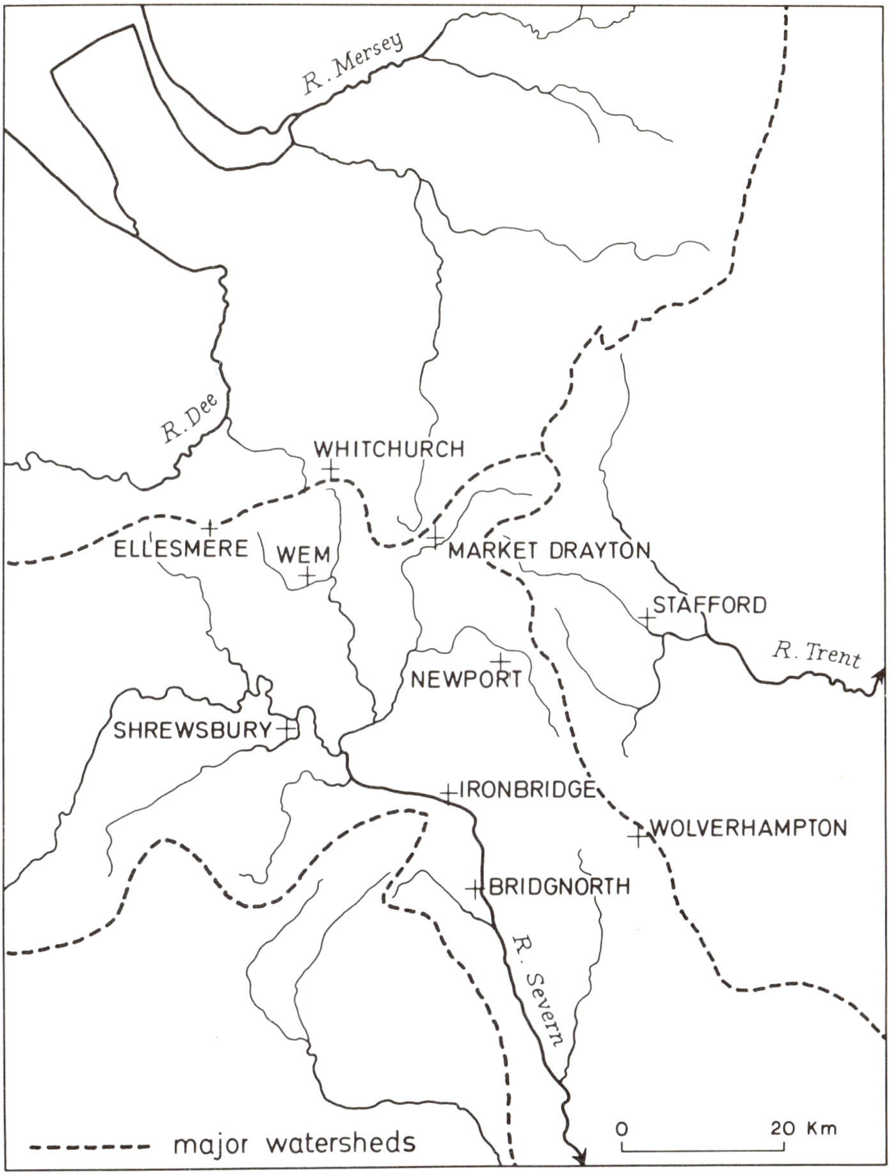

Fig. 4.1 Drainage and major watersheds in the Cheshire–Shropshire region

marginal features could be traced at many points. Further, it was stated that the strandline in the glacial sediments was difficult to follow due to postconstructional modification but that on the 'solid' rocks the notch was more definite and clearly independent of rock type.

Interest in Lake Lapworth was revived by Poole and Whiteman (1961) in an interpretation of glacial events in the west Midlands–Mersey area.

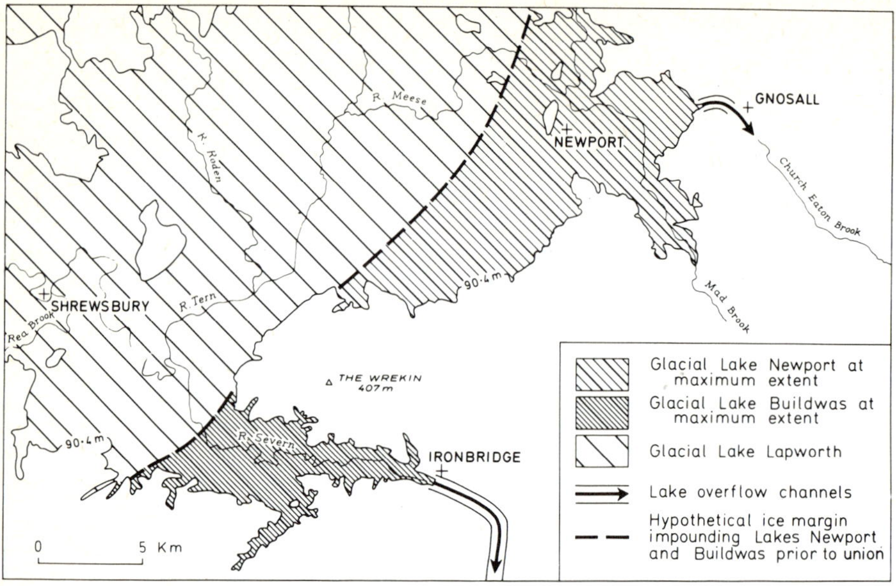

Fig. 4.2 Glacial lakes Newport, Buildwas and part of Lapworth

They extended the original definition of the lake to include a huge water body which they claimed extended from Ironbridge as far north as Chorley in Lancashire. This lake had a height of 99.5 m, that is coextensive with Dixon's Lake Newport, and was described as having a generally well developed shoreline feature throughout its extent. Following this phase, they recognised a slightly smaller lake at the 91.9 m level with a less well marked shoreline. After this episode, south of the Mersey–Dee–Severn watershed, rapid and uniform downcutting at the Ironbridge Gorge was seen as being responsible for the draining of the lake without any stillstand periods during which shorelines may have developed. In a review of Pleistocene events in the Cheshire–Shropshire lowland region Worsley (1970) considered that the data available gave general support to the glacial lake hypothesis; it was made clear that a wide range of geomorphological evidence was consistent with the concept of a former glacial lake having the general dimensions of the classic Lake Lapworth.

The shoreline problem

A substantial body of opinion has favoured the Lake Lapworth idea, although the amount of unambiguous field evidence remains limited. Comprehension of this situation may be easier if a theoretical consideration of the problem is attempted. Following the extinction of a glacial lake certain erosional and depositional phenomena might be expected to remain in the landscape which would help to establish the lake's former presence. These comprise: (1) deltas produced where either extraglacial or direct meltwater runoff entered the lake basin with the deposition of bedload sediments; (2) bottom sediments derived from the suspension load material settling out over time; (3) glacially rafted material including erratic blocks and masses of till, provided that there was a sufficient depth of water; (4) a channel or threshold which regulated the lake level; and (5) strandlines or littoral features produced by the interaction of waves and lake ice against the shore. Of all these features the most enigmatic is undoubtedly the last, since many workers have been struck by the rarity of strandlines in various areas of Britain. It seems generally true that strandline features are often absent where other criteria strongly suggest their presence.

There are several reasons for the restricted development of possible former shorelines and for their poor state of preservation. First, to develop a marked shoreline the level of a waterbody must remain stable for a certain period. Such a situation is unlikely to have been common in some ice marginal environments since the discharge into the lake may be highly variable, being largely controlled by the ablation rate which, in turn, responds to the vagaries of climatic change. The water reservoir of the lake itself may be subject to evaporation loss, and this factor alone is capable of causing significant changes in water level. The principal factor likely to affect water level changes is variation in the altitude of the lake outlet, caused by erosion and ice jamming. In a situation such as that postulated for Lake Lapworth the isostatic instability factor is not likely to have been more than a minor influence because of its position almost at the glacial limit. However, the progressive erosion of the outlets may well have been considerable. All the potential thresholds are underlain by Carboniferous or Permo-Triassic rocks, lithologies which are relatively unresistant and thus vulnerable to enlargement by high meltwater discharges. It should be stressed that this common cause of glacial lake instability, namely the periodic draining of marginal lakes through an impounding ice barrier, is not realistic in this particular environment. This is because the large ice cap to the north, which is envisaged as having a thermal regime conducive to freezing the glacier sole to the rockhead, would make this type of catastrophic release impossible. The postulated Lake Lapworth is one in which a progressive lowering of the exit channel might be expected, producing a steady reduction in the lake level associated with a regressive strandline lacking any significant still-stands. Regressive shores are invariably allied with poorly developed strand features, in marked contrast with the more massive structures often found characterising transgressive beaches.

Second, the area of Lake Lapworth as defined by the 90 m contour was sufficient to permit long fetches, generating powerful wave activity in normal circumstances and on face value this is one of the most difficult factors to reconcile with the paucity of shoreline development. By analogy with modern glacial lakes fronting cold based glaciers (Conn Lake in Baffin Island), the lengths of the partial and of the full open water seasons are critical in determining the amount of wave activity. In the event of the lake supporting an ice cover, morphological activity along the strand will be severely limited. In contrast, during the partial open water season the grounding of ice floes and icebergs is important in moving material up the beaches to build ice-push ramparts. Clearly, the degree of annual breakup of the winter ice is critical, together with the amount of shore ice surviving the melt season and protecting the shore from encroachment by wind-powered ice rafts.

Third, on the assumption that the lake level was lowered progressively through its existence, the ever-increasing areas of newly exposed shore would be at the mercy of subaerial processes. It is apparent from palaeoclimatic work that, following the ice withdrawal, there was a period of perhaps 5 000 years of periglacial conditions of sufficient severity to produce permafrost beneath the seasonally frozen layer. Under this kind of climatic regime, even land surfaces with quite gentle slopes would be susceptible to solifluction and frost-creep processes, which would tend to obliterate rapidly any existing shore features. Indices of magnitude from the period in question (*c.* 15 000–10 000 years BP) are difficult to establish, but radio carbon dating results from within the region at Church Stretton (Rowlands and Shotton, 1971) and Wybunbury (Yates and Moseley, 1967) illustrate the significant effects of periglacial processes in the short climatic deterioration of the British pollen zone III at the end of the last glacial period.

In the light of this discussion it appears reasonable to argue that any shoreline evidence in the area is likely to have been preserved imperfectly and thus it is not surprising that uncertainty surrounds its recognition today. It is on the basis of other criteria that confirmation or refutation of the Lake Lapworth hypothesis should be sought. Although most workers have maintained that true shoreline features may be recognised, others have been more sceptical. It is possible that some of the supposed shorelines are local accumulations produced in small marginal lakes (Shaw, 1972*a*). In such a case, consistency in the altitude of the strandlines would not be expected.

The first substantiated account of an area considered inundated by Lake Lapworth appeared in the Stafford and Market Drayton memoir of the Geological Survey (Whitehead *et al.*, 1927), in which Dixon described the critical area east of Newport. This account is significant since he was one of the few survey officers who had a particular interest in the interpretation of fluvioglacial landforms. This interest had been stimulated by his mapping experience in Cumberland where the evidence for ice contact deposition was particularly well preserved. In the

Newport area, however, the ice had retreated earlier and consequently there had been a greater time for postdepositional modification, especially under severe periglacial conditions. Dixon's earlier experience was of value in recognising and interpreting these more degraded forms.

Since Dixon's account was the most comprehensive discussion of field data interpreted as evidence for Lake Lapworth, this ground was re-examined by Worsley (1967*b*) as part of a wider study of glaciation in the Shropshire plain area. The district east of Newport as far as the Gnosall col on the Severn–Trent watershed was mapped at the 1 : 10 560 scale. Before the fieldwork, the areas identified by Dixon as showing a glacial lake strandline were inserted on the field maps from the reference maps available at the Geological Survey Library in London. Following this, it was apparent that the shoreline as identified by Dixon was not of a constant height, for it repeatedly crossed the 90.4 m contour. It is important to note that this accurately levelled contour was not on the maps available at the time of Dixon's work. In his memoir it is specifically stated that the strandline was marked by a notch or change in slope, most frequently at a till over bedrock boundary; that the strandline was composite, having a height range of 7.6 m; and that the different strandlines were not differentiated. Although it was possible to delimit extensive areas of flat ground which abutted against the solid outcrops forming a boundary rising at a steep angle, no feature was positively located which could be objectively identified as a shoreline. On the basis of this evidence, the case for the former presence of a glacial lake is inconclusive.

An alternative approach, first used by Dixon, is available and is based on his classification and interpretation of ice marginal depositional landforms (Fig. 4.3). This scheme enables interpretation of landforms in terms of the palaeoenvironment at the time of formation and in particular the distinction between ice margins terminating on land and in water. In the Wolverhampton memoir (Whitehead *et al.*, 1928), a distribution of ice stands between Wolverhampton and Newport, as identified by the Newport esker chain and the Penkridge esker, is illustrated. This map has been reproduced so often it is now established as a type example of how former ice marginal positions may be reconstructed from esker morphology (Sparks, 1972), even though one of the authors of the memoir questioned the validity of the ice stands. Even a cursory examination of the esker outcrops on the map gives rise to doubts of the significance of the total distribution and a field check supported this view since much of the area shown as an esker system consists of sand and gravel without any distinctive morphology. Nevertheless the survey officers recognised and mapped some very fine esker systems as their written accounts testify, but it is remarkable that no map has been published showing the distribution of true eskers as specific landform features.

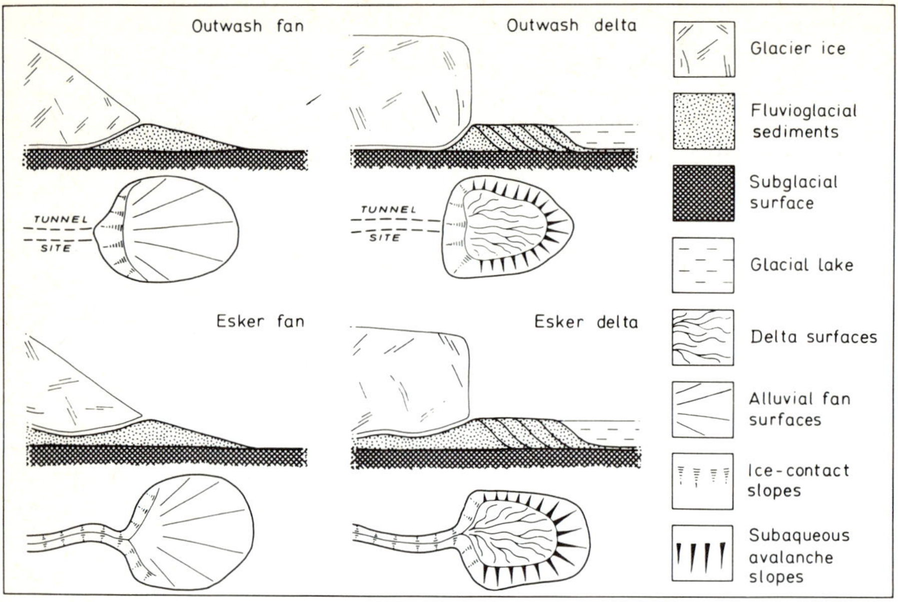

Fig. 4.3 Classification of ice-marginal depositional forms. Based on Whitehead *et al.*, 1928

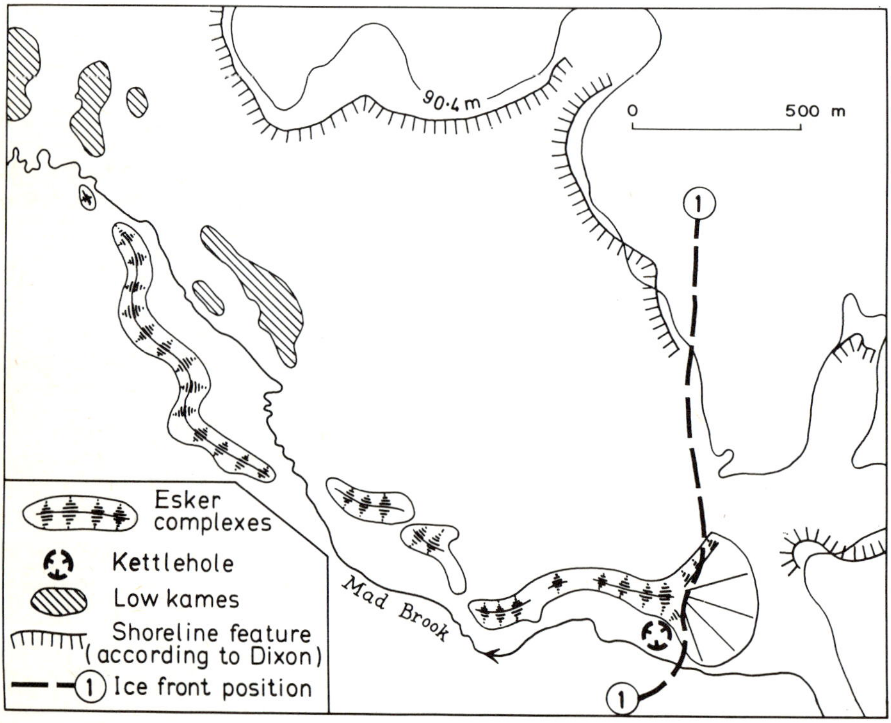

Fig. 4.4 Esker-fan system along part of the Mad Brook valley

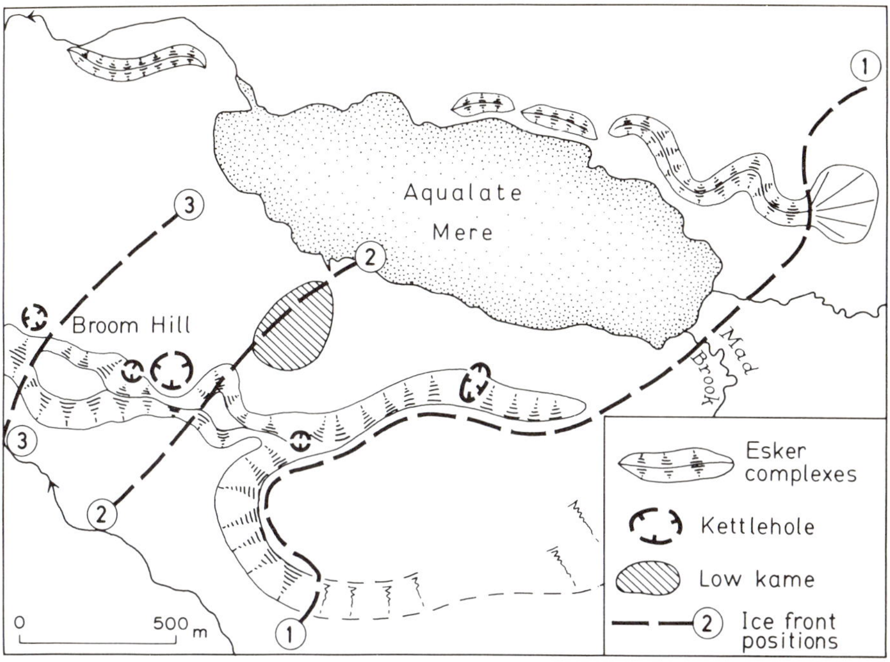

Fig. 4.5 Esker-fan system north of Aqualate Mere and the Broom Hill complex esker-delta

Three esker systems have been mapped morphologically (Figs 4.4 and 4.5), although at the time of examination no sections were available which might have revealed their sedimentary structures. One system follows the axis of the Mad Brook valley, although it is not a continuous feature. Parts have been removed by man but it seems reasonably certain that it never was a single depositional structure. At the southern (downstream) terminus of the esker there is a very prominent fan without any vestige of a flat-topped upper surface. Since the stream which produced the esker drained into a closed valley head from which the ice margin was retreating, it is highly probable that the meltwater was ponded in the proglacial area. However, in this instance the sediment supply was so limited that aggradation failed to rise above the water level of the lake. A similar esker system occurs north of Aqualate Mere, again with a fan-shaped terminus. South of Aqualate Mere is the Broom Hill esker which originates just east of Newport. This does not have the simple, sinuous form of the others: it is essentially straight, and has three beads in its course and a sizeable kettlehole on the northern side. The terminus is complex and appears to have been produced by a set of ice marginal fans emanating from a prominent ice contact slope which runs from southwest to northeast. The morphology of the overall feature surviving today is difficult to comprehend, although it does have a southeasterly fall from the 'feeding' esker. The recognition of ice contact slopes and esker termini demonstrates the likelihood that the ice front withdrew to the northwest across the area with a well defined margin. The gross topography in relation to such a type of ice retreat suggests that, as the ice moved back, the heads of pre-existing valleys emerged. Undoubtedly the meltwater drainage from the ice flowed into the recently vacated valleys and, since the local slope morphology would have prevented free drainage from the ice margin, any meltwaters produced must have been ponded. The distribution of the various esker fans demonstrates that, as the ice retreated, the meltwater lake became progressively larger. The col at Gnosall, which would have been the lowest point on the lake perimeter, is associated with a valley outwash train which runs along the valley of the Church Eaton Brook. This feature confirms that the Gnosall col formerly functioned as an overflow channel and thus Dixon's basic hypothesis of a glacial lake in the Newport area is confirmed by several types of field evidence.

Sedimentological data

The conclusion reached by Shaw (1972*a*) clearly requires further consideration since, if the Lake Lapworth concept is to remain valid, the sedimentological evidence must be reconciled to the theoretical framework. To meet this difficulty, Shaw investigated the sedimentological characteristics of the glacigenic sequence in the Shrewsbury district. The interpretative model finally adopted by Shaw (1972*b*) was the well established one of mass stagnation, with the underlying bedrock playing a major influence in the pattern of ice wastage. Within the decaying ice sheet Shaw envisaged an integrated drainage system flowing through crevasses and basal tunnels. As the ice stagnated in the deeper hollows in the bedrock surface, shear planes formed at the contact zone between the stagnant ice and the more active ice lying further to the north. Along these shear planes basal debris was carried to higher levels and eventually released on the ice surface by ablation. From outcrops sited above the ice margin this debris would have been subject to gravity-controlled mass movement producing flow tills.

On the basis of observations from two sand and gravel pits, Shaw concluded that the identification of ice contact and fluvioglacial sediments occurring close to the present level of the Ironbridge Gorge threshold indicated that, at the time of deposition and therefore deglaciation at the two localities, the gorge was already functioning as a subaerial outlet for the glacial meltwater. He argued from this proposition that the gorge must have originated from a previous, unspecified glacial episode and accordingly it was not possible to form a glacial Lake Lapworth at this later period of deglaciation. At this juncture it may be noted that the suggestion that the gorge may date from a time prior to the last ice advance is not new. Robertson (Whitehead *et al.*, 1928) suggested that Wills's (1924) interpretation need not assume the non-existence of the bedrock-cut gorge, for if it were blocked by iceborne material the same hypothesis would be valid. Furthermore, Horton (1969) speculated that the rapid downcutting at Ironbridge was related to the re-excavation and deepening of an older valley which had been infilled by sediments similar to those remaining at Buildwas, the implication being that the Buildwas sequence was older than the last major ice advance.

On the basis of sedimentological data alone, the reasoning which leads to the rejection of the glacial lake hypothesis seems convincing and as a result the existence of yet another meltwater lake would seem to have been disproved. In view of this a close examination of the nature of the evidence which supports this alternative interpretation is necessary.

Recognition of ice contact landforms and sediments

The sedimentological and geomorphological relationships in the area leave little doubt that many of the glacial landforms were created in an ice contact environment. This factor alone does not invalidate a glacial lake explanation, since it is widely recognised that dead ice landform assemblages often occur in association with glacial lakes (Lundqvist, 1972).

Palaeocurrent data

It is evident that certain bedform structures found at two localities indicate a palaeocurrent flow towards the east and it may be accepted that this is towards the presentday gorge at Ironbridge. However, the use of this data in support of a freely flowing drainage by way of the Ironbridge Gorge must be tempered with caution on the following grounds. First, there is an unstated basic assumption that the palaeocurrent was flowing freely, behaving as an unconstrained gravity-powered flow. One alternative possibility is that the sediments in question are the products of a confined flow situation and that they were subglacially deposited. Although this concept is not favoured on anything but a most limited scale, it should be recorded that some authorities would support it (Hoppe, 1963). Indeed, Shaw (1972*b*) favours a similar mechanism to explain some of the sedimentary sequences south of Shrewsbury, where an integrated subglacial drainage system has been postulated. The meltwater could have been flowing under hydrostatic pressure towards the Ironbridge Gorge, then lying at a higher elevation.

Second, the discussion so far has assumed that the presentday distribution of sediments and landforms indicates directly the level of deposition within the palaeoenvironment with respect to Ordnance Datum. Such an assumption makes no allowance for the possibility that much of the deposition originally occurred over an extensive sheet of stagnant basal ice in a supraglacial position. The case for this mode of deposition operating on an extensive scale during the wastage of the late Pleistocene lowland ice sheets in Britain has been argued persuasively by Boulton (1972). It was primarily field experience in the Cheshire–Shropshire area, where the sedimentary landform relationships could not be accounted for satisfactorily using the former generally accepted deglacial interpretative models, that led Boulton to seek a modern analogue which, as far as possible, would be more in accord with local glacial geological evidence. This search led Boulton to Spitzbergen where subpolar glaciers terminate in a relatively maritime climatic environment in a lowland situation underlain by soft rocks. This type of area exhibits some of the characteristics which might have been prevalent in Shropshire during the last deglaciation. On the basis of field experience in Spitzbergen, Boulton postulated that the generalisation that the landform sedimentary association, which is common in the Cheshire–Shropshire region, may have arisen from the wastage of a subpolar glacier. Supplementary evidence that the glacier may have been frozen to its bed is provided by the widespread occurrence of fossil permafrost structures in the region (Worsley, 1966, 1967*a*; Morgan, 1971). This suggests that the broad climatic environment before, during and immediately following the Irish Sea glacier advance supported widespread permafrost and, as a consequence, the mean annual temperature must have been at least 15° C lower than at present for much of the time.

The assemblage of sedimentary relationships recorded by both Wills in describing the Buildwas Sands exposures and by Shaw at other localities are of a nature currently considered to be diagnostic of the wastage of subpolar ice. It follows that much of the deposition may have been supraglacial and that this, by definition, occurred over extensive areas of stagnant ice. Subsequent to final deposition, the buried ice slowly wasted, gently lowering the supraglacially-deposited sequences on to the subglacial surface without inducing any widespread deformation. As a result, the sedimentary directional properties inherited from an earlier period of deposition at a higher level may well have been preserved. Application of this concept to the two localities with the eastward-flowing palaeocurrents, and bearing in mind the altitudinal relationships between Lake Lapworth and Ironbridge Gorge, requires sediment deposition over an ice body some 50 m thick. This figure must necessarily be an estimate but is consistent with the thicknesses of buried glacier ice in contemporary Arctic environments undergoing deglacierisation.

It follows from these arguments that any claim that, at the time of deposition at the two localities in question, the Ironbridge Gorge was necessarily functioning at its present level is premature. Further, supra-glacial deposition in the area has already been invoked by Shaw (1972*a*, p. 628), stating that 'large areas would have been below the level of the ice-supported streams beds', a view supported by his illustration (1972*b*) of sediments deposited over basal ice prior to being let down on to the subglacial surface without any *extensive* deformation.

Third, another difficulty encountered by the hypothesis that the over-flow threshold of the Ironbridge Gorge approximated to its present height during deglaciation is the implication that relatively little bedrock erosion has been achieved by the Severn during the 16 000 years or so since deglacierisation. The gorge today is scarred by dormant and active landslips, the former Shrewsbury to Bridgnorth railway was bedevilled by subsidence throughout its existence, and even the historic iron bridge is suffering from the encroachment of the opposed river banks. All this type of evidence testifies to the comparative 'youthfulness' of the gorge and the inequilibrium condition of the river channel and the surrounding valley slopes.

The glacial chronology

On the basis of the acceptance of the hypothesis that the gorge is an old feature antedating the last glacial episode, and that the base of the Buildwas Sands was linked directly to the threshold height of the gorge at the time of deposition, it is necessary to postulate a substantial aggra-dation phase, as represented by the sand sequence, so that the gorge was choked with sediment at least to the upper height of the Buildwas sequence as seen today. Such an event has implications for the Severn terrace chronology. According to Wills (1937), the Main Terrace was formed when the gorge was partly cut, having an origin within the gorge itself. This means that the gorge must date from before the last glacia-tion: on the other hand the stratigraphic data, reviewed above, indicate consistently that it was created during the phase of ice withdrawal from the advance limit of the Bridgnorth to Wolverhampton line. To question the correlation of the ice retreat with the Main Terrace and to appeal for more surveying above and below the gorge (Shaw, 1972*a*), simply to account for some sedimentary structures in two exposures, reject by implication the mapping previously done by the Geological Survey and Wills. In the circumstances, the overwhelming weight of evidence is firmly in favour of the earlier workers.

Relation of the Whitchurch moraine complex to the lake

The northern boundary of Lake Lapworth is formed by a west–east trending bi-arcuate end moraine complex which bisects the Cheshire–

Shropshire plain. The relationship of this feature to the lake has been commented on previously (Worsley, 1970), when it was proposed that some of the morphological characteristics were influenced strongly by a standing waterbody lying immediately to the south during its construction. Since then, Boulton (1972) has made some broad generalisations about the feature, seeking to demonstrate that many of the moraine complexes in areas of late Pleistocene glaciation could be better understood in terms of an essentially supraglacial origin. He observed that fluvioglacial sediments are often a dominant constituent of moraine complexes and suggested that this situation can be matched by many modern supraglacial environments where fluvial sediments are aggrading between and above ice-cored moraines. On this basis he postulated that the attribution of a moraine ridge to the direct action of an ice margin is often inappropriate and that an inverted fluviolacustrine mound of an ice-cored moraine is a frequent occurrence resulting from the total wastage of a former ice core. He maintained that, in this type of situation, the ice cores inhibit the free outward flow of meltwaters and thereby promote the deposition of the outwash bedload in close proximity to them. Thus he suggested that the lack of extensive outwash sandurs is consistent with his hypothesis and that the very limited development of outwash trains associated with the Whitchurch feature exemplifies this relationship. Whether Boulton rejects the idea that the lack of an extensive sandur is also indicative of the former presence of a glacial lake with outwash rivers rapidly entering standing water to deposit their bedload is not clear.

The application of Boulton's general idea to the genesis of parts of the moraine complex in question is justified although caution is essential when making genetic interpretations of glacial landforms solely on the basis of their morphology. One of the rare exposures in the moraine complex is afforded by the sand and gravel workings near Ellesmere. There the dominant lithology is indeed outwash material, including some flow tills, but near the northern limit of the exposure is a glacial tectonic structure consisting of thrusted slices of till and overturned outwash lithologies signifying active rather than passive ice determining the development of the landform. The area with the tectonic structures corresponds to an arcuate ridge which has a large kettlehole at the focus on the northwestern side. As with many generalisations in glacial geomorphology, the inverted fluviolacustrine ice-cored moraine hypothesis, involving collapse over an ice core, encounters difficulties in particular instances.

One important difference occurs between the situations described by Boulton in Spitzbergen and the palaeoglacial environment in north Shropshire during the last glaciation. This concerns the overall physiographic framework, since the drainage from the Spitzbergen glaciers is usually free without the impounding of large ice-dammed lakes. A further difference is the comparative scale and Boulton admitted that one of the major problems is the relatively restricted extent of the contemporary areas of active deglacial phenomena (in Spitzbergen the maximum amount of glacial retreat being given as 12 km). He also noted that, where the main meltwater channels are incised into deep gorges, they tended not to interfere with the processes of ice core collapse and allied mass movement. The incision is a response to a falling base level – usually the sea – due to isostatic rebound near to the glacial termini.

The most likely situation to have prevailed in north Shropshire was one of impounded proglacial drainage between the ice margin and the watershed lying to the south. It is assumed on the basis of the arguments above that the glacier was cold based, at least in the marginal zones, for part of the period it occupied the area. This implies that the glacier would have been an efficient impounder of the meltwater and the only escape routes would have lain to the south through the low cols in the watershed. Given this situation, the very presence of impounded meltwater would inhibit the development of extensive fringing sandurs and induce deposition at or very near the ice margin. As Boulton agreed, the Whitchurch moraine complex does not possess any major sandur areas and it is suggested that an important element in determining this situation was the existence of Lake Lapworth.

Welsh ice advance to Shrewsbury

This event is enigmatic, to judge from the literature, since its precise limits have never been defined. Despite this there is a general consensus among various workers in the locality that the limit lies in or immediately adjacent to Shrewsbury. Urbanisation has hindered its delimitation, a task further hampered by the fact that no readily recognisable landform such as a terminal moraine ridge marks its limit. The two major sources of data available on the Shrewsbury area are Geological Survey maps and the various descriptions of the exposures in the Mousecroft Lane sand and gravel pit in the southwestern suburbs of the town. It is clear from the stratigraphy of the pit that, following the deposition of a glacigenic sequence derived from the Irish Sea containing sizeable masses of glacier ice, braided outwash streams emanating from glacier ice of Welsh upland provenance laid down an aggradation sequence before the melting of the buried ice within the Irish Sea sediments was complete. Finally, the Welsh ice overrode the outwash gravels and a till was emplaced. On the limited basis of small two-dimensional structures, Shaw (1971) suggested that the Welsh ice was cold based, a conclusion supporting the general concept that the glacial episode under discussion was one with a mean annual temperature significantly below 0° C. Clearly such conditions would serve to perpetuate any buried ice masses after the Irish Sea ice had withdrawn and during the later Welsh advance. The remarkable impression obtained from the Shrewsbury area is that the hummocky terrain is the product of the melt-out of the ice masses in the Irish Sea sediments and that the subsidence associated with this process occurred after the vacation of the area by the later Welsh ice. Consequently, both the Irish Sea and Welsh glacigenic sequences are deformed by the same melt-out period.

In their mapping, the officers of the Geological Survey demonstrated that the highest and what they interpreted as the oldest terrace in the Severn Valley west of the Ironbridge Gorge could not be recognised above Shrewsbury, although they thought that it extended for some distance up the valley of the Rea Brook, a right bank tributary joining the Severn at Shrewsbury (Pocock *et al.*, 1938). The preliminary results of mapping the Quaternary sediments and morphology in the Shrewsbury area in connection with a study of the Severn terraces upstream of Ironbridge indicate that, of the three terraces identified by the survey, only the two lower terraces are true river terraces – the Atcham and Cressage Terraces. An examination of the survey map reveals that, save for the area immediately east of Shrewsbury, their highest terrace (the Uffington) has a restricted extent. On investigation, these small tracts have appeared more likely to be *remanié* patches of glacigenic sediments rather than 'normal' fluvial aggradations. In the vicinity of Shrewsbury the surface of the Uffington Terrace grades imperceptibly into glacial sands and gravels as mapped by the survey. Despite the built-up nature of the area, many temporary exposures occur, revealing that the fluvioglacial sediments are indistinguishable from the Welsh outwash gravels beneath the Welsh till at the Mousecroft Lane sections. This applies also to the Uffington Terrace sediments in the type area around Uffington village.

Bearing in mind the restricted extent of the Uffington Terrace, the possibility must be entertained that in this district the feature originally mapped as a river terrace is, in fact, the distal equivalent of the Welsh glacier outwash, and that the nearer representatives have become disguised in the course of mapping as lithologically undifferentiated glacial sands and gravels. Few exposures of the Welsh till have been located to the east of the Rea Brook and in this area the sequence is Welsh ice outwash overlying tills and fluvioglacial materials of Irish Sea provenance. The capping of Welsh outwash gravels may be traced eastwards across a plateaulike area to a point where there is a sharp break of slope into the Severn valley proper. This break is attributed to the postglacial incision of the main river, although this does not preclude the possibility that some of the downcutting occurred during deglaciation. To the southeast, this break in slope swings away from the main river, where it appears that the slope is not erosional but depositional. In this area a former quarry showed the Welsh gravels overlying sediments derived from the Irish Sea but the site has been filled and levelled. However, the Welsh gravels are not encountered further east and the distinct possibility emerges that the depositional slope may be in part a delta front and so represent the termination of the Welsh outwash sandur.

These relationships suggest that the fringing sandurs in front of the ice at both Shrewsbury and Whitchurch were controlled by waterbodies lying in their paths. Again, this situation induced deposition and thereby prevented the bed load material from extending any further towards the Ironbridge exit. Thus, landform sedimentary relationships are consistent with the presence of an extensive glacial lake beyond the ice margins. Within the Shrewsbury area the field worker is confronted by a major problem in that the modern physiography is essentially the fossilised equivalent of an extensively deformed glacial depositional landscape, involving areas of both till and fluvioglacial deposits. As a consequence, fragments of the original surface are not readily recognisable and this partial concealment explains why the limit of the Welsh ice in the area has been so difficult to define. The major contributing factor in this case appears to have been the survival of masses of ice preserved in the Irish Sea sediment suite when the debris of Welsh provenance was laid down. The widespread but relatively thin capping of Welsh sediments suggests that, at the time of deposition, the area had a fairly low relief permitting the meltwaters from the Welsh ice to anastomose across the area as the sandur was developed. The impressive array of kettleholes which litter the Shrewsbury region today are the result of former melt-out processes allied to the presence of the buried ice, ultimately giving rise to the hummocky relief.

Although the Welsh glacier sandur probably entered a glacier lake lying near to its advance limit, it should be emphasised that this event occurred some time after the creation of Lake Lapworth at 90 m OD. The general deltaic terminal surface level at Shrewsbury lies at approximately 66 m, demonstrating that by this time the controlling outlet of Lake Lapworth at Ironbridge had been incised by some 24 m. It follows that the lake area must have been reduced. Despite this, a sizeable meltwater lake remained. The outwash in the Tern valley (Worsley, 1970), which terminates near Hodnet, has a minimum height of *c*. 65 m. This similarity in altitude suggests that the two deposits are almost synchronous features controlled by the same lake level, further supporting the concept of a single extensive lake. If this relationship is valid, it also serves to increase the probability that the Welsh Readvance occurred almost at the same time as the Irish Sea ice sheet withdrew from the Whitchurch moraine complex.

Conclusions

Considerable support remains for the former presence of Lake Lapworth. On balance, it is inadvisable to reject the concept on the basis of limited sedimentological evidence, especially when this is amenable to interpretation in a variety of ways, none of which necessitates the elimination of a glacial Lake Lapworth. It is fair to say that while the former existence of the lake is not absolutely proven, very much more of the evidence is explained by accepting the lake hypothesis than by rejecting it. Many problems remain, including the nature, origin and age of the Buildwas Sands and the nature of the underlying bedrock surface. Perhaps the most perplexing question is why the gorge seems to have been formed so late in the Pleistocene when a similar overflow situation is likely to have prevailed during earlier glaciations. One possibility is that only in the last glaciation did the Irish Sea ice penetrate so far into the northwest Midlands, and that previously the ice emanating from the Welsh uplands was dominant in the area, with a retreat pattern from east to west rather than from south to north. Thus the precursor of Lake Lapworth may have drained into the Trent system as did Lake Newport prior to its capture by Lake Buildwas.

It has been suggested that the Ironbridge Gorge is essentially an exhumed feature. This view leads to severe difficulties with the Severn terrace chronology unless there was considerably more ice marginal variation allied to the last glaciation maximum than is generally assumed. Certainly, experience of contemporary ice margins unequivocally demonstrates a general instability and it may be that reconstructions of ice marginal variation do not incorporate a sufficient degree of randomness. This variation is much easier to recognise theoretically than to demonstrate in practice. Thus, on balance, Lake Lapworth survives for the present.

References

BOULTON, G. S. (1972) 'Modern Arctic glaciers as depositional models for former ice sheets', *Q. J. Geol. Soc. Lond.*, **128**, 361–93.

CLAPPERTON, C. M. (1968) 'Channels formed by the superimposition of glacial meltwater streams, with special reference to the east Cheviot Hills, north-east England', *Geogr. Annlr.*, **50**, 207–20.

GREGORY, K. J. (1965) 'Proglacial lake Eskdale after 60 years', *Trans. Inst. Br. Geogr.*, **36**, 149–62.

HARMER, F. W. (1907) 'On the origin of certain cañon-like valleys associated with lake-like areas of depression', *Q. J. Geol. Soc. Lond.*, **63**, 470–514.

HOPPE, G. (1963) 'Subglacial sedimentation with examples from northern Sweden', *Geogr. Annlr.*, **45**, 41–9.

HORTON, A. (1969) 'Pleistocene and Recent deposits', in B. A. Hains and A. Horton, eds, *The Central England District*, British Regional Geology, pp. 89–103.

KENDALL, P. F. (1902) 'A system of glacier-lakes in the Cleveland Hills', *Q. J. Geol. Soc. Lond.*, **58**, 471–571.

LUNDQVIST, J. (1972) 'Ice-lake types and deglaciation pattern along the Scandinavian mountain range', *Boreas*, **1**, 27–54.

MANNERFELT, C. M. (1945) 'Några glacialmorfologiska formelement och deras vittnesbörd om inlandsisens avsmältningsmekanik i svensk och norsk fjällterräng', *Geogr. Annlr.*, **17**, 3–239.

MORGAN, A. V. (1971) 'Polygonal patterned ground of Late Weichselian age in the area north and west of Wolverhampton, England', *Geogr. Annlr.*, **53A**, 146–56.

POCOCK, R. W. and WRAY, D. A. (1925) 'Geology of Wem', *Mem. Geol. Surv. England and Wales*.

POCOCK, R. W., WHITEHEAD, T. H., WEDD, C. B. and ROBERTSON, T. (1938) 'Shrewsbury District', *Mem. Geol. Surv. England and Wales*.

POOLE, E. G. and WHITEMAN, A. J. (1961) 'The glacial drifts of the southern part of the Shropshire–Cheshire Basin', *Q. J. Geol. Soc. Lond.*, **117**, 91–130.

ROBINSON, H. (1969) *Morphology and Landscape*, University Tutorial Press.

ROWLANDS, P. H. and SHOTTON, F. W. (1971) 'Pleistocene deposits of Church Stretton (Shropshire) and its neighbourhood', *Q. J. Geol. Soc. Lond.*, **127**, 599–622.

SAWYER, K. E. (1970) *Landscape Studies: an introduction to geomorphology*, Edward Arnold.

SCHUMM, S. A. and SHEPHERD, R. G. (1973) 'Valley floor morphology: evidence for subglacial erosion?', *Area*, **5**, 5–9.

SHAW, J. (1971) 'Mechanism of till deposition related to thermal conditions in a Pleistocene glacier', *J. Glaciol.*, **10**, 363–73.

SHAW, J. (1972*a*) 'The Irish Sea glaciation of north Shropshire – some environmental reconstructions', *Fld Stud.*, **3**, 603–31.

SHAW, J. (1972*b*) 'Sedimentation in the ice – contact environment with examples from Shropshire (England)', *Sedimentology*, **18**, 23–62.

SISSONS, J. B. (1961) 'Some aspects of glacial drainage channels in Britain, part 2', *Scott. Geogr. Mag.*, **77**, 15–36.

SPARKS, B. W. (1972) *Geomorphology*, 2nd edn, Longman.

STENBORG, T. (1969) 'Channel formation and glacier drainage', *Geogr. Annlr.*, **51A**, 112–13.

WATTS, W. W. (1898) 'Long excursion to the Birmingham district', *Proc. Geol. Ass. Lond.*, **15**, 417–28.

WHITEHEAD, T. H., DIXON, E. E. L., POCOCK, R. W., ROBERTSON, T. and CANTRILL, T. C. (1927) 'The country between Stafford and Market Drayton', *Mem. Geol. Surv. England and Wales*.

WHITEHEAD, T. H., ROBERTSON, T., POCOCK, R. W. and DIXON, E. E. L. (1928) 'The country between Wolverhampton and Oakengates', *Mem. Geol. Surv. England and Wales*.

WILLS, L. J. (1924) 'The development of the Severn valley in the neighbourhood of Ironbridge and Bridgnorth', *Q. J. Geol. Soc. Lond.*, **80**, 274–314.

WILLS, L. J. (1937) 'The Pleistocene history of the west Midlands', *Rep. Brit. Ass.*, Sec. C, pp. 71–94.

WORSLEY, P. (1966) 'Some Weichselian fossil frost wedges from east Cheshire', *Mercian Geol.*, 1, 357–65.

WORSLEY, P. (1967*a*) 'Fossil frost wedge polygons at Congleton, Cheshire, England', *Geogr. Annlr.*, **48A**, 211–19.

WORSLEY, P. (1967*b*) 'Some aspects of the Quaternary evolution of the Cheshire Plain and adjacent areas', unpublished PhD thesis, University of Manchester.

WORSLEY, P. (1970) 'The Cheshire–Shropshire Lowlands', in C. A. Lewis, ed., *The Glaciations of Wales and Adjoining Regions*, Longman, Geographies for Advanced Study, pp. 83–106.

YATES, E. M. and MOSELEY, F. (1967) 'A contribution to the glacial geomorphology of the Cheshire Plain', *Trans. Inst. Br. Geogr.*, 42, 107–25.

Monica M. Cole

5 Recent developments in biogeography

In 1952, under the direction of Professor Beaver, Keele became one of the first centres to develop research and provide courses in biogeography. Some aspects of this work were summarised in 1965 in an attempt to indicate the value of biogeographical studies in bringing about a more rational use of natural resources. In this account the importance of fundamental investigations of the relationships of vegetation and soils to environmental parameters and their applications in conservation, the assessment of land potential for agriculture and other uses, in mineral exploration and in studies concerned with disease incidence and the maintenance of health were all cited (Cole, 1965). Today, in a world suddenly aware of the danger of pollution and wild life deterioration, it seems appropriate to outline the spectrum of the subject and examine recent developments in a tribute to one who was among the first to recognise its importance.

The spectrum of biogeography

Biogeography includes a range of discrete but related studies which may be grouped into six major categories. Those concerned with the structure, composition and distribution of plant and animal communities, at all levels of classification, are basically ecological and were among the first to receive attention. Related geobotanical and biogeochemical studies which employ quantitative field analytical and computing techniques in an evaluation of the effects of specific environmental parameters, including bedrock geology and atmospheric pollution, on plant distribution and on certain aspects of plant physiology and morphology are more recent. The information they yield is of importance in the third category of studies, those related to historical plant and animal geography, in which interest centres on past environments and on the influence of geomorphological evolution and geological and climatic changes on floral and faunal distributions. The fourth category embraces soil studies which cover pedogenesis in relation to landscape evolution and cultural practices. A fifth category, of recent growth, measures productivity and energy exchange within ecosystems. The sixth includes integrated and applied studies concerned with problems of resource use and conservation, pollution and reclamation, and disease incidence and control.

The breadth of this spectrum poses many problems, including that of defining biogeography itself. This difficulty may be illustrated by reference to two recent books which treat the subject differently: Seddon (1971), in bringing together new and interesting material, concentrates on distributions in the classical style of plant and animal geography; Watts (1971), in dealing with principles, ranges over energy controls, biogeochemical cycles and environmental limitations within ecosystems and devotes more attention to the physical nature of the environmental parameters than to the plants and animals whose interrelationships with the environment constitute the subject of biogeography.

Ecological studies of vegetation communities

In Britain plant geography began with the mapping of vegetation units (Moss, 1907, 1910, 1913). This was followed by studies in which the types of vegetation were related to individual habitat factors (Crompton, 1911; Tansley, 1911; Adamson, 1918; Salisbury, 1925; Woodhead, 1929) and later to the complex of interacting environmental factors which produced the concept of the ecosystem, expressed in Tansley's classic work *The British Islands and their Vegetation* (1939). In turn such work provided the basis for studies of ecological processes and cyclic phenomena (Watt, 1947; Barclay Estrup and Gimingham, 1969; Barclay Estrup, 1970) and, following the introduction of methods of association analysis (Williams and Lambert, 1959, 1960, 1961, 1966; Lambert and Williams, 1962, 1966) and of ordination (Curtis and McIntosh, 1951; Bray and Curtis, 1957; Curtis, 1959), of the multivariate analysis of vegetation in terms of species and habitat. In areas of complex vegetation cover such approaches have elucidated the relationships between individual plant communities and environmental conditions and have been applied with particular success to studies of heathland communities (Ivimey-Cook and Procter, 1966; Harrison, 1970).

Until recently studies of the forests, savannas and semidesert communities in the tropics focused on the structure, physiognomy and floristic composition of the vegetation units, and on a subjective assessment of the respective roles of environmental parameters and man's activities, in determining the major boundaries, notably that between forest and savanna where fire was considered of paramount importance. The recognition of repetitive mosaic patterns of vegetation and soils in the Sudan (Morrison *et al.*, 1948) and of the associations between vegetation, and geomorphology and climatic change, with their influence on soil genesis, soil moisture and soil nutrient status in the southern continents (Cole, 1960, 1963, 1964), invited examination of the precise relationships using quantitative and analytical techniques. In the last decade investigations of these relationships have elucidated the distribution patterns and in some cases discounted the role of fire. Thus, combining soil profile studies with precise measurements of soil moisture conditions in the Rupununi area of Guyana, Eden (1964, 1970) has demonstrated that the boundary between the herbaceous savanna and the savanna woodland is dictated by soil moisture variations, which reflect relief and drainage conditions. Within the savanna woodlands, while accepting that soil compaction and grass fires might account for compositional differences, he found no evidence that fire had caused a retreat of the forest and noted that where fire had extended into the forest in very dry years, regeneration of forest species occurred during subsequent wet years.

In Venezuela comparable studies undertaken at the biological station at Calabozo, in the dry savanna region of the central Llanos, produced similar conclusions and underlined the importance of geomorphology as the main single factor operating edaphically through water availability (Sarmiento and Monasterio, 1969). The investigations carried out in the savanna-forest boundary zone of the northeastern Mato Grosso in Brazil established the vegetation boundary as coincident with one of soil texture, being probably affected by moisture availability; the role of fire was dismissed (Askew *et al.*, 1970; Daultrey, 1970). In western Nigeria Moss and Morgan (1970) established the stability of the forest-savanna boundary over a ten-year period despite increasing population density and increased cultivation and confirmed the overriding importance of environmental factors in controlling vegetation distributions. By contrast, in those areas of West Africa and New Guinea, where environmental factors favour forest, Adejuwon (1970) and Flenley (1969) have demonstrated the role of fire and human occupance in bringing about its destruction and replacement by grassland.

Studies of animal communities

Among recent analyses those concerned with exploding populations of pests and with vectors of disease illustrate the importance of biogeographical research. Studies of geographical distributions and migrations have been basic to the development of the control strategy of the Centre for Overseas Pest Research, which has grown out of the Anti-Locust Research Organisation. These have indicated the outbreak areas of the red locust and the African migratory locust and assisted the elucidation of the processes which lead to upsurges and new plagues of the desert locust.

In Tanzania ecological studies of locust populations led to the conclusion that permanent populations were found only in areas of valley grassland with impeded or closed drainage, notably the Rukwa valley (Vesey-Fitzgerald, 1955). Later investigations, embracing biogeographical and geomorphological considerations, disclosed that locust dynamics, behaviour and control are influenced by the distribution of grass species changes in which occur as a result of fire and flooding, the latter being connected with fluctuations in the level of Lake Rukwa. A twenty-year interval occurs between the peaks of the lake level and during this period the vegetation alternates between the species of wet and dry habitats with consequent repercussions for the locust populations (Dean, 1967, 1968).

Recognition of the influence of weather on the migration of bird and insect pests has prompted investigation of the weather systems which at times may either concentrate or disperse pest populations. The use of radar and of satellite imagery has assisted studies of night-flying species and facilitated forecasting of movements and breeding of locusts, army worm moths and other pests, particularly in sparsely populated areas. Recent studies of the spread of bovine trypanosomiasis carried by tsetse flies (*Glossina* spp.) in southwest Uganda have demonstrated a dynamic relationship with the growth and extension of *Acacia gerrardii* and *A. hockii* to the stage where they form a suitable environment for the flies (Ford and Clifford, 1968). The spread of these tree species is related in turn to changes in the environmental parameters of which increased dissection of the terrain may be of critical importance.

Historical biogeography

Histories of the flora and fauna of Britain and continental Europe are among the earliest biogeographical works. Forbes (1846) first explained the discontinuous distribution of Arctic-Alpine species on the mountain ranges of Europe as relicts of an Arctic flora which occupied the lowland plains of Europe during the Great Ice Age. Subsequently these species ascended the mountains and retreated northwards with ice recession. Thereafter attention focused on an explanation of the distribution of floras and faunas in terms of migrational and evolutionary shifts induced by past climatic changes and geological processes. Distributional data which could be correlated with Pleistocene geology and archaeological remains were accumulated particularly by Reid for his book *The Origin of the British Flora* (1899). The great advance in historical plant geography, however, came with the advent of radiocarbon dating and the development of pollen analysis and its application in Quaternary investigations.

Pioneered by van Post, this development was introduced by Erdtman (1924) into Britain and other English-speaking countries in the mid-1920s. The publication of Wulff's *Historical Plant Geography* (1943), summarising much of the European and American work on the flora of postglacial and interglacial Quaternary deposits and of older deposits extending back to the Tertiary, provided a landmark in the literature. Of equal significance was Godwin's *The History of the British Flora* (1956), which re-examined the distributional data of Reid in the context of evidence from pollen analysis. More recently, palynological studies have been extended to the southern continents, where van der Hammen and his collaborators in South America have established a Quaternary chronology in which the glacial periods of the northern hemisphere are equated with pluvial periods in Colombia and Guyana (van der Hammen and Gonzalez, 1960, 1964; van der Hammen, 1963; Gonzalez *et al.*, 1966). In Africa van Zinderen Bakker and his co-workers have produced evidence of temperature changes in the Pleistocene parallel to those in Europe (Morrison, 1961, 1966, 1968; van Zinderen Bakker, 1962, 1964; Coetzee, 1964).

Soil studies

Concerned primarily with properties and qualities of soil for agriculture, early studies concentrated on relationships with bedrock geology as the provenance of parent material or on climate as the agent of pedogenesis. The recent reciprocal movement of geographers into the soil surveys of many countries and of soil scientists from such organisations into university geography departments has led to the recognition that profile features and soil distribution are the result of the interplay of all environmental factors operating throughout pedogenesis. This concept, which is essentially biogeographical, has contributed to an orderly understanding of soils. Soil classification has proceeded from the early Russian and American systems, which respectively defined the major soil groups and soil series, to the recent European systems, notably that of Belgium (Tavernier and Marechal, 1962) which uses the features of the entire profile in a genetic context (Duchaufour, 1963), and to the Seventh Approximation formulated by the United States Department of Agriculture (1960), which is based on the diagnostic features of the profile.

Within the classification framework the first significant advance was the acceptance of Milne's catenary concept, which explains the sequence of soil profiles repeated over a regular succession of relief features developed over one rock type (Milne, 1935, 1947). Subsequent contributions have included Prescott's (1947) recognition of profile features which could not have arisen under the prevailing climatic conditions of the areas of Australia in which they are now found, studies of soil cycles by Butler (1959), van Dijk (1959) and Stephens (1958) in Australia, and Watson's analysis (1964, 1965) of soils of polycyclic and polygenetic origin in Rhodesia. These works have stimulated further research in which soil morphology has been examined in the context of environmental conditions existing when the soils were first developed and of those which have subsequently modified that morphology.

Recognition of the impact of changing climatic conditions on soil profile characteristics has encouraged renewed investigations into the origin and subsequent changes in laterites and lateritic soils, which have commanded attention and controversy since those of southern India were first described by Buchanan (1807). Prescott and Pendleton's (1952) review of the numerous existing field and analytical studies stimulated further interest in the nature and distribution of laterite and lateritic soils in the context of geomorphology and of climatic change, particularly in Africa and Australia (Playford, 1954; Terrill, 1956; Simonett, 1957; Valentin, 1959; Mulcahy, 1960; Mulcahy and Hingston, 1961; Pullan, 1967). These studies have elaborated the development of the characteristic features of the lateritic profile under hot humid conditions, with a fluctuating watertable to promote the formation as a B horizon of the laterite (plinthite in the Seventh Approximation) layer. They have shown that where this laterite, indurated by exposure to air, occurs at or near surface today, it represents the truncated relict of a former mature profile from which, following degradation of the vegetation in a subsequent drier period, the A horizon material has been removed to form the sand plains of the continental interiors.

Current thought and field investigations are tending towards the establishment of a comparable explanation in respect of calcrete or caliche whose distribution, nature and mode of precipitation has been described in the western part of the USA (Breazeale and Smith, 1930; Nikiforeff, 1941; Bretz and Horberg, 1949; Brown, 1956; Gile, 1961) and in North Africa (Moseley, 1965). Attention has also been directed to the relationship between calcrete, geomorphology and climatic change by Durand (1954, 1959), Boulaine (1958, 1966) and Gigout (1960). Durand compared the calcrete crusts with laterite, suggesting that during a period of equilibrium between soil, vegetation and climate, bedrock limestone was dissolved under the influence of forest vegetation, the calcium carbonate then being precipitated from transported solutions in the form of thick layers of powdery calcrete. Subsequently, under drier conditions causing the progressive degeneration of the forest, nodular calcrete and later sheet calcrete was formed. More recent work in Australia and southern Africa, particularly Botswana, lends support to this theory.

On a smaller scale, within the British Isles, the importance of geomorphology and climatic change in soil development has been highlighted by the identification of loess as a component in the soil profiles developed over limestone bedrock in the southern Pennines (Piggott, 1962, 1970; Cazalet, 1968) and in the soils occurring on the exhumed sub-Eocene erosion surface which carries clay-with-flints over the chalk of the South Downs (Hodgson *et al.*, 1967). The introduction of the dynamic concepts associated with these changing environmental conditions have brought new dimensions to studies of pedogenesis and a greater understanding of the nature and distribution of soils within a biogeographical framework.

Biogeographical-geobotanical and biogeochemical investigations of vegetation communities

In the last decade the use of analytical, remote sensing and computing techniques has permitted more precise investigations of the influence of individual environmental parameters on the distribution of vegetation communities and individual plant species than hitherto possible. Such techniques were embodied in studies of the vegetation, soils geomorphology and bedrock geology of two areas initiated when the author was at Keele.

The southern Pennines

The plant communities of this area were first mapped using the units within the moorland formation originally proposed by Moss (1910), the soil profiles being examined within each unit. The resulting distributions were examined with reference to the major environmental parameters, particularly climate, relief and drainage, geomorphology and geology. More detailed pedological, geomorphological, geobotanical and biogeochemical investigations were then undertaken over selected areas, concentrating particularly on those with distinctive plant assemblages and underlain by mineralised bedrock or polluted from industrial or vehicular sources.

The vegetation map of the Goldsitch Moss syncline and relief basin, with its bounding rim of the Roaches and Ramshaw Rocks, provides an

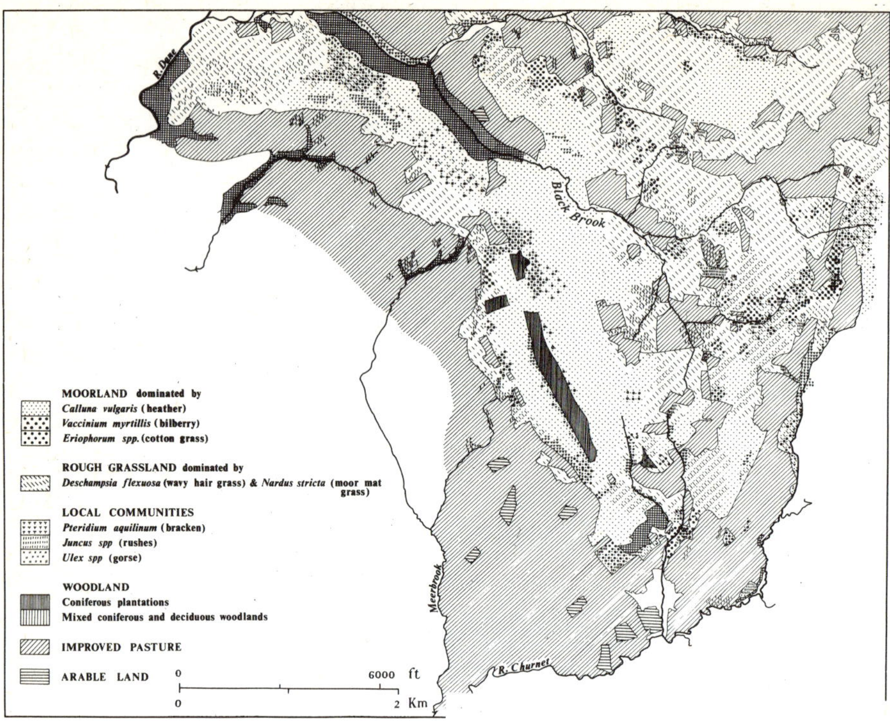

Fig. 5.1 Goldsitch Moss syncline: vegetation and land use

example of the initial survey (Fig. 5.1). The distribution of the plant communities is closely related to relief and drainage, soils and geomorphology (Fig. 5.2). While the relief is a reflection of geology (Fig. 5.3), with the resistant elements of the Millstone Grit Series forming the ridges bounding the basin and the covering Coal Measures its lower lying centre, the drainage is superimposed. This pattern is inherited from an earlier surface at a higher level, and the streams which meander in the centre of the basin penetrate its rim in steepsided gorges. In places in the middle of the syncline the watertable is at or only a few metres below the peat-covered surface; this factor largely accounts for the occurrence of plant communities dominated by *Eriophorum* spp., which form the cotton grass bogs. The better drained land carries *Calluna vulgaris* (L.) Hull. which dominates moorland communities; where reclaimed this land soon reverts to moorland unless heavily and regularly fertilised. On the steeper slopes along the river gorges the common bracken, *Pteridium aquilinum* (L.) Kuhn, the forerunner of woodland, is extending. The level tracts below the scarp face on the outer margins of the syncline are occupied by *Eriophorum* spp. communities whose distribution is related to the presence of poorly drained peaty gley soils, the profiles of which

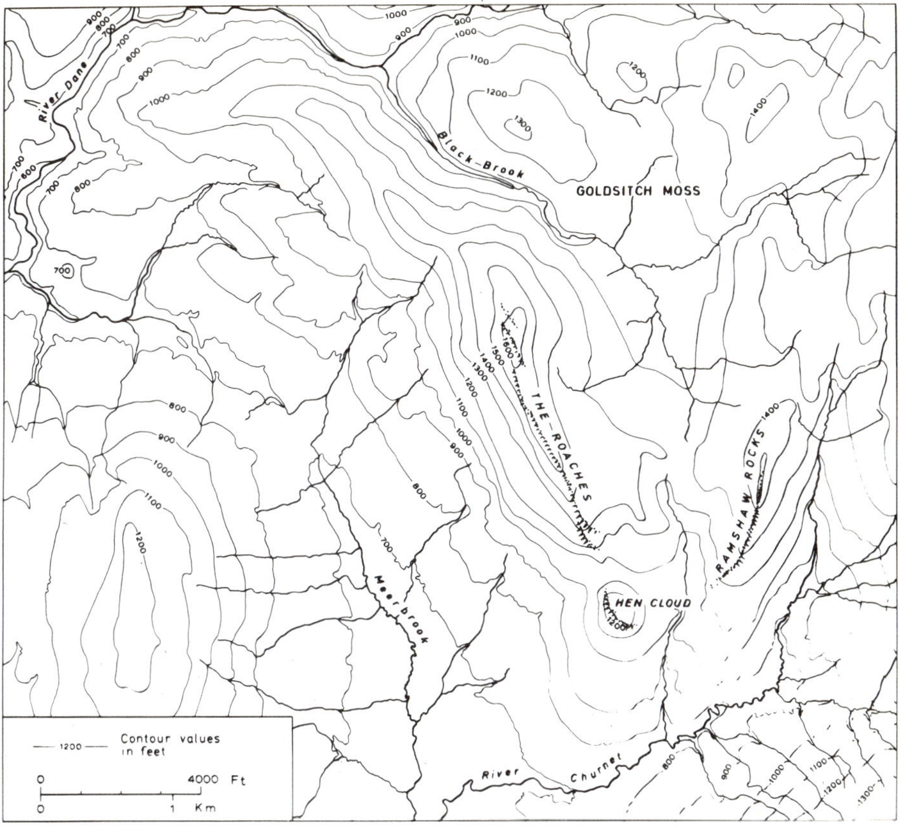

Fig. 5.2 Goldsitch Moss syncline: relief and drainage

show about 0.3 m of peat overlying several metres of stiff plastic gleyed clay before the watertable is reached. By contrast the gently sloping land under improved grass is characterised by multiple podzol profiles which have developed most probably in successive layers of colluvial material deposited over gritstone bedrock during past climatic oscillations. In summary, the distribution of the vegetation communities is closely related to soil and drainage features, which themselves are the product of geomorphological and climatic change over a long period.

The discovery of the multiple podzol profiles stimulated detailed investigation of the relationships between soils and geomorphology (Cazalet, 1968). Field studies of soil profiles were supported by mechanical and chemical analysis of samples, and by studies of thin sections to ascertain mineral composition and clay translocation. Overall the investigations established the relationships between pedogenesis, parent material, relief and drainage. They substantiated the close link between soil type and parent material and highlighted the importance of

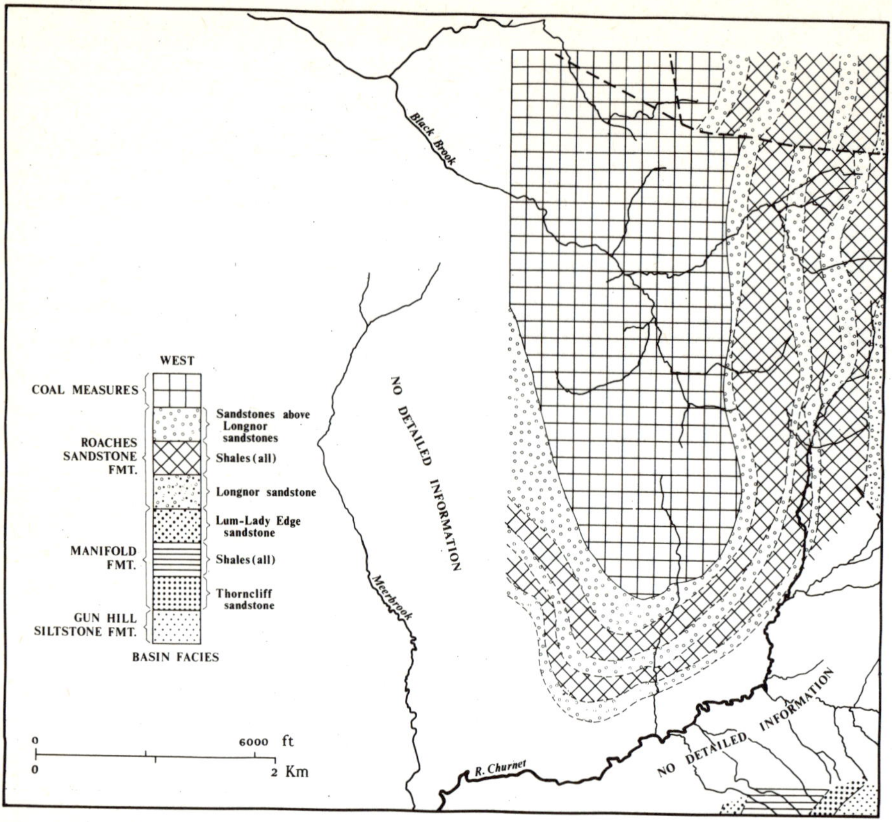

Fig. 5.3 Goldsitch Moss syncline: geology

the combination of colluvial redistribution of material and of groundwater impedance. The discovery of loess and white plastic clay overlying limestone bedrock in a number of profiles, together with high silt levels in many samples, and the evidence of clay translocation in thin sections led to an understanding of a wider distribution of loess and glacial till than had hitherto been suspected. In turn, the establishment of the relationship between soils and morphological features, including planation surfaces, furnished a better appreciation of the sites and soils for agriculture.

Analysis of the soil for trace elements revealed concentrations of copper, lead and zinc in some areas and of molybdenum in others. This prompted geobotanical, biogeochemical and geochemical investigations in several areas of known and potential bedrock mineralisation and in areas suspected of being contaminated from mining, industrial and vehicular sources. The distribution of the copper, lead and zinc indicator plants, *Viola lutea* Huds var *calaminaria* Lej, *Minuartia verna* (L.) Hiern

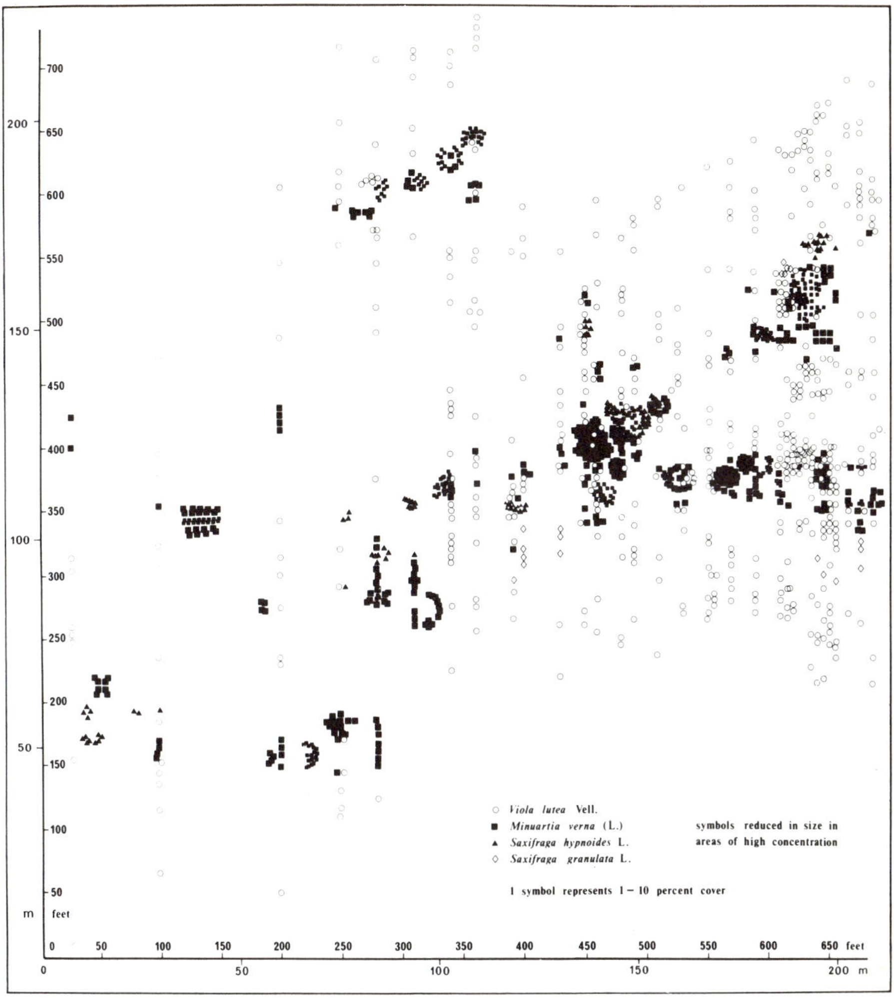

Fig. 5.4 High Low. Distribution of *Viola lutea* **Huds var** *calaminaria* **Lej,** *Minuartia verna* **(L.) Hiern and** *Saxifraga hypnoides* **L.**

and *Saxifraga hypoides* L. over High Low was mapped (Fig. 5.4) and the occurrence of all species relative to the copper, lead and zinc values in the soil along one sample transect line is shown (Fig. 5.5). Both reveal the striking changes which occur in the composition of the vegetation over the toxic ground, where tolerant strains of *Festuca ovina* L., associated with *Koeleria cristata* Pers. and, over peak lead values, with *Festuca rubra* L., dominate the grass cover, replacing the mixed sward with *Agrostis tenuis* Sibth., *Poa pratensis* L., and a variety of other grasses and herbs over unmineralised ground. *Viola lutea* var *calaminaria* clearly delineates the geochemical anomaly which extends well beyond the

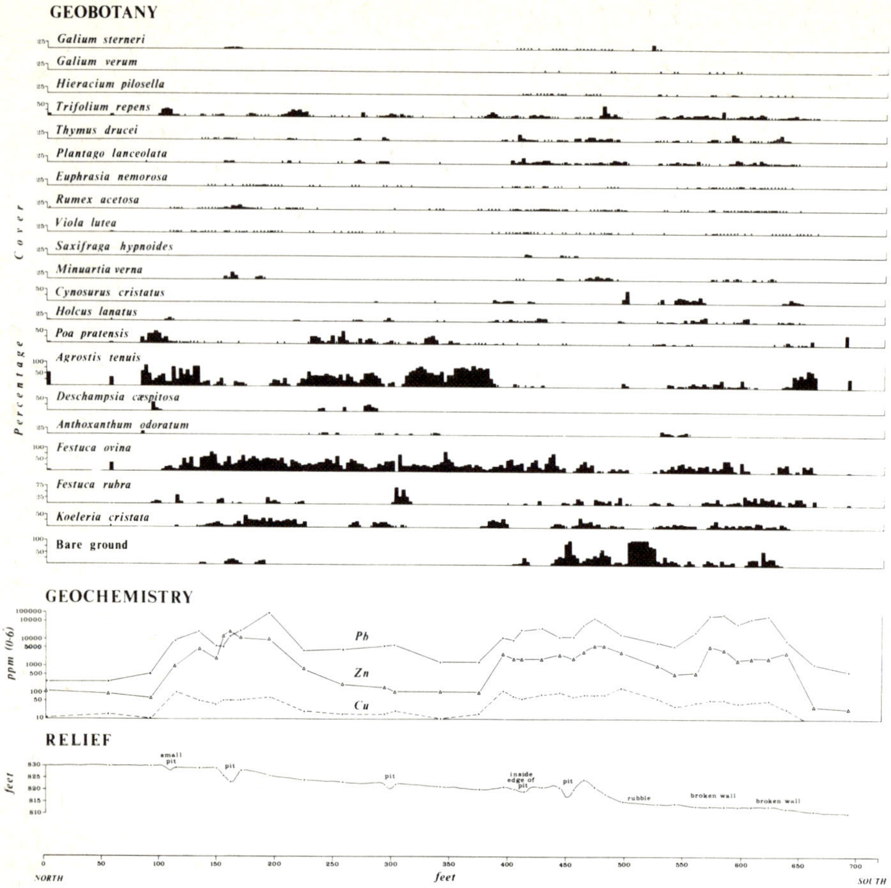

Fig. 5.5 Transect across High Low showing relationship between the distribution of the most common plant species, and the copper, lead and zinc levels in the –80 mesh fraction of the soils sampled at 10–15 cm

vicinities of some old pits around which *Minuartia verna* is common. Biogeochemical analysis of selected species (Fig. 5.6) reveal high uptakes of lead and zinc and indicate that this ground should not be used for grazing.

Analysis of plant and soil samples from other areas characterised by the indicator plants, notably Ecton Hill, and the vicinity of the old Magpie mine also revealed high levels of copper, lead and zinc. Examination of *Calluna vulgaris* samples collected along transects across Ramshaw Rocks showed high lead levels near the Leek–Buxton road, where vehicle exhaust fumes may be responsible. Overall the results of the studies of the vegetation and soils in the southern Pennines confirmed the close relationships between plant distributions and environmental parameters, revealed the nature of the problems for agricultural

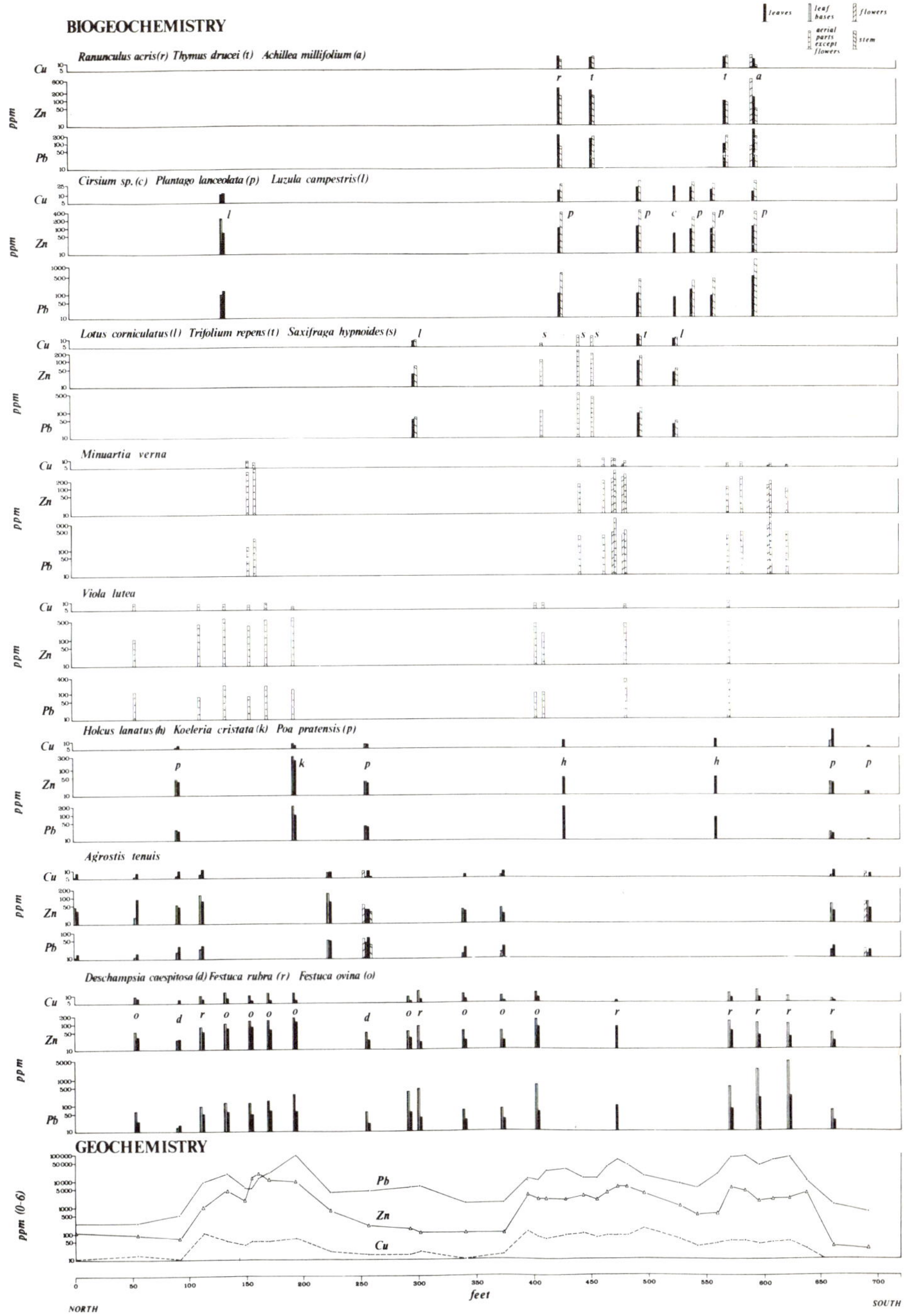

Fig. 5.6 **Transect across High Low showing the copper, lead and zinc contents of selected plant species**

land use, and identified those associated with contamination from bedrock and other sources.

The Mount Isa–Cloncurry area of Australia

Investigations in this area of western Queensland in 1962 established a hierarchical relationship between vegetation units and environmental parameters and disclosed the presence of an anomalous plant community composed mainly of the grass *Eriachne mucronata* R. Br. and the small shrub *Polycarpaea glabra* C. T. White et Francis. This community replaces the low tree and shrub savanna characterised by *Eucalyptus brevifolia* F. Muell. and *Eucalyptus argillacea* W. V. Fitzg. trees and *Triodia pungens* R. Br. grass, where high levels of lead, zinc and copper in the soil emanating from the underlying Dugald River lode create conditions too toxic for most species (Nicholls *et al.*, 1965).

In 1970 the area was chosen for a remote sensing programme which investigated the use of multispectral photography and thermal line scan imagery in the identification of vegetation units, soil types, physiographic features and bedrock geology including rock types, structural features and mineralised zones. The main object was the recognition of the spectral signatures produced by the vegetation as indicative of environmental features. Parallel studies of the imagery from the air survey and of the ground truth information from field investigations undertaken during and following the flying programme show that the false colour photography, which covers the visible and near infrared part of the spectrum in which plant reflectivity is most sensitive, yields most information on the vegetation cover. This is particularly true of the ground cover, where individual grass species and combinations of species produce distinctive spectral signatures. The tree species, mostly *Tristania grandiflora* Cheel and *Bauhinia carronii* F. Muell, occupying river banks and stream courses are readily distinguished by their strong red hues from the species of the drier country, mostly *Eucalyptus* spp. and *Acacia cambagei* R. T. Baker which have pink spectral tones. Similarity of tone however precludes the identification of individual *Eucalyptus* species.

On the broad scale within the low tree and shrub savanna, the savanna grassland, composed mainly of the 'soft' grasses *Astrebla squarrosa* C. E. Hubb, *Enneapogon avenaceus* (Lindl.) C. E. Hubb., *Brachyachne convergens* (F. Muell.) Stopf. and *Iseilema macratherum* Domin, which produce spectral signatures in the red purple range, is readily distinguished from the harsh spinifex grassland characterised by tussocks of *Triodia pungens* which produce a blue spectral signature. The former occupy the black soil plains and the latter areas of near surface bedrock. Within the savanna grassland of the black soil plains differences of spectral tone produce remarkable lozenge-shaped areas whose overall distribution pattern is suggestive of subsurface drainage conditions.

These patterns are caused by repetitive sequences of grass and herb communities which occupy discrete zones characterised by particular soil moisture conditions associated with the length of time water lies after rains. The revelation of these patterns is of considerable importance to an understanding of the vegetation–soil–moisture relationships of the black soil plains, which in Australia and Africa constitute vast areas of potentially valuable grazing land inaccessible after rains when the black soils become exceedingly sticky.

On the more detailed scale the anomalous plant communities over the Dugald River lode produced a unique spectral signature in the blue green range in the infrared photography. Studies of the imagery for all the areas flown revealed a number of similar anomalies which were investigated in the field. Interpretation of the imagery and field data for one such area, the north of Roseby Homestead, and south of the Dugald River lode, is given in Fig. 5.7.

On the infrared colour photography sixteen spectral signatures produced by discrete plant communities associated with particular soil, relief and drainage and geological features were identified (Table 5.1), and the areas exhibiting these either uniquely or in combination of two or more were outlined on overlays of the air photographs enlarged to the same linear scale as the scan lines of the microdensitometer. Some of the signatures occurring in combination were then grouped with reference to ground truth information to produce the interpretation depicted in Fig. 5.7. In broad outline this shows an area in the centre of the frame devoid of trees and characterised by an anomalous community similar to that over the Dugald River lode and composed of *Eriachne mucronata*, *Bulbostylis barbata* (Rottb.) C. B. Clarke and *Fimbristylis* sp. nov., giving a blue-green reflectance (123C and 123D) in the colour infrared (spectral signature 1). This is flanked to the north by communities with scattered *Eucalyptus* trees and a ground cover characterised by *Enneapogon polyphyllus* (Domin) Burbidge grass giving a red purple reflectance (65C, spectral signature 4), with variable amounts of *Eriachne dominii* Hartley and *Sporobolus australasicus* Domin (spectral signatures 3, 2 and 29 giving respectively reflectances in the red-purple 65C and violet 87D and 85D range) and bare ground (spectral signatures 16 and 17 with white and blue green 121C reflectances) on the gently sloping terrain of the interfluves and with *Acacia chisholmii* F. M. Bail. shrubs (spectral signature 30 in the red-purple 62D and 68D) along the creeks. In turn these are succeeded northwards by communities with *Acacia cambagei* trees and a ground cover of *Sporobolus actinocladus* F. Muell. and *Dactyloctenium radulosum* (Desf.) Beauv. or *Sporobolus australasicus* with variable amounts of bare ground giving a combination of spectral signatures over the lower terrain with more varied soils derived from colluvial and alluvial material. South of the anomalous vegetation zone,

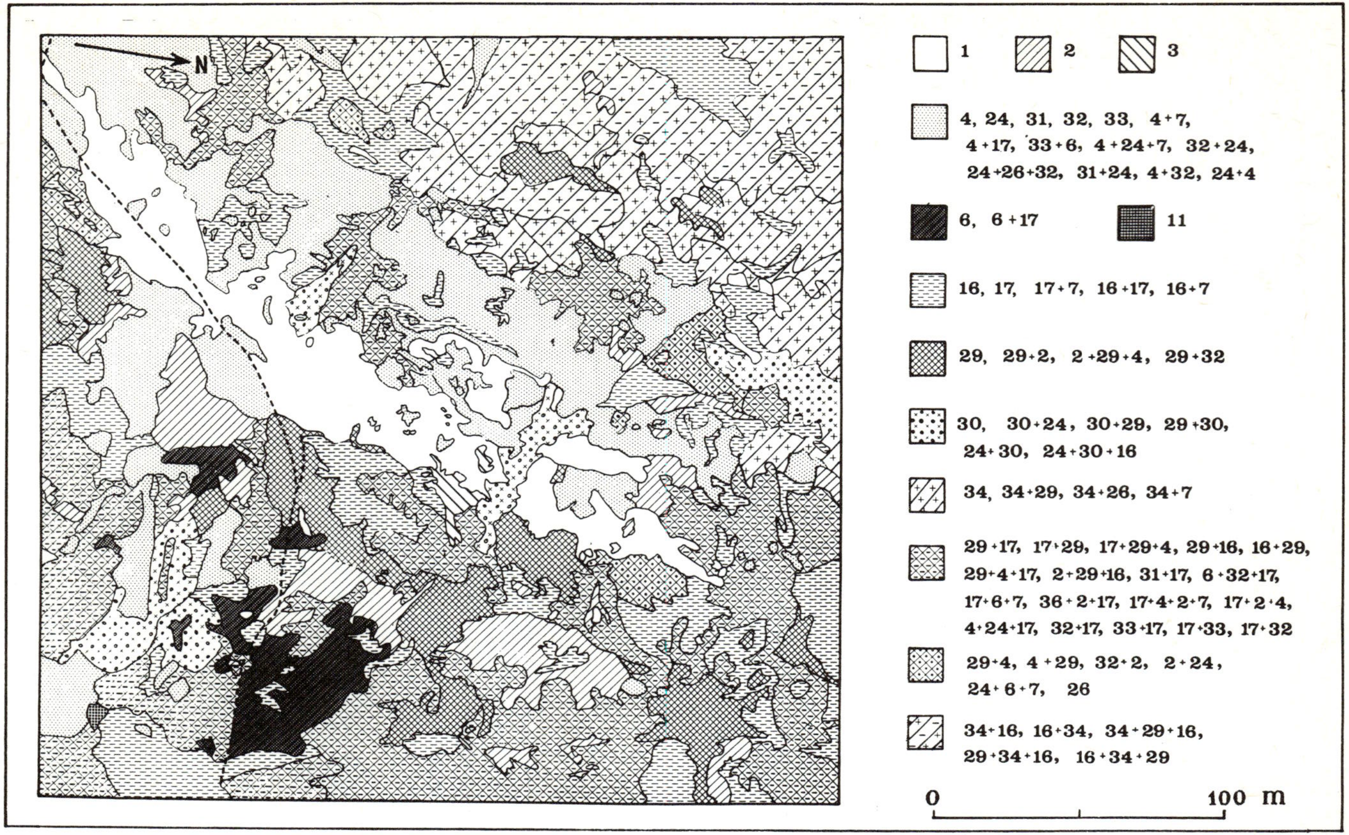

Fig. 5.7 Roseby Homestead anomaly. Spectral signatures displayed on the infrared colour film 28 frame 277 flown at a scale of 1 : 5 000 and grouped with reference to ground truth information

Table 5.1 Roseby Homestead anomaly. Interpretation of the spectral signatures displayed on the infrared colour film 28 frame 277 flown at a scale 1:5 000

Spectral signature	*Composition of vegetation association plant community*	*Soils, relief, geology*
1. Blue green 123D and 123C	*Eriachne mucronata, Bulbostylis barbata and Fimbristylis* sp.	Thin red sandy soils over calc-silicate rocks with some evidence of copper mineralisation, some quartz gravel at surface. Level to gently sloping terrain
2. Violet 87D	*Sporobolus australasicus* dominant	Red sandy soils over mixed colluvial and residual material of interfluves (items 2, 3, 29, 4)
3. Red purple 63C	*Eriachne dominii* dominant in ground layer	
29. Violet 85D	*Eriachne dominii* and *Sporobolus australasicus* co-dominant in ground vegetation	
4. Red purple 65C	*Enneapogon polyphyllus* dominant in ground layer	
6. Greyed purple 186D	*Enneapogon polyphyllus* and *Sporobolus australasicus* co-dominant in ground layer	Red sandy clay loam over mixed colluvial and residual material
31. Purple 76D and blue green 123D	*Enneapogon polyphyllus* and *Cleome viscosa* forming sparse ground cover with bare patches between or *Bulbostylis barbata*, forming sparse ground cover with much bare ground	Red sandy clay loams over mixed colluvial and residual material of slightly dissected terrain at margins of interfluves
32. Red 49C and 49D with some violet 87D and 92C	*Eriachne dominii, Aristida contorta* and *Enneapogon polyphyllus* forming fairly dense ground cover	Red sandy clay loams with mixed colluvial and residual material of slightly dissected terrain at margins of interfluves (items 32, 33)
33. Red purple 73C and 73D	*Bulbostylis barbata, Cleome viscosa* and *Enneapogon polyphyllus* forming ground layer	
30. Red purple 62D and 68D	*Aristida contorta, Enneapogon polyphyllus* and *Eriachne dominii* dominant in ground layer. *Acacia chisholmii* shrub layer	Gently sloping dissected terrain parallelling stream courses

Continued on p. 136

Table 5.1 – *continued*

Spectral signature	*Composition of vegetation association plant community*	*Soils, relief, geology*
11. Blue green 123C with red purple 65C	*Triodia pungens* and *Enneapogon polyphyllus* co-dominant in ground layer	Red sandy soils over mixed colluvial and alluvial material of interfluves
16. White	Largely bare ground	
17. Blue green 121C	Largely bare ground	
24. Red purple 62C	*Cenchrus pennisetiformis* dominant	Alluvial clays; sands and gravels near stream course
26. Violet 87D	*Chrysopogon fallax* dominant	Poorly drained dark brown to black clays subject to waterlogging
34. Blue green 120B and 114B with purple 78D and black and violet 84C and violet blue 91D	Much bare ground studded with *Acacia cambagei* trees with some ground cover of *Sporobolus actinocladus* and *Dactyloctenium radulosum*	Red to reddish brown sandy loams over colluvial and alluvial material

communities with infrequent *Eucalyptus* trees and a ground vegetation of *Eriachne dominii* and *Sporobolus australasicus* with much intervening bare ground again characterise the interfluves, with soils derived from an admixture of residual and colluvial material. These give way to those in which *Cenchrus pennisetiformis* Hochst & Steud ex Steud (spectral signature 24 giving a bright red purple 62C reflectance), *Enneapogon polyphyllus* and *Cleome viscosa* L. become more important and produce varied combinations of spectral signatures over the sands, sandy clay loams and clays derived from colluvial and alluvial material near the Dugald River (south of the frame).

When the infrared colour imagery was projected through a red filter and superimposed on the panchromatic imagery with a yellow filter projected through either a yellow filter or a green filter, the number of visually identifiable spectral signatures was increased to nineteen but the recognisable units were larger and the boundaries between them simpler. In both cases more than one spectral signature was identifiable within the anomalous plant community and also within the communities whose ground layer is dominated by *Enneapogon polyphyllus* grass to the north. By contrast the numerous spectral signatures and combinations of spectral signatures within the areas of covered ground characterised by

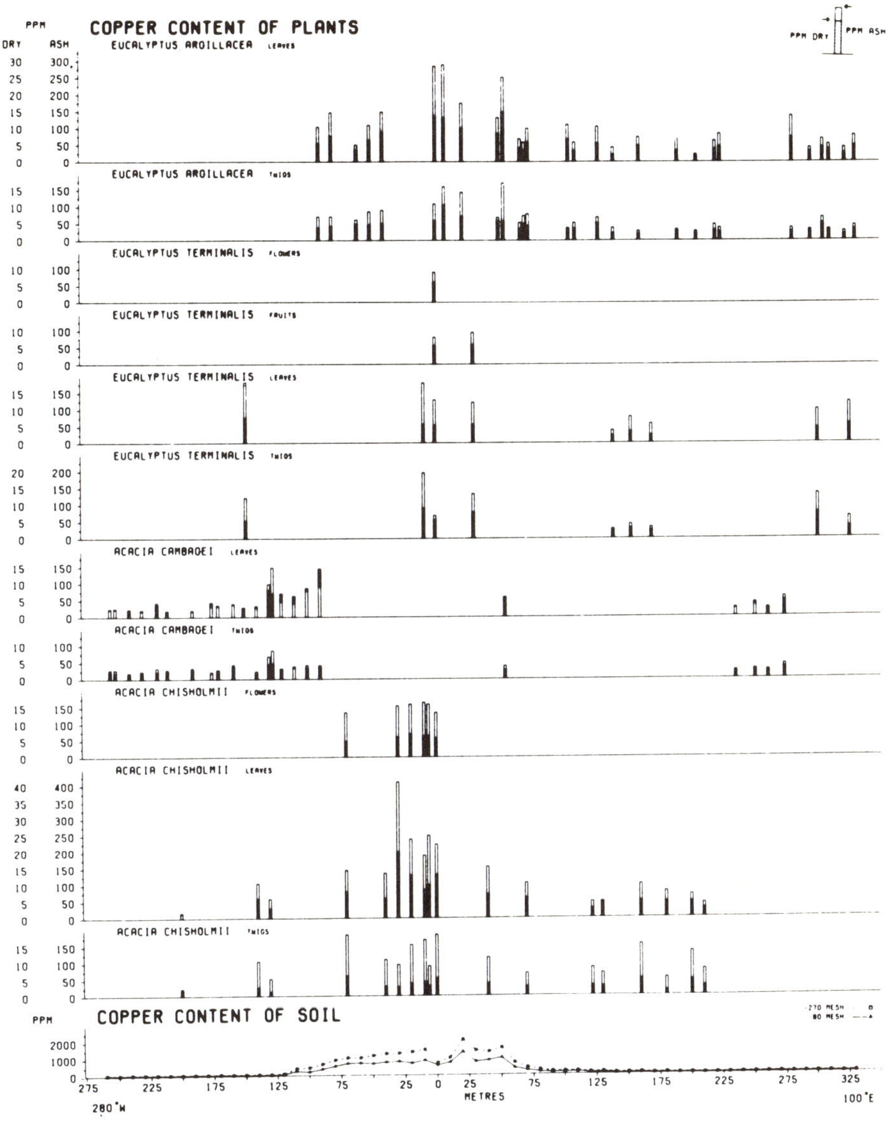

Fig. 5.8 **Computer print-out histogram showing the copper contents of the leaves and twigs of trees and shrub species and of the –80 fraction of the soils sampled at 10–15 cm along a transect orientated across the spectral/geobotanical anomaly, Roseby Homestead area**

Eriachne dominii and *Sporobolus australasicus* were reduced greatly, mainly because the combination of imagery accentuated the contrast between bare ground and ground with some plant growth.

In order to ascertain the reason for the anomalous plant community, the soils and plants were sampled along a transect line. The results of the analyses were computed to produce the histogram print-out (Figs 5.8

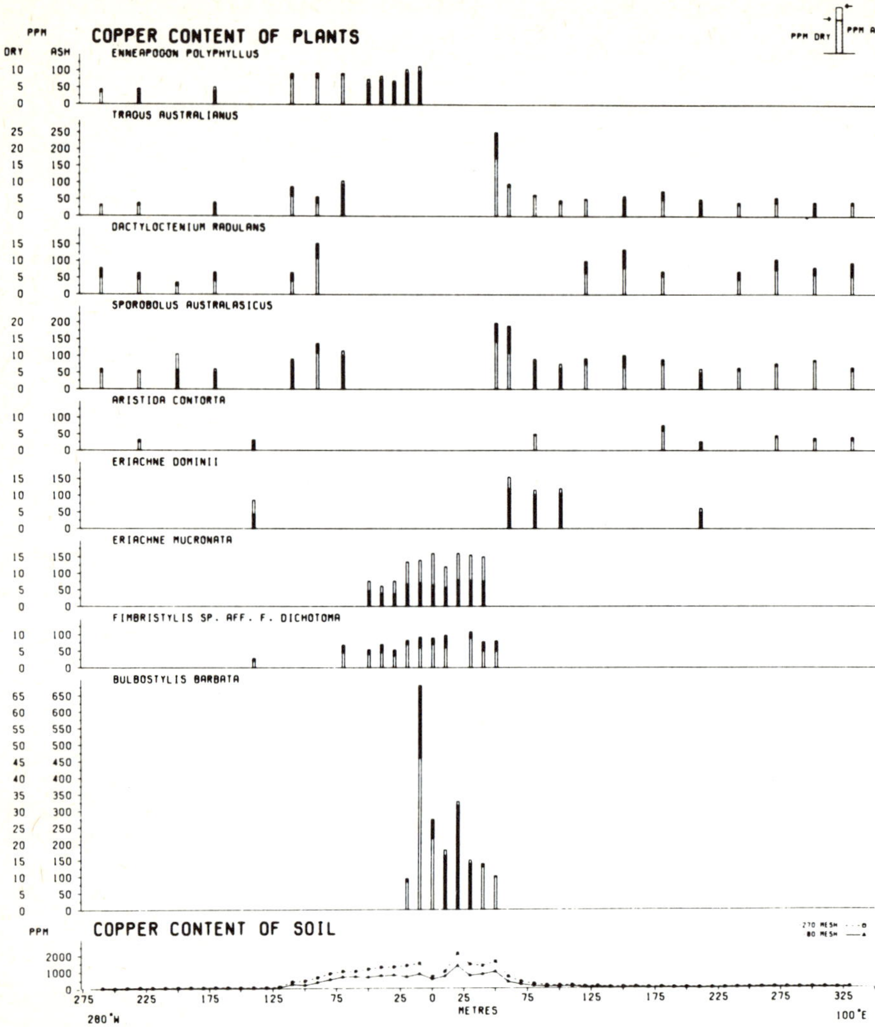

Fig. 5.9 Computer print-out histogram showing the copper contents of grass and sedge species and of the –80 mesh fraction of the soils sampled at 10–15 cm along a transect orientated across the spectral/geobotanical anomaly, Roseby Homestead area

and 5.9). Produced more accurately and rapidly than could have been achieved by a draughtsman, the histograms reveal that the anomalous plant community is associated with relatively high copper levels in the soil, whose relationship to possible bedrock mineralisation, however, is unknown in the absence of trenching and drilling.

The individual photographic frames are being scanned currently by a microdensitometer (which measures density on a grey scale), and given the necessary parameters of spectral units from visual interpretation of selected frames it is hoped to proceed to automatic print-out of vegetation maps of frames for which visual interpretation and ground truth data are not available. If and when this is achieved, the rapid mapping of large areas may become practicable.

Applied biogeographical studies

In the last decade numerous studies have been undertaken using biogeography, geobotany, geochemistry and biogeochemistry in the assessment of land potential for agriculture and other uses (notably by the Directorate of Overseas Surveys from the United Kingdom and by the Commonwealth Scientific and Industrial Research Organisation in Australia), in mineral exploration (Cole *et al.*, 1968; Cole, 1971*a* and *b*, 1973), and in the analysis of disease incidence and control. In addition, increasing concern for the quality of environment has prompted research into the effects of pollution and measures of reclamation.

Studies of environmental pollution span a range of subjects, only some of which are biogeographical in scope. Some problems, such as those concerned with the impact of oil slicks on wild life in coastal areas (Nelson Smith, 1968), of china clay effluent on the bottom fauna of St Austell and Mevagissey bays (Howell and Shelton, 1970) and of pollution from a sulphate cellulose factory on the aquatic flora in Finland (Eloranta, 1970), are readily defined. The effects on plant growth of old mining activities and of industrial processing is apparent in many parts of the world, and the accumulation of poisonous substances in human tissues is of increasing concern, as for example the high lead levels in milk teeth of children from certain areas in American cities. Here, however, the problems are complex and before it is possible to set any standards for chemical quality that could serve as a basis for legislation to control pollution of the physical environment it is necessary to establish the range of metal contents in plants, soils and waters in natural uncontaminated and in polluted environments (Cannon and Anderson, 1971). In the USA studies of environmental geochemistry relative to health and disease have been initiated by the United States Geological Survey. In this country, Webb and his collaborators at the Centre of Applied Geochemistry at Imperial College are compiling a valuable geochemical atlas of the British Isles. Bradshaw and his collaborators at Liverpool have been evolving strains of metal-tolerant grasses for reclaiming old mine dumps (Antonovics, 1966; Bradshaw, 1970*a* and *b*; Antonovics *et al.*, 1971; Smith and Bradshaw, 1972), while Goodman and his associates have been engaged in reclaiming that area of the Swansea valley despoiled by industrial processes (Goodman and Roberts, 1971).

There remains a vast field in which integrated research involving biogeographical-geobotanical, biogeochemical and geochemical studies in the context of all environmental parameters is required. A need exists to establish the range of contents of individual elements in plants, soils and waters over different geological formations and under differing climatic conditions in natural uncontaminated and in polluted environments, in order to assess the hazards, to formulate controls and to effect suitable reclamation measures where pollution has occurred.

Conclusions

In the past twenty years biogeography has developed from a traditional plant and animal geography into a scientific discipline using quantitative analytical, sensing and processing techniques. At the same time the spectrum of studies has been extended. In the last five years the use of remote sensing techniques, coupled with the enhancement of imagery and computerisation of data to automatic print-out form, has opened new vistas for biogeographical studies, particularly where imagery from satellites and rockets surveying earth resources and providing repetitive views of terrain at different seasons can supplement the data from air survey and ground investigation. The mapping of vegetation units over large areas, the monitoring of fires and of vegetation changes consequent on fire and on clearing for cultivation may become possible, while movements of animal populations including locust swarms may be recorded.

These new techniques assist biogeographical studies in ways comparable to those provided by the use of biogeochemistry, geochemistry and quantitative data handling which followed the earlier development of spectrophotometers and computers. The essence of the subject remains the elucidation of the interrelationships between plants, animals and the environment through field and laboratory investigations but the new techniques enlarge the scale of research and make possible the study of increasingly complex problems. In a world now conscious of environmental quality the subject offers contributions to a wiser use of resources.

References

ADAMSON, R. S. (1918) 'Relationships of some associations of the southern Pennines', *J. Ecol.*, **6**, 97–109.

ADEJUWON, J. O. (1970) 'The ecological status of coastal savannas in Nigeria', *J. Trop. Geogr.*, **30**, 1–10.

ANTONOVICS, J. (1966) 'The genetics and evolution of differences between closely adjacent plant populations with special reference to heavy metal tolerance', unpublished PhD thesis, University of Wales.

ANTONOVICS, J., BRADSHAW, A. D. and TURNER, R. G. (1971) 'Heavy metal tolerance in plants', *Adv. Ecol. Res.*, **7**, 1–85.

ASKEW, G. P., MOFFATT, D. J., MONTGOMERY, R. F. and SEARLE, P. L. (1970) 'Inter-relationships of soils and vegetation in the savanna-forest boundary zone of north-eastern Mato Grosso', *Geogr. J.*, **136**, 370–6.

BARCLAY ESTRUP, P. and GIMINGHAM, C. H. (1969) 'The description and interpretation of cyclical processes in a heath community: I. Vegetation change in relation to the *Calluna* cycle', *J. Ecol.*, **57**, 737–58.

BARCLAY ESTRUP, P. (1970) 'The description and interpretation of cyclical processes in a heath community: II. Changes in biomass and shoot production during the *Calluna* cycle', *J. Ecol.*, **58**, 243–50.

BOULAINE, J. (1958) 'Sur la formation des carapaces calcaires', *Travaux des collaborateurs service de la carte geologique de l'Algerie, Bull.*, 20(ns), 7–19.

BOULAINE, J. (1966) 'Sur les relations entre les carapaces calcaires et les sols isohumiques de climat xerotherique', *Sci. Sol.*, **1**, 3–16.

BRADSHAW, A. D. (1970*a*) 'Plants and industrial waste', *Trans. Proc. Bot. Soc. Edinb.*, **41**, 71–84.

BRADSHAW, A. D. (1970*b*) 'Pollution and plant evolution', *New Scient.*, 17 Dec., pp. 497–500.

BRAY, J. R. and CURTIS, J. T. (1957) 'An ordination of the upland forest communities of southern Wisconsin', *Ecol. Monogr.*, pp. 325–49.

BREAZEALE, J. F. and SMITH, H. V. (1930) 'Caliche in Arizona', University of Arizona, College of Agriculture, *Agric. Exper. Bull.*, **131**, 419–41.

BRETZ, J. and HORBERG, L. (1949) 'Caliche in south-eastern New Mexico', *J. Ecol.*, **57**, 491.

BROWN, C. N. (1956) 'The origin of caliche on the north-east Llano Estacado, Texas', *J. Geol.*, **64**, 1.

BUCHANAN, F. (1807) *A Journey from Madras through the countries of Mysore, Canara and Malabar etc. in 1800–1*, London.

BUTLER, B. E. (1959) 'Periodic phenomenon in landscape as a basis for soil studies', *CSIRO Soil publication*, No. 14.

CANNON, H. L. and ANDERSON, B. M. (1971) 'The geochemist's involvement with the pollution problem', in H. L. Cannon and H. C. Hopps, eds, *Environmental geochemistry in health and disease, Geol. Soc. Am. Mem.*, p. 123.

CAZALET, P. C. D. (1968) 'Investigation of the profile characteristics, development and distribution of the soils of part of the southern Pennines', unpublished PhD thesis, University of London.

COETZEE, J. A. (1964) 'Evidence for a considerable depression of the vegetation belts during the upper Pleistocene on the East African mountains', *Nature, Lond.*, **204**, 564–6.

COLE, M. M. (1960) 'Cerrado, caatingan and pantanal; distribution and origin of the savanna vegetation of Brazil', *Geogr. J.*, **126**, 168–79.

COLE, M. M. (1963) 'Vegetation and geomorphology in Northern Rhodesia: an aspect of the distribution of the savannas of central Africa', *Geogr. J.*, **129**, 290–310.

COLE, M. M. (1964) 'Vegetation nomenclature and classification with particular reference to the savannas', Presidential address 1963, South African Geographical Society, *S. Afr. Geogr. J.*, **45**, 3–14.

COLE, M. M. (1965) *Biogeography in the Service of Man*, Inaugural lecture, Bedford College, London.

COLE, M. M. (1971*a*) 'Biogeographical–geobotanical and biogeochemical investigations connected with exploration for nickel–copper ores in the hot, wet summer-dry winter savanna woodland environment', *S. Afr. Instn. Min. Metall. J.*, **71**, No. 10, 199–209.

COLE, M. M. (1971*b*) 'The importance of environment in biogeographical-geobotanical and biogeochemical investigations', *Geochemical exploration, Canadian Institution of Mining and Metallurgy*, Special volume 11, pp. 1–12.

COLE, M. M. (1973) 'Geobotanical and biogeochemical investigations in the sclerophyllous woodland and shrub associations of the Eastern Goldfields area of Western Australia, with particular reference to the role of *Hybanthus floribundus* (Lindl.) F. Muell. as a nickel accumulator and indicator plant', *J. Appl. Ecol.*, **10**, 269–320.

COLE, M. M., PROVAN, D. M. J. and TOOMS, J. S. (1968) 'Geobotany, biogeochemistry and geochemistry in mineral exploration in the Bulman–Waimuna Springs area, Northern Territory, Australia', *Applied Earth Science Trans. Section B, Institution of Mining and Metallurgy*, **77**, 81–104.

CROMPTON, C. B. (1911) *The Vegetation of Caithness considered in relation to Geology*, Committee for Survey and Study of British Vegetation.

CURTIS, J. T. (1959) *The Vegetation of Wisconsin: an ordination of plant communities*, University of Wisconsin Press.

CURTIS, J. T. and McINTOSH, R. P. (1951) 'An upland continuum in the prairie-forest border region of Wisconsin', *Ecology*, **32**, 476–96.

DAULTREY, S. G. (1970) 'An analysis of the relation between soil moisture, topography and vegetation types in a savanna area', *Geogr. J.*, **136**, 399–405.

DEAN, G. W. J. (1967) 'Grasslands of the Rukwa Valley', *J. Appl. Ecol.*, **4**, 45–57.

DEAN, G. W. J. (1968) 'Studies of factors affecting the formation of hopper bands of the red locust [Nomodacris septemfasciator] in an outbreak area', *J. Appl. Ecol.*, **5**, 273–90.

DUCHAUFOUR, P. L. (1963) 'Soil classification, a comparison of the American and the French systems', *J. Soil Sci.*, **14**, 149–55.

DURAND, J. H. (1954) *Le sols d'Algérie*, Government general de l'Algérie, Direction du service de la Colonisation et de l'Hydraulique, Pedology No. 2.

DURAND, J. H. (1959) *Les sols rouges et les croutes calcaires en Algérie*, Direction de l'Hydraulique et de l'Equipment rural, Service des études scientifiques, Etude générale 7, Clairbois-Birmandreis.

EDEN, M. J. (1964) *The savanna ecosystem – northern Rupununi, British Guiana*, McGill University, Savanna Research Project, Savanna Research Series No. 1.

EDEN, M. J. (1970) 'Savanna vegetation in the northern Rupununi, Guiana', *J. Trop. Geogr.*, **30**, 17–28.

ELORANTA, P. (1970) 'Pollution of aquatic flora of water by sulphate cellulose factory at Mantla, Finnish Lake District', *Ann. Bot. Fenn*, **7**, 63–141.

ERDTMAN, G. (1924) 'Studies in the micropalaeontology of post-glacial deposits in northern Scotland and the Scottish Isles, with special reference to the history of the woodlands', *J. Linn. Soc.* (EN), **46**.

FLENLEY, J. R. (1969) 'The vegetation of the Wabag region, New Guinea highlands, a numerical study', *J. Ecol.*, **57**, 465–90.

FORBES, E. (1846) 'On the connexion between the distribution of the existing fauna and flora of the British Isles and the geological changes which have affected the area, especially during the epoch of the Northern Drift', *Mem. Geol. Surv. G.B.*, **1**, 336–432.

FORD, J. and CLIFFORD, H. R. (1968) 'Changes in the distribution of cattle and of bovine trypanosomaisis associated with the spread of tse-tse flies [*Glossina*] in south-west Uganda', *J. Appl. Ecol.*, **5**, 301–38.

GIGOUT, M. (1960) 'Sur la genese des croutes calcaires pleistocene en Afrique du Nord', *C.R. Soc. Geol. Francais*, **1**, 8–10.

GILE, L. H. (1961) 'A classification of the horizons in soils of a desert region, Done Ana County, New Mexico', *Proc. Soil Sci. Soc. Am.*, **25**, 52.

GODWIN, H. (1956) *The History of the British Flora*, Cambridge University Press.

GONZALEZ, E., VAN DER HAMMEN, T. and FLINT, R. F. (1966) 'Late Quaternary glacial and vegetational sequence in valle de Lagunillas, Sierra Nevada del Cocuy, Columbia', *Leid. Geol. Meded.*, Deel 32, pp. 157–82.

GOODMAN, G. T. and ROBERTS, T. M. (1971) 'Plants and soils as indicators of metals in air', *Nature, Lond.*, **231**, 287–92.

HARRISON, CAROLYN M. (1970) 'The phytosociology of certain English heathland communities', *J. Ecol.*, **58**, 573–89.

HODGSON, J. M., CATT, J. A. and WEIR, A. H. (1967) 'The origin and development of clay-with-flints on the South Downs', *J. Soil Sci.*, **18**, 85–97.

HOWELL, R. R. and SHELTON, R. G. J. (1970) 'The effect of China Clay on the bottom fauna of St Austell and Mevagissey Bays', *Mar. Biol. Assoc. UK*, **50**, 593–607.

IVIMEY-COOK, R. B. and PROCTER, M. C. F. (1966) 'The application of association analysis to phytosociology', *J. Ecol.*, **54**, 179–92.

LAMBERT, J. M. and WILLIAMS, W. T. (1962) 'Multivariate methods in plant ecology IV. Nodal analysis', *J. Ecol.*, **50**, 775–802.

LAMBERT, J. M. and WILLIAMS, W. T. (1966) 'Multivariate methods in plant ecology VI. Comparison of information analysis and association analysis', *J. Ecol.*, **54**, 635–64.

MILNE, G. (1935) 'Composite units for the mapping of complex soil associations', *Trans. Third International Congr. Soil Scientists*, **1**, 345.

MILNE, G. (1947) 'A soil reconnaissance journey through parts of Tanganyika Territory, December 1935–February 1936', *J. Ecol.*, **25**, 192–264.

MORRISON, M. E. S. (1961) 'Pollen analysis in Uganda', *Nature, Lond.*, **190**, 483–6.

MORRISON, M. E. S. (1966) 'Low altitude vegetation history with special reference to Africa', *Proceedings International Symposium on World Climate 8000–0 B.C.*, Royal Meteorological Society, pp. 142–8.

MORRISON, M. E. S. (1968) 'Vegetation and climate in the uplands of south-western Uganda during the later Pleistocene period. I. Muchoya swamp, Kigezi district', *J. Ecol.*, **56**, 363–84.

MORRISON, C. G. T., HOYLE, A. C. and HOPE-SIMPSON, J. F. (1948) 'Tropical soil-vegetation catenas and mosaics. A study of the south-west part of the Anglo–Egyptian Sudan', *J. Ecol.*, **36**, 1–84.

MOSELEY, F. (1965) 'Plateau calcretes of Libya', *Z. Geomorph.*, **9**, 167–85.

MOSS, C. E. (1907) 'Geographical distribution of vegetation in Somerset, Bath and Bridgewater district', *Geogr. J.*, **16**, 35–64.

MOSS, C. E. (1910) 'Fundamental units of vegetation', *New Phytol.*, **9**, 18–53.

MOSS, C. E. (1913) *The Vegetation of the Peak District*, Cambridge.

MOSS, R. F. and MORGAN, W. B. (1970) 'Soils, plants and farmers in West Africa', in J. P. Garlick, and R. W. J. Keay, eds, *Human Ecology in the Tropics*, Pergamon Press, pp. 242–86.

MULCAHY, M. J. (1960) 'Laterites and lateritic soils in south-western Australia', *J. Soil. Sci.*, **11**, 206–25.

MULCAHY, M. J. and HINGSTON, F. J. (1961) 'The development and distribution of the soils of the York–Quairading area, western Australia in relation to landscape evolution', *CSIRO Soil publication* No. 17.

NELSON SMITH, A. (1968) 'The effects of oil pollution and emulsifier cleansing on shore life in south-west Britain', *J. Appl. Ecol.*, **5**, 97–108.

NICHOLLS, O. W., PROVAN, D. M. J., COLE, M. M. and TOOMS, J. S. (1965) 'Geobotany and geochemistry in mineral exploration in the Dugald River area, Cloncurry district, Australia', *Trans. Instn. Min. Metall.*, **74**, 695–799.

NIKIFOREFF, C. C. (1941) 'Hardpan and micro relief in certain soil complexes of California', *U.S. Dept. of Agriculture Tech. Bull.* No. 745.

PIGOTT, C. D. (1962) 'Soil formation and development on the Carboniferous Limestone of Derbyshire: I. Parent materials', *J. Ecol.*, **50**, 145–56.

PIGOTT, C. D. (1970) 'Soil formation and development on the Carboniferous Limestone of Derbyshire: II. Relation of soil development to vegetation on the plateau near Coombs Dale', *J. Ecol.*, **58**, 529–42.

PLAYFORD, P. E. (1954) 'Observations on laterites in western Australia', *Aust. J. Sci.*, **17**, 11–14.

PRESCOTT, J. A. (1947) 'A soil map of Australia', *CSIRO Bull.*, **177**.

PRESCOTT, J. A. and PENDLETON, R. L. (1952) *Laterite and lateritic soils*, Commonwealth Bureau of Soil Science Technical Communication 47.

PULLAN, R. A. (1967) 'A morphological classification of lateritic ironstones and ferruginised rocks in Northern Nigeria', *Nigerian J. Sci.*, **1**, 161–73.

REID, C. (1899) *The Origin of the British Flora*, London.

SALISBURY, E. J. (1925) 'Notes on the adaptive succession in some dune soils with special reference to the time factor', *J. Ecol.*, **13**, 322–8.

SARMIENTO, G. and MONASTERIO, M. (1969) 'Studies of the savanna vegetation of the Venezuelan Llanos: I. The use of association analysis', *J. Ecol.*, **57**, 579–98.

SEDDON, B. (1971) *Introduction to Biogeography*, Duckworth.

SIMONETT, D. (1957) 'Observations on laterite and other ironstone soils in north Queensland', *J. Proc. R. Soc. N.S.W.*, **91**, 23–35.

SMITH, R. A. H. and BRADSHAW, A. D. (1972) 'Stabilization of toxic mine wastes by the use of tolerant plant populations', *Trans. Sect. A Mining Industry, Instn. Min. Metall.*, **81** (791), 230–7.

STEPHENS, C. G. (1958) 'Phrenology of Australian soils', *Trans. R. Soc. S. Aust.*, **81**, 1–12.

TANSLEY, A. G. (1911) *Types of British Vegetation*, London.

TANSLEY, A. G. (1939) *The British Islands and their Vegetation*, Cambridge University Press.

TAVERNIER, R. J. F. and MARECHAL, R. (1962) 'Soil survey and soil classification in Belgium', *Trans. Int. Soc. Sci. Commn.*, **IV–V**, 298–307.

TERRILL, S. E. (1956) 'Laterite and materials of similar appearance in south-western Australia', *J. R. Soc. W. Aust.*, **40**.

UNITED STATES DEPARTMENT OF AGRICULTURE, SOIL SURVEY STAFF (1960) *Soil Classification: a comprehensive system, 7th approximation.*

VALENTIN, H. (1959) 'Geomorphological reconnaissance of the north west coast of Cape York peninsula (northern Australia)', *Second Coastal Geomorphology Conference, Louisiana State University*.

VAN DER HAMMEN, T. (1963) 'A palynological study of the Quaternary of British Guiana', *Leid. Geol. Meded.*, Deel 29, 125–80.

VAN DER HAMMEN, T. and GONZALEZ, E. (1960) 'Upper Pleistocene and Holocene climate and vegetation of the Sabana de Bogatá (Colombia, South America)', *Leid. Geol. Meded.*, Deel 25, pp. 261–315.

VAN DER HAMMEN, T. and GONZALEZ, E. (1964) 'A pollen diagram from the Quaternary of the Sabana de Bogatá (Colombia) and its significance for the geology of the northern Andes', *Leid. Geol. Meded.*, Deel 29, pp. 147–90.

VAN DIJK, D. C. (1959) 'Soil features in relation to erosional history in the vicinity of Canberra', *CSIRO Soil publication*, No. 13.

VAN ZINDEREN BAKKER, E. M. (1962) 'A late glacial and post-glacial climate correlation between East Africa and Europe', *Nature, Lond.*, **194**, 201–3.

VAN ZINDEREN BAKKER, E. M. (1964) 'A pollen diagram from Equatorial Africa; Cherangani, Kenya', *Geologie Mijnb.*, **43**, 123–8.

VESEY-FITZGERALD, L. D. F. (1955) 'The vegetation of the outbreak areas of the Red Locust (Nomadacris septemfaciata Serv), in Tanganyika and Northern Rhodesia', *Anti-Locust Bull.*, **20**, 1–31.

WATSON, J. P. (1964) 'A soil catena on granite in Southern Rhodesia: I. Field observations, II. Analytical data', *J. Soil Sci.*, **15**, 238–57.

WATSON, J. P. (1965) 'A soil catena on granite in Southern Rhodesia: III. Clay minerals; IV. Heavy minerals; V. Soil evolution', *J. Soil Sci.*, **16**, 158–69.

WATT, A. S. (1947) 'Pattern and process in the plant community', *J. Ecol.*, **35**.

WATTS, D. (1971) *Principles of Biogeography*, McGraw-Hill.

WILLIAMS, W. T. and LAMBERT, J. M. (1959) 'Multivariate methods in plant ecology: I. Association analysis in plant communities', *J. Ecol.*, **47**, 83–101.

WILLIAMS, W. T. and LAMBERT, J. M. (1960) 'Multivariate methods in plant ecology: II. The case of an electron digital computer for association analysis', *J. Ecol.*, **48**, 689–710.

WILLIAMS, W. T. and LAMBERT, J. M. (1961) 'Multivariate methods in plant ecology: III. Inverse association and analysis', *J. Ecol.*, **49**, 717–29.

WILLIAMS, W. T. and LAMBERT, J. M. (1966) 'Multivariate methods in plant ecology: V. Similarity analysis and information analysis', *J. Ecol.*, **54**, 427–45.

WOODHEAD, T. W. (1929) 'History of the vegetation of the southern Pennines', *J. Ecol.*, **17**, 1–34.

WULFF, E. V. (1943) *Introduction to Historical Plant Geography*, Dawson.

Heather M. Tinsley

6 The vegetation of upper Nidderdale: Man's impact in the post-Romano–British period

Numerous studies in historical ecology have established the importance of man in the modification and destruction of the natural plant communities of the British Isles. The most significant activity has been the removal of trees, either for use as timber, or for the provision of cultivated land. Clearance has occurred not only in the lowlands, since Dimbleby (1962) and others have shown that man has influenced the formation of upland heath and moorland habitats. The first major modification of British woodlands dates from the arrival of the early Neolithic people during the fourth millennium BC. From this period cleared areas were enlarged steadily as population increased. Only with the introduction of the iron plough, however, did permanent cultivation begin to form entirely new habitats where the soil was regularly disturbed.

This pattern of vegetational change can be found in the central Pennines around Nidderdale, in the West Riding of Yorkshire. Suggestions that this moorland area was once wooded were first made by Lucas in 1872. In working for the Geological Survey he observed the remains of trees in the peat and made detailed comparisons between the distributions of living trees and the altitudes at which remains of the same species were buried. From these observations he deduced that on the highest land 'the birch must have formed a thick and almost universal forest by itself' with oaks predominating at lower altitudes, and spreading down into the dale (Lucas, 1872, p. 75). Later studies of buried tree remains made during a vegetation survey of the entire Pennine region produced conclusions similar to those of Lucas on the former distribution of trees (Smith and Rankin, 1903).

Recent palynological work in the Nidderdale region (Tinsley, 1972) established that the reduction of this upland woodland began about 4 000 years ago in the late Neolithic period, and that the influence of man on woodland decline, since that time, has been continuous. Pollen diagrams from a number of peat deposits on the moors immediately east of upper Nidderdale show clearance phases which suggest that these uplands were exploited by successive cultural groups for timber and pasture. The first archaeological evidence in the central Pennines of the adoption of more advanced agricultural techniques dates from the late Iron Age and Romano–British period. At the beginning of the Roman period almost all the present moorland area of upper Nidderdale was clear of trees, woodland surviving only on the flanks of the upland and in the valleys. The purpose of this study is to demonstrate how, in the Nidderdale area, a region ignored in previous palaeoecological work, evidence from pollen diagrams, place names and documents may be combined in a detailed analysis of vegetation history since the Romano–British period.

The physical background

Upper Nidderdale is a sheltered, steep-sided valley with a flat floor. Three reservoirs have occupied a large area of the valley bottom since 1900 and the remaining land is used mainly as permanent pasture (Fig. 6.1). The fertile appearance of the dale contrasts sharply with the

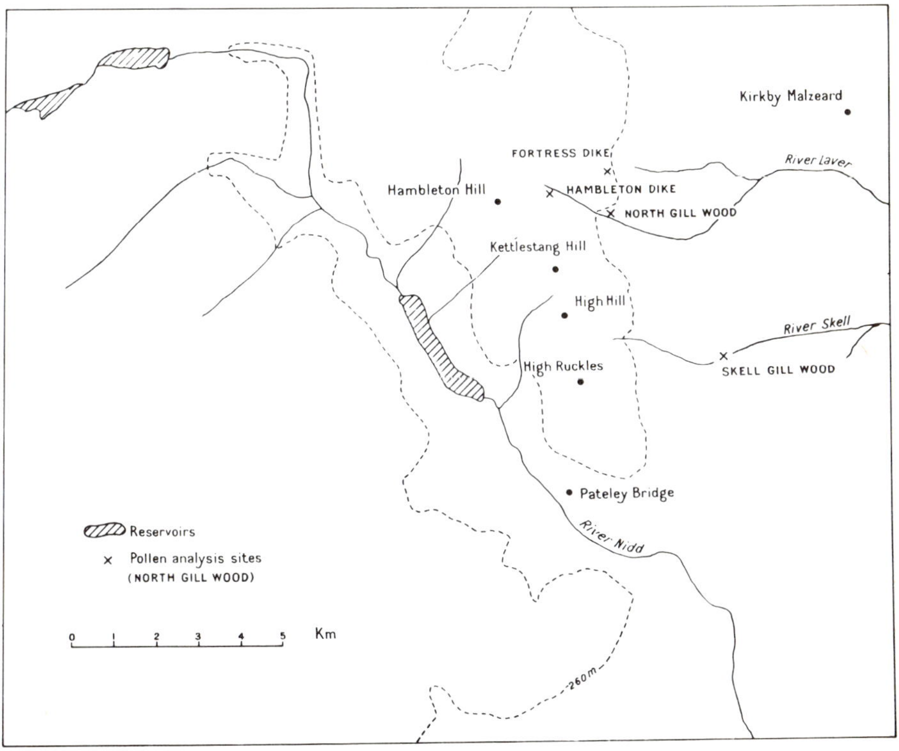

Fig. 6.1 **Upper Nidderdale**

surrounding bleak, gritstone uplands, which today are devoted to the monoculture of *Calluna vulgaris* (ling), being highly valued as grouse moors. On the moors current management policy involves burning on a ten-year rotation to stimulate new *Calluna* growth and to reduce competition by other plants. The upper limit of improved land is at approximately 260 m, and below this rough upland pasture grades into better quality grassland. Natural woodlands dominated by *Quercus petraea* (oak) are restricted to the upper parts of some of the narrow, deeply-incised valleys which cut into the moors. Along the sides of the dale and in some of the valleys plantations of mixed conifers are now widespread.

The interfluve between the rivers Nidd and Laver, on which pollen sampling sites were established, is a broad peat-covered plateau at about 290 m with four isolated hills, High Ruckles, High Hill, Kettlestang Hill and Hambleton Hill rising to over 360 m. On the western and southern edges of the plateau there is an abrupt break of slope and a steep descent to the valley of the Nidd. To the east the land gradually slopes to the Vale of York, while to the north the plateau adjoins the area of high land which forms the Nidd–Ure watershed. The principal drainage is to the southeast towards the rivers Skell and Laver *via* Skell Gill and North Gill Beck, but on the west side of the watershed three short streams flow into the Nidd through deeply incised valleys. The soils of the summit plateau are typically acidic. Shallow, well humified peat covers most of the surface but in places deeper deposits of topogenous peat have developed. Rapidly accumulating flush peat deposits are typical of the narrow stream valleys, and clayey, alluvial soils occur in the valley of the Nidd.

Pollen sampling sites

Detailed investigations of vegetation history were first made possible by the development of pollen analysis at the start of this century, but in the British Isles palynological studies have become widespread only since about 1950. As techniques have become more refined, the amount of information deduced from pollen diagrams has increased, and recently several workers have demonstrated detailed correlations between fluctuations in certain pollen types and the activities of man (Oldfield, 1963; Turner, 1964; Moore and Chater, 1969). Correlations of this type are most evident in pollen diagrams from sediments which have accumulated rapidly. In this study, therefore, attention was directed primarily to the flush peat deposits in the narrow gills and two sites were chosen within oakwoods in the valleys of Skell Gill and North Gill Beck (Fig. 6.1).

The flushes are colonised by species of *Sphagnum* (bog moss) and *Juncus* (rushes). At the North Gill Wood site (SE 16727256, altitude

267 m) 180 cm of peat have accumulated in an abandoned stream meander. The upper 65 cm of peat consist primarily of *Sphagnum* remains, while the lower part is a well humified monocotyledonous peat, with wood of *Betula* (birch) becoming increasingly abundant towards the base. At the Skell Gill Wood site (SE 19536852, altitude 229 m) 80 cm of *Sphagnum* peat have accumulated in a depression. The tenurial history of these two sites differs. The Skell valley forms part of the High and Low Bishopside parish boundary, where, during the Middle Ages, land was occupied by smallholders, tenants of the Archbishop of York. North Gill Wood was part of a land grant to Fountains Abbey in the eleventh century and was managed for over 500 years as an element in a large monastic estate.

A third site, Hambleton Dike (SE 15177342, altitude 366 m), was chosen in an area of open moorland formerly belonging to Fountains Abbey where a pocket of topogenous peat has accumulated in a depression. The bulk of the deposit is pale brown, monocotyledonous peat in which wood remains are abundant, but at 15 cm from the surface it is overlain by black, partly oxidised *Calluna* peat. Only the top part of the pollen diagram from this site is illustrated in this essay. Additional evidence is drawn from other sites for which pollen diagrams have been published. In particular, organic deposits from an earthwork attributed to the Romano–British period have yielded interesting palynological data (Tinsley and Smith, 1974).

The pollen diagrams

All the pollen diagrams are based on a count of at least 500 pollen grains excluding spores. The peat stratigraphy at each site is indicated in a column to the left of each diagram (Fig. 6.2). The pollen diagrams have been divided into zones and subzones on the basis of changes in frequency of pollen of those taxa considered to be indicators of agricultural activity. These zones are of local significance only; they are discussed in this paper with particular reference to North Gill Wood, as an absolute

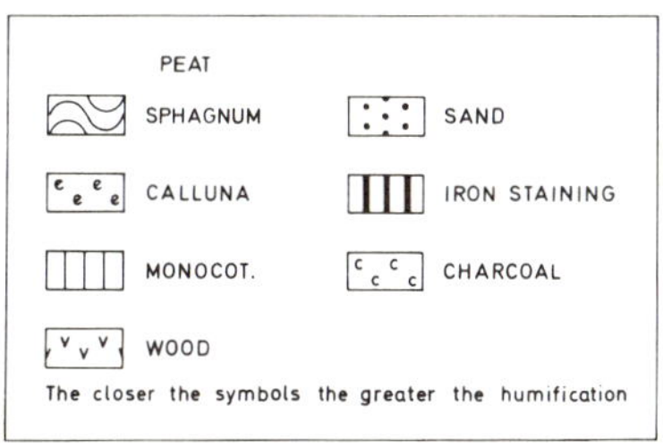

Fig. 6.2 Key to stratigraphic symbols

chronology has been established for this diagram by radio carbon dating. Zones E, F and G fit into a wider scheme of zonation which has been devised for the Nidderdale area (Tinsley, 1972).

Zone E

The earliest pollen records are from Hambleton Dike (Fig. 6.3), where the base of the *Calluna* peat has a radio carbon date of 250 BC ± 80.[1] Because of the compact and highly humified peat at this site the pollen sequence is contracted compared with those in the valley sites and zonation of this diagram is impossible. In addition, as environmental conditions on the open moors contrasted with those in the sheltered valleys, different ecological responses resulted, preventing comparison of the pollen assemblages from moorland and valley sites. The upper few centimetres of peat at Hambleton Dike, however, must have formed in the same time period as the valley flush peats, since the base of the North Gill Wood diagram (Fig. 6.4) has a radio carbon date of AD 900 ± 90.[2] At Hambleton Dike the pollen assemblage is dominated by Ericaceae (heaths) and tree pollen values are low throughout the diagram.

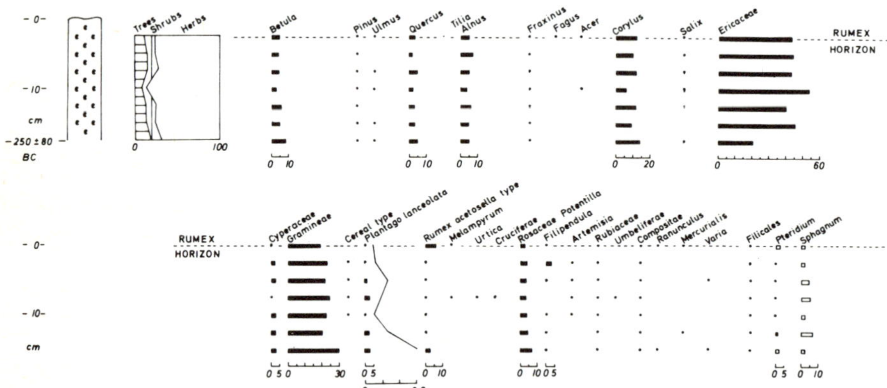

Fig. 6.3 Hambleton Hill: total pollen diagram

The pollen assemblage in zone E at the valley sites has relatively high percentages of tree and shrub pollen. *Corylus* (hazel) is well represented at both sites; *Quercus* is the dominant tree pollen type at Skell Gill Wood (Fig. 6.5), and *Betula* at North Gill Wood. Pollen values for the Gramineae (grasses) and the agricultural weeds are relatively low at both sites compared with subsequent zones. The pollen assemblage indicates open rather than closed woodland, and the high proportion of *Betula* at North Gill Wood suggests that around this site the woodland may have been secondary at the opening of zone E. Evidence from these three sites supports the hypothesis that the summit plateau was clear of trees by the

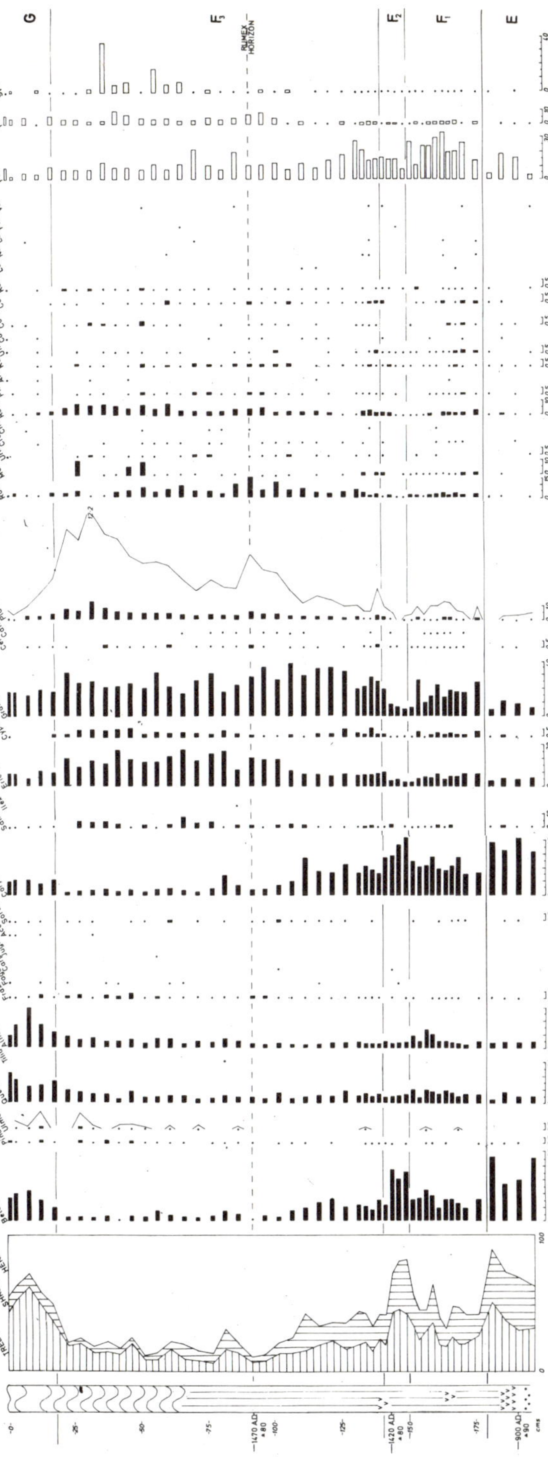

Fig. 6.4 North Gill Wood: total pollen diagram

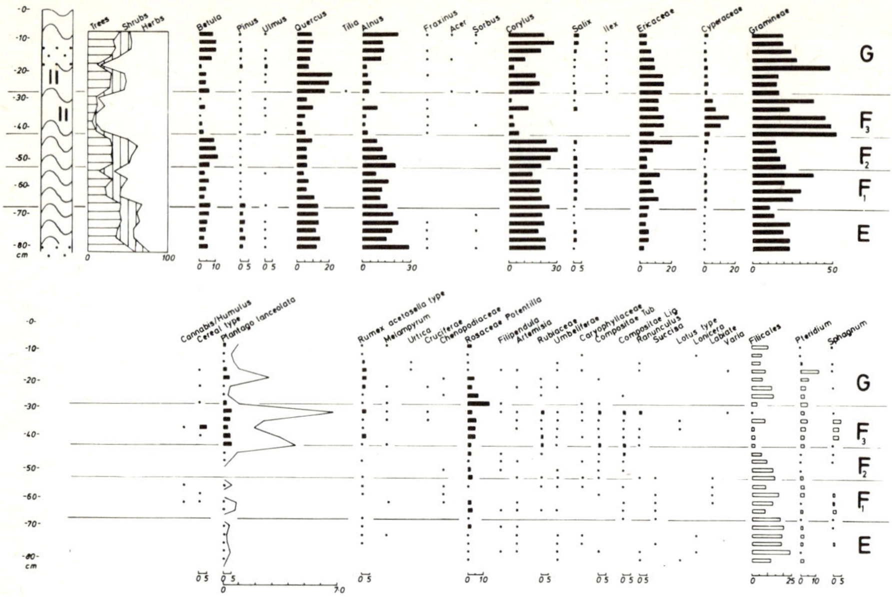

Fig. 6.5 Skell Gill Wood: total pollen diagram

Romano–British period. Any clearances on the lower slopes before this appear to have been followed by regeneration, and thus the upland flanks remained wooded until a much later date than the plateau.

Zone F

The opening of this zone is distinguished by an increase in non-arboreal pollen, particularly of the grasses, and in the incidence and variety of herb pollens. There are small peaks in the pollen of *Plantago lanceolata* (ribwort plantain), which is associated with pastoralism (Turner, 1964), and in the pollen of the Rosaceae (*Potentilla* type) which has been related to high grazing pressures (Moore and Chater, 1969). Pollen of cereals and of either *Cannabis* (hemp) or *Humulus* (hop) are also present in this subzone. The distinction between *Cannabis* and *Humulus* pollen on the basis of morphology is difficult, but Godwin (1967) suggested that most of the pollen of this type in British diagrams could be assigned to the genus *Cannabis*, and that it was usually associated with arable cultivation. The subzone F pollen assemblage implies clearance of the gill woodlands in response to both arable and pastoral activity.

Subzone F_2 appears to be a phase of regeneration, as tree and shrub pollen percentages, in particular *Betula* at North Gill Wood, show marked increases. Pollen of grasses and the agricultural weeds correspondingly declines. At North Gill Wood this phase has been radio carbon dated to AD 1420 ± 80.[3]

The pollen assemblage of subzone F_3 is typical of a period of further clearance and agricultural activity, arboreal pollen being reduced to low levels while pollen of first the Gramineae, and then the Ericaceae, increases. *Plantago lanceolata* is prominent throughout this subzone, and is associated with cereal pollen and a wide variety of agricultural weeds. At North Gill Wood, *Rumex acetosella* type pollen (sheep's sorrel) is present at values greater than 4 per cent of the total with a marked peak at 92.5 cm. Small peaks in the pollen of cereals and of *Plantago lanceolata* also occur at this level; *Cannabis/Humulus* type pollen is present in adjacent samples with the pollen values for *Corylus* and the trees being at a minimum. A sample of peat from the *Rumex* horizon has a radio carbon date of AD 1470 ± 80.[4] Near the surface of the Hambleton Dike diagram there is a marked peak in *Rumex* pollen, a feature noted in the upper horizons of other pollen diagrams from sites on former monastic land (Tinsley, 1972). Iversen (1941) commented that in pollen diagrams increases in the pollen of *Rumex acetosella* and *Rumex acetosa* (sorrel) occurred only in conjunction with agriculture, and that these weeds were characteristic of poor, sandy fields. At sites in north Lancashire, Oldfield and Statham (1963) postulated an association between increases in *Rumex acetosella/acetosa* pollen and the creation and extension of acidic pasture. Though pollen of *Rumex acetosella* type occurs in subzone F_3 at Skell Gill Wood, there is no marked *Rumex* horizon at this site.

The radio carbon dates from 145 cm and 92.5 cm in subzones F_2 and F_3 differ by only fifty years, indicating a rapid growth of the flush in North Gill Wood. Statistically these two dates cannot be separated as both have been quoted with an error of eighty years, but they are internally consistent and therefore there is no suggestion of disturbance in the profile.

Zone G

A marked increase in the pollen of *Quercus* and *Alnus* (alder) in zone G occurs at Skell Gill Wood and North Gill Wood, with the arboreal pollen component forming at least 50 per cent of the total.

Historical evidence

The Romano–British period

The period from 250 BC to about AD 600 spans the late Iron Age, Romano–British times, and the early Dark Ages. In the Pennines it can be termed the Brigantian period after the native British inhabitants. The pollen analytical evidence from Hambleton Dike and from other sites in this area establishes that the summit plateau around Nidderdale was heathland at this time. In the lower layers of the *Calluna* peat on the moors are high proportions of pollen of pastoral indicators – *Plantago lanceolata, Rumex acetosella* type and Rosaceae (*Potentilla*) type. In general the Brigantians have been thought of as pastoralists (Piggott, 1958), although the occurrence of querns indicates that cereals also were produced in Nidderdale. Cleared fields were probably sited on the better lowland soils within woodland, with the open uplands used for pasture.

The end of the organised Brigantian kingdom in AD 43 was a result of the Roman occupation. The rural economy of the central Pennines (Hartley, 1967) was little affected by the Romans, although there was probably increasing movement and trade in the Nidderdale area. Important Roman lead mines were situated at Greenhow Hill adjacent to the Nidd valley, and the influx of population to the area as a result of mining must have been considerable, hoards of Roman coins and ingots of lead having been found in a number of places in the upper dale.

Detailed palynological evidence for the impact of man on vegetation in this period is lacking, but it is reasonable to suppose that charcoal for smelting was derived from local timber, causing some destruction of woodland. In this context it is significant that the woodland surrounding the North Gill Wood site at the inception of peat growth appears to have been secondary, suggesting a phase of woodland clearance before AD 900.

Fortress Dike Camp (altitude 259 m) on the moors east of Nidderdale has been attributed to the Romano–British period (B. R. Hartley, pers. comm.). Palynological investigations of buried soil and peat deposits connected with the site suggested that the earthwork was constructed in a heathland environment. However, cereal pollen grains occur in the organic horizons associated with the occupation of this site, indicating that arable cultivation was taking place in the Nidderdale region in the Romano–British period (Tinsley and Smith, 1974). Some cleared areas must have existed in the generally wooded dale, as the open heathlands surrounding the earthwork would have been unsuitable for cultivation. Production of meat and grain may have been increased in the Nidderdale area at this time to provide food for the mining population.

The Dark Ages

From radio carbon evidence it appears that Fortress Dike Camp was abandoned finally some time after the beginning of the seventh century. There is historical evidence that at this time invasions of first the Anglians and then the Danes disrupted the Brigantian way of life. Angles may have reached Ripon in lower Nidderdale as early as AD 500 (Stenton, 1947). Certainly York was in the hands of the English by the start of the seventh century (G. R. J. Jones, pers. comm.). The Danish invasion of Northumbria in 865 instituted a further period of unsettled times in the dales. Exploratory expeditions were followed by permanent Danish settlement in east Yorkshire, and during the same period Viking settlers of Norse–Irish descent penetrated the western dales (Butler, 1967). Place names provide the chief evidence for the impact of the new settlers on the vegetation, although the information derived from such a source must be treated with caution as its exact significance is debatable. On the summit plateau Anglian and Norse names referring to moorland are both widespread (Smith, 1961): *moor* (Anglian *mōr*, Norse *mór*, both meaning a high tract of barren and uncultivated land), *Jordan's Moss, Woodale Moss* (Anglian *mos*, Norse *mosi*, an area of bog moss) and *Fleet Seaves* (Anglian *sef*, a sedge). In contrast, Nidderdale and some of its tributary valleys were wooded when the Anglians arrived. Names in *-ley* (Anglian *-leah*, a clearing in a wood) are common and indicate the extent of the lowland clearances which must have taken place in this period. Thus, 'the line of *-leys* up the valley, Whipley, Winsley, Darley, Bewerley and Pateley marks the progress of the English in carving their settlements and fields out of woodland' (Jennings, 1967, p. 24). Similarly *-thwaite* (Norse, a clearing or meadow) occurs frequently, as in Bouthwaite, Braithwaite and Gouthwaite (Fig. 6.6).

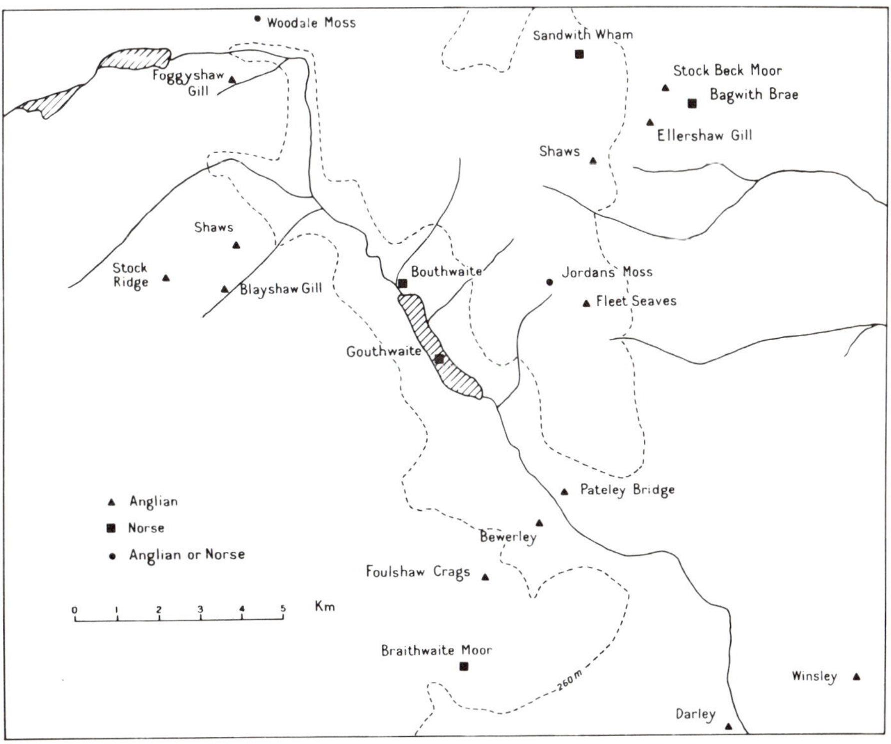

Fig. 6.6 Location of place names mentioned in text. Based on Smith, 1961

Other place names suggest that woodland still existed in the Dark Ages in areas now moorland. *Shaws* (Anglian, a copse) occurs on Carle Moor between 260 and 320 m, at the edge of Stean Moor between 320 and 360 m and in *Foulshaw* at Foulshaw Crags on Flat Moor at 320 m. The distribution of this element indicates that woodland once occupied some of the narrow valleys which dissect the summit plateau: Ellershaw Gill, Blayshaw Gill, Foggyshaw Gill. The element *-with* (Norse, a wood) occurs well above the present moorland limit at Sandwith Wham (305 m) and Bagwith Brae (190 m).

On the basis of place-name evidence, the greater part of the present moors was probably deforested by the seventh century, but the lowlands were well wooded and trees extended up the steep sides of the Nidd valley. In certain areas tongues of woodland probably spread on to the moors to about 300 m, between cleared areas with heath vegetation, at least until after the initial phase of Norse settlement in the ninth century. There is some evidence of felling in these upland woodlands, the element *stock* (Anglian, a stump) occurring in Stock Beck Moor (228–275 m) and in Stock Ridge (450 m).

The pollen data from the peat deposits at Fortress Dike Camp support the conclusions based on place-name evidence. The upper part of the organic layer in which cereal pollen was found has a radio carbon date of AD 630 ± 80.[5] Above this level a marked increase in the pollen of *Betula* occurred and fragments of birch bark and wood were found within the profile, both suggesting an extension of woodland on this part of Carle Moor. The earthwork is adjacent to the part of Carle Moor named *Shaws*, thus providing a link between the palynological and place-name evidence.

The Norman conquest and the monastic period

The period of Norse authority ended with the Norman conquest in 1066. Following the Norman conquest resistance was strong in Yorkshire and the area suffered heavily from William I's punitive action following the Northumbrian rebellion in 1069. All fourteen places mentioned for Nidderdale in Domesday Book (1086) were described as 'waste', as were most of the settlements in the surrounding area. An entry of waste for a West Riding vill seems to imply that such a holding had no population, and that a large amount of agricultural land must have been abandoned at this time (Maxwell, 1962). The Domesday record yields little information on the distribution of woodland in this area in the eleventh century, perhaps because most of the entries are brief owing to the wasted condition of many of the holdings. At Bewerley and Dacre, woodland three leagues long and three wide is recorded, but the exact significance of these measurements is not clear. As twelfth- and thirteenth-century documents record extensive areas of woodland in places not mentioned as wooded in 1086 the Domesday record may be incomplete.

The revival of agriculture and the renewal of woodland clearance in this area were due primarily to the activities of Cistercian monks from the abbeys of Byland and more particularly of Fountains. Early in the twelfth century much of upper Nidderdale was owned by the de Mowbray family. They created a hunting area, known as the Chase of Nidderdale, in which the rights to cut timber and cultivate land were strictly controlled to foster game. By the middle of the twelfth century almost the whole of the Chase had passed into monastic hands through land grants. Evidence of widespread woodland in the Nidderdale area at the time when the abbeys first received their land charters is considerable. Rights to timber are mentioned frequently: for example, de Mowbray granted Fountains Abbey 'dead wood in his forest of Kirkeby (*Kirkby Malzeard*) standing or lying whatsoever bears no leaf' (Lancaster, 1915, p. 97). Woods are noted in land grants at Bewerley, Lofthouse, Poppleton (later Covel House Grange), Backstone Gill and on the eastern margins of the moors at Carlesmoor, Galphay, and Grewelthorpe (Fig. 6.7). As with Domesday, the detail in these charters is insufficient to establish the area or exact positions of these woodlands. However, it seems likely that the woods chiefly occupied the steep valley sides in

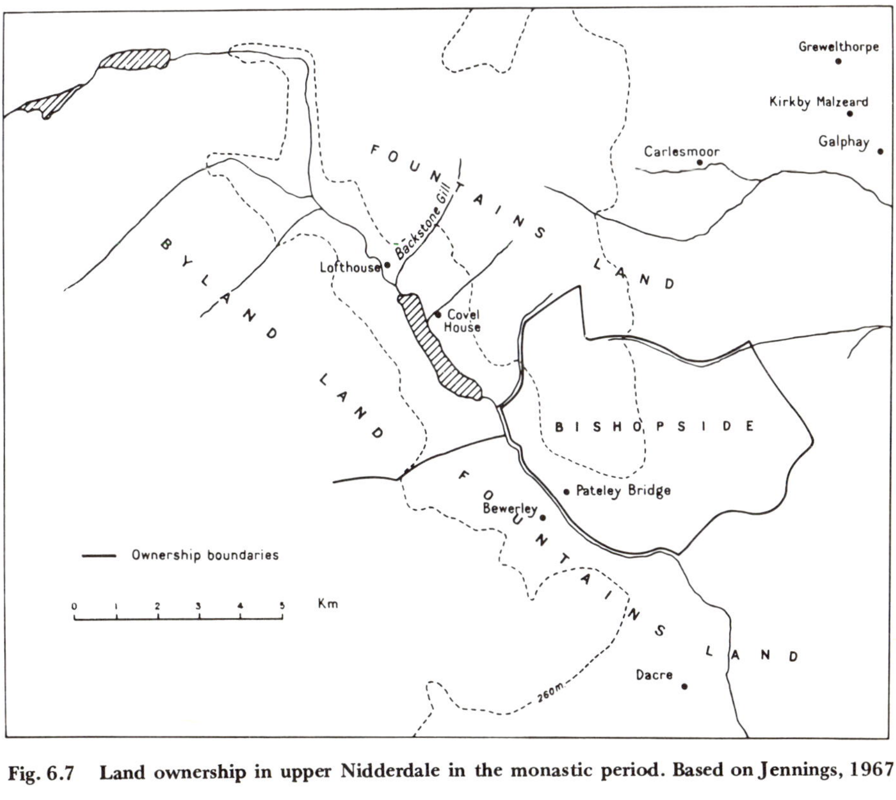

Fig. 6.7 Land ownership in upper Nidderdale in the monastic period. Based on Jennings, 1967

Nidderdale and the slopes below 230 m. There is no positive evidence to indicate that at this date there was woodland on the present moors.

The establishment of monastic activities in the Nidderdale area can be seen in the pollen diagram from the North Gill Wood site, where land was granted to Fountains Abbey in 1175 by Roger de Mowbray. The basal peat has a radio carbon date of AD 900 ± 90 and therefore the early wooded phase of zone E may be related to the latter part of the Dark Ages. From this basis the agricultural phase of subzone F_1 seems to represent clearance carried out by the monks of Fountains.

With the growth of population and assarting in Nidderdale in the latter part of the thirteenth century the area of valley woodlands must have been considerably depleted. However, the main influence on woodland reduction in this period was the demand for charcoal for smelting iron, the mines near Pateley Bridge being extensively worked. A thirteenth-century agreement relating to the Chase of Nidderdale allowed the Abbot and Convent of Fountains 'to burn charcoal within the said Chase from every kind of wood when and as they chose' (Lancaster, 1915, p. 82).

The clearance phase of subzone F_1 is more marked in the pollen diagram from North Gill Wood than in that from Skell Gill Wood. This contrast may be a consequence of different tenurial history, the monastic lands possibly being subjected to a more organised exploitation than the Bishopside area, which was occupied by smallholders.

The regeneration phase recorded in the valley flush pollen diagrams in subzone F_2 suggests a general reduction in agricultural activity in the area, although the phase was not long enough to allow primary woodland to develop. The radio carbon date obtained for peat from subzone F_2 at North Gill Wood places this phase between 1340 and 1500. Population expansion and agricultural activity was interrupted in the first half of the fourteenth century when raids by the Scots took a severe toll of the population. A series of bad harvests followed, and finally Nidderdale was ravaged in 1349–50 by the Black Death. Population numbers fell drastically and the cultivated area contracted. An overall adult mortality in the plague year in Nidderdale in the region of 40 per cent has been estimated (Jennings, 1967) and the woodland regeneration of subzone F_2 was probably a direct reflection of these events. A similar regeneration phase in pollen diagrams associated with the period of the Black Death has been recorded from Weardale (Roberts *et al.*, 1973).

The recovery of the Nidderdale area from these catastrophes is illustrated in the pollen diagrams by the decline of tree pollen and by the increase in the pollens of agricultural indicators which occur in subzone F_3. A series of large farms or granges were created by monks on their Nidderdale estates, under the management of lay brothers or tenants. The agricultural system was almost entirely pastoral, much of the valley being devoted to cattle, though small areas were used for cereals on some of the granges (Donkin, 1962). The occurrence of *Cannabis* type pollen at this level in the flush peat pollen diagrams, and in subzone F_1, is of great interest. Records of hemp growing in this area are common for the fifteenth and sixteenth centuries (Jennings, 1967), but in the monastic period documentary evidence for the cultivation of this crop is scanty. There is one reference to the purchase of hemp seed by Fountains Abbey in 1457–58 (Fowler, 1918), but no indication as to where this was planted. The pollen evidence suggests that hemp was cultivated in Nidderdale throughout the monastic period. There is also a continuous record for cereal pollen throughout subzones F_1 and F_3.

In 1491 the Fountains granges in Nidderdale supported 612 dairy cattle, 1 200 sheep and 380 steers and bullocks (Jennings, 1967). The sheep may have been pastured on the moors in spring but then driven to richer pastures on the limestone of the Craven estates, the steers and bullocks spending summer on the Nidderdale moors. The influence of these pastoral activities on the moorland vegetation is not easy to trace through the condensed pollen record from the plateau peats. Certainly peat was cut for fuel on both the monastic estate and the common land of Bishopside, rights to turbary being mentioned in some charters. Seasonal firing of the moorland pasture also seems to have been carried out to promote new growth, a practice continued to the present. On the lowlands the destruction of woodland must have continued in the period after the Black Death. A valuation of the woods belonging to the Fountains estate at its dissolution in 1539 suggests that along the east side of upper Nidderdale only isolated patches of scrub woodland remained. Contemporary documents also record that to the east of the moors clearance resulted in an extension of the moorland vegetation and in the sixteenth century the unenclosed commons stretched as far as Kirkby Malzeard.

From palynological evidence in subzone F_3, agricultural activities in the period after the Black Death seem more important on the monastic land than in Bishopside. The pronounced peak in pollen of *Rumex acetosella* type at 92.5 cm at North Gill Wood can be correlated with similar peaks in the surface horizons of peat at Hambleton Dike, Fortress Dike Camp and other sites on former Fountains land. This horizon is far less marked at sites in the Bishopside area. At North Gill Wood the *Rumex* horizon has been radio carbon dated to AD 1470 ± 80, a period associated with the later phase of monastic activities in the Nidderdale area. As *Rumex acetosella* is regarded as a particular problem of old and poorly managed pasture, the *Rumex* horizon could represent deteriorating grazing conditions, a product of overstocking of upland pasture on the monastic granges. On the other hand, *Rumex acetosella* currently colonises bare, eroding peat on the Nidderdale moors, and the *Rumex* horizon could indicate the initiation of peat erosion in medieval times. The smallholders farming in Bishopside may have made less intensive use of the commons and thus avoided these effects. Cistercian influence on the vegetation of this area would seem considerable and the palynological and documentary evidence suggest that the landscape by the beginning of the sixteenth century must have looked very similar to that of the present.

The post-monastic period

The dissolution of the monasteries in 1539 profoundly changed the social organisation of areas formerly controlled by the Cistercians. Monastic land passed into the hands of private owners, although the agricultural system continued much as before. Cattle were the main source of income in the seventeenth century and cultivation of hemp and flax became more important. As the population of the area grew large parts of the commons on the margins of the moors were enclosed by Act of Parliament or private agreement, producing eventually the present moorland limits. The final destruction of the lowland woodland came in the late sixteenth and seventeenth centuries as lead smelting expanded.

In the pollen diagrams this period of continued agricultural activity and clearance is represented by the upper part of subzone F_3. Pollen percentages of the Gramineae and the agricultural weeds are high, but to the end of this subzone there is a gradual increase in tree pollen at the valley sites. In zone G the relatively high values for *Betula, Alnus* and *Quercus* pollen indicate the establishment of the woodlands which now exist in the valleys of Skell and North Gill. The date of this regeneration phase is difficult to determine. It may stem from the late nineteenth century, when the decline in the economic prosperity of Nidderdale led to a reduction in farm numbers on marginal land, and when landlords found it more profitable to let shooting rights on the moors than to keep the land in pasture. However some of the oaks appear to be between 120 and 200 years old and the abandonment of grazing in these valleys could date from an earlier period.

The implications of current moorland management practices

During the twentieth century the contraction in upland farms has continued, and many deserted farmsteads near the limit of improved land testify to a formerly more intensive use of the moors. The tenant farmers are limited in the numbers of sheep which they are allowed to keep, as the interests of sheep and grouse conflict; at present the moors have a stocking density of 3.2 sheep per hectare.

As woodland clearance has changed the vegetation of the lower land, so burning has altered the character of the moorland. There is a long tradition of firing the moors, medieval documents recording the value of this practice for pasture improvement. In the nineteenth century regular and systematic burning was carried out widely to encourage new growth of *Calluna* shoots on which grouse feed. Burning discourages species which are not fire-tolerant, and the creation of an almost pure cover of *Calluna* has resulted, in place of a more mixed association of moorland plants. The high frequencies of Ericaceae pollen in the surface horizons of the Hambleton Dike pollen diagram and the high charcoal content of the peat at this site reflect current management policies.

The practice of burning, even when controlled, dries out the surface peat which, as a result, is easily eroded, laying bare the sterile subpeat soils. In addition, repeated burning increases the acidity of the peat cover. Should human pressure on these moors be reduced, it is debatable whether any tree regeneration could now occur. Lucas's 'thick and universal' birch forest may be lost forever. The present productivity of the moors is carefully maintained, but the balance is precarious. Accidental fires occur with increasing frequency as recreational pressure increases. Clearly, the management practices of recent centuries have produced an inherently unstable environment.

Conclusions

Human influence on the vegetation of the Nidderdale area since Romano–British times has been considerable. In this period the woodlands which once clothed the flanks of the uplands and covered the valley floors have been almost completely cleared, and the nature of the moorland vegetation substantially changed. The grassy heath community, which replaced the upland birch woodland in the time between the later Neolithic and the Romano–British period, has in turn been replaced by an almost pure *Calluna* sward as a consequence of heavy grazing and firing. These changes appear to have been most marked on the land owned by Fountains Abbey, and certainly the Cistercians were of importance in shaping the contemporary landscape of this area.

This picture of vegetation history over the past 2 000 years for the Nidderdale area is substantially similar to that described by Roberts, Turner and Ward (1973) for the northern Pennines. Palynological evidence for widespread clearances by Cistercians has been obtained from sites near Strata Florida Abbey in Wales (Turner, 1964) and in Furness (Oldfield, 1963). Further research of this nature could shed considerable light on the role of monastic institutions in shaping the vegetation of upland Britain.

Notes

1. The radio carbon dating in this essay was carried out by Professor K. Kigoshi, Gakushuin University, Tokyo, Hambleton Dike code number Gak. 3670.
2. North Gill Wood (base) code number Gak. 3673
3. North Gill Wood (140 cm) code number Gak. 3672.
4. North Gill Wood (90 cm) code number Gak. 3671.
5. Fortress Dike Camp code number Gak. 3851.

References

BUTLER, L. A. S. (1967) 'Prehistoric archaeology', in M. W. Beresford and G. R. J. Jones, eds, *Leeds and its Region*, British Association, Leeds, pp. 89–93.

DIMBLEBY, G. W. (1962) 'The development of British heathlands and their vegetation', *Oxford For. Mem.*, p. 23.

DONKIN, R. A. (1962) 'Cattle on the estates of medieval Cistercian monasteries in England and Wales', *Econ. Hist. Rev.*, second series, **15**, 31–53.

FOWLER, J. T. (1918) *Memorials of the abbey of St Mary of Fountains*, vol. 3, Surtees Society.

GODWIN, H. (1967) 'Pollen-analytical evidence for the cultivation of *Cannabis* in England', *Rev. Palaeobot. Palynol.*, **4**, 71–80.

HARTLEY, B. R. (1967) 'The Roman period', in Beresford and Jones, *Leeds and its Region*, pp. 94–7.

IVERSEN, J. (1941) 'Land occupation in Denmark's Stone Age', *Dann. Geol. Unders.*, 3, 1–23.

JENNINGS, B. (1967) *A History of Nidderdale*, Huddersfield, Advertiser Press.

LANC STER, W. T. (1915) *Abstracts from the Chartulary of Fountains Abbey*, Leeas.

LUCAS, J. (1872) *Studies in Nidderdale*, London.

MAXWELL, I. S. (1962) 'Yorkshire: the West Riding', in H. C. Darby and I. S. Maxwell, eds, *The Domesday Geography of Northern England*, Cambridge University Press, pp. 1–84.

MOORE, P. D. and CHATER, E. H. (1969) 'Studies in the vegetational history of mid-Wales: the post-glacial period in Cardiganshire', *New Phytol.*, **68**, 183–96.

OLDFIELD, F. (1963) 'Pollen-analysis and man's role in the ecological history of the south-east Lake District', *Geogr. Annlr.*, **45**, 23–40.

OLDFIELD, F. and STATHAM, D. C. (1963) 'Pollen-analytical data from Urswick Tarn and Ellerside Moss, north Lancashire', *New Phytol.*, **62**, 53–63.

PIGGOTT, S. (1958) 'Native economies and the Roman occupation of north Britain', in I. A. Richmond, ed., *Roman and Native in North Britain*, Nelson, pp. 1–27.

ROBERTS, B. K., TURNER, J. and WARD, P. F. (1973) 'Recent forest history and land use in Weardale, northern England', in H. J. B. Birks and R. G. West, eds, *Quaternary Plant Ecology*, Oxford, Blackwell, pp. 207–21.

STENTON, F. M. (1947) *Anglo-Saxon England*, 2nd edn, Oxford University Press.

SMITH, A. H. (1961) *The Place-names of the West Riding of Yorkshire*, parts I–VIII, English Place-Name Society, Cambridge.

SMITH, W. G. and RANKIN, W. M. (1903) 'Geographical distribution of vegetation in Yorkshire: part 2, Harrogate and Skipton district', *Geogr. J.*, **12**, 149–78.

TINSLEY, H. M. (1972) 'A palynological study of changing woodland limits on the Nidderdale Moors', unpublished PhD thesis, University of Leeds.

TINSLEY, H. M. and SMITH, R. T. (1974) 'Ecological investigations at a Romano–British earthwork in the Yorkshire Pennines', *Yorks. Archaeol. J.*, **46**, 25–34.

TURNER, J. (1964) 'The anthropogenic factor in vegetation history', *New Phytol.*, **63**, 73–89.

Elizabeth M. Shaw

7 Hydrology – a multidisciplinary subject

'Hydrology is the science that treats of the waters of the Earth, their occurrence, circulation and distribution, their chemical and physical properties and their reaction with their environment, including their relation to living things. The domain of hydrology embraces the full life history of water on the Earth' (Price and Heindl, 1968, p. 529). This comprehensive statement contains the essence of Meinzer's (1942) classical definition but is considerably broader in scope, embracing the chemical and physical properties of water, and taking into consideration the time dimension. These additions are of increasing importance as the development of the world's water resources expands and their pollution is further appreciated and understood. Hydrology is a dynamic subject, and just as the basic element water is in a continuously changing state, so is knowledge of it constantly developing.

Hydrology is a pure science, but while the pure scientists have made significant advances in specialised facets of the subject, civil engineers have contributed most to its general development. The engineer has used the subject to solve practical problems and hence it could properly be considered an applied science, a view supported by Wolf (1964). While appreciating the value of specialised studies in the subdivisions of hydrology, such as precipitation, surface flow or groundwater, he considered that the Hydrology Group of the Institution of Civil Engineers should concentrate on the application of the subject to engineering, despite its interdisciplinary membership.

No fully acceptable definition of hydrology yet exists. There is a limit to the breadth and depth of knowledge that a simple practitioner can assimilate and comprehend. As the subject expands the specialist will be required increasingly to contribute detailed knowledge. The hydrologist remains concerned mainly with the occurrence, circulation and distribution of water together with its direct interaction with man and his environment. Hydrology is at once both a pure and an applied science.

To understand the physical principles, a good foundation in the pure sciences is a primary requisite, while in applying the subject to problem solving a keen insight into the use of engineering judgment is an asset.

Evolution of hydrology

The establishment of mankind depended on a favourable climate and more particularly on a plentiful water supply. Many of the earliest traces of protoman have been found in equatorial regions, but it is at the margins of the tropics that mankind developed. Here, although temperatures were favourable, water was limited and as he progressed from a nomadic hunting existence, man became obliged to live in groups near reliable water sources. The increase in population resulting from successful cultivation of crops and domestication of animals led to the congregation of people in sites with constant supplies.

Darlington (1969) has suggested that one reason for the pause in expansion from 5000 to 3500 BC, before which settlements seemed to reach an optimum of 2 000 inhabitants, was the need for larger supplies of water. Not until men learned to harness the water of major rivers (Tigris and Euphrates) for irrigation could they occupy valley lands and develop the civilisations of Sumeria, Babylonia and Egypt. Comparable cultures grew independently in Central America, where progress was limited by difficulties in communications, and more successfully in China which benefited from the Hwang-Ho and Yangtze-Kiang.

Knowledge of water supply developed on a pragmatic basis to fulfil the primitive needs of man. Changes in climate resulting in persistent drought brought famine, and communities perished; floods swept away fields and villages. These natural catastrophes halted population growth and economic development but they contributed to the fund of experience on which hydrology is based. The advance of the subject up to the nineteenth century has been catalogued by Biswas (1970). Historical accounts begin with the first written records of Egypt. From the earliest days of this civilisation, the seasonal fluctuations in discharge of the Nile were recognised; the river stage was measured by the famous Nilometers from about 3000 BC and these records have formed a significant contribution to the understanding of discharge variations (Hurst, 1951). The Egyptians exploited the economic advantages of the river, damming its flow and diverting its floods into storage basins and irrigation channels, evolving complex laws for water rights, all with a minimum of understanding of the processes involved.

The Greek philosophers may be considered the first serious students of hydrology (Gk *hudor*, water; *logia*, a discourse), though they by no means thought of the subject as a whole. Thales (sixth century BC) who had visited Egypt and been inspired by the Nile, believed that water was the primary element. Hippocrates (fifth century BC) added to the knowledge of water by experiment when he demonstrated loss of water from a vessel by evaporation. By the time of Plato (fifth to fourth century BC) there were two principal theories on the origin of springs and rivers. One invoked a vast underground reservoir within the earth's interior, which fed rivers, the water eventually being returned to the underground depths by a circuitous route. The second explanation involved rainfall sinking into high ground, accumulating and emerging as springs to feed rivers. Aristotle (fourth century BC) ranged more widely over the subject. In addition to elaborating on the four fundamental elements, earth, air, fire and water, his text *Meteorologica* contained discourses on astronomy, geology, chemistry and physics as well as on the topics usually understood to belong to meteorology, such as precipitation, the wind, the ocean and again the origin of rivers and springs. He recognised the role of the sun's energy in causing evaporation although he did not differentiate between air and water vapour. His explanation of the rivers was also realistic: if cooling air over mountains changes into water (rainfall) then there is also conversion of air into water inside the earth; the waters collect together and emerge in springs in mountainous areas and flow to form rivers.

The Romans adopted the theories of the Greek philosophers, developing a more practical approach. Vitruvius described methods of finding groundwater and gave details of the availability of water in various soil types. He also clarified the atmospheric part of the hydrological cycle as outlined by Aristotle. The Romans were renowned for their many engineering works for water supply and sewage disposal but, for all their expertise in aqueduct construction, had no real understanding of the basic hydraulics. Hero of Alexandria (first century AD) proposed the concept that the discharge of a stream was equivalent to the cross-sectional area multiplied by the velocity but it was never accepted by his contemporaries.

Not until the Renaissance were further contributions made to the knowledge of hydrology. A comprehensive explanation of the hydrological cycle has been attributed to Leonardo da Vinci (late fifteenth century) although his understanding of the groundwater contribution to springs and rivers was incomplete. Palisey (early sixteenth century) was convinced that rainfall must be the source of rivers since gravity would not allow seawater to be returned to the subterranean reservoirs. From the limited fund of knowledge and experience available to these early scholars it is remarkable that the concept of the hydrological cycle should have been so well developed. However, controversy over the origins of subterranean waters continued for another century before the age of scientific inquiry and measurement brought proof that the quantities of rainfall were sufficient to sustain river flow.

Measurement in hydrology

The form of the hydrological cycle which is currently accepted (Fig. 7.1) owes its origins to the work of Perrault, the probable author of *De l'origine des fontaines*, published in Paris in 1674 (Dooge, 1959) in which measurements of rainfall and runoff in the catchment of the upper Seine at Aignay-le-Duc are described. Although the measurements of stream flow are known to be crude the results obtained for the area of some six square leagues over three years of observations give the runoff as one-sixth of the rainfall. In 1686 Mariotte published his results from the Seine catchment above Paris, 3 000 square leagues, and his improved rainfall and runoff calculations confirmed Perrault's work. Thus the conjectures of the early philosophers on the sources of springs and rivers were made redundant by this work, though disagreement on the detailed mechanisms remained for many more years.

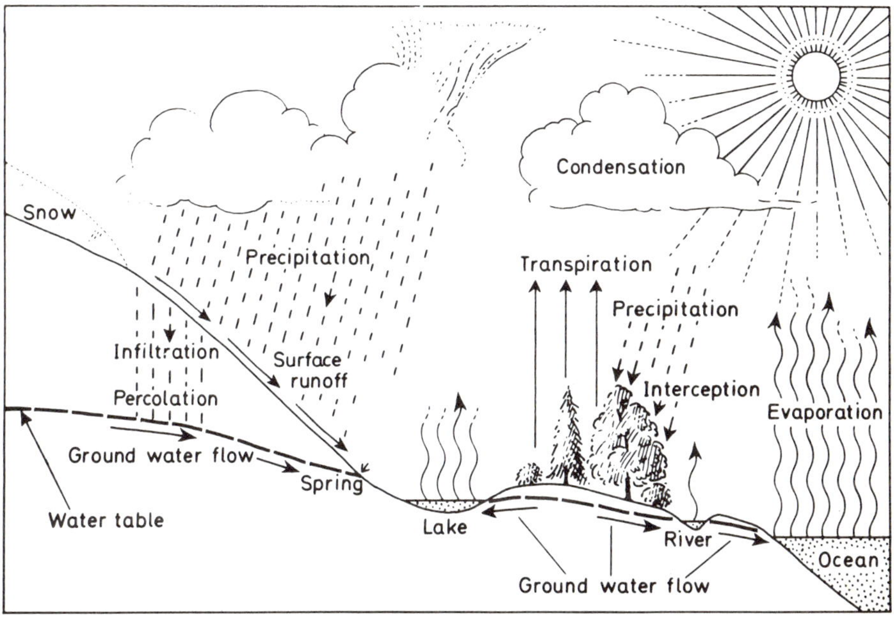

Fig. 7.1 The hydrological cycle. Based on Todd, 1959

These were among the first purely hydrological measurements but instrumental observations of individual components of the hydrological cycle had been made much earlier. First mention of rainfall measurement comes from India as early as 400 BC. Records of rainfall were made in Palestine during the first century AD, and rainfall gauges were designed in Korea in the fifteenth century. Kurtyka's *Precipitation Measurements Study* (1953) brings together material of worldwide origins on methods of observation and on the variety of instruments for the measurement of rain and snow. Over 1 000 annotated references document the historical development of all the principal standard gauges and observational procedures now adopted.

Special mention must be made of British contributions to development of rainfall measurement, since this country possesses one of the world's densest network of rain gauges and the most efficient and advanced system of rainfall data processing. A notable pioneer was Symons, a Board of Trade clerk, who in 1860 published the first of the annual collections of rainfall data, and later devoted his life to the encouragement of research, design of instruments and the assemblage and publication of rainfall observations through the British Rainfall Organisation. One of his successors, Mill, a geographer, summarised the progress in the science of rainfall measurement (Mill, 1901). Assumption of the responsibility for the British Rainfall Organisation by the Meteorological Office in 1919 placed the work on an official footing. The increase in demand for rainfall data by hydrologists led to a re-thinking of the basis of rainfall measurement founded by Symons. New equipment was evaluated at several selected Meteorological Office test sites (Bleasdale, 1959). Instead of the standard 5-inch diameter gauge, with its Snowdon funnel and brass rim, set level 1 foot above a short turf surface, gauges were set in the ground with their rims at ground level and experiments made with a variety of antisplash surrounds. The studies spread to further fields and numerous egg-box grids were devised and tested (Plate 5). The increased 'catch' has been reported in the results of the experiments at the Institute of Hydrology (Rodda, 1967, 1968). At the Water Research Association the effects of sloping sites with increased exposure and varying wind velocities have been demonstrated (Green, 1970). The efforts to obtain satisfactory measurements of rain falling on the ground surface have been accompanied by trials of cheaper gauges.

The traditional copper gauge has become expensive and both plastic and fibreglass gauges have been considered as possible replacements. A much-tested new system has been recommended by the Meteorological Office (Maidens, 1965). Two funnel sizes with 150 cm^2 and 750 cm^2 apertures can be attached to storage vessels or be fitted with a tipping bucket mechanism to give records of rainfall on a digital counter or continuously on magnetic tape. Commercial instrument makers have developed independently other rain gauging systems, some in conjunction with automatically recording climatological stations, but extensive and prolonged field trials, particularly in remote mountain regions, have encountered practical electronic difficulties and the systems are proving expensive. Although autographic records are essential for many hydrological purposes and remote sensing and telemetric instruments are invaluable for flood warning, the network of simple standard gauges continues to form the basic foundation of most rainfall information. For

5. A new rain gauge installation, Plynlimon experimental catchment

rainfall amounts over short durations, the Dines Tilting Syphon Rain Recorder, particularly with its open time scale chart, will continue in favour until the new self-recording systems become more reliable and economical. The increase in automated chart reading facilities and computer processing of rainfall intensities supports this view.

While rainfall is the source component of the land phase of the hydrological cycle, streamflow is of ultimate interest to the civil engineer and measurement of river discharge is thus of prime importance. The Egyptians recorded river levels, but quantities beyond direct volumetric measurement had to await the development of greater skills. The Renaissance experimentalists Leonardo da Vinci and Castelli appreciated the need for velocities; much later, Mariotte mastered the measurement of open channel flow, combining cross sectional area and changes of velocity with depth (Rouse and Ince, 1963). The mathematical scholars of the seventeenth and eighteenth centuries advanced the theories of hydromechanics and the engineers, with their contributions to hydraulics, made practical use of these findings. Although the techniques of river discharge measurement have been available for three centuries the need for continuous monitoring of streamflow has only been felt in the last century.

In the USA, the Geological Survey began the systematic collection of records of streamflow in conjunction with the study of the use of water for irrigation. From the establishment in 1888 of an experimental and training camp at Embudo, on the Rio Grande in New Mexico, a nationwide organisation of over 4 000 stream gauging stations developed. Annual federal funds ensured efficient coordination of the work and results were published in the renowned series of Water-Supply Papers. A recommended code of practice of river measurement techniques was developed. The resulting manual, completed in 1935, was used in the survey and finally published as 'Stream-gaging procedure' in the *Geological Survey Water-Supply Paper 888* (Corbett, 1945).

This widespread practice of current meter gauging in the USA had also been pioneered in the United Kingdom. However, a fundamental difference was that the government did not consider the measurement of river flow essential policy. McClean, the lone advocate of a practical survey of the resources of British rivers, persisted in his aims and set up a self-financed organisation, River Flow Records (McClean, 1936). In contrast to Symons's views on instruments and collative work, many of the river flow measurements were made by McClean himself. However, the larger proportion of the practical work was supervised by the organisation's Chief Surveyor, Chapman, who was also responsible for the training and supervision of temporary surveyors. Most of McClean's pioneering work was carried out on Scottish rivers, first during the 1912–15 period and later from 1929. The most notable gauging stations were those of the Ness basin, and the rivers Garry, Moriston, Spey and Aberdeenshire Dee. McClean's work in the early 1930s was instrumental in the government's establishment of the Inland Water Survey Committee of the Ministry of Health and Secretary of State for Scotland in 1935. While public recognition of the value of river flow measurements had been obtained little financial help was given. Official publication of river discharge records began with the *Surface Water Year Book of Great Britain 1935–36* and *1936–37* and the regular gauging of rivers gradually expanded with the initiative of many land drainage engineers using their optional powers under the River Boards Act 1948. Statutory obligation to measure river discharges did not become operable in England and Wales until the Water Resources Act 1963.

British hydrometric practice has since made rapid progress. The basic current meter gauging method has been supplemented by the use of numerous calibrated structures, from the initially favoured Crump Weir (Plate 6) to the recently more popular flat-vee weir does not require wing walls to gauge the same range of flows (Crump, 1952; White, 1968). In this field of measurement, too, the advance in instrumental techniques is apparent. The continuously recording pen charts for river stage are still advocated by practical engineers, but data processing is facilitated by the use of punched tape recorders and sometimes by magnetic tape recorders. Research is also being directed towards the

6. A Crump Weir, Hollinsclough gauging station, River Dove

development of even more sophisticated techniques of flow measurement. Ultrasonic waves and laser beams have been tried with varying degrees of success. An increasingly acceptable method is the technique of chemical gauging using various non-toxic compounds and radioactive tracers. These are particularly valuable methods for measuring the turbulent flow of upper reaches of rivers where the application of traditional methods is difficult or even impossible.

Scientific hydrology

If the cycle can have a beginning, then the oceans, containing 95 per cent of all water resources, can be considered as the main origin. The initiation of the cycle requires energy and in this respect Monteith, a physicist and practising hydrologist, has contributed greatly to the understanding of the role of solar radiation. In addition to noting the atmospheric pollution effects on the incoming short wave radiation, he devised a portable solarimeter, thereby providing the means for measuring in the field the energy available near the earth's surface for effecting water loss (Monteith, 1959). The physics of evaporation, the first step from the oceanic source but also the cause of serious loss from surface reservoirs, has been one of the principal interests of Penman, another physicist, at Rothamsted. Again, the scientist has provided a useful 'tool' for the engineer. Although owing allegiance to the agriculturalist, his studies resulted in a semi-empirical but physically based formula for estimating evaporation from an open water surface (Penman, 1948, 1956). The scientific literature devoted to the problem of evaporation is almost as immeasurable as evaporation itself although the 'evapotron', designed to measure directly the upward vapour flux and developed in Australia (Dyer, 1961), has considerable potential. The study of evaporation continues but although there is still no universally agreed method of measurement, the hydrologist knows that with the data sources and well tried instruments available, a satisfactory estimate of water loss can be obtained.

The atmospheric processes concerned with water have been investigated in detail. Once more a physicist, Mason, in his studies of cloud physics, has made significant contributions, and many meteorologists, especially the Scandinavians such as Bergeron, have added to the understanding of the processes which convert water vapour into precipitation. In addition to his scientific publications Mason has produced a layman's explanation of the atmospheric conditions in which precipitation is formed (Mason, 1962). Weather forecasting, of great benefit to all aspects of civil engineering but vital for the hydrologists concerned with flood prediction, has made great advances in recent years based on the work of a physicist-mathematician, Richardson (1922). His numerical methods of weather prediction, demonstrated before the days of digital computers, have been developed into the complex mathematical models of the atmosphere used today on a routine forecasting basis by the Meteorological Office.

The next phase in the cycle, the role of interception and transpiration in reducing total precipitation to the engineer's 'effective' precipitation, is the concern of the botanists. In the United Kingdom, the experimental sites at Bagley Wood (Department of Forestry, Oxford) and at Crowthorne (Department of Botany, Imperial College London) have acted as scientific laboratories in which total rainfall, throughfall, stemflow and transpiration by trees were measured. The initial results from Crowthorne showed that loss by interception was 27 to 35 per cent of the rainfall and that the rate of evaporation of intercepted water far exceeded the transpiration rate (Rutter, 1959). These studies were complemented by Law (1957), a water engineer, who added to his experiments the measurement of the runoff from an instrumented forested plot. While not undertaking the detailed botanical measurements of transpiration, Law obtained water balances for a Sitka spruce covered lysimeter and thereby obtained estimates of interception and total evaporation losses.

The complex interactions between the various components of the hydrological cycle were recognised by the team of scientists led by

Pereira, a physicist, which was set up with the special purpose of studying the hydrological effects of changes in land use in selected catchments in East Africa. Physicist and soil physicist, botanist and agronomist cooperated to organise the measurements of the several hydrological variables and to analyse their interrelationships with a view to identifying the major contributions to the water balance equation (Pereira *et al.*, 1962). This highly successful project inspired the rapid development of the catchment studies being undertaken in this country by the Institute of Hydrology. Particular attention is being paid to the physical processes of the hydrological cycle (for example at Thetford Forest Site) and the techniques of measurement are benefiting greatly from the latest improvements in electronic sensors, continuous data recording and automated analysis. It is debatable, however, whether in a comparable number of years, the detailed data obtained from the automatic weather stations operated on a routine basis (Strangeways and McCulloch, 1965) will be as continuously reliable as the personal observations made at a standard climatological station (Plate 7).

7. Duty observer, Keele climatological station, Christmas vacation 1961–62

The advent of the International Hydrological Decade encouraged many more catchment studies in participating countries all over the world. In this country the representative basins of the Trent River Authority are outstanding and the active work of the Department of Geography of the University of Hull is well reported (Ward, 1967).

Infiltration and the dynamics of soil moisture have also been the preserve of the physicist. The Agricultural Research Council's Unit of Soil Physics directed by Childs has been the scene of intensive fundamental research in connection with drainage for cultivated land. The numerous scientific papers that have stemmed from this Cambridge team since its inception in 1948 demonstrate their expertise and the eminence of the late director is recognised internationally (Childs, 1967). This complex part of the hydrological cycle has been studied especially in arid and semi-arid lands where the availability of soil moisture for plant growth needs to be determined. Again in such areas, where irrigation is practised, adequate drainage must be installed to prevent water logging and consequent soil sterility. Agricultural engineers have been the main contributors in the practical field of soil moisture management.

For knowledge and comprehension of the hidden groundwater sector of the water cycle the hydrologist must turn to the geologist. The vast storages below the watertable are now so significant that the specialist hydrogeologist has evolved from the pure scientist, as an authority on the disposition and character of rock strata and in particular of sedimentary aquifers. One of the greatest contributors to groundwater hydrology in this country was Ineson, who directed the compilation and publication of the first *Groundwater Year Book of England and Wales, 1964–66*. The vital aquifer for the supply of water to densely populated southeast England is the chalk. Its variable composition and hence changing permeability gave rise to a detailed study of its water-yielding properties (Ineson, 1959). Valuable research work is continued by the geologists of the Water Department of the Institute of Geological Sciences and by the groundwater hydrologists of the Water Research Centre.

The increase in scientific endeavour in hydrology, together with the use of improved techniques of measurement on a denser network of sampling stations, must result eventually in a more complete appreciation of the phases of the hydrological cycle. For the hydrologist seeking urgent answers to practical problems the access to more data will be welcomed, and the facility for more sophisticated methods of analysis offered by the computer should stimulate the researcher in engineering hydrology.

Engineering hydrology

Although claims have been put forward for Plato, Leonardo da Vinci and Perrault as the first hydrologists, hydrology as such has been established only recently as an academic subject and its acceptance as a recognised specialism by the engineering profession in the United Kingdom is only in its infancy. Nevertheless, both water supply and land drainage engineers have been obliged to pursue hydrological studies as requisite preliminaries to their schemes. An impetus to the assemblage of hydrological knowledge and the development of analytical techniques came in the mid-nineteenth century with the statutory duty of public authorities to provide piped potable water and to organise the disposal of waste water and sewage. Immediate concerns were the determination of the yield of mountain catchments and the capacity of the sewer systems for increased storm runoff as paved areas expanded. Both these problems are today still being presented to hydrologists for solution in developing areas.

In their search for values of runoff or discharge on which to base their designs, early civil engineers depended on the records of the two major components of the hydrological cycle, namely rainfall and streamflow. As the relatively simple rainfall measurements were much more plentiful than discharge records it became obvious to relate rainfall to runoff. For the water engineer responsible for ensuring constant supplies, the simple linear relationship on an annual basis sufficed. Once the regression constants were obtained from a few years of annual runoff values, the greatest concern was to estimate the sequence of the three driest years likely to occur within the imagined life time of an impounding scheme. The sewer designer, however, had a much more critical job, as failure could occur within a very short time. The resultant runoff from an intense rainstorm can reach a peak within minutes in the drainage system of an urban area and within hours in an average-sized British catchment.

Irish civil engineers pioneered the attempts to estimate flood flows. At first only daily totals of rainfall were available but the use of an automatic rain gauge (Mulvaney, 1851) ensured the success of their efforts for small areas. The result of their research was the 'rational formula' of worldwide acclaim and varied nomenclature:

$$Q_{max} = C \, . \, I \, . \, A$$

where Q_{max} represents peak discharge, I rainfall intensity, A catchment area and C the runoff coefficient which varies with catchment and can be related to catchment characteristics. The urgent need for estimating flood peaks in all kinds of conditions and catchments led to the evolution of many empirical flood discharge formulae. Eventually a measure of the response of a catchment to given rainfall was developed by an American consulting engineer (Sherman, 1932). The hydrograph from 1 inch of 'effective' rainfall of duration T distributed uniformly in time and space over a catchment was defined as the 'unit hydrograph' (*UH*) of duration T (*TUH*). There was a considerable timelag before the new concept was appreciated, and engineers continued to apply the rational formula, although researchers adopted the unit hydrograph. Led by Nash

(1958), the full implications of the linear properties of the mathematical representation of the hydrograph were explored. A flood hydrograph from a storm rainfall is determined by the summation of discrete increments of the rainfall applied to the discrete ordinates of the unit hydrograph. The hydrograph resulting from a single instantaneous pulse of rainfall became known as the 'instantaneous unit hydrograph' (*IUH*) which can be described by the integral:

$$q(t) = \int_{t^*=0}^{t} h(t - t^*) \, . \, i(t^*) \, . \, dt^*$$

where q and i are runoff and rainfall and $h(t)$ is the *IUH* function. The derivation of the *IUH* has been demonstrated by several workers (O'Donnell, 1966), and these methods have been applied successfully to the solution of engineering problems. The response of the catchment in producing runoff from rainfall is far from linear however and non-linear response functions are now being studied.

Conceptual mathematical models represent an alternative approach to the problem of catchment behaviour. A simple overall catchment model involves four storage elements and interconnecting inputs and outputs (Dawdy and O'Donnell, 1965). Another more comprehensive model is the Stanford Watershed Model which is being used successfully on a commercial basis to give the answers to water resource, irrigation and flood protection problems (Crawford and Linsley, 1966). Once the parameters of the models are determined to a satisfactory standard of accuracy from series of data they can be used to obtain single storm discharges or sequences of discharge data as required. Another type of conceptual model is the component model in which the several physical processes of the hydrological cycle are represented mathematically and linked. These are more complex but, since the days of the eminent American consulting engineer Horton (1945) who studied most of the components, recent advances in computing could result in a renewal of interest in such comprehensive representations of catchment hydrology. Apart from mathematical models, physical modelling has widespread use in most branches of civil engineering. For modelling of the natural catchment, there are complex scaling factors to be overcome, but these can be circumvented if the 'models' in the laboratory are treated as catchments in their own right (Hall and Wolf, 1968).

So far, the techniques considered for estimating runoff have been deterministic in form but it has long been recognised that hydrological events have a tendency to display a randomness in their occurrence. Hence within the last ten years, there has grown up a strong school of stochastic hydrology. This aspect of the subject, based on the statistics of time series analysis, has recently been reviewed (O'Connell, 1972). The pure research aspect of stochastic hydrology seeks to explain and quantify the natural trends and persistences in the sequences of hydrological records but the mathematical engineers of this specialism are well aware of its value in flow forecasting, determining flood frequencies and as an aid in the decision-making necessary in the design of major civil engineering schemes.

Conclusions

An attempt has been made to recount the development of hydrology and the fields of endeavour which are having most impact today. Nevertheless there are aspects of the subject accepted by definition that have not been mentioned. Of these, the chemical and physical properties of waters of the earth have been the most neglected by hydrologists, but are now being re-evaluated. Water quality is of utmost concern to the hydrologist and although its measurement may be left to the chemist, account must be taken of factors of turbidity, dissolved oxygen content, conductivity and temperature in the assessment of water resources. In addition, suspended sediment and bed load are physical characteristics of open channel flow, whose importance is now being increasingly appreciated by hydrologists as indeed are all aspects of river mechanics (Walling and Gregory, 1970). Future developments in hydrology are seen to be the further fusion of the contributions of scientific and engineering hydrology. The greater understanding and expression of the physical processes involved will be assimilated into the conceptual models of the engineers concerned with providing answers to practical problems. In the land phase of the hydrological cycle, it is expected that a greater use of catchment characteristics will be made in defining model response functions. Indeed a landform frame or a stream network of a catchment may be contrived through which the rainfall input passes to become a runoff output. Catchment characteristics introduced into hydrology by Horton have been studied and modified by many physical geographers, and engineers once more are taking an active interest in this field.

The study of hydrological techniques has received a considerable impetus with the current emphasis on water resources development. In both theoretical and practical fields hydrologists are tending to become water resource managers. However, there is still ample scope for research in all branches of the subject and the prospects of greater enlightenment on the vagaries of the hydrological cycle are ensured by the increase in availability of continuous series of basic measurements. The extension of cooperation, first stimulated by the International Hydrological Decade in the International Hydrological Programme, should result in the interchange of ideas and in the further advance of the subject.

References

BISWAS, A. K. (1970) *History of Hydrology*, Amsterdam; London, North Holland Publishing Co.

BLEASDALE, A. (1959) 'The measurement of rainfall', *Weather, Lond.*, **14**, 12–18.

CHILDS, E. C. (1967) 'Soil moisture theory', *Adv. Hydrosci.*, **4**, 73–117.

CORBETT, D. M. (1945) 'Stream-gaging procedure', *Geological Survey Water-Supply Paper 888*, United States Department of the Interior, Washington.

CRAWFORD, N. H. and LINSLEY, R. K. (1966) 'Digital simulation in hydrology: Stanford Watershed Model IV', *Technical Report 39*, Dept. of Civil Engineering, Stanford University.

CRUMP, E. S. (1952) 'A new method of gauging stream flow with little afflux by means of a submerged weir of triangular profile', *Proc. Instn. Civ. Engrs*, **1**, Pt 1, No. 2, 223–42.

DARLINGTON, C. D. (1969) *The Evolution of Man and Society*, Allen & Unwin.

DAWDY, D. R. and O'DONNELL, T. (1965) 'Mathematical models of catchment behaviour', *Proc. Am. Soc. Civ. Engrs*, **91**, HY4, 123–37.

DOOGE, J. C. I. (1959) 'Un bilan hydrologique au XVIIe siècle', *Houille Blanche*, 14e Année No. 6, 799–807.

DYER, A. J. (1961) 'Measurements of evaporation and heat transfer in the lower atmosphere by an automatic eddy-correlation technique', *Q. J. R. Met. Soc.*, **87**, 373, 401–12.

GREEN, M. J. (1970) 'Some factors affecting the catch of rain-gauges', *Met. Mag. Lond.*, **99**, 1170, 10–20.

HALL, M. J. and WOLF, P. O. (1968) 'Design criteria for laboratory catchment experiments, with particular reference to rainfall simulation', *Int. Ass. Scient. Hydr.*, **76**, 395–405.

HORTON, R. E. (1945) 'Erosional development of streams and their drainage basins: hydrophysical approach to quantitative morphology', *Bull. Geol. Soc. Am.*, **56**, 275–370.

HURST, H. E. (1951) 'Long-term storage capacity of reservoirs', *Trans. Am. Soc. Civ. Engrs*, **116**, 770–808.

INESON, J. (1959) 'Yield-depression curves of discharging wells, with particular reference to chalk wells, and their relationship to variations in transmissibility', *J. Instn Wat. Engrs*, **13**, 119–63.

KURTYKA, J. C. (1953) *Precipitation Measurements Study*, Report of Investigations No. 20, State Water Survey Division, Urbana, Illinois.

LAW, F. (1957) 'Measurement of rainfall interception and evaporation losses in a plantation of Sitka Spruce trees', *Int. Ass. Scient. Hydr. General Assembly, Toronto*, **44**, 397–411.

MAIDENS, A. L. (1965) 'New Meteorological Office rain-gauges', *Met. Mag. Lond.*, **94**, 1114, 142–4.

MASON, B. J. (1962) *Clouds, Rain and Rainmaking*, Cambridge University Press.

McCLEAN, W. N. (1936) *A note on the work of 'River Flow Records'*, London, privately published.

MEINZER, O. E., ed. (1942) *Physics of the Earth, vol. 9: Hydrology*, New York (National Research Council, Committee on Physics of the Earth), McGraw-Hill.

MILL, H. R. (1901) 'The development of rainfall measurement in the last 40 years', *British Rainfall 1900*, pp. 23–45.

MONTEITH, J. L. (1959) 'Solarimeter for field use', *J. Scient. Instrum.*, **36**, 341–6.

MULVANEY, T. J. (1851) 'On the use of self-registering rain and flood gauges in making observations of the relations of rainfall and flood discharges in a given catchment', *Proc. Instn. Civ. Engrs Ir.*, **4**, 18–31.

NASH, J. E. (1958) 'Determining runoff from rainfall', *Proc. Instn. Civ. Engrs*, **10**, 163–84.

O'CONNELL, P. E. (1972) 'Stochastic and statistical models', *Proceedings of the International Symposium on Uncertainties in Hydrologic and Water Resource Systems*, III, 1363–403.

O'DONNELL, T. (1966) 'Methods of computation in hydrograph analysis and synthesis', *Recent trends in hydrograph synthesis; Proc. of Technical Meeting 21*, TNO 13, The Hague, pp. 65–103.

PENMAN, H. L. (1948) 'Natural evaporation from open water, bare soil and grass', *Proc. R. Soc.*, **193**, 120–46.

PENMAN, H. L. (1956) 'Estimating evaporation', *Trans. Am. Geophys. Un.*, **37**, 1, 43–50.

PEREIRA, H. C., McCULLOCH, J. S. G., DAGG, M., HOSEGOOD, P. H. and PRATT, M. A. C. (1962) 'Assessment of the main components of the hydrological cycle', *E. Afr. Agric. For. J.*, **27**, Special Issue, 8–15.

PRICE, W. E. and HEINDL, L. A. (1968) 'What is hydrology?', *Trans. Am. Geophys. Un.*, **49**, IHD Bull. No. 6, 529–33.

RICHARDSON, L. F. (1922) *Weather Prediction by Numerical Process*, London.

RODDA, J. C. (1967) 'The systematic error in rainfall measurement', *J. Instn. Wat. Engrs*, **21**, 173–7.

RODDA, J. C. (1968) 'The rainfall measurement problem', *IUGG General Assembly of Bern IASH Pub. No. 78*, pp. 215–31.

ROUSE, H. and INCE, S. (1963) *The History of Hydraulics*, New York, Dover Publications.

RUTTER, A. J. (1959) 'Evaporation from a plantation of Pinus Sylvestris in relation to meteorological and soil conditions', *Int. Ass. Scient. Hydr. Symposium of Hannoversch-Münden*, **48**, 101–10.

SHERMAN, L. K. (1932) 'Streamflow from rainfall by unit-graph method', *Engng News. Rec.*, pp. 501–5.

STRANGEWAYS, I. C. and McCULLOCH, J. S. G. (1965) 'A low priced automatic hydrometeorological station', *Bull. Int. Ass. Scient. Hydrol.*, **10**, 57–62.

TODD, D. K. (1959) *Ground Water Hydrology*, Wiley.

WALLING, D. E. and GREGORY, K. J. (1970) 'The measurement of the effects of building construction on drainage basin dynamics', *J. Hydrol.*, **11**, 129–44.

WARD, R. C. (1967) 'Water balance in a small catchment', *Nature, Lond.*, **213**, 123–5.

WHITE, W. R. (1968) 'The flat-vee weir', *Wat. & Wat. Engng*, 72, 13–19.

WOLF, P. O. (1964) 'What is hydrology? How is it applied?', *Proc. Instn. Civ. Engrs*, 27 (Informal discussion reported by D. G. Hall), 662–4.

L. M. Cantor

8 The medieval landscape of Staffordshire

In this essay the medieval period is defined as that from the Domesday survey in 1086 to the accession of Henry VII in 1485. The rate of change in the landscape over these 400 years was slow and the most distinctive attribute of medieval technology was its immobility (Postan, 1972). As a consequence, changes in land use were gradual and, for the most part, on a small scale. In attempting to portray the development of the Staffordshire landscape over this long period it is inevitable that the result will be incomplete. This is partly because the evidence is variable and fragmentary. With the exception of the remarkable Domesday survey relatively few documents exist for the period before 1300. In part it is also a reflection of the limited amount of research undertaken into the evolution of the Staffordshire landscape. The aim of this essay is to provide an inventory of the information already available for a reconstruction of the medieval landscape.

The early medieval landscape

Any attempt to describe the Staffordshire landscape at the beginning of the Middle Ages must rely almost entirely on the Domesday Book but it must be appreciated that 1086 does not represent an important datum in terms of landscape evolution as a whole. Moreover, the limited nature of the survey and its method of compilation require care in interpretation of its data. Nevertheless, it is invaluable in providing a general idea of the nature of the Staffordshire countryside in the late eleventh century.

At this time, the county was poor and largely unsettled, consisting for the most part of extensive areas of forest and upland containing scattered, small settlements (Midgley, 1958; Page, 1908). Most of the settlements and population were concentrated in the central lowlands, along the valleys of the Dove, Trent and lower Tame, and in the western part of the county. In contrast, Cannock Forest and the southern plateau, Needwood Forest and the Pennine fringe were all thinly settled (Pinnock, 1971). The number of settlements listed in Domesday is limited and many contemporary villages do not appear by name. However, as pointed out in the context of Dorset (Taylor, 1970), the Domesday Book's greatest weakness is in its inadequate settlement record; settlements which it does not list may well have existed in or before the eleventh century. The Domesday settlement pattern exhibited characteristic elements of both upland and lowland England, as to be expected from Staffordshire's position within the country. Lowland parishes took the form of a single nucleated township, while those of the uplands were dispersed, consisting of two or three hamlets (Johnson, 1964; Sylvester, 1969). Of those manors listed the survey records that almost all contained some woodland. In addition, there must have been extensive areas of forest and woodland which escape mention because their timber was not being exploited. Although they are not mentioned by name, the existence at this time of both Needwood and Cannock Forests can be detected.

The relatively low value of the Staffordshire manors, as compared with those in other counties, leads to the assumption that the county was poor, with subsistence agriculture dominant. Outside the few towns the rural population recorded in the survey was only a little over 3 100, most of whom lived in settlements of less than ten recorded inhabitants. Town life was probably on a small scale and only three boroughs are mentioned: Stafford, Tutbury and Tamworth. Stafford was a creation of the conquest and possibly had a population of about 750 (Midgley, 1958). Tutbury a possible post-conquest creation of the Ferrers family (Palliser and Pinnock, 1971) may have had a population of about 200, while there are no indications as to the size of Tamworth.

In the early medieval period most of the county was owned by a few great landowners, notably the crown, the Earl of Chester, Robert de Stafford and Henry de Ferrers. Throughout the Middle Ages, the part played in the development of the Staffordshire landscape by such individuals was important but as yet little known and a study of the geography of medieval landholding in the county is overdue. The largest landowner in the county in 1086 was the crown which held, among others, a group of manors east of Lichfield extending from King's Bromley beyond Tamworth to Drayton Bassett. Among the next most important landholders were the de Staffords whose manors included Stafford and Madeley. In the northern part of the county, the Earls of Chester held extensive estates and the Audleys and the Verdons, though not mentioned in Domesday Book, became prominent in this area by the thirteenth century. The de Ferrers family owned large estates around Needwood Forest, based on their castle at Tutbury while, in the south-western part of the county, the de Somerys became important at the end of the twelfth century when they held a number of manors, including Dudley, where in the thirteenth century they built a castle on the site of an earlier Norman stronghold.[1] Elsewhere in the county, the Bishops of Coventry and Lichfield, as they later became, held large areas of land. Their estates were gradually increased in extent until about 1300, after which they remained unchanged to the end of the Middle Ages, and included the manors of Cannock, Rugeley, Longdon, Lichfield, Eccleshall, Haywood, Baswich and Brewood. Monastic landowners, who were less important than in other counties, included the Benedictines with houses at Burton and Tutbury, and in the thirteenth century the Cistercians with an abbey at Dieulacres, near Leek (Hilton, 1970; Wagstaff, 1970). As a feature of the landscape, the most visible token of Norman power was the castle. In the Domesday Book three are mentioned: Dudley, Stafford and Tutbury. Other Norman castles subsequently erected include Audley, Newcastle-under-Lyme, Tyrley, Chartley, Weston Jones, Lichfield and Tamworth (Cantor, 1966; Palliser, 1972). The castles belonged either to the crown, as at Stafford, Tutbury and Newcastle, or to Norman barons like William FitzAnsculph at Dudley, the Earl of Chester at Tyrley and the Marmions at Tamworth. Over the period as a whole, some twenty-four castles or fortified houses were in existence (Fig. 8.1).

The development of the medieval landscape

The record of the next 400 years is one of the gradual 'filling-up' and development of this late eleventh-century landscape. In this process the significant themes are population change, the disafforestation of large areas from the late twelfth century, the creation of hunting parks, the growth of settlement, the development of field systems and the slow but gradual growth in trade and industry.

Any attempt to estimate the population of medieval English counties, whether based on the Domesday Book or on later documents like fourteenth-century poll tax returns, is subject to margins of error on a heroic scale (Postan, 1972). However, while the documentary evidence is perhaps of no great reliability, it does provide sufficient indication of the directions in which the total English population moved. From this, it may be concluded that the population of the country as a whole greatly increased, perhaps doubled, between 1086 and the first part of the fourteenth century. Against this general background, developments in Staffordshire may be judged. The county records for the first part of the Middle Ages are scanty. In the twelfth century, for example, the only document of importance is the Burton Cartulary, a collection of surveys

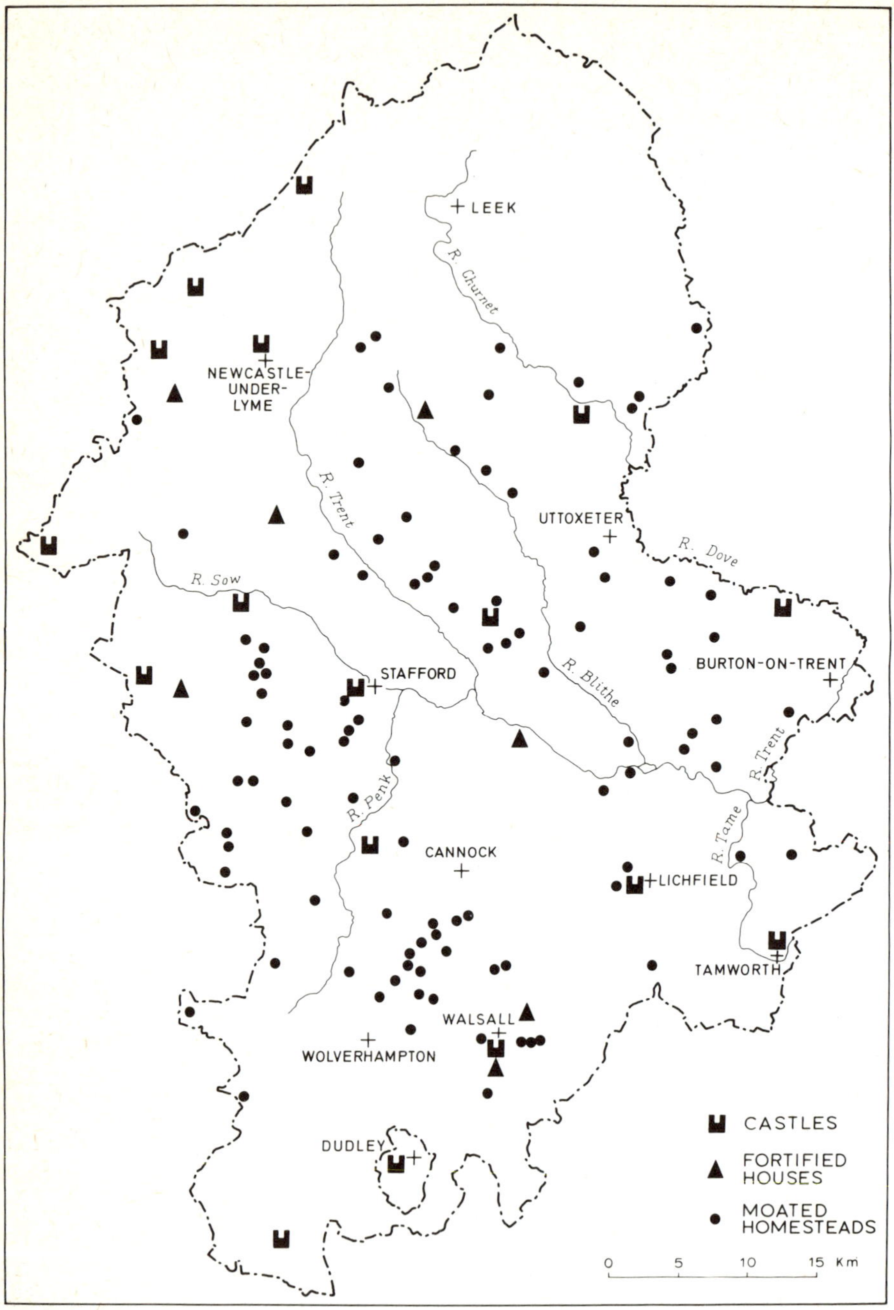

Fig. 8.1 Medieval castles, fortified houses and moated homesteads in Staffordshire. Based on Emery, 1962; Cantor, 1966; Palliser, 1972

of the manors belonging to the abbey of Burton-on-Trent, which date from between 1100 and 1133 and which cast some light on the agricultural economy of the period (Wrottesley, 1884; Page, 1908). However, there is sufficient evidence to indicate that, by the thirteenth century, the population was growing and was creating land hunger and, consequently, pressure on the land.

One way in which this pressure manifested itself was in the disafforestation of royal forests. In the twelfth century virtually the whole of the southern half of the county was occupied by royal forests and by chases of great landowners while in the north was the extensive 'New Forest' (Fig. 8.2). Forests which belonged to the crown were legally areas in which the forest laws applied. These laws, more or less strictly applied by royal forest officials, were restrictive of settlement and were designed to protect the 'beasts of the forest' (the red and fallow deer, the roe deer, and the wild boar) for the king's hunting. The chase was similar, being in effect a private forest. Some of the greater nobility and clergy were given royal permission to create hunting grounds, or chases, on their own estates. Like the king, they appointed their own officials to protect the chases against depredations and infringements. Both these areas therefore inhibited settlement and farming and, after the end of the eleventh century, were under pressure from a growing, land-hungry population. Partly in response to this demand and partly to raise revenues the crown from time to time permitted the disafforestation of all or parts of royal forests (that is, the forest laws and the officials who enforced them were withdrawn) in return for substantial sums of money.

This process of disafforestation can be seen clearly at work in Staffordshire (Cantor, 1968). Although forests and chases were not necessarily wooded, in Staffordshire a wooded aspect would seem to have been invariably the case. In the areas of the county concerned the soils and climate would have supported woodland and the trees would have been left largely undisturbed as they provided food and 'covert' for the beasts of the forest and chase. It may therefore be assumed that in the latter part of the twelfth century, the county was still very largely wooded. As legal disafforestation occurred and settlement followed in its wake, there was a slow and gradual clearing of the woodland. The first royal forests to disappear were the 'New Forest', probably towards the end of the twelfth century, and Brewood Forest, early in the thirteenth. The Shropshire section of Brewood Forest, approximately one-third of the total area, was disafforested in 1209 for a payment to the crown of 200 marks (Greenslade and Jenkins, 1967); in 1204 the Staffordshire inhabitants of the forest, occupying the other two-thirds, had obtained a charter of disafforestation, presumably by paying a somewhat larger sum. Gradually, over the next century, disafforestation took place in other parts of the county and, as a consequence, by 1300 Cannock and

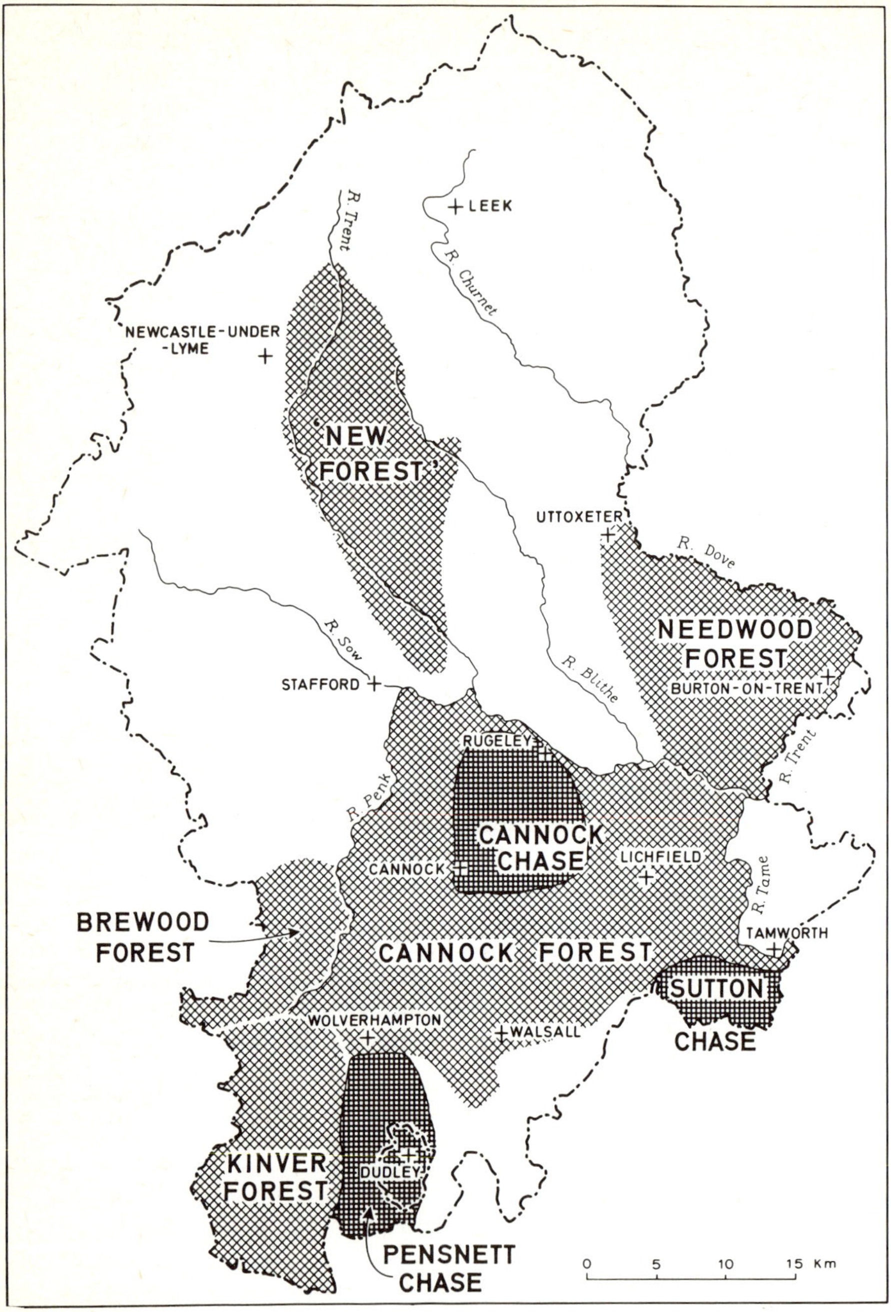

Fig. 8.2 The medieval forests and chases of Staffordshire. Based on Cantor, 1968

Kinver Forests were much reduced in extent. The Bishop of Lichfield's chase at Cannock, which is not to be confused with the royal forest, continued in existence until the dissolution.

While the growth in population between 1200 and 1350 was marked by the shrinkage of the royal forests, it was also characterised by a rapid increase in the numbers of another form of hunting ground, namely the park. A common feature of the English medieval landscape, it consisted of an area of woodland or open country, enclosed by an earth bank and topped by a wooden paling fence. Rarely of great extent, it was an enclosure stocked with deer, which provided hunting and meat for the lord of the manor. Part of the lord's desmesne, it was usually sited on poor ground on the edge of the manor and consequently did not encroach on its agricultural land. It made a contribution to the manorial revenues in that it was commonly a source of timber; cattle and horses were pastured on it; swine were fed from the pannage from the trees; it might be stocked with partridges and pheasants; and it frequently contained fishponds (Cantor and Moore, 1963). In the course of the Middle Ages, the county contained more than 100 parks (Fig. 8.3). Not all of them existed at the same time, and many of them were relatively short-lived. However, they formed one of the relatively few significant elements in a landscape that was still largely forest and woodland, and were frequently used to march on and as points of location in indicating boundaries and limits (Cantor, 1962, 1964, 1965). Following the plagues of the mid-fourteenth century, the number of parks in use declined considerably. A great deal of labour was required to maintain them and after about 1350 this was either not available or too expensive. As a result, many parks either fell out of use or were leased as pastures.

The growth of population between 1200 and 1350 was inevitably reflected in the growth of settlement: for example, in the spread of moated homesteads and in the creation of new towns. Moated homesteads appear to date from the twelfth, thirteenth and fourteenth centuries, with a concentration between 1250 and 1320. Described as 'working-units in the land use pattern of the times', they were basically farmsteads whose only defence was a moat and, rarely, minor earthworks (Emery, 1962, p. 378). They reflected the need for a dry building site: by digging a moat, by piling debris within it and then levelling, the farmer created an artificial island. Built for the most part by the smaller feudal landowners, their moats provided a supply of water, especially useful for livestock in summer; being stocked with fish they also supplemented the sometimes meagre diet. While providing no great barrier to marauders, they helped to guard livestock against the depredations of wild animals. There were at least 100 such moats in Staffordshire in the Middle Ages (Fig. 8.1), concentrated mainly along the river valleys in the central part of the county, west of Stafford and between Cannock and

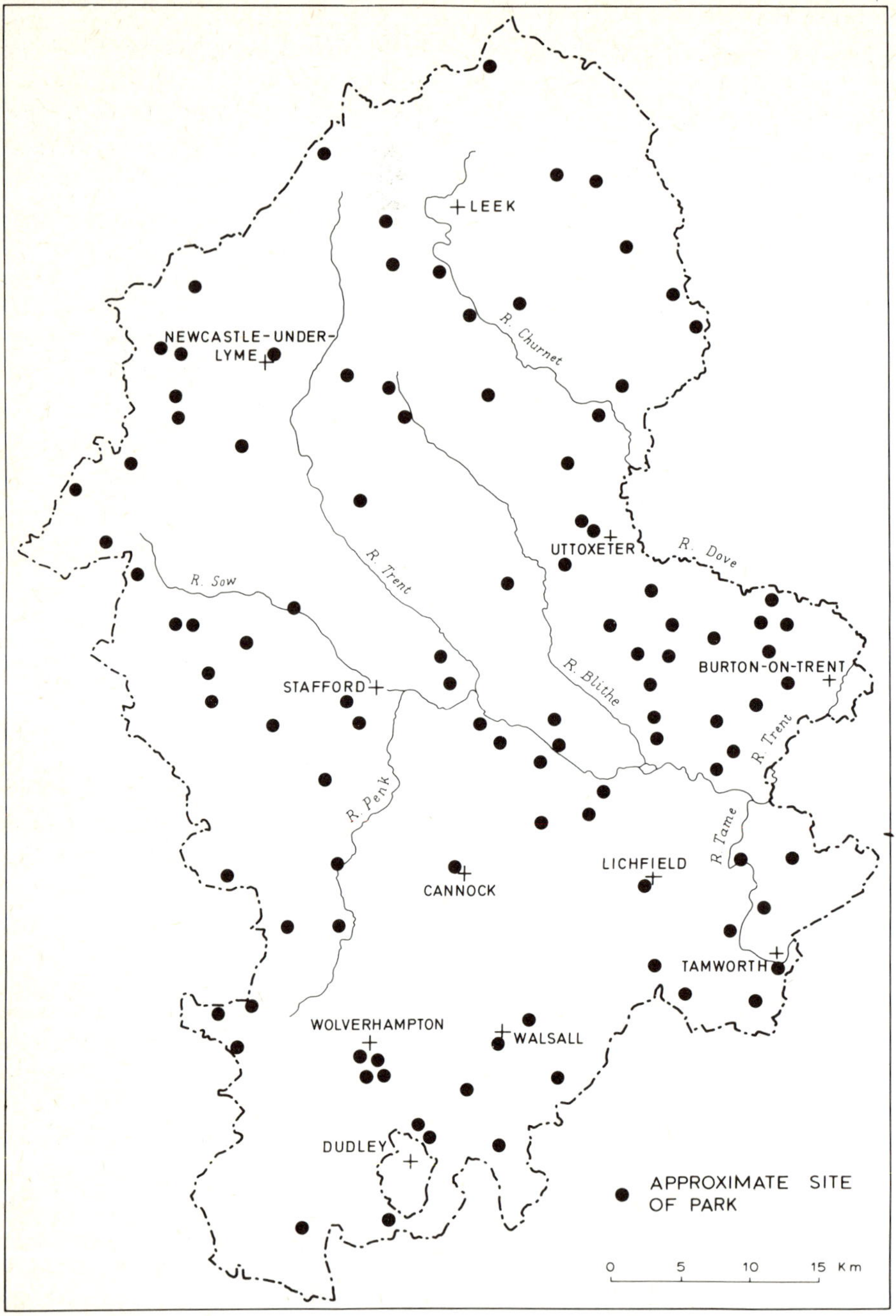

Fig. 8.3 The medieval parks of Staffordshire. Based on Cantor, 1962, 1964 and 1965; Cantor and Moore, 1963

Wolverhampton (Page, 1908). As might be expected, they were largely absent in Cannock and Kinver Forests and in the northeast moorlands.

One of the problems facing the student of medieval urban settlement is that of definition. In Staffordshire Palliser and Pinnock (1971) defined a town as a community which included a majority, or a large minority, of inhabitants who gained their living by means other than agriculture. This is a broader definition than that of a borough which, in legal terms, was distinguished by having burgage tenure and chartered privileges. The chief function of the medieval town was to act as a market and distribution centre for its agricultural hinterland, by means of shops, markets and fairs. Palliser and Pinnock identified at least forty-five market centres in Staffordshire in the Middle Ages. Of these the most important were possibly the boroughs which, in addition to the three Domesday ones, included Lichfield, Newcastle, Newborough, Burton-on-Trent, Leek, Walsall, Stone, Uttoxeter, Wolverhampton, Cannock and Dudley. Created by Bishop Roger de Clinton between 1129 and 1149, Lichfield was a 'new town' whose population in 1377 has been estimated as approximately 1 400 (Thorpe, 1950–51). Newcastle-under-Lyme was another twelfth-century creation; built around its castle between 1154 and 1162, it had become a royal borough by 1173 (Palliser, 1973). Newborough was established in Needwood by the Earl of Derby between 1100 and 1139 (Beresford, 1967). The Church was instrumental in developing a number of settlements, of which Lichfield is the most obvious. At Burton, the town most probably grew up round the abbey while at Leek the migration of Dieulacres Abbey from Cheshire in the early thirteenth century resulted probably in the expansion of an existing settlement (Beresford, 1967). It was given a grant in 1214, when timber was made available to newcomers and building plots were allocated rent free for three years. In the southern part of the county, Dudley may well have been like Newcastle in that it also had a castle nucleus (Wrottesley, 1906).

In general terms, the market towns, like the moated settlements, were concentrated in the valleys of the Trent and its tributaries and in the western part of the county, and were absent in Cannock Forest and in the northeast moorlands. This distribution was a reflection of economic opportunities; the richer agricultural soils and the more workable lands (based on the clays, sands and gravels of the river terraces) supported a relatively dense population by the standards of the time, while the poorer soils were thinly populated. The exception was Kinver Forest which, despite its relatively rich soils, was inhibited from becoming well populated by its royal status. In the county as a whole towns were not important in the Middle Ages. Staffordshire was predominantly rural and most of its towns were to all intents and purposes 'villages in disguise' (Palliser and Pinnock, 1971, p. 54).

During the first half of the period the gradual growth in population must have been accompanied by a growth in the agricultural economy. Staffordshire in the Middle Ages was characterised by a 'mixed' agricultural economy: that is, arable farming developed in the more fertile river valleys of the centre and west of the county, while pastoral farming was dominant in the infertile northeast moorlands and on the heavy Keuper marls of Needwood Forest. From various evidence it can be deduced that by the thirteenth century the open field system had developed on a moderately extensive scale. Gray asserted that the three field system was important in the county and that it was best developed in the Trent valley (Gray, 1915). Later the Orwins stated that there was evidence that medieval open field farming was the rule in the county, except in the north (Orwin and Orwin, 1938). Tate, on the other hand, considered that there was always relatively little open field farming within the county and certainly there is plenty of evidence showing that, when enclosure occurred, it was frequently direct from the waste (Roberts, 1973).

Staffordshire was probably one of those counties which exhibited great regional and local variations in respect of the development of open field systems. Not only were some regions, like the northeast moorlands, largely devoid of open fields, but even in areas where they were dominant, like the Trent valley, the proportion of land devoted to open fields could vary greatly from one manor to the next. Roberts (1973) designated such open field systems as being within the 'woodland zone' as distinct from the 'champion zone' where the open field systems were more extensive and homogeneous. However, there is evidence to support Gray's contention that, where the open field system existed, three fields were most common. Certainly, it would seem that Newcastle-under-Lyme and Madeley both possessed three field systems in the Middle Ages, and the same was true of Wednesbury in 1315, with three fields known as Monway Field, Church Field and Hall Field (Ede, 1962; Jenkins, 1963; Kennedy, 1970). Many manors, like Newcastle and Wednesbury, for example, added further fields at a later date. Another medieval manor with a three field system was Stafford, where in 1467 there were three fields known respectively as Great Hyde Field, Castle Field and Maynor's Green Field, and precise indications to their extent exist.[2] Over the county as a whole, arable farming would seem to be expanding in the thirteenth and fourteenth centuries. The late thirteenth-century rentals and surveys of the Bishop's estates demonstrate this expansion in his manors at Lichfield, Brewood, Eccleshall, Rugeley, Cannock, Haywood and Berkswich[3] and there was assarting on the edge of Needwood Forest early in the fourteenth century (Birrell, 1962).

Pastoral farming was important and extensive throughout the Middle

Ages. Needwood Forest, for example, was famous for its grassland, and during the fourteenth and fifteenth centuries supported prosperous non-seigneurial stock farming, particularly in the demesne manors on the edge of the forest. Stock farming was also very common in the north-eastern part of the county. In 1319, for example, Sir Thomas Furnival, lord of the manor of Alton, impounded into his park 160 sheep from near Wootton-under-Weaver (Lynam, 1911) and there was a vaccary in existence at Quarnford, near Alstonefield in 1307 (Wrottesley, 1908). Sheep were numerous in Cannock Forest in the thirteenth century (Wrottesley, 1884) and in 1342 a majority of the inhabitants of Stafford and Newcastle appear to have obtained a living by sheep farming (Vanderzee, 1807).

According to Baker (1966) the high tide of medieval land colonisation in England may well have begun to turn by the early decades of the fourteenth century, and the arable acreage, which had been expanding, now began to decline. Among the reasons suggested for this decline are soil exhaustion, climatic hazards and shortage of seed corn. However, if the recession of arable land, and a consequent increase in pastoral farming, had already begun by the mid-fourteenth century, it was greatly accelerated by the effects of the plagues. These are graphically illustrated by an inquisition post-mortem of 1349 which says of Wootton-under-Weaver that it contained 'no messuages, demesne lands, nor meadows . . . places and perquisites of the courts there are worth by the year two shillings and no more, on account of the pestilence' (Wedgwood, 1913). This compares with twenty years earlier when the demesne includes 120 acres of arable (Wedgwood, 1913). Although the severity of the plagues in Staffordshire was perhaps not as great as in other counties, it must have suffered, for villages like Wootton-under-Weaver took a long time to recover. Johnson (1964) estimated that this village did not redevelop an open field economy until Tudor times. Other villages like the nearby Blore and Okeover did not recover and it is from this period that most of the Staffordshire lost villages date. Bate and Palliser (1970–71) list over 100 suspected lost village sites, most of which probably vanished in the later Middle Ages when, already declining because of epidemics and migration, they were destroyed by landlords enclosing land either for sheep pastures or parks. How much of the county was affected by enclosure of this sort cannot be stated at present. However, there are some indications of the processes involved: in the case of Wednesbury, Ede (1962) estimated that by 1500, the amount of enclosed land equalled that of the open fields and Johnson (1964), in his study of northeast Staffordshire, suggested that enclosure was taking place in that part of the county after 1350.

Enclosure had long predated the recession of the second half of the Middle Ages. In Staffordshire, direct enclosure from the waste was taking place at least as early as the twelfth century and piecemeal enclosure of the open fields occurred from 1250. This process gathered momentum in the next century and was a common feature of the rural economy well into the modern period. The extent to which enclosure occurred in a given manor or region depended on many factors including the period of initial colonisation and settlement, the character of seigneurial control and the availability of reserves of common pasture within a manor (Roberts, 1973).

Throughout the Middle Ages Staffordshire was primarily an agrarian county and industry was little developed, a situation that continued until the Industrial Revolution. But the beginnings of industrial activity can be detected in the Middle Ages. Pottery-making, for example, was quite common throughout the county, where conditions were suitable, although no one district specialised in its manufacture. There are documentary references to pot-making in north Staffordshire in the late thirteenth and in the fourteenth centuries (Greenslade and Jenkins, 1967) but it is clear that this activity predates them. As with pottery, small-scale, scattered coalmining was found in the county throughout the Middle Ages. The first recorded reference to coalmining appears to be at Sedgley in 1273; other references include Cannock in 1298 and Beaudesert in 1367 (Wrottesley, 1888). According to an account of 1423, the Bishop of Coventry and Lichfield was leasing his 'mines of sea-coal' at Beaudesert for an annual rent to John Rouse and Henry Bullock.[4] In north Staffordshire, coal was mined as early as 1282, though the quantities were small both compared to later centuries and to contemporary outputs in northeast England and in Scotland (Greenslade and Jenkins, 1967). There appears to have been a little iron-mining in the county and also some iron-working, based on forges using charcoal. In 1293, for example, the Earl of Stafford had a 'factory' for making iron in Madeley Park (Wrottesley, 1885), and Newcastle had developed as a centre for marketing iron by the mid-fourteenth century (Jenkins, 1963). In contrast with the modern period, the northern part of the county seems to have been the chief centre for the mining and working of iron during the later Middle Ages. Other mining and quarrying included that of stone, used principally as a building material, and of gypsum, the main centre being Tutbury where it was mined as early as the twelfth century. The oldest Staffordshire textile industry appears to have been the making of woollens which, although widespread, was a small-scale activity serving local markets (Greenslade and Jenkins, 1967). Evidence exists of glass-making by the fourteenth century, mainly in the centre of the county around Abbots Bromley, Rugeley and Wolseley. There were certainly glasshouses in operation in this area by the late 1470s (Greenslade and Jenkins, 1967), and glass furnaces existed at the end of the period in Blore and Tyrley, near Market Drayton (Pape, 1933–34).

Conclusions

The population of the country appears to have declined dramatically during the fourteenth century, perhaps by as much as one-half. This decline may have preceded the Black Death of 1348, but if so it was certainly accelerated by this catastrophe, and further outbreaks of pestilence occurred in the early 1360s and in the mid-1370s (Postan, 1972). Population did not increase substantially until the end of the fifteenth century (Hollingsworth, 1969). Pinnock (1971) has suggested that there is no reason to suppose that Staffordshire differed from the rest of the country in this respect. However, he concludes that in the period from 1327 to 1377 the decline in population was accompanied by a shift in distribution. By the end of the fourteenth century the central lowlands and the western part of the county ceased to be the most populous areas. Analysis of the 1377 poll tax shows a radical change whereby the south and east had a relatively large population compared to other parts of the county and by the end of the fifteenth century, the south and east were dominant. On the southern plateau, industry was beginning to develop, the forerunner of the Black Country, and towns like Walsall, Wolverhampton and Birmingham were all growing, while in the east a more prosperous agriculture was developing. Thus to a considerable extent the Domesday pattern of population distribution had been reversed by the end of the medieval period and the roots of a rapid modern growth of population and industry on the southern plateau had been established.

Notes

1. PRO, Calendar patent rolls, 1258–66, 307.
2. SRO, D641/1/2/68, Stafford receiver general's accounts, 1467–68.
3. SRO, D1734/J2298, Rentals and surveys of the Bishop of Coventry and Lichfield, 1297–98.
4. SRO, D1734/3/2/1, Bailiff's accounts of the estates of the Bishop of Coventry and Lichfield.

References

BAKER, A. R. H. (1966) 'Evidence in the "Nonarum Inquisitiones" of contracting arable lands in England during the early fourteenth century', *Econ. Hist. Rev.*, second series, **19**, 518–32.

BATE, P. V. and PALLISER, D. M. (1970–71) 'Suspected lost village sites in Staffordshire', *S. Staffs. Arch. Hist. Soc.*, **12**, 31–6.

BERESFORD, M. W. (1967) *New Towns of the Middle Ages*, Lutterworth Press.

BIRRELL, J. R. (1962) 'The forest economy of the honour of Tutbury in the fourteenth and fifteenth centuries', *Univ. Birmingham Hist. J.*, **8**, 114–34.

CANTOR, L. M. (1962) 'The medieval deer parks of north Staffordshire', *N. Staffs. J. Fld Stud.*, **2**, 72–7.

CANTOR, L. M. (1964) 'The medieval deer parks of north Staffordshire', *N. Staffs. J. Fld Stud.*, **4**, 61–6.

CANTOR, L. M. (1965) 'The medieval parks of south Staffordshire', *Trans. Birmingham Arch. Soc.*, **80**, 1–9.

CANTOR, L. M. (1966) 'The medieval castles of Staffordshire', *N. Staffs. J. Fld Stud.*, **6**, 38–46.

CANTOR, L. M. (1968) 'The medieval forests and chases of Staffordshire', *N. Staffs. J. Fld Stud.*, **8**, 39–53.

CANTOR, L. M. and MOORE, J. S. (1963) 'The medieval parks of the Earls of Stafford at Madeley', *N. Staffs. J. Fld Stud.*, **3**, 37–58.

EDE, J. F. (1962) *A History of Wednesbury*, Wednesbury Corporation.

EMERY, F. V. (1962) 'Moated settlements in England', *Geography*, **47**, 378–87.

GRAY, H. L. (1915) *English Field Systems*, Harvard University Press.

GREENSLADE, M. W. and JENKINS, J. G. (1967): see *Victoria History*, vol. 2.

HILTON, R. H. (1970) 'Lord and peasant in Staffordshire in the Middle Ages', *N. Staffs. J. Fld Stud.*, **10**, 1–20.

HOLLINGSWORTH, T. H. (1969) *Historical Demography*, Hodder & Stoughton.

JENKINS, J. G. (1963): see *Victoria History*, vol. 8.

JOHNSON, F. G. (1964) 'The settlement pattern of north-east Staffordshire', unpublished MA thesis, University of Wales, Aberystwyth.

KENNEDY, J., ed. (1970) *Madeley, a History of a Staffordshire Parish*, Keele University Library.

LYNAM, C. (1911) *St Mary's Abbey, Croxden*, London.

MIDGLEY, L. M. (1958): see *Victoria History*, vol. 4.

ORWIN, C. S. and C. S. (1938) *The Open Fields*, Oxford.

PAGE, W. (1908): see *Victoria History*, vol. 1.

PALLISER, D. M. (1972) 'Staffordshire castles: a provisional list', *Staffs. Arch.*, **1**, 5–8.

PALLISER, D. M. (1973) 'A town is born', in J. Briggs, ed., *Newcastle-under-Lyme 1173–1973*, Newcastle-under-Lyme Corporation, pp. 1–12.

PALLISER, D. M. and PINNOCK, A. C. (1971) 'The markets of medieval Staffordshire', *N. Staffs. J. Fld Stud.*, **11**, 49–63.

PAPE, T. (1933–34) 'Medieval glassworks in north Staffordshire', *Trans. N. Staffs. Fld Club*, **68**, 74–121.

PINNOCK, A. C. (1971) 'Population and prosperity in Staffordshire 1086–1539', unpublished MA thesis, University of Keele.

POSTAN, M. M. (1972) *The Medieval Economy and Society*, Weidenfeld & Nicolson.

ROBERTS, B. K. (1973) 'Field systems of the west Midlands', in A. R. H. Baker and R. A. Butlin, eds, *Studies of Field Systems in the British Isles*, Cambridge University Press, pp. 188–231.

SYLVESTER, D. (1969) *The Rural Landscape of the Welsh Borderland*, Macmillan.

TAYLOR, C. (1970) *Dorset*, Hodder & Stoughton.

THORPE, H. (1950–51) 'Lichfield: a study of its growth and function', *SHC*, 1950–51, pp. 137–211.

VANDERZEE, G., ed. (1807) *Nonarum inquisitiones in curia scaccaria*, Record Commission.

Victoria History of the County of Staffordshire, Dawsons Pall Mall and Oxford University Press.
Vol. 1, ed. W. Page (1908); facsimile edn, 1966.
Vol. 2, ed. M. W. Greenslade and J. G. Jenkins (1967).
Vol. 4, ed. L. M. Midgley (1958).
Vol. 8, ed. J. G. Jenkins (1963).

WAGSTAFF, J. M. (1970) 'The economy of Dieulacres Abbey 1214–1539', *N. Staffs. J. Fld Stud.*, **10**, 83–101.

WEDGWOOD, J. C., ed. (1913) 'Inquisitions post mortem', *SCH*, 1913, pp. 1–154.

WROTTESLEY, G., ed. (1884) 'An abstract of the contents of the Burton Chartulary', *SHC*, **5**, 1–104.

WROTTESLEY, G., ed. (1885) 'Extracts from the plea rolls, 1272–1294', *SHC*, **6**, 37–300.

WROTTESLEY, G., ed. (1888) 'Assize rolls and De Banco rolls of the reign of Edward II', *SHC*, **9**, 1–119.

WROTTESLEY, G., ed. (1906) 'The chartulary of Dieulacres Abbey', *SHC*, new series, **9**, 291–365.

WROTTESLEY, G., ed. (1908) 'A history of the Bagot family', *SHC* new series, **11**, 1–224.

E. M. Yates

9 Aspects of open field agriculture in Staffordshire

In his study of English field systems, Gray (1915) considered that Staffordshire lay on the western boundary of the province within which the two or three field system was the normal form of husbandry. Within the county he found the system to be most perfectly developed in the Trent valley, for example at Rolleston. Elsewhere, irregularities were present in the field systems, attributed by Gray to the county's forested nature. As Gray studied the whole of England, his Staffordshire evidence was limited, but in addition to Rolleston he listed seventeen other places for which there was evidence of open field and two places, Rocester and Wootton-under-Weaver, with evidence of irregular or more complex field systems.

Since Gray's work, additional evidence for open field arable in Staffordshire has been recognised, confirming and extending his list. Thomas (1933) gathered evidence for enclosure of open field arable and commons from the 1517 inquisition into depopulation, Star Chamber records, early maps, court rolls and enclosure awards. Much of this relates to enclosure of common grazing, but there is also a substantial body of data for open field arable, including some twenty-one parliamentary enclosures.

Further manuscript and published evidence of open arable is also available. Claims for pasture for two years after the corn had been cut and carried provide an important source, since the claim implied that the land was or had been in common. Such claims, recorded in Plea Rolls, are numerous in Staffordshire and extend from the thirteenth century to the sixteenth, being made when land was separated from the common arable. Examples relate to Milwich, Bushbury, Norbury, Billington, Oakley, Elleford, Eccleshall, and Lapley.[1] Among the later material are references in the 1623 court rolls of Tunstall to selions (normally the strips in open field arable) at Norton[2] and to open field arable at Leek.[3] Certain rent and account books of Staffordshire farmers also contain evidence for open field arable: examples are for 1646 at Fauld and Coton in the rental of Cassibilan Burton,[4] at Streethay in an account book of Richard Pyott beginning in 1723, and at Kinvaston in an account book of Edward James[5] beginning in 1692. Plot (1686) described most of the heavy soils as still being open field, but Pitt (1794) recorded only Stone, Stafford, Bloxwich and Cheddleton as having open field. The papers of the Sneyd family refer to some open field surviving at Keele in 1729.[6] Thomas's map evidence of open field arable can also be supplemented as in the cases of Caldon, 1631, and Marston, 1722 (Figs 9.1 and 9.2). Further evidence of open field arable in Staffordshire awaits inspection, but the lists of Gray and of Thomas and the other examples already quoted establish that this form of husbandry was widely distributed.

Physical background

Any analysis of the distribution of open field arable in Staffordshire must take into account the pattern of soil types and physical subdivisions. The northeast of the county, extending from the edge of the Cheshire plain in the west, across the Potteries coalfield and the gritstone moorlands to the limestone upland in the east, is Pennine in character. The southern part is dominated by the south Staffordshire coalfield, separated from the Pennines by an extensive belt of Keuper Marl. This belt is divided into an eastern and a western section by the outcrops of the Bunter and the Keuper Sandstones which ring the south Staffordshire coalfield, extending north into Cannock Chase. This complexity in the geological bases for farming was recognised by Plot, who suggested a division into northeast and southwest, separated by a central area termed the woodlands (because of the presence of Needwood and other forest areas). Pitt (1808) refined this division, distinguishing the limestones of the northeast as a separate soil region. He also recognised as other regions the meadows along the major streams, notably the Dove, and the pastures as in the moorlands. Three other soil classes are distinguished: strong, light and mixed. Strong soils are those of the Keuper Marls, forming an almost continuous belt across the county; light soils derive from the Triassic sandstones, particularly east and west of the Black Country; and mixed soils are associated with the coalfields. Evershed (1869), citing the geological map, subdivides the county into a dairy farming area, corresponding with Pitt's region of strong soils, a light soil area, Cannock Chase, the limestones, the moorlands and the streams. All three writers adopt a regional division that, with varying degrees of refinement, is based substantially on a simplified geology.

The distribution of townships for which there is evidence of open field

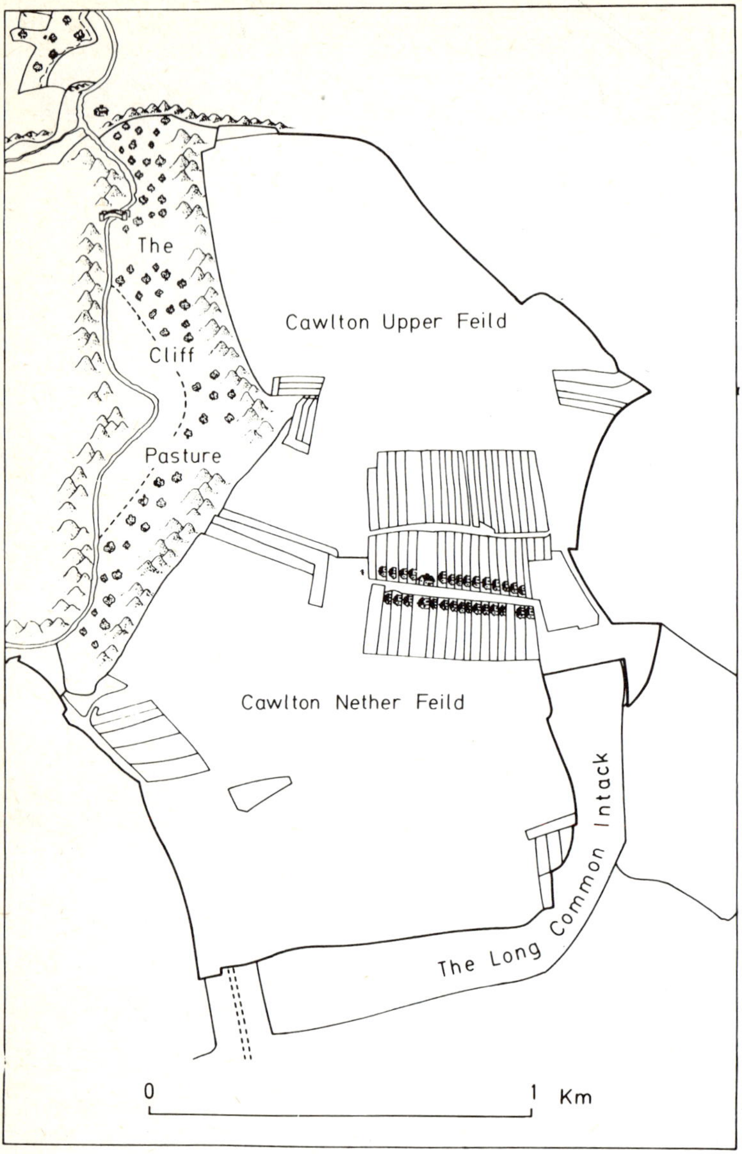

Fig. 9.1 Two field system at Caldon, 1631. Based on Portland MSS, Duke of Newcastle's estate book, 1631 map

arable involves all the regions distinguished by these writers (Fig. 9.3). The overall distribution set against the diversity of soils, relief and climate is anomalous. The ratio of cultivated to uncultivated land varied from region to region, being lowest in the gritstone moorlands, but in all regions at least some of the arable land was farmed on an open field

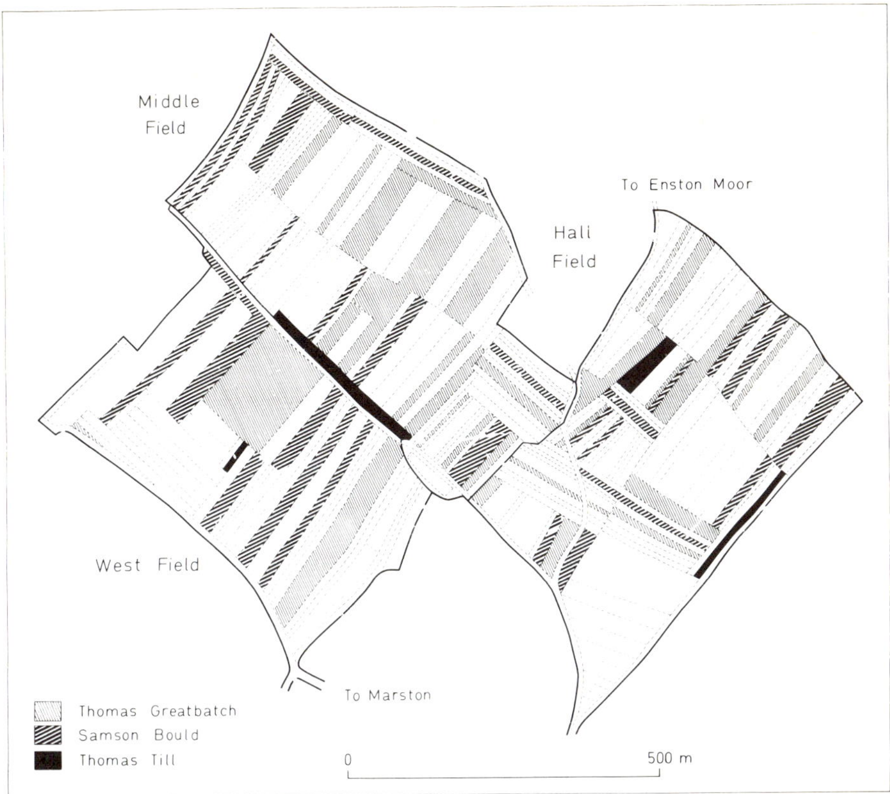

Fig. 9.2 Three field system at Marston, near Stafford, 1722. Based on SRO, Giffard MSS, D590/349, 369c

system. The details of the system and other aspects of the Staffordshire open field husbandry require further comment.

Terminology

'Field', as used by Gray, referred mainly to an administrative subdivision of arable land. The fundamental land unit was the bundle of strips or selions, sometimes lying as ridges (Plot, 1686), and constituting the furlong, shott or platt, separated by access tracks from other such units. Across the head of the ridges ran the headlands on which the plough team turned. Ridges not running the whole length of the field were termed butts, and were sometimes separated by grassy baulks. Where strips ran in ridges, deep furrows assisted drainage. In 1537 at Wolverhampton these drainage ditches continued across the headlands and carried away so much water that a nearby meadow was damaged.[7] The

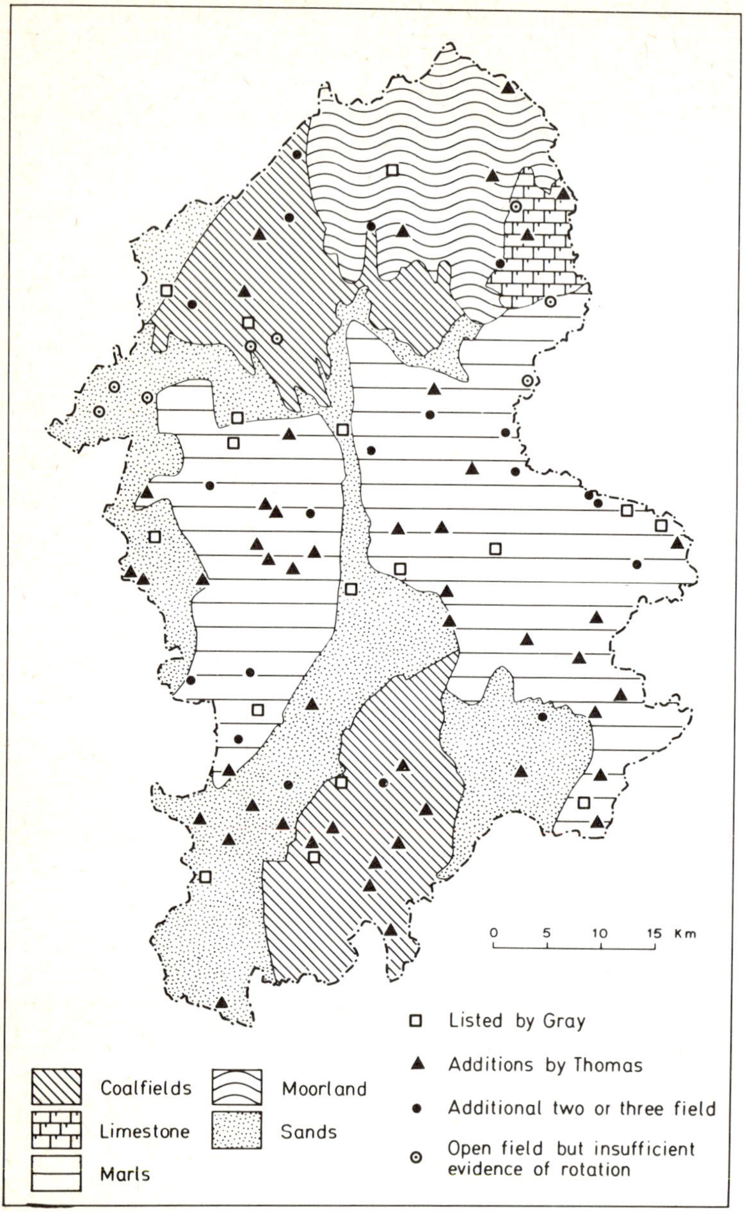

Fig. 9.3 Open field arable with two or three course rotation in Staffordshire

fencing of groups of furlongs and the application of a rotation to such groups gave them a permanency and promoted their development into specific fields, such as north or south field. The strips, however, were often identified merely by reference to the furlong and the whole area of arable land was known as 'the field' without further subdivision. A Trentham charter of 1330 illustrates this point: 'a messuage in Hanchurch held in bondage and four acres of arable land . . . in the field of Hanchurch lying in Colnefurdying'.[8] There is no mention of any division apart from the furlong. In counties such as Berkshire were open field agriculture still existed at the time of the first OS maps, 'the field' was the normal descriptive term. This twofold use of the term 'field' has led to some confusion. It was used for the whole of the open arable belonging to a village community, and for a division of that arable land, consisting of a group of furlongs, for rotation purposes.

The three shift rotation could be employed on open arable land, with common grazing rights attached, without the subdivisions for the rotation being named. This is exemplified at Stafford where Foregate, Marston and Port Fields were enclosed in 1807 under an Act of 1800.[9] The preamble to the Act leaves no doubt that a three shift rotation was being practised, but the fact that three fields are named is deceptive, as each was itself divided into three. Common was permitted over one-third of a field from harvest time until 12 November of the following year, and on the other two-thirds until 13 February. This is the normal three course of fallow, winter-sown corn, and spring-sown corn. But for the evidence of the enclosure Act, the three fields named would generally be interpreted as the 'three fields' constituting one functional unit.

On the other hand, the two or three shift rotation was applied to enclosed land. Heathcot Grange, belonging to the Augustinian Priory of Ranton, had at the time of the dissolution 32.5 acres of pasture and 65 acres of arable.[10] This ratio suggests a three shift rotation but no open arable was involved. The two or three field system is taken generally to imply subdivided open field arable with a two or three shift rotation. It follows that evidence of a three shift rotation is not in itself evidence of this system. Similarly open subdivided arable need not imply the existence of the two or three field system, for the infield in the infield–outfield system (according to Gray the Celtic system) was subdivided. The existence of the two or three field system can only be established with evidence of open subdivided arable of considerable acreage together with common grazing on the fallows, and the two or three shift rotation (Thirsk, 1964). Such evidence is provided for example by the claims to common, as land was separated from the open field.

There are several problems of terminology and interpretation (Baker, 1969). Even the degree of 'openness' is controversial, since in some areas numerous small permanent enclosures, each subdivided into strips, could be grouped for a three course rotation (Yates, 1961; Baker, 1966). There is the additional question of the degree to which the open field arable with a three shift rotation dominated the husbandry of an individual township or vill. In the two or three field system, as described by Gray, the major part of the land within the township formed a functional entity: elsewhere only a small part of the land was organised in this manner.

In considering the distribution of open field arable in Staffordshire, the evidence for a two or three shift rotation in some of the townships listed by Thomas is meagre or lacking, despite the existence of open field, although Plot's comment on the widespread usage of the three shift rotation should be remembered in this context. With these reservations, those townships within which some arable land lay open, subdivided into strips, subject to a two or three shift rotation, and open to common grazing during the fallow period, have been mapped (Fig. 9.3). This recording does not imply that such lands dominated or that such a husbandry system was preponderant in the township.

Two shift rotation on open arable land

Many of the places, including Stafford, to which reference is made in claims for pasture over fallow and stubble, are in the central lowlands of the county, where Pitt's 'strong soils' could sustain the high demands of the three shift rotation. Conversely, it would be expected that the less demanding two shift rotation, with half the ploughland rested each year, would be present on poorer soils (Gray, 1915). Although there are extensive areas of poor soils in Staffordshire, evidence for the two shift rotation is meagre. Even on some of the poor soils derived from the Bunter Sandstone outcrops, the three shift rotation was practised. Thus, in an extent of 1349 for Bentley, a low valuation was quoted 'because the land is very poor and sandy, and a third part lies fallow every year and in common'.[11] Without doubt this is the three shift rotation on open arable. The best example of the two shift rotation on open arable is shown by a 1631 map of Caldon in the northeastern uplands of the county, where altitude and thinness of soil might have inhibited the introduction of the three shift system.

The meagre evidence for the two shift rotation in the county is puzzling. A possible explanation, although only conjectural, is to be seen in the ample rough grazing and good meadows available throughout the county. The existence of these resources is undeniable and they would have made it possible to carry a large head of cattle to a given acreage of arable land, and, by heavy dunging, to apply the three shift rotation to mediocre soils (Pitt, 1794). Paradoxically, the abundance of grazing in upland areas is also reported as a reason for the abandonment of arable land. At Alverton demesne lands were uncultivated in 1339 due to their poor and marshy nature and because 'the herbage is worth nothing by the year on account of the abundance of pasture in those parts'.[12]

Importance of the two or three field system

Evidence for open field with the two or three shift rotation rarely establishes the degree of dominance of the system within the vill. In the two or three field system the land of the vill was arable, meadow, or waste, all common according to season. Furthermore, the system was predominantly associated with grain production and the arable lands formed a large part of the total area. A Staffordshire example of this emphasis on the open field arable as a facet of the two or three field system is the manor of Crakemarsh in Uttoxeter, where a plea roll of 1469 reads 'the place where it [trespass by cattle] was alleged to have taken place was a great field containing 500 acres and parcel of the manor of Crakemersshe and the said field from time out of memory had been without any fence between it and the fields of the vill and the manor of Uttoxeter'.[13] The open fields at Caldon were of comparable size, being over 400 acres (Fig. 9.1).

In the two and three field system only the lands of the demesne lord and some of the freeholders might lie outside the area subject to common rights, although there are many examples of demesne land lying within the system. In 1337 an extent of Madeley showed 180 acres of demesne, of which 120 were sown each year while the other 60 acres were fallow and in common.[14] This implies a three shift rotation, with the demesne contiguous to or intermixed with the open field arable of the vill. In other Staffordshire examples of vills with open field arable, such arable occupied only part of the cultivated land. At Marston in 1722, the 150 acres of open field formed only a small part, approximately one-sixth, of the total area of the vill.[15] The three fields, Middle, West and Hall Fields, were divided between ten farmers all of whom held other land beside that in open field. All had land in each of the three fields although none had the same amount in each, an unequal distribution which would in itself have made for difficulty if the farmers were dependent solely on open field land. These features are demonstrated from a redrawn version of the original map, showing only field divisions and the land of three farmers, Sampson Bould, Thomas Greatbatch and Thomas Till (Fig. 9.2), and from the distribution of the holdings of each of these farmers within Marston (Table 9.1).

Table 9.1 Distribution of sample holdings in Marston (acres)

			Distribution among open fields		
Tenant	*Total holding*	*Total in open field*	*Middle field*	*West field*	*Hall field*
Sampson Bould	118.5	19.3	6.0	8.0	5.3
Thomas Greatbatch	134.5	30.0	12.8	6.2	11.0
Thomas Till	4.8	1.5	0.8	0.1	0.6

Source: SRO, D590/349, 369c, Giffard MSS.

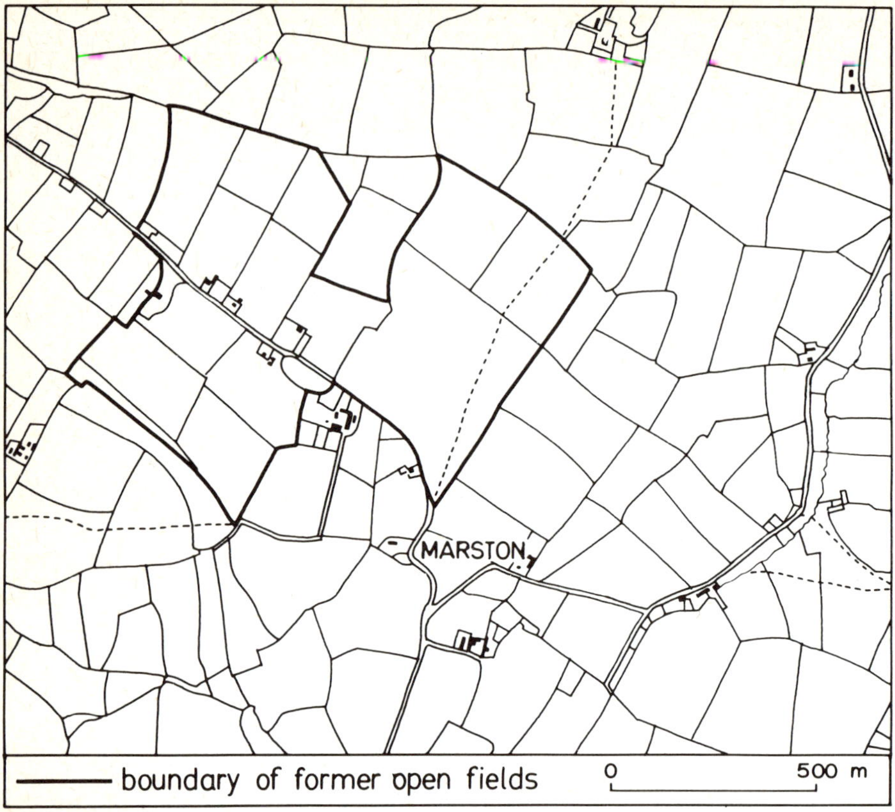

Fig. 9.4 Modern field pattern at Marston. Based on 6-inch OS sheet SJ 92 NW (*Crown copyright reserved*)

As Thomas Greatbatch had two fields, both called Westfield Piece, of 8 and 13 acres, presumably removed from the open field, it might be argued that enclosure had begun and that the small area of open field in Marston in 1722 was a result of this process. However, the compactness of the three fields does not support this interpretation. Furthermore, the field boundaries on the 6 inch OS map for the area of the open field and its surrounds (Fig. 9.4) do not reveal evidence for any substantial enclosure from open field. The irregular field boundaries, particularly to the north of the open field area imply enclosure from the waste. Only in the southwest is there some suggestion in the field boundaries of early enclosure of open field. The later enclosure of that area still open field in 1722 is revealed by the regularity of the field boundaries. Gray argued that the two and three field system was characterised by the open arable situated in two or three fields, the right to common, the even division of each peasant's land among the fields, and the bulk of each peasant holding being open field. At Marston, the bulk of each holding lay outside the open field, and the distribution of each holding across the three fields was uneven. Marston, therefore, cannot be regarded as an example of this system, although Gray himself commented on one of his examples of a three field system, Tutbury, as being of small acreage and having few tenants.

Small open field acreage

The explanation for this lesser role of the open field arable within a vill may be sought perhaps in the wooded nature of the county to which Gray made reference and which has been noted in connection with the meagre evidence for the two shift rotation. Staffordshire was still well wooded in Domesday times (Wheatley, 1954; Midgley, 1958) and there are many references to pannage. Indeed pigs figure in a remarkable way in the place names of the county: Kingswinford, Swinfen, Swynnerton, and Swinshead (now Swinchurch). Many of the first post-conquest documents of any relevance to agriculture deal with pigs: in 1156, 140 pigs were bought; in 1166 payment was made for the transport of pigs to Chester; and in 1170, 100 pigs were sent to Ireland for the campaign of that year.[16] At the same time, the woodland and waste which provided the pannage was vigorously cleared and many fines were paid for assarting. In 1130, for example, the men of Chillington and Swinford paid one mark for assarting without permission, and the forest offences for which so many places in Staffordshire were fined in 1166 probably included assarting. In 1170, a fine of £109 4*s* 10*d* was laid on the whole of Staffordshire for encroachment on forest lands.[17] The late twelfth century was clearly a period when considerable extension of farming took place at the expense of forests and wastelands. In Yorkshire, where owing to William I's harrowing the amount of waste land was very great, such new or reclaimed land was added to the open fields and divided into strips (Bishop, 1935).

In certain other counties, assarts were made and held individually without becoming part of the common arable land, a practice giving rise to a patchwork pattern of fields and lanes. Staffordshire has examples of both processes. The second process would lead to the open field land forming only a proportion of the arable lands of the vill. Early evidence of clearance by individuals is given in a charter of the Bishop of Coventry of about 1150; two assarts at Broom Hall in Brewood are specified as those of Siwardus Sutor and Ailric Berley.[18] The smallness of the assarts recorded in late thirteenth-century pleas of the forest for Kinver and Cannock may also be interpreted as evidence for individual clearance.[19] Many were between one-half and two acres, which suggests individual rather than communal effort. Documents dealing with the exchange of assarted lands in which the assarts are treated as units also suggest that such lands had not been merged with open fields. At the end of the twelfth century, Bernard de Francelega granted Nicholas, son of William the Cook, 'an assart which lies between Ludebroc and the assart of Walter the Deacon of Womborn', and about 1211 Isabel de Pattingham gave to the nuns of Brewood an assart in Chillington in exchange for half a virgate in Pattingham.[20] Such evidence, meagre though it is, points to individually held assarts, probably around the original open field of the community.

Later the evidence becomes fuller and more definite. At Sedgley in 1272 the rents of assarted tenants were given separately from those of customary tenants, and there are marked differences between the labour dues.[21] There can be little doubt that the newly won land had not been added to the open fields. In Tunstall, 25 acres of land surrendered to the lord of the manor in 1326 included 'rudinge Willi doune-in-the-tourne, Solinhurst rudinge, Grene delves and two small rudings lying near Chaderleyes delves'.[22] 'Rudinge' is cognate with riddance and means land cleared of trees. There are no furlong names, as is usual in areas of two or three open fields, and it is clear that the 25 acres lay in small separate fields: the assarts had been kept separate and not merged with any area of open field. An extent of King's Bromley in 1330 describes the rents from forty-six sokemen, each holding half or a whole virgate.[23] Those rents are followed by those of twenty-nine 'foreign tenants' who held land newly taken from the waste. Their rents varied from ¼*d* to 1*s* 8¼*d* and there is little likelihood that they were part of a community associated with a three field system.

This practice of holding assarted lands separate was not uniform throughout the county. In some areas the assarts were divided into strips and formed part of the open arable fields. An instance is afforded by a grant of land in Hanford made to Trentham Priory about 1280.[24] Included in the grant were 3½ acres in the field called Le Cruftes: one acre lay between the land of John, son of Agnes, and the land of John, son of Richard; one acre between the King's highway and the land of William Griffin; half an acre lay by the land of John, son of John; and one acre lay by the ditch by the assart of Achard. The grant leaves no doubt that there was at Hanford a large area of open arable, since four furlongs are named: Longfurlong, Piletfurlong, Drifurlong and Scyrewell le furlong. Le Cruftes and the assart of Achard were presumably enclosures from the waste on the periphery of this open arable, but Le Cruftes, at least, was divided into strips. There is no indication as to how this land was organised in a rotation system.

Another thirteenth-century charter mentions 'the ploughlands of the old assarts of Hanchurch', and this also suggests assarted land belonging to the community and forming part of its open arable land. Again, there is no evidence as to how these lands were organised for rotation.[25] More detail on this latter point is contained in a thirteenth-century land grant in Field in the parish of Leigh to the Cistercian Abbey of Dieulacres.[26] The selions and headlands lay in two fields, Middle Field and Standich Field, and in named furlongs which appear to have been part of a third. The presence of headlands, furlongs and selions indicates the existence of open field arable; the name Middle Field and the distribution of the lands in the grant show that the arable was grouped in a three field system. Mixed with the furlongs of the third field, but similarly divided into strips and headlands, was 'Le Haut Assart'. Furthermore four selions in 'Le Guys', one of the named furlongs, are described as held as new assarts. There can be little doubt that the assarts had been incorporated into open arable with the application of the three shift rotation. What governed an emphasis on incorporation or non-incorporation is difficult to establish. Time was an important factor since, as the open field system decayed, any newly asserted land would be less liable to incorporation.

Arable land outside the two or three field system

The origins of open field with a two or three shift rotation are still obscure but Meitzen's suggestion that it was introduced into England at the time of the Anglo-Saxon colonisation is no longer widely held (Meitzen, 1895). There is now a body of evidence showing that the spread of the system took place in medieval times (Schwarz, 1966). Staffordshire was thinly populated in the eleventh century, as shown by the Domesday returns, the presence of large areas of woodland and the small number of hundreds. Few large estates existed in the county and therefore the organisation necessary for the introduction of a more advanced agriculture was lacking (Hilton, 1970). Even in the relatively well populated part of the county, many of the Domesday manors were small. Many lacked a demesne and at others the population was so low that nothing approaching Gray's two or three field system could have existed. In the north of Pirehill Hundred, for example, of the eight manors held by Richard the Forester only two, Whitmore and Clayton, had demesne (Midgley, 1958). The demesne at Whitmore had only one plough and the recorded population was three villeins and two bordars; at Clayton, with half a plough in demesne, there were four villeins and six bordars.

In general evidence for a lack of manorial organisation can be found in the existence of a high number of freemen or men holding land by some form of non-servile tenure. Their presence suggests that some holdings or some arable land were not integrated fully into the open field system.

For Staffordshire, few freeholders are listed in the Domesday returns, although it is noteworthy that in Totmanslow Hundred, with its moorland, and in northern Pirehill Hundred Anglo-Saxons were still holding small estates in 1086.

However, there is evidence from the Burton Abbey surveys of the late eleventh century that the villein class in the Domesday survey included some men who before the conquest had been freemen, although holding lands by non-military tenure (Round, 1906; Bridgeman, 1917). The land in the majority of the Burton manors was divided between the inland, probably consolidated demesne, and the geldable land. The latter was shared between two groups: those who held for rent and those who held for work. At Leigh, 42 bovates were held for rent and 24 for work; at Branston 15 for rent and 11 for work; and at Stretton, 32 bovates for work and 32 for rent, indicating that a considerable group of rent-paying farmers existed on the abbey lands. Furthermore, although this group (the censarii) was distinguished from that of the work-rent farmers (the villani), among the rent-payers there was a great range in the size of holding. At Stretton, for example, the rent-payers had holdings varying from two to eight bovates of geldable land. Similarly, the rent-payers held widely varying shares of the inland. Only one freeman was recorded on all the manors of the abbey in 1086, so that these minor rent-paying tenants must have been descendants in many cases of the men described in 1086 as villeins (Midgley, 1958).

Villein in the Domesday survey covers a range of status and a variety of tenures. Vinogradoff (1892, p. 210), discussing the increase of freeholders in the thirteenth century, considered that 'it is at least probable that the thirteenth-century freeholders were originally customary freeholders entered as villeins in Domesday and rising to freedom again in spite of the influence of feudalism'. Stephenson (1943, p. 245) also refers to the legal debasement which resulted from the establishment of the rule that 'a rustic unless he could offer an acceptable proof to the contrary' was an unfree villein. Many of the rent-paying farmers on the Burton Abbey estates were possibly in this category. Certainly in thirteenth- and fourteenth-century records large numbers of freeholders appear. In 1305, for example, when Joan, widow of Robert de Huggeford, lord of Hilderstone, sued her son William for her dower in Hilderstone, she had also to sue some fifty-four other individuals. Since villein land belonged to the land of the manor, the fifty-four must all have held land freely. There are similar cases from the same period, as at Wolverhampton with twenty-six people involved in 1299.[27] The holdings of these men must have been small, as borne out by the inquisitions post-mortem. At Weston-under-Lizard in 1305 there were six free tenants, four of whom each held half a virgate, William Bagot held a nook, and Richard le Smyth nine acres. William Bagot also held half a virgate at

Newton in Blithfield where there were four other freeholders.[28] Hilton (1970, p. 9), after studying a variety of records, came to the conclusion 'that certainly more than half of the peasant families of early fourteenth-century Staffordshire were tenants of freeholdings'. Some of these freeholdings were undoubtedly newly formed as the result of clearance, but it is contended that others were of considerable antiquity and, in this thinly populated county, never part of an open field system.

Application of the three shift rotation to lands outside the open field

Lands lying outside the open field could be subject, nevertheless, to the three shift rotation. Evidence of the rotation alone is insufficient to indicate the presence of open field. Separate demesne lands are an instance where the three shift rotation could be employed and are usually distinguished in the records by lack of reference to common rights. Numerous fields could be grouped for rotation, however, and an example of this is given in the survey made in 1297–98 of the manors of the Bishop of Coventry and Lichfield.[29] Thus the fields in Brewood were in three groups for purposes of rotation. The first group comprised 46 acres in Heyruding, 4 acres in two furlongs in Henheth, 56 acres in Prestesruding, 5 acres in Kilmondesruding and 3 acres in Undereruding; the second group 38 acres in Michelenotehurst, 42 acres in Halenotehurst, 33 acres in Levencheruding and 50 acres in Hatton; and the third group 85 acres with 10 acres fallow in Bentley Field and 33 acres in Steynashruding and Priesbrok. The rough equality of area and the description of each grouping as *saisona* are evidence for the three shift rotation, but the field names render it unlikely that a two or three field system is involved. Indeed, the ruding names suggest a series of assarts around perhaps an original core (Bentley Field). Nevertheless, in 1358 it was stated that the lands in Brewood lay in common every third year:[30] either the assarted fields had been reorganised in the interval into a three field system or common right did attach to all or most of the various fields named. Gray cited Brewood as an example of the three field system.

A similar grouping of numerous fields to apply a three shift rotation is revealed in a 1307–08 survey of lands belonging to Burton Abbey.[31] At Stretton in 1308, 76 acres were under wheat, 58 under oats and 10 under peas, while 83 lay fallow. The field and furlong names noted show that more than three open fields were involved. The three shift rotation could also be applied within the bounds of the various granges established in north Staffordshire by the Cistercians and Augustinians. The example of Heathcot Grange in Ranton has been discussed. Similarly, Rushton Grange of Hulton Abbey was made up of 216 acres of which

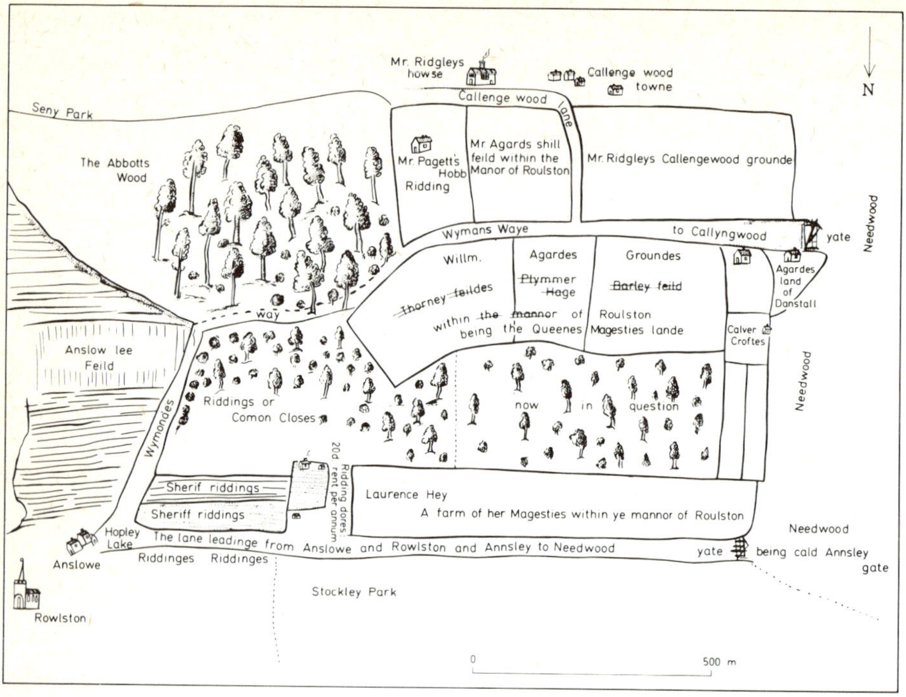

Fig. 9.5 Lands near Anslow, on edge of Needwood Forest, 1596. Based on PRO, MPC/241

166 were pasture and 24 meadow. There were 16 acres of arable, all in Wallfield, together with 8 acres of pasture. These acreages suggest that the nucleus of the grange was a 24-acre field, used as arable on a three shift rotation, and surrounded by pasture closes.[32] There is also tentative evidence for a two shift rotation applied to enclosed land at 'Sheriff riddings' as shown on a 1596 map of Anslow and Needwood (Fig. 9.5).

Increase in field numbers

Any increase in the number of fields into which the furlongs were grouped entailed a reduction in the period in which the land lay fallow. The success of such changes in the rotation must have been conditioned by the fertility of the soil and the availability of dung. According to Gray the grouping into four or more fields was the response of some open field vills to the growth of convertible husbandry.

In Netherstone Hall, a manor in Shenstone, four open fields were recorded in a 1567 survey, while in Mosshouse manor in the same parish, there appear to have been five fields in 1574.[33] At Acton Trussel there were four open fields in 1574, while four existed at Wednesfield in 1652.[34] Brewood, in which numerous fields are named in 1297–98 and which Gray regarded as an example of the three field system, had four fields in the later seventeenth century (Pugh, 1959).

Infield–outfield and temporary breaks

At the other extreme, some of the poorer soils never became part of even a biennial cropping system. Where the area of fertile soil was limited, only a small part of the lands belonging to the vill, the infield, was continuously cropped. The major part, the outfield, was left as pasture, occasionally cropped until the yield fell below the minimum return, and then allowed to revert. Gray found evidence of this system mainly in the north and west of England, but there are traces elsewhere. The existence of this practice is suggested at Baswich, where the terrain was described in 1298 as too poor for rotation, one of the fields being named Hunger Hill.[35]

There are various references to temporary breaks made in the areas of outfield and other rough grazing. In a 1306 dispute over commoning in the manor of Clifton Campville, it was stated that 'Geoffrey, the lord of the manor had frequently ploughed and sown parts of the waste [Howden Heath] in different places and the jury found against him'.[36] In 1428, several pieces of land in Wednesfield had 'break' names indicating temporary inclosure: Newbreche, Bembreche, and Knollebreche.[37] Plot (p. 341) refers to such soils as those around Swynnerton 'which will naturally bear nothing but rye, french wheat, and oats, nor these unless it be well mucked and then not above three years together but it must rest again'. He continued by describing the crop sequence on the heaths, which were seldom enclosed, any enclosure for cultivation being temporary. After a period of grazing temporary enclosures were made, the ground broken-up, limed and then planted with a rotation of rye, barley, peas and oats, before relapsing to common grazing. The court rolls of Tunstall for 1610 show a different but related use on poor soils. Licence was given to Thomas Danyell to lease for twenty-one years a pasture called Brownhills and another called the Hill in Burslem, containing 3 acres, and to dig clay there 'filling up the pit after him'. Danyell covenanted that he would not plough Brownhills after fourteen years unless he 'do well sufficiently muck the same and then to till it but for three years'.[38]

Conclusions

After studying the evidence relating to Staffordshire field systems, the initial reaction is renewed admiration for Gray's formative work. The field systems of Staffordshire were undoubtedly not typical of those of the English Midlands. Away from the major valleys, the common open

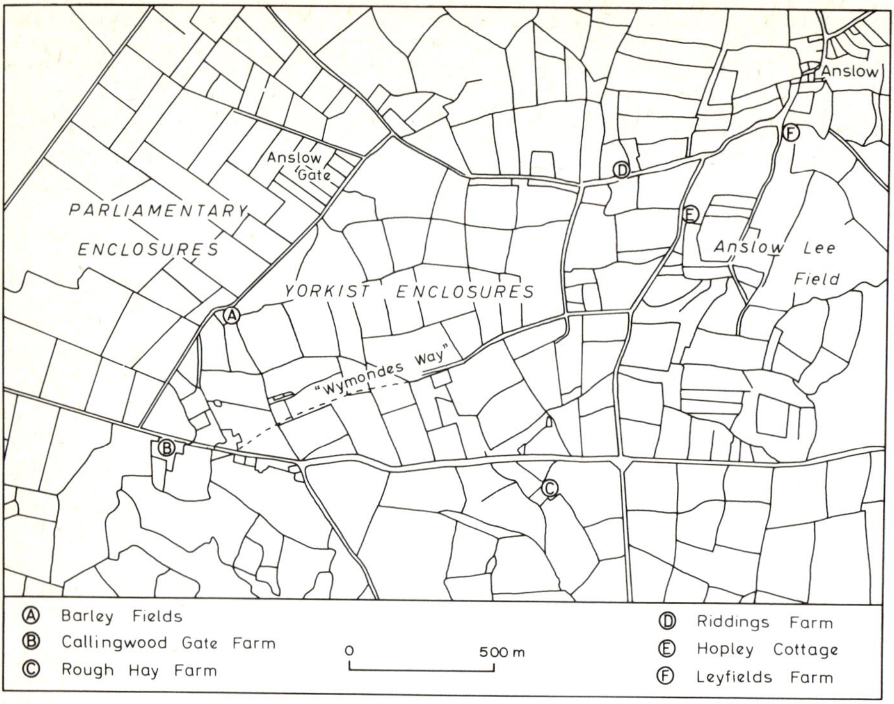

Fig. 9.6 Modern field pattern near Anslow, on the edge of Needwood Forest. Based on 6-inch OS sheets SK 12 NE and SK 22 NW (*Crown copyright reserved*)

field in many townships formed only part of the arable land. This, as Gray suggested, was due in part to the character of the county, which was wooded, thinly populated and in many ways a frontier land. The newly assarted lands could be grouped in a three shift rotation and, then in some instances, recast as an open field system. This appears to have been the sequence at Brewood, when the manorial lord was the Bishop of Coventry and Lichfield. In other examples the land was kept separate in freeholdings. Such land may have been subject to a three shift rotation, as was the case in some church lands.

On poorer soils less demanding rotations were employed. The similarities are with Cheshire rather than with the Midlands, and the resemblance between the agricultural system at Brewood and Wynbunbury is striking, particularly as the latter manor was also held by the Bishop of Coventry and Lichfield (Sylvester, 1958). Generally, Sylvester considered Cheshire to be a county characterised by the one field system, the single field of a township being the town field, and this was likened tentatively to the Celtic field system (Sylvester, 1949, 1956). Town fields are similarly in evidence in Staffordshire, as at Winnington near Mucclestone, mentioned in a 1577 survey as the town field or common field, and at Bloor in 1722.[39] It is not clear, however, whether these town fields were infields. At Stafford, the unit fields (Foregate, Marston, and Port Fields) were each cultivated with a three shift rotation. They differ from the two and three field system of the Midlands only in scale, whereas the infield in the infield–outfield system was cultivated as a whole each year (Dodgshon, 1973).

The landscape created by assarting, where such lands were not recast as part of an open field system, is shown by a 1596 map of Anslow and Needwood (Fig. 9.5), drawn in relation to a dispute over certain ridings. This demonstrates the difference between the open field land and the fields assarted in the late fifteenth century. Scattered settlement and irregular fields had been created and the pattern persists today (Fig. 9.6), enhanced by the piecemeal enclosures of Anslow and the parliamentary enclosure of Needwood. One of the landowners noted on the 1596 map is William Agard. Agardsley, on the western side of the forest, was the only cleared land in Needwood (one virgate) recorded in Domesday, in itself too small to constitute an open field system but cleared before the medieval assarting. Held in 1066 by Ulvic, a freeman (Midgley, 1958), it is perhaps representative of the individual, separate farms which were increased in number by medieval clearance but never incorporated into any open field system.

Appendix

Open field townships in Staffordshire

Open field townships with a three shift rotation listed by Gray

Abbots Bromley	Clayton Griffith	Mere End
Barton	Essington	Norbury
Baswich	Heywood	Swynnerton
Bonehill	Hilderstone	Trysull and Seisdon
Bradley	Leek	Tutbury
Brewood	Madeley	

Open field townships with irregular field systems listed by Gray

Rocester	Wooton-under-Weaver

Open field townships listed by Thomas*

A. Evidence mainly from the inquisition into depopulation and Star Chamber records

Adbaston	Cumberford	Perton and Trescott
Biddulph	Drayton Basset	Rushall
Bridgford	Gnosall	Throwley
Cheadle	Loxley	Tunstall
Chillington	Meertown	Wrottesley

B. Early evidence for open field

Ashley
Barton
Bentley
Bradley
Coton
Derrington
Eccleshall†
Essington
Forton
Hammerwich†
Hilderstone
Hinley†
Madeley
Mere End
Norbury
Rocester
Rolleston
Stretton on the Moors
Swynnerton

C. Evidence from manorial court rolls for open field in 1700 but enclosed piecemeal by 1800

Alrewas
Bilston
Codsall
Colton
Darlaston
Hatherton
Pattingham
Pilsall
Rugeley
Shenstone
Trysull
Tutbury
Walsall
Wednesbury
West Bromwich
Whittington
Wolverhampton

D. Evidence from parliamentary Acts for enclosure of open field

Abbots Bromley
Alstonefield
Barton under Needwood
Basford and Cheddleton
Butterton
Castlechurch
Checkley
Cumberford and Whittington
Edingale
Elford
Handsworth
Hixon and Drointon
Longnor
Newcastle
Newton Regis
Pattingham
Stafford
Stone
Trysull and Seisdon
West Bromwich†
Whitgreave

Open field townships with two or three shift rotation, supplementary evidence

Anslow[5]
Billington[7]
Bloor[8]
Bloxwich[6]
Bushbury[7]
Caldon[5]
Cheddleton[6]
Crakemarsh[7]
Eccleshall[7]
Fauld and Coton[1]
Grindon[5]
Hanchurch[2]
Hanford[2]
Keele[8]
Kinvaston[1]
Lapley[7]
Leigh[2]
Marston[4]
Milwich[7]
Norton-in-the-Moors[3]
Oakley Elleford[7]
Streethay[1]
Uttoxeter[7]
Winnington[8]

* Evidence for a three shift rotation is not given for all townships.
† Probable examples.
1. Account book; 2. chartulary; 3. court rolls; 4. Giffard MSS; 5. map; 6. Pitt; 7, plea roll; 8. Sneyd MSS.

Notes

1. G. Wrottesley, ed. (1893) 'Plea roll for 1335', *SHC*, **14**, 39; (1892) 'Plea roll for 1369', *SHC*, 13, 74; (1889) 'Plea roll for 1311', *SHC*, **10**, 10; (1890) 'Plea rolls for 1332, 1333 and 1337', *SHC*, **11**, 20, 38 and 74; W. K. Boyd, ed. (1906) 'Chancery proceedings Elizabeth I, *SHC*, new series, **9**, 219–20.
2. J. C. Wedgwood, ed. (1927) 'Tunstall court rolls', *Trans. N. Staffs. Fld Club*, **62**, 71.
3. W. F. Challinor, (1923) 'The mystery of the Leek town lands', *Trans. N. Staffs. Fld Club*, **58**, 79–84. It is difficult to see, however, how the Domesday entry for Leek can be taken as evidence of common fields: J. M. Wagstaff (1970) 'The economy of Dieulacres Abbey 1214–1539', *N. Staffs. J. Fld Stud.*, **10**, 83–101.
4. BM, Add. MS. 31,917.
5. SRO, William Salt Library, SMS 431 and HM 27/2.
6. Sneyd MSS, Keele University, Box S.65. The three fields were named as Blast Field, Smith Field and Hall Field in 1729.
7. W. K. Boyd, ed. (1910) 'Court of Star Chamber proceedings', *SHC*, 1910, pp. 69–70.
8. F. Parker, ed. (1890) 'Trentham Chartulary', *SHC*, **11**, 327.
9. SRO, Q/RDC 16, Act for enclosing lands in or adjoining the parish of St Mary Stafford (39 and 40 George III *c*. 70, private Act).
10. PRO, SC6/Hen.VIII/3352.
11. J. C. Wedgwood, ed. (1913) 'Inquisition post mortem', *SHC*, 1913, p. 129.
12. *Ibid.*
13. G. Wrottesley, ed. (1901) 'Plea rolls', *SHC*, new series, **4**, 154.
14. J. C. Wedgwood, ed. (1913) 'Inquisition post mortem', *SHC*, 1913, p. 61.
15. SRO, D590/349, 369c, Giffard MSS.
16. R. W. Eyton, ed. (1880 and 1881) 'Pipe rolls', *SHC*, **1**, 23, 46, 63 and 135; and **2**, 31.
17. R. W. Eyton, ed. (1880) 'Pipe rolls', *SHC*, **1**, 2, 48 and 61.
18. G. Wrottesley, ed. (1882) 'Episcopal charters', *SHC*, 3, 182.
19. G. Wrottesley, ed. (1884) 'Forest pleas', *SHC*, **5**, 144.
20. G. Wrottesley, ed. (1882) 'Staffordshire chartulary, Wombourne and Basset charters', *SHC*, 3, 219 and 193.
21. J. C. Wedgwood, ed. (1911) 'Inquisition post mortem', *SHC*, 1911, p. 154.
22. J. C. Wedgwood, ed. (1924) 'Tunstall court rolls', *Trans. N. Staffs. Fld Club*, **59**, 42.
23. J. C. Wedgwood, ed. (1911) 'Inquisition post mortem', *SHC*, 1911, p. 257.
24. F. Parker, ed. (1890) 'Trentham chartulary', *SHC*, **11**, 306.
25. *Ibid.*, p. 325.
26. G. Wrottesley, ed. (1906) 'Chartulary of Dieulacres Abbey', *SHC*, new series, **9**, 319–20.
27. G. Wrottesley, ed. (1886) 'Plea rolls', *SHC*, **7**, 142 and 52.
28. J. C. Wedgwood, ed. (1911) 'Inquisition post mortem', *SHC*, 1911, p. 282.
29. SRO, D1734/J2268.
30. BM, Add. MS, 6165ff, 95–104.
31. SRO, D1734/2/3/112a.

32. PRO, SC6/Hen.VIII/3352.
33. BM, Add. MS, 25297.
34. BM, Add. MS, 24822.
35. SRO, D1734/J2268.
36. G. Wrottesley, ed. (1886) 'Plea rolls', *SHC*, **7**, 169.
37. G. Wrottesley, ed. (1896) 'Plea rolls', *SHC*, **17**, 121.
38. J. C. Wedgwood, ed. (1927) 'Tunstall court rolls', *Trans. N. Staffs. Fld Club*, 62, 78.
39. BM, Add. MS, 25,297; Sneyd MSS, Keele University, Box 31.

References

BAKER, A. R. H. (1966) 'Field systems in the vale of Holmesdale', *Agric. Hist. Rev.*, **14**, 1–24.

BAKER, A. R. H. (1969) ' Some terminological problems in studies of British field systems', *Agric. Hist. Rev.*, **17**, 136–40.

BISHOP, T. A. M. (1935) 'Assarting and the growth of open fields', *Econ. Hist. Rev.*, **6**, 13–29.

BRIDGEMAN, C. G. O. (1917) 'Burton Abbey surveys', *SHC*, 1916, pp. 209–300.

DODGSHON, R. A. (1973) 'Infield–outfield in Scotland', *Trans. Inst. Br. Geogr.*, **59**, 1–24.

EVERSHED, H. (1869) 'The agriculture of Staffordshire', *J. R. Agric. Soc.*, new series, **5**, 293–317.

GRAY, H. L. (1915) *English Field Systems*, Harvard University Press.

HILTON, R. H. (1970) 'Lord and peasant in Staffordshire in the Middle Ages', *N. Staffs. J. Fld Stud.*, **10**, 1–20.

MEITZEN, A. (1895) *Siedling und Agrarwesen der Westgermanen und Ostergermanen, der Kelten, Romer, Finnen und Slawen*, Berlin.

MIDGLEY, L. M., ed. (1958): see *Victoria History*, vol. 4.

PITT, W. (1794 and 1808) *General View of the Agriculture of Staffordshire*, London.

PLOT, R. (1686) *The Natural History of Staffordshire*, Oxford.

PUGH, R. B., ed. (1959): see *Victoria History*, vol. 5.

ROUND, J. H. (1906) 'Burton Abbey surveys', *SHC*, new series, **9**, 271–89.

SCHWARZ, G. (1966) *Allgemeine Siedlungsgeographie*, Berlin.

STEPHENSON, C. (1943) 'Feudalism and its antecedents in England', *Am. Hist. Rev.*, **48**, 245–65.

SYLVESTER, D. (1949) 'Rural settlement in Cheshire', *Trans. Hist. Soc. Lancs. Ches.*, **101**, 1–37.

SYLVESTER, D. (1956) 'The open fields of Cheshire', *Trans. Hist. Soc. Lancs. Ches.*, **108**, 1–33.

SYLVESTER, D. (1958) 'A note of medieval three course arable systems in Cheshire', *Trans. Hist. Soc. Lancs. Ches.*, **110**, 183–6.

THIRSK, J. (1964) 'The common fields', *Past and Present*, **29**, 3–25.

THOMAS, K. R. (1933) 'The enclosure of open field and commons in Staffordshire', *SHC*, 1931, pp. 60–9.

Victoria History of the County of Staffordshire, Dawsons Pall Mall and Oxford University Press.
Vol. 4, ed. L. M. Midgley (1958).
Vol. 5, ed. R. B. Pugh (1959).

VINOGRADOFF, P. (1892) *Villeinage in England*, Oxford.

WHEATLEY, P. (1954) 'Staffordshire', in H. C. Darby and I. B. Terrett, eds, *The Domesday Geography of Midland England*, Cambridge University Press, pp. 160–214.

YATES, E. M. (1961) 'A study of settlement patterns', *Fld Stud.*, 1, 65–84.

R. A. Butlin

10 Rural change in Northumberland, 1600–1880

The county of Northumberland at the time of the accession of James I and the union of the crowns still bore the hallmarks of a frontier or 'march' area, a condition reflected in many aspects of rural society, economy and settlement. After 1603, however, the rate of rural change accelerated as lawlessness and violence, formerly endemic in the border regions, were gradually eliminated, and structural, tenurial and economic innovations paved the way for revolutionary improvements in agriculture which reached their peak during the nineteenth century. The purpose of this essay is to identify and analyse the causes and effects of rural change in the period between about 1600 and 1880.

Northumberland in the early seventeenth century

Regional contrasts

Distinctive regional contrasts in patterns of land use, settlement and social structure are evident in Northumberland in the early seventeenth century with major differences between the uplands of the west and the coastal plain and adjacent areas in the east. The former, comprising the high moorland and fell areas of the Cheviots and Border Fells and drained by the Allen, Aln, Rede, upper North Tyne and upper Coquet, was described by Camden (1695, p. 487) as a region

> where the country itself is mostly rough and barren, and seems to have harden'd the very carcasses of its inhabitants: whom the neighbouring Scots have render'd yet more hardy, sometimes inuring them to war, and sometimes amicably communicating their customs and ways of living; whence they are become a most warlike people and excellent horsemen.

Within the remote border dales a distinct form of economy had developed, involving small patches of arable and meadowland, frequently in common, and extensive unenclosed pastures for cattle rearing, in combination with a variety of nefarious activities including sheep- and cattle-stealing (Butlin, 1973). Arable land was used intensively for the cultivation of oats, barley and rye, and meadowland for hay. The extensive upland pastures were used only during the summer, when the animals were grazed on the shieling grounds. The other activities of the clans of the border dales received much attention in contemporary accounts – the 1604 border survey, for example, stated that the young and active people of Tynedale continually 'spoiled' or stole in both England and Scotland for lack of a living, attributed to the overpopulation of the region (Sanderson, 1891). Violence and lawlessness were certainly an integral part of everyday life among the kinship groups or clans of the border dales, being a reflection of the inadequacy of the land to support many people, of the insecure condition of the border regions and of 'border tenures', and of the peculiar nature of their social groupings.

The predominant pattern of settlement in the western regions was of dispersed farmsteads and small hamlets, the latter usually inhabited by members of the same clan. These clans were in fact extended families, bearing the same surnames, the most common being Charlton, Rede, Hall, Hedley, Brown, Anderson, Pott, Fletcher, Foster, Dodd and Ridley. The dwellings of the more important members of these families often took the form of 'peles' or 'bastles' – fortified tower houses which occur most frequently in the western border regions of the county. The presence of this type of building appears to have been influenced by the form of settlement (not highly conducive to successful communal defensive tactics, as in the villages of the coastal plain), the frontier location, the availability of stone for building, the declining influence of the border barons, the relative freedom of border tenants because of their occupation, in many cases, of crown land, and the supposed operation of 'tenant-right'. The latter three factors enabled tenants to retain profits which could be employed for building constrution (Ramm *et al.*, 1970). An additional feature of clan life was the operation of both partible inheritance and gavelkind in some of the dales.

Totally different from the border dales and uplands of the west were the regions in the east, centre and south of the county, particularly the coastal plain, the cementstone vales, the low plateaus and the lower Tyne valley, where the environment favoured the development of a balanced form of mixed farming. Villages were common in these areas, and the arable area in the townships was frequently about 700 acres and comprised, for the most part, open, subdivided, commonable fields which occurred mainly in groupings of two, three or four, and between which the holdings of freeholders and customary tenants were evenly distributed (Butlin 1964, 1973). The relatively large area of common pasture in many townships permitted a flexibility of management of common arable, which was not possible in more densely populated counties, and also the keeping of substantial numbers of livestock. The inhabitants of the eastern lowlands appear to have been more peaceable by nature than their western counterparts, but frequent raids, both by the Scots and by raiders from Tynedale and Redesdale, necessitated communal defensive measures, including the driving of livestock into the village at night. Land tenures varied in these regions, but included forms of border tenure. An important feature of the lowland townships was the bondland or 'husbandland', the tenure of which entitled the holder to a share of the common land of the township.

Border politics

Although there were distinct regional variations in land use, society and landscape in Northumberland, it is equally clear that certain political and administrative factors were prevalent throughout the county. The most obvious of these was the frontier or border conditions and the consequent necessity for protection and defence. These needs had been for centuries the responsibility of a small number of powerful feudal families such as the Percies, Dacres, Cliffords and Nevilles. Their power was extensive, and in practice frequently exceeded that of the crown. Their castles were defensive strongholds that 'represented the acme of power and magnificence, and were the sources from which favours and patronage flowed . . . but behind the decorous façade all too often the great lords were bound in a close *mafia* with the upland thieves, and patronised and protected border lawlessness' (James, 1965, p. 7). The independence of these northern magnates led to increased attempts by the crown to control the affairs of the border, notably by the appointment to the wardenship of the marches of men who were not members of these great northern families and who, if they survived the turbulent political intrigues of the late Tudor period, could build up large estates.

The direct influence of the magnates and the newer gentry on change in the rural economy was limited by the necessity to maintain adequate forces for the defence of the border, and they were thus discouraged from enclosing land and from extorting higher fines from their tenants, who sheltered behind the apparent security of border tenures and tenant-right. The border situation, however, merely delayed the wave of reform and change until the seventeenth century, though some indications of future developments were already apparent in the sixteenth century.

The processes of change

Changes in land tenure

One of the most obvious and immediate effects of the union of the crowns in 1603 was the review of tenures relating to border service. The crown, for its part, established a commission in 1604 to survey the border realms, whose duties included a determination of the customs, ownership, and value of crown lands. One fact which emerged from their report is the frequency of the claim by tenants in the border regions to hold by 'tenant-right':

> The Tenantes challenge to holde their Tenements by title of tenante right, payeinge, their rents and doinge their services upon the Border, wch services were to bee at the commande of the Keeper to serve in field on horse or foot; for the defence of the Border land in wch many of th' inhabitants have lost their lives (Sanderson, 1891, p. 51).

Tenant-right comprised 'a special constellation of manorial customs which varied slightly from manor to manor' (Watts, 1970, p. 64), and differed from other forms of customary border tenure in that the tenants assumed that their border service entitled them to favourable customs, including low and fixed rents or fines, and transferability and widow inheritance of the right. The unstable and insecure conditions, particularly of the border dales, had militated formerly against an orderly review and renewal of tenures, and the tenants had benefited, a fact recognised by the commissioners in 1604:

> All the Customary Tennants have anciently held by coppie of Courte roule, payinge to the lord at every alienation one yeres rent for a fine; but by reason the courtes have not there byn orderly kept, manie and most part of the tennants have no coppies, by wch means the Prynce hath lost his fines, but the tennants in generall claime to be Coppie-houlders of inheritance, though in examininge their coppies, the fines in them are very different (Sanderson, 1891, p. 124).

Border tenure became obsolete with the union of the crowns, and James I, and later the northern lords, took immediate action to force tenants to compound for tenures which were less secure for them and more profitable for the landlords. James contended that tenant-right ceased to exist with the death of Queen Elizabeth and therefore could not be accepted as an adequate basis for tenure in the north. His contention was supported by the judges of the Star Chamber in 1625, and other landlords had followed successfully James's advice on the matter: in 1621 the Reads of Harbottle had their tenant-right claim to the use of the summer pastures in Redeshead rejected by the Court of the Exchequer, being forced to compound as tenants-at-will or accept leases (Watts, 1970). Other landlords did not question titles but compounded for large sums of money. Similar changes took place on the Percy estates. Henry Percy (1564–1632), ninth Earl of Northumberland, embarked (partly through the necessity of paying a large fine for his suspected involvement in the Gunpowder Plot) on a policy of estate management and revision of tenancies. Wherever possible he replaced customary tenure with leases, and after 1613, when a ruling was made that copyhold estates were for the life of the lord and tenant, he forced those of his copyhold tenants who held by the custom of Cockermouth to accept leases (Batho, 1962; Watts, 1970). To this end, he issued a general certificate which enabled his agents to convert copyhold to leasehold tenure, and his income showed a marked increase as a result (Batho, 1957).

This change to leasehold, which facilitated enclosure, did not take place without dispute. The tenants at Lesbury for example refused at first to acquiesce and tried to defend and maintain their position as copyholders. Thomas Fotherley, one of the Earl's agents, writing from Newcastle in 1614, reported that

> in those trialls for some coppieholdes in Newham, Lucker and Lesburye, the judges making some haste awaie, because they would not ride upon Sondaies, and that I feared a packt jurie, the matters being of very great consequent to your lordship. I forbare the trialls of them for this present, and besides in mine own hearing two causes concerninge tennants rights passed on their behalf (Northumberland County History Committee (NH), 1895, ii, p. 427).

Such tenurial changes paved the way for profound structural changes, of which the principal was enclosure, manifested in a variety of forms.

Enclosure

Enclosure in Northumberland, as elsewhere, had been continuous since the early Middle Ages, expressly in the form of the engrossment and enclosure of demesne and freehold land, emparkment, and enclosure from the waste. Some depopulating enclosure took place in the second half of the sixteenth century, and continued into the early years of the seventeenth. Examples of townships experiencing this type of enclosure included Hartley, about 1597, Seaton Delaval, about 1597, Tughall, about 1567, Seghill, about 1594, Newham, about 1597, and Little Houghton, about 1598 (Tate, 1942; Butlin, 1967; Kerridge, 1969). At the muster of 1580 men said that much ground had been turned into pastures and the commissioners of 1584 complained of this and of the taking of towns into demesne, though no details are given. The process continued in the seventeenth century: a report of Hetton in 1627 stated that in the 'peaceful time of King James I' a Mr William Carve came from Sleaford in Lincolnshire 'and decayed all ye said farms, Dixons onlie excepted, and took all ye lands that belonged to them into his own hands' (NH, 1935, xiv, p. 229). An account of Brotherwick in 1616 stated that 'Lancelot Ogle hath converted all the arable grounds into pasture'.[1] An interesting aspect of this type of enclosure in Northumberland is that it occurred later than in the Midlands and other parts of England – a reflection, perhaps, of the late abolition of border tenures and the existence of extensive wastes and pastures which, employed in a mixed farming economy, was an 'immunising factor' (Beresford, 1954, p. 371). The contribution of this period to the deserted villages of the county, of which there are many (Beresford and Hurst, 1971), is still uncertain.

In spite of these enclosures the greater part of the county appears to have remained unenclosed at the beginning of the seventeenth century. That is not to say that no efforts had been made to improve the old systems of communal land management. During the late sixteenth and in the seventeenth century, such attempts were made by the division of the lands of many Northumberland townships into halves, thirds or quarters, thereby reducing the area over which a tenant's lands were distributed (Butlin, 1967). The township of Long Houghton, for example, had been divided into two in the late sixteenth century because the location of parcels of land within the township had become intolerable:

> The arable land lyeth for the most part nigh the sea and is donged with the sea wracke, and, because of the greatnes of the said town, the town is now devidit in two partes, for that there were xxvii tenants besyde cotteagers, havynge alwayes and in every place every one tenant one ryge by himsellfe, and so consequently, from ryge to ryge that everie tenant had one rig, then the first did begin to have his ryge for his lot agayne, and so by ryge and ryge it was in every place devidit among them to the great chardge and laboure of every one of the said tenants; although the same partition did give to every tenant like quality of all sortes of lands, yet it was so paynefeule to them and ther cattell that for the most part the said tenants did never manure ther ground threwghly, whereby they did fall in great povertie; and also ther severall ground, called ther oxen pasture . . . was in breaffe tyme over eatying and made bare of fedying.[2]

The township was divided into a north end and a south end. Other townships so divided in the late sixteenth century were Rock and Thirston.

It is peculiar that this process of division should have continued into the seventeenth century, when enclosure became easier. Nevertheless, there are examples of divisions. In 1685 the copyhold lands of the township of South Charlton were divided into two parts. The land of four of the tenants was grouped together on the eastern side of the township and the rest of the land allocated to the remaining customary tenants. It was discovered that the two parts had been very roughly and unequally allocated, and that no survey had been made (NH, 1895, ii, p. 308). At Birling in 1640 the tenants divided the township into three parts, the north, west, and south sides, and apportioned the arable land to each of the sectors. This proved to be unsatisfactory in the longer term; at the end of the century a further division took place, consolidating the tenant holdings which were then enclosed.[3]

The division of townships represented a progressive step, but the advantages of full enclosure were recognised by landowners and their agents. A letter from the Earl of Northumberland's receiver, Robert Delaval, in 1617, points out the advantages of enclosure:

> Where there are 20 tenants, more or less in a town, that occupy in common it follows that they are forced to keep in tillage for their cornfields two-thirds part of the whole town, by which means that part that lies furthest from the situation of their homes, is the poorest land, being by reason it is the furthest off, seldomest manured, and so proves neither kindly for grass nor corn. Therefore to divide these farms, into four severall quarters, would be the best means to cause enclosure . . . and in time may cause tenants to remove their houses to the midst of their several farms, and that part they now have most barren may be thereby as fertile and good as their best. And as leases may expire, by the means of this division, your lordship may keep one part divided or more to demesne, which is then fit for your lordship's use, either to be employed with a store for your necessary provisions for corn grazing or cattle feeding or increase of rent (Batho, 1962, p. li).

With this type of process and the new leaseholds, some of the larger landowners could effect enclosures by unity of possession (Kerridge, 1969) which, as in cases cited above, could cause depopulation.

The major instrument of enclosure in Northumberland in the seventeenth and early eighteenth centuries was, however, enclosure by agreement, usually in the form of enclosure by commission (Kerridge, 1969). Over fifty examples of enclosure by agreement can be cited for the period between 1600 and 1758. The sequence of events leading to and effecting enclosure was similar in many cases – dissatisfaction with the existing state of affairs, proposals for enclosure and the drafting of articles of agreement, petition (where necessary) to the lord of the manor, the appointment of commissioners and surveyors, the allotment and its implementation, and occasionally, ratification by other authorities.

Petitions for enclosure usually list the disadvantages of the existing arrangements; for example, a 1617 petition refers to 'the daylie inconvenience which they [the freeholders and customary tenants] suffer by reason of the intermixture of their lands and tenements which do now lie in common each with others' (NH, 1899, v, p. 212). Almost 100 years later the freeholders and tenants of Rennington were similarly petitioning the Duke of Somerset for enclosure of common lands 'because they were lyeing dispersedly in and throughout the lordship, by reason whereof your petitioners are very much demnified and cannot make any improvement' (NH, 1895, ii, p. 159). The basic themes of such petitions are desire for improvement and the elimination of disputes over land holdings. The progress of enclosure by agreement was sometimes hindered by opposition. A proposed enclosure of the common lands at Tynemouth, for example, was delayed by the opposition of three freeholders, Thomas Dunn, John Bowe and Gilbert Otway, who in 1649 claimed that the action would lead to depopulation.[4]

Commissioners and surveyors were appointed to make and implement awards. The appointment of 'some cunning expert and trusty surveyor of lands'[5] was adjudged to be vital to the operation (though such experts were not always employed), and the later awards refer to well-known local surveyors, such as John Fryer of Newcastle.[6] The number of commissioners or arbitrators appointed by the interested parties varied, but there were usually four involved in the enclosure of large sections of townships. Thus in the case of the Wall enclosure of 1691, the arbitrators were William Errington of Todderidge, Arthur Shaftoe of Gunnerton, John Simpson of Ovingham Hall and Thomas Dobson of Humshaugh, who took upon themselves 'the charge and burden of the said division'.[7] The basis for the allocation of land varied, but the normal practice was to allocate one-twelfth of the land to the lord of the manor for his abstract rights over it, the rest being divided between the freeholders and other tenants. Provision was made in the awards for the erection of hedges, fences and gates, the creation of rights of way and wayleave, the phasing out of pre-existing rights of common, and for the defrayment of expenses.

The extremely varied sources of information on enclosure by agreement (Beresford, 1961) make it difficult to build up a completely accurate picture of its chronology and spatial occurrence. A study of locally registered agreements, however, indicates the following trends for Northumberland: an even flow of enclosure agreements in the seventeenth century and an intensification in the first two decades of the eighteenth. The latest date for such an enclosure agreement in the county was 1851. The townships affected most frequently were those of the coastal plain, plateaus and river valleys, where arable land in the form of large, open, subdivided fields had been a characteristic ingredient

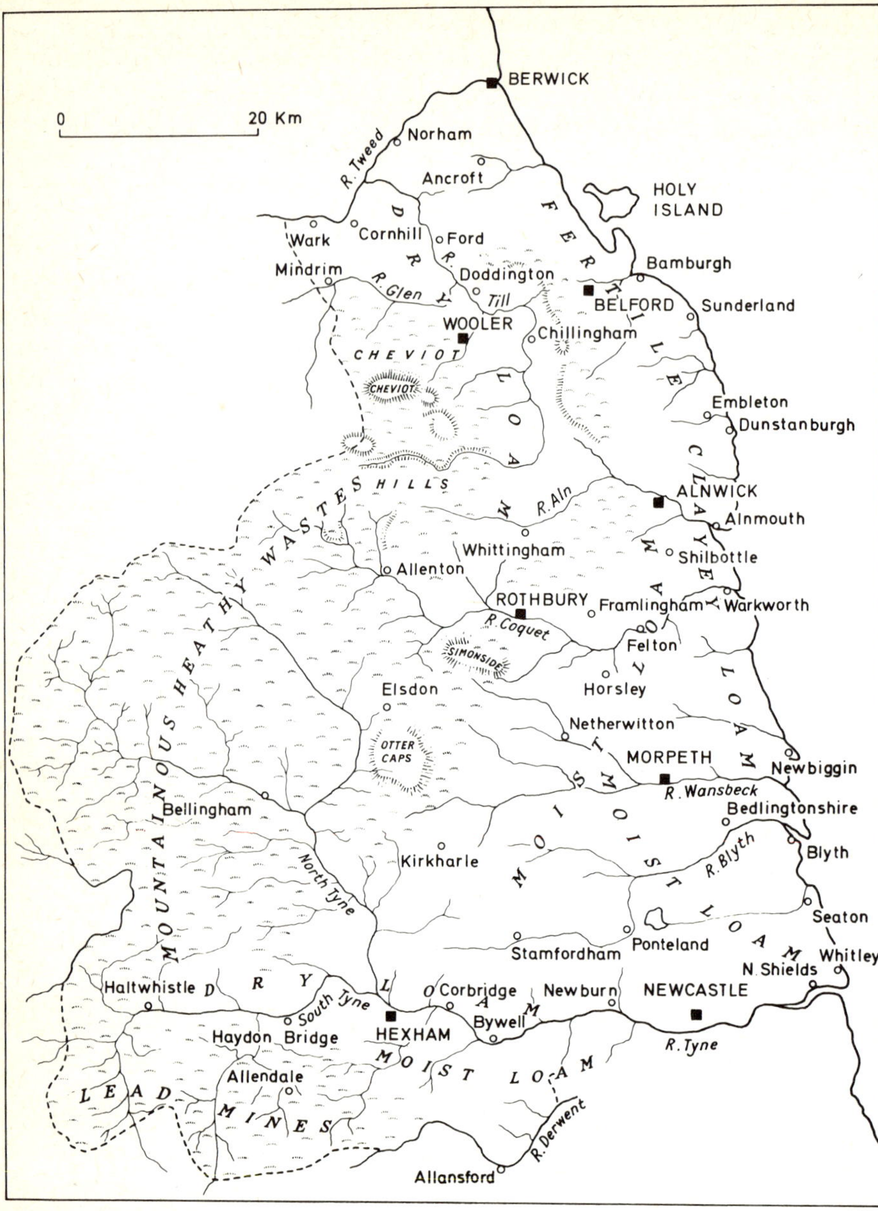

Fig. 10.1 The regions of Northumberland, 1794. Based on Bailey and Culley, 1794

of the agrarian organisation (Fig. 10.1). It appears certain that the open arable fields in Northumberland were enclosed principally by agreement. Substantial areas of meadow, pasture and waste were included in such agreements and the amount of waste included increased in the eighteenth century.

The motives for enclosure included the possibility of increased income for landlords from rents, the rationalisation of the size and distribution of holdings, and the recognition that general improvement of agricultural standards could be expedited by this type of structural change. Progress could, of course, be achieved within the older system of communal management, but in the late sixteenth and early seventeenth centuries the desire for further improvement is clearly expressed in the documents relating to enclosure. The 1707 petition of the tenants at Rennington to the Duke of Somerset is a good example, claiming that dispersion of holdings prevented improvement. Enclosure by agreement took place between 1720 and 1760, and in 1727 a surveyor stated that 'the tenants were very poor and scarce able to pay their rents, but since they divided their farms, having both coal and lymestones within their grounds they have improved their tenements to be worth £25 per annum' (NH, 1895, ii, p. 159).

The effects of this type of enclosure on the landscape are illustrated by contemporary maps. Rennington was surveyed in 1782,[8] and the resulting map shows that the new fields had been fenced and hedged in a regular and geometric pattern, with many fields of similar size. There was no conspicuous contrast between the size and shape of fields enclosing areas of former common arable and those on the former common waste, testifying to the rapidity of enclosure of both arable and waste. Similar features are evident on maps of Tynemouth, Bilton and High Buston.[9] Greater variations of field size and shape occurred in those townships with high proportions of common waste and relatively small quantities of common arable and meadow, with the fields enclosed from the waste being larger. Barrasford in the North Tyne valley, for example, was enclosed by agreement in 1717, and a map made shortly afterwards reveals long narrow fields as large as 40 acres extending to the fell area, with the former arable land now arranged in regular square fields.

With respect to changes in land use and agricultural improvement, there was some conversion from former arable to pasture as a result of enclosure, but the scale on which this took place is difficult to establish. Enclosure before the 1770s did not effect major improvement in agriculture, save by consolidation of holdings, and by providing a means of separating herds of animals and of keeping livestock from growing crops. The cropping system within the enclosed fields remained unaltered, with a three-course rotation of wheat, oats and fallow, or variations thereon, predominating; in some cases closes were farmed by more than one tenant. Nevertheless, the agriculture of the enclosed townships of the lowland areas was more advanced than that of the upland regions, which were the target of much adverse comment from agricultural writers. Arthur Young was particularly scathing in his comment on the periodic cultivation of moorlands. This practice was described by John Grey (1841, p. 153):

> Great tracts of land were not brought regularly under the plough, most farms being divided into what was called the *infield* and *outfield* land; the former being subjected to a severe course of cropping, with very inadequate cultivation; the latter left to its native produce as a pasture for sheep and young cattle, or, if choice portions of it were occasionally broken up, it was to rob it of its fertility by the production of a succession of corn crops, and then to leave it to recruit itself by several years of pasture.

Young's (1770, ii, p. 199) comment on moorland agriculture referred to the various practices as

> strokes of barbarism, which tend to damp and even extinguish the spirit of improvement from the infallible want of success, and to the leaving a country, after what is here called improvement, in as miserable and waste a state as before it was begun.

The age of improvement, 1770–1850

The last thirty years of the eighteenth century saw the beginnings of an era of conspicuous improvement in the agriculture of Northumberland, starting in the northern part of the county and gradually affecting other areas, though the chronology, scale and spatial effects varied widely. The main ingredients of change were enclosure by parliamentary Acts, the introduction of new farming techniques, general improvements in estate management (including new tenancy agreements and greater investment in buildings and houses), and a limited degree of underdraining of land. The principal stimuli included desire for improvement, and higher income from rents and sales, the latter being facilitated by increasing demands for agricultural produce at local and national levels in the late eighteenth century.

Enclosure

Although the process of enclosure by agreement continued in Northumberland throughout the whole of the eighteenth century (the last such enclosure took place in 1851), it did so with diminishing significance after 1775. Difficulties were experienced in some of the larger townships in securing unanimous agreement for enclosure, and for a time decrees in Chancery were sought. Eventually parliamentary sanction had to be applied for and the first private enclosure Acts were introduced. Considerable expense was involved in this type of enclosure, and in Northumberland the Acts related principally to marginal land which could be readily improved. The first of the private Acts in 1729 applied to Elsdon, proposing the enclosure of 10 500 acres of common waste, and the last in 1844 to the enclosure of Haltwhistle Commons. In all there were fifty-four Acts for the county from 1729 to 1844, of which only nine related to arable fields (Tate, 1942). An area of approximately 10 per cent of the county was affected by private enclosure Acts before 1845. The occurrence of such Acts between 1729 and 1844 is evenly distributed with some marked intensities in the periods 1770–80 and 1790–1810 (Fig. 10.2).

Nine of the related awards contain reference to common arable, but apart from the fields of Corbridge, the amount of arable affected by the awards is small. Enclosure of common arable fields was therefore almost an incidental feature of this particular movement, the main purpose and effect of which was to enclose common pastures and wastes, particularly those capable of improvement. The enclosure of such areas by Act of Parliament in the eighteenth and early nineteenth centuries was extensive. Bailey and Culley (1794, p. 110) state that 'the commons in the county capable of being converted into profitable tillage land are now very trifling, the greatest part having been enclosed within the last 30 years; the whole amounting to near 120 000 acres'. The amount of land affected by an individual award varied from a few hundred to several thousand acres, the fell enclosures constituting the largest areas. The major districts enclosed were the Fell Sandstone uplands, the moors of the upper North Tyne and Redesdale regions, the western scarplands and the north Pennines – in effect the lower moorland areas of the county.

Allotments and fields resulting from these enclosures were, on the whole, large, a direct result of the allocation of extensive areas with relatively few inhabitants. The new fields were regular, often square in shape, and delimited by hedges and walls, the latter being more common on the higher moorland areas. The new hedges were quickset on an earth-mound base, and the stone walls were usually

> two feet and a half at bottom, fifteen or sixteen inches at top, and four to four feet and a half high: about half way up a row of through stones are put, at the rate of nine or ten in a rood of seven yards, and on the top a coping of sods, or stones, *set edgeways*; the latter is preferable, as being the most lasting and presenting a more awful aspect, to deter the mountain sheep from attempting to leap them (Bailey and Culley, 1805, p. 62).

New roads were frequently laid out, adding to the planned appearance of an enclosed landscape. Enclosure was a necessary prerequisite for agricultural improvement, but the degree of improvement which followed varied within the county, as Bailey and Culley (1794, p. 111) state: 'The increased value of such enclosed commons depends . . . entirely upon the systems of cultivation pursued.' The introduction of substantial improvements to the agriculture of Northumberland was effected largely by a

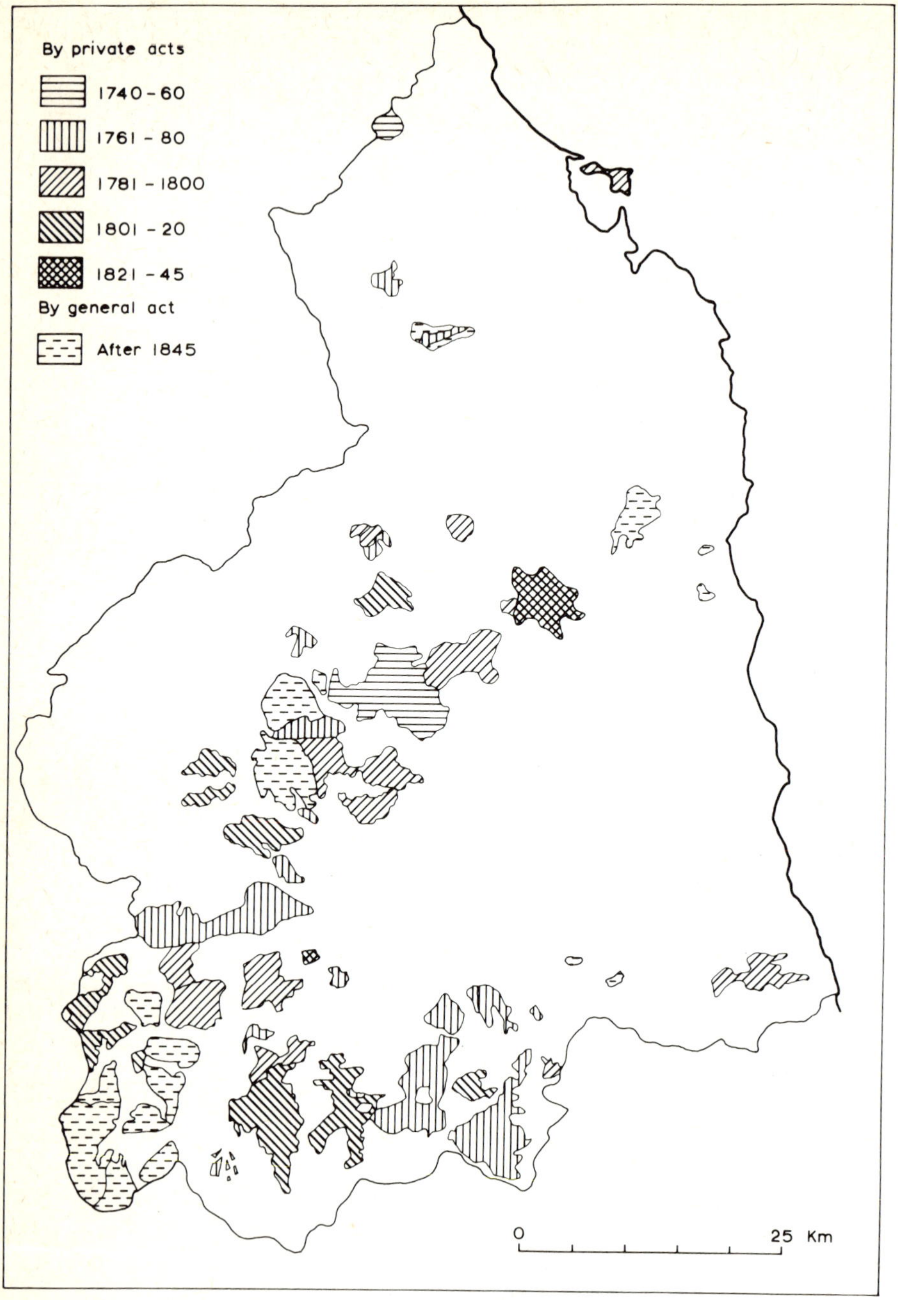

Fig. 10.2 Enclosure by Acts of Parliament. Based on Butlin, 1961

few men, and their innovations diffused at varying rates over a period of a century or more.

The improvers and their improvements

The introduction of new farming techniques took place when, in the words of one of the major improvers, John Grey of Dilston (1841, p. 153):

> Men of intelligence, activity and industry were attracted from other quarters to settle in the fertile vales in the northern part of the county, of whom none bear a more distinguished name in the annals of agricultural improvement, are more deserving of the praise and gratitude of their countrymen, than the late Messrs. Culley. These able men, who had traversed the provinces of the land which were considered to possess the most approved modes of husbandry and the best descriptions of stock, migrated in the year 1767 from their paternal property on the banks of the Tees into the district in Northumberland called Glendale, carrying with them superior knowledge and intelligence, which they brought at once to bear in their extensive undertakings with unremitting application and perseverance. The example set by these and other energetic and spirited agriculturalists, together with the signal success which attended their exertions, gave a stimulus to the surrounding district, and in a few years the inexpert operation and languid system of husbandry which had previously prevailed gave place to others of extraordinary expedition and efficacy.

The Culleys, Mathew and George, had been encouraged to learn the tenets of improved crop and animal husbandry by their father, himself an improver, and in 1767 they leased a farm at Fenton in north Northumberland. This area, until they arrived, was characterised by backward agriculture and very large farms. These often exceeded 500 or 1 000 acres, and were rarely less than 100 acres, possibly a reflection of the lasting effect of border warfare and a declining population (Hughes, 1963). The Culleys' main innovation was the introduction of a new system of convertible husbandry which eventually became known as the Northumberland five-course system, although the five-year rotation only became common after 1800 (Rowe, 1971). Initially, a six- or seven-year rotation was introduced, of oats, turnips, barley, then clover and other artificial grasses for three years. The five-course rotation was oats, turnips with some potatoes, spring wheat and barley, clover and other grasses, and pasture for the fifth year, though with some variation in years under grass in areas of poor land or high altitude. The most important innovation in their new rotation was turnips, which thrived on the light soils, provided winter feed for sheep who could be folded on the fields, and assisted in the elimination of fallows, for 'the practice of making naked fallows on all kinds of soils, once in three or four years, was general throughout this county till the introduction of turnips; in a few years the fallow of the dry lands were covered with this valuable plant' (Bailey and Culley, 1805, pp. 67–8). The cultivation of turnips was improved by the introduction of sowing by drilling, and the practice of sowing the seeds on heavily manured ridges.

Considerable improvements in animal husbandry were made by the Culleys, who introduced the shorthorn cattle into Northumberland, and bred the Border Leicester sheep by crossing the new Leicester breed with Cheviots. These new crop and animal husbandries were introduced in an area of light sand and gravel loams, the 'dry' or 'turnip' soil regions of the Till and Breamish valleys, but spread to other 'dry' areas, such as parts of the Tyne valley, and eventually to some of the areas of heavier soil. Among other innovators in Northumberland were John Grey of Dilston, whose main contribution was initially in Glendale and later on the Greenwich Hospital estates, particularly in the Tyne valley, and John Bailey, the co-author of the Board of Agriculture report.

In addition to the new breeds and crop rotations, there were improvements in farm machinery. Threshing machines were becoming common in the northern part of the county by the end of the eighteenth century. They had been developed in East Lothian in the late 1780s using the principle of the flax mill and thence spread to the village areas of north-east England. Before the introduction of steam power, rotary power sources were used to operate the new threshers, and the horse-powered wheel was incorporated into a new type of building – the wheelhouse, formerly very common in Northumberland (Hellen, 1972). The new turnip drills were a great improvement on broadcast sowing, but corn drills were not in widespread use. The main type of plough was a swing variety, an imitation of the Rotherham plough, but made to varying specifications.

Land draining was a further improvement in Northumberland farming, though tile drains were not used until the nineteenth century. Bailey and Culley (1805, p. 128) refer to land draining as

> one of those improvements that has lately made its way into Northumberland, and is now mostly practised in the middle and northern parts of the county . . . hollow-drains are generally used, filled with stones where they can be got; where they cannot be obtained (but at a great expense), sod-drains are the only resource, especially in the northern parts, where there is little wood. Of late years, great improvements have been made upon the sheep farms of the Cheviot Hills, by cutting surface drains, about one-foot wide, and as much deep, in an oblique direction to the declivity of the ground.

Embanking and draining of low-lying land were carried out in the northern vale of the county, and the Culleys became interested in the creation of water-meadows (Rowe, 1971).

The role of leases

Throughout this period, there was constant change in the form of leases granted to tenants on Northumberland estates, and in the process of renewal. Bailey and Culley (1805, p. 32) record that the leases 'for twenty-one years are lett on most of the principal estates, especially in the northern parts of the county', though 'some proprietors of land in the other districts, lett only for nine, twelve, or fifteen years; and a few lett no leases'. Leases for twenty-one years were most common on the estates of the improving landlords, and John Grey (1841, p. 157) believed that 'to the custom of letting land on long leases, more than to any other cause, though of course in conjunction with other causes, this county is indebted for its rapid improvement and high state of cultivation'. The mechanism of this process, according to Grey, was that long leases offered the necessary security and inducement to expend energy and capital on improvement. The role of leases has been disputed by Chambers and Mingay (1966, p. 48), who note that in general the lease cannot be regarded 'as a very important instrument in agricultural improvement, nor its absence as a great obstacle to efficiency'. Given this view, the lease may be seen as a direct instrument of change only when it contains clauses directly relating to improvement: leases on the Greenwich Hospital estates in the Tyne valley after 1833 did contain such clauses which compelled tenants to adopt new methods (Hughes, 1963). In addition the lease can be interpreted as a symbol of improvement when it related to estates where landlords were investing large amounts of capital and where tenants responded quickly to the spirit of improvement. At the other extreme, there are examples of estates in Northumberland in the eighteenth century where there were no leases, or, as in the case of the Greenwich Hospital estates, where on renewal the farms were advertised for tender and the highest bidder was normally accepted (Hughes, 1963).

Prices and markets

Although broadly based agricultural improvement was taking place in several clearly identified regions of the county by the late eighteenth century, further change, albeit of limited variety and of relatively short duration in some cases, was stimulated by the increase in grain prices between 1793 and 1815. During this wartime period, the last remaining portions of land accessible to the plough were requisitioned: 'Large stones were dug up and removed from the sides of mountains to procure an arable surface; bogs were drained, and lands hitherto unproductive were inclosed, and, by the application of lime and good husbandry, made to wave with golden harvests' (Grey, 1841, p. 155). This large-scale ploughing-up campaign, accompanied by a remarkable extension of wheat-growing, is a good example of a market and price stimulus to structural and sectorial change in agriculture. The results of this campaign varied: the enclosures remained, the high prices and rents did not, and the effect of such cultivation on marginal land was in many cases disastrous, with the land being exhausted by continuous cropping. Similar, but less spectacular and consequential, price fluctuations had been experienced earlier in the eighteenth century, though none of them had the universal effect of the high prices of the Napoleonic period and their subsequent fall. Price fluctuation also occurred at local levels, and could affect the fortunes of small farmers who derived part of their income from lead-mining, or who supplied the mining areas with fodder for the pit-ponies (Hughes, 1963).

The spatial incidence of improvement

The spatial distribution of the effects of rural improvement by 1850 was relatively limited. Despite the fact that by the 1840s 'the improved farming of Northumberland and the Lothians was considered a pattern for all England' (Fussell, 1948, p. 296), the only improved areas were the north, the coastal plain, and parts of the Tyne valley, for in 1851 the agriculture of much of the southeast of the county was deemed to be

> as little drained and as badly farmed as any district we have yet seen in England, and that the occupiers of the small farms can only eke out a scanty subsistence by careful parsimony, and by employing no labour except that of themselves and their families.[10]

Real improvement, therefore, had taken place, mainly on large farms where landlords and tenants had shared a spirit of progress, where large amounts of capital had been invested in enclosure, draining and machinery, and where the new forms of crop rotation and animal husbandry had been introduced. Agricultural development had thus been initiated largely by the enthusiasm of a relatively small number of men, whose model farms and estates were characterised by such innovations. To meet these new levels of production, spacious farmhouses and farm buildings were provided, together with farm cottages of good quality and, in some areas, planned villages, as at Blanchland, Stamfordham, Ford, Simonburn and Rock (Newton, 1972). The landscapes of improvement were described by William Cobbett, who refers to the Tyne valley at Hexham as containing

> as fine corn-fields as those of Wiltshire, and fields of turnips of both kinds, the largest, finest and best cultivated that my eyes ever

> beheld . . . the farms are all large, and the people who were on them either live in the farmhouse, or in buildings appertaining to the farmhouse; and they are all well fed, and have no temptation to acts like those which sprang up out of the ill-treatment of the labourers in the south (Cobbett, 1830, p. 290).

Rural change, 1850–1880

Enclosure continued in the second half of the nineteenth century, mainly under the provisions of the General Enclosure Acts of 1845 and 1852, whereby some twenty-two enclosures were made, involving about 80 000 acres of land, mainly moorland and barren wastes which would have been too costly to enclose by private Acts. The degree of improvement of the economy of the upland areas so enclosed was probably limited, though the standards of quality and management of the sheep which grazed the pastures of these remote areas were increased.

Further changes took place in the new agricultural system of the county, which 'by 1880 had been improved by the extended use of the Leicester Cross in sheep; by the increased use of cake feeding and the new artificial manures; by drainage and by the use of steam cultivation' (Fussell, 1948, p. 298). Land draining took a new form in this period, with the increased use of tile drains, though open drains were still employed, and large areas of major estates were drained effectively in the 1850s and 1860s (Phillips, 1969). The beginnings of changes in land use are discernible by the 1870s, with an increase in the overall acreage of grassland and a decrease in the amount of tillage, a trend which was to continue until 1914.

The overall effects of continued improvement are difficult to estimate, for very little research has been done on Northumberland in the late nineteenth century. The fact remains that while improvement did take place before the onset of the agricultural depression (by which the county does not appear to have been greatly affected),

> there is no statistical measure whatever of the area on which these improvements were made and their expansion is a matter of belief rather than knowledge. It is probable that the additional improvements were added to the great farms already in the hands of the great capitalist farmers who must advance with the time or drop out of the race (Fussell, 1948, p. 299).

Conclusions

The pace, processes and effects of rural change in Northumberland from the seventeenth to the nineteenth century were varied and complex, and are still understood only in general terms. At the beginning of the seventeenth century rural society and economy were not far removed from a backward and feudal state. By the early nineteenth century, substantial improvements had been effected through enclosure and many agricultural innovations. Nevertheless, the impression remains of a 'feudal' area, in the sense that rural life and livelihood were still dominated by influential landowners and their large estates. The most significant single factor in the whole period may well have been the influence of the border barons and their descendants, responding initially to the possibilities of increased income from land afforded by the termination of border warfare and therefore of border tenures. At a later date they again reacted to financial stimuli, but now afforded by the improved husbandries, enclosure, market changes and rising prices for farm products.

Notes

1. Alnwick Castle MSS, A II, ii.
2. Alnwick Castle MSS, A IV, 5.
3. Alnwick Castle MSS, J XI (7).
4. Alnwick Castle MSS, J X (le).
5. NRO, ZMD/117b.
6. NRO, ZBL/68/2.
7. NRO, ZBM/18/19.
8. Alnwick Castle MSS, O VIII, 6.
9. Alnwick Castle MSS, O VI, 33; O VIII, 6; O XVI, 6(b); O XXII, 12.
10. *The Times*, 28 November 1851.

References

BAILEY, J. and CULLEY, G. (1794, 1805) *A General View of the Agriculture of Northumberland*, London.

BATHO, G. R. (1957) 'The finances of an Elizabethan nobleman: Henry Percy, ninth Earl of Northumberland', *Econ. Hist. Rev.*, second series, 9, 433–50.

BATHO, G. R. (1962) *The Household Papers of Henry Percy, ninth Earl of Northumberland*, Camden Society.

BERESFORD, M. W. (1954) *The Lost Villages of England*, Lutterworth Press.

BERESFORD, M. W. (1961) 'Habitation versus improvement: the debate on enclosure by agreement', in F. J. Fisher, ed., *Essays in the Economic and Social History of Tudor and Stuart England*, Cambridge University Press, pp. 40–69.

BERESFORD, M. W. and HURST, J. G. eds. (1971) *Deserted Medieval Villages*, Lutterworth Press.

BUTLIN, R. A. (1961) 'The evolution of the agrarian landscape of Northumberland, 1500–1900', unpublished MA thesis, University of Liverpool.

BUTLIN, R. A. (1964) 'Northumberland field systems', *Agric. Hist. Rev.*, 12, 99–120.

BUTLIN, R. A. (1967) 'Enclosure and improvement in Northumberland in the sixteenth century', *Archaeologia Aeliana*, fourth series, 45, 149–60.

BUTLIN, R. A. (1973) 'The field systems of Northumberland and Durham', in A. R. H. Baker and R. A. Butlin, eds, *Studies of Field Systems in the British Isles*, Cambridge University Press, pp. 93–144.

CAMDEN, W. (1695) *Britannia*, ed. E. Gibson, London.

CHAMBERS, J. D. and MINGAY, G. E. (1966) *The Agricultural Revolution 1750–1880*, Batsford.

COBBETT, W. (1830) *Rural Rides*, Dent, Everyman edition, 1912.

FUSSELL, G. E. (1948) 'High farming in the north of England', *Econ. Geogr.*, 24, 296–309.

GREY, J. (1841) 'A view of the past and present state of agriculture in Northumberland', *J. R. Agric. Soc.*, 2, 151–92.

HELLEN, J. A. (1972) 'Agricultural innovation and detectable landscape margins: the case of wheelhouses in Northumberland', *Agric. Hist. Rev.*, 20, 140–54.

HUGHES, M. E. (1963) 'Lead, land and coal as sources of landlord income in Northumberland between 1700 and 1850', unpublished PhD thesis, Newcastle division, University of Durham.

JAMES, M. E. (1965) *Change and Continuity in the Tudor North*, Borthwick Papers 27, York.

KERRIDGE, E. (1969) *Agrarian Problems in the Sixteenth Century and after*, Allen & Unwin.

NEWTON, R. (1972) *The Northumberland Landscape*, Hodder & Stoughton.

NORTHUMBERLAND COUNTY HISTORY COMMITTEE (1893–1940) *A History of Northumberland*, 15 vols, Newcastle upon Tyne.

PHILLIPS, A. D. M. (1969) 'Underdraining and the English claylands, 1850–80: a review', *Agric. Hist. Rev.*, 17, 44–55.

RAMM, H. G., MACDOWALL, R. W., and MERCER, E. (1970) *Sheilings and Castles*, Royal Commission on Historical Monuments. London.

ROWE, D. J. (1971) 'The Culleys, Northumberland farmers, 1767–1813', *Agric. Hist. Rev.*, 19, 156–74.

SANDERSON, G. ed. (1891) *A Book of the Survey of the Debatable and Border Lands adjoining the realm of Scotland and belonging to the crown of England, 1604*, London.

TATE, W. E. (1942) 'A handlist of English enclosure Acts and awards: part 26, Northumberland', *Proc. Soc. Antiq. Newcastle-upon-Tyne*, fourth series, 10, 119–40.

WATTS, S. J. (1970) 'Tenant right in early seventeenth-century Northumberland', *Nth. Hist.*, 5, 64–87.

YOUNG, A. (1770) *A Six Months' Tour through the North of England*, 4 vols, London.

J. R. Walton

11 The residential mobility of farmers and its relationship to the parliamentary enclosure movement in Oxfordshire

Parliamentary enclosure has long attracted attention as one of the fundamental elements in the series of changes which contributed to the transformation of English agriculture and rural society during the eighteenth and nineteenth centuries, yet its causes and consequences deserve further investigation. The purpose of this essay is to examine one significant aspect of rural change – the residential mobility of farm occupiers – and to consider the relationship between occupier turnover and the incidence of parliamentary enclosure.

Early opinions on enclosure as it affected both agriculture and society owed much to the pioneering work of Lord Ernle (1912) and of the Hammonds (1911). Ernle, stressing the fundamental role of the East Anglian landowners Townshend, Coke and Walpole as innovators, saw enclosure as the means whereby their innovations, particularly the Norfolk four-course system, gained widespread acceptance. The Hammonds, equally convinced of the importance of enclosure to social change, emphasised its crucial significance in dispossessing the small owner occupier and creating a large reserve of landless labour to work the newly created large farms of the post-enclosure era and to man the emergent factory industries. Subsequently, both these views have been modified substantially. The prime function of the Norfolk landowner is now seen as that of publicist rather than innovator, while enclosure is known not to have been a crucial precondition of agricultural innovation. It is now widely accepted that many of the innovations thought previously to be dependent directly on the reorganisation of agriculture through enclosure could have been and were introduced on common field holdings during the seventeenth century or earlier (Kerridge, 1967; Havinden, 1961).

The Hammonds' findings have undergone more drastic reconsideration. Their conclusions about distress and hardship were made on the basis of a number of individual case studies (the Otmoor enclosure riots in Oxfordshire being one) which are now recognised as atypical of the general situation. Revision of their ideas has been based on the examination of two types of different but complementary sources. On the one hand, at an aggregate or national level, use has been made of census and registration data. The lack of national census data before 1801, and of registration data before 1837, creates difficulties in analysing population movement at an aggregate level in the period of most intense enclosing activity during the last two decades of the eighteenth century. Nevertheless, on the basis of a detailed examination of such material, Redford (1926) demonstrated that there was no appreciable movement of population from those counties with considerable enclosing activity in the Midlands, the east and the south to the industrialising districts of the north, and confirmed that throughout the nineteenth century migratory movements were largely short-distance, occurring in the fashion suggested by Ravenstein (1885, 1889). More recent work by White (1969), using simple correlation and regression techniques on county statistics of migration rates and enclosure acreages, has lent further support to these conclusions.

The second approach in the scrutiny of the Hammonds' view of agricultural change has been the single county study, based usually on the land tax assessments. This type of study has attracted more attention than aggregate analyses, possibly because only at this level can the land tax assessments be handled with ease. These offer, at a time of intense enclosing activity, an insight into rural social change not obtainable from census and registration data. The first studies using land tax assessments to present a view of enclosure and its social consequences different from that suggested by the Hammonds date from the same period as their work, but made a much smaller initial impact than the Hammonds's far more colourful conclusions. Johnson (1909) and Gray (1910) suggested that the small occupying owner was not a victim of the enclosure movement. Gray, for example, showed that in Oxfordshire the numbers of small occupiers increased until 1815, even in enclosing parishes, and such subsequent decline as occurred appeared no greater than rates of decline before the parliamentary enclosure period.

Since 1910 Gray's method of approach has been applied to other enclosing areas, and broadly similar conclusions reached. Davies (1927) studied the land tax assessments for a large sample of English parishes and concluded that parliamentary enclosure after 1780 led to an increase rather than a decline in all types of occupying owner, though such owners had ceased to be the dominant element in England long before 1780. More recent work, including that of Chambers (1940), Hunt (1958), Martin (1967), Grigg (1963) and Mingay (1964), mainly reinforces the findings of Johnson, Gray and Davies. Martin (1964) is alone in not minimising the ill-effects of enclosure on the small occupier, having drawn attention to the relatively high cost of enclosure during the last two decades of the eighteenth century and the early years of the nineteenth. In Warwickshire, at least, this seems to have increased the vulnerability of the smaller occupier, particularly in those parishes where ownership structure was dominated by a large number of freeholders ('open' parishes), and conflicting interests delayed the movement toward enclosure until this later period. Such consequences appear to be relatively uncommon in relation to the overall position of the freeholder during the enclosure period which, in numerical terms, was one of stability or, at least, of no more rapid decline than previously.

As a result of the increasing use of land tax data, much attention has been directed to the validity of the assessments as an index of landholding size. It has long been appreciated that, because of variations in the level of assessment and the infrequency of reassessment, size of payments could not be related directly to holding size throughout a county, but that the total assessment for each parish had to be related to the parish's acreage, and a parish acreage equivalent determined to serve as the basis for calculating the size of an individual holding. This procedure was followed by Davies in his work on the tax, and has been applied subsequently by most other researchers. However, comparisons of ownership units calculated by this method with the independent evidence from manorial and estate surveys led Mingay to conclude that there was no constant relationship between the level of tax assessment and the size of holding, while Grigg demonstrated that there were obvious difficulties in assuming a parish acreage equivalent to be constant when land quality could vary more within a parish than between parishes. Further difficulties arise from voluntary redemption of the tax after 1798, the payment of tax on houses alone and on other non-agricultural uses, the problem of occupiers and owners with blocks of land scattered throughout several parishes, tax evasion, and the possibility that some long-term lessees or tenants paying tax themselves may have been returned as owners. Although, as Martin (1966) has argued, acreage equivalents are normally relatively accurate, and many other problems do not arise in practice, it is important to bear in mind these possible limitations of the evidence when undertaking a land tax study.

In spite of the extent and detail of previous work, there are some aspects of enclosure and its social consequences which still remain to be examined. Throughout earlier research, there has been a tendency to relate enclosure only to changes in the prosperity of the owner occupier. Since it is now accepted that the freeholders' dominance of agricultural society had ceased before the middle of the eighteenth century, it seems irrelevant to continue examining enclosure in relation only to owner occupiers. Few previous studies have devoted any attention to other types of occupier. This essay adopts a different approach in two respects: first, it considers the consequences of the enclosure process for all farmers, not simply the owner occupiers; and second, it attempts to assess the impact of enclosure on the individual farmer rather than on particular groups of farmers, whether defined by farm size or tenure. Hitherto most studies have examined only the small owner occupier, and then at an aggregate level. As the smaller size groups of owner occupier did not generally decline any more rapidly as a consequence of the enclosure process, it would seem appropriate to pose different questions. Though enclosure may not have caused the numerical reduction of any particular group of farmers, so far it has not been demonstrated that the process did not increase the rate of turnover of farmers within each size group. Clearly, examination of changes in the rate of residential mobility of farmers during enclosure gives a much better impression of the impact of that movement on the individual than an assessment of changes before and after enclosure in the numerical strength of any particular group of agriculturalists. Measures of residential turnover are recognised as important in the analysis of social change in cities, but their value in the study of rural society, especially in a historical context, has not been examined.

Mechanisms and evidence of farmer turnover

Rental was one of the most important of the mechanisms which might produce increases in farmer turnover during enclosure. An increase in the rental value of land was one of the first results of enclosure. One contemporary observer, Sir John Sinclair, estimated that enclosure during the 1790s yielded an average rental increase of 9 shillings per acre (Thompson, 1907). Young (1813) refers to much greater increases than this for some Oxfordshire enclosures: for example, a threefold increase at Fringford and Stoke Lyne, a fivefold increase at Stratton Audley and Wootton, and an anticipated increase of from 12 or 15 to 40 shillings per acre in the enclosure of Deddington. The evidence of farm and estate accounts suggests that increases of this magnitude were probably extreme cases, but rental increases were, nevertheless, considerable. Income on rentals from land belonging to the Harcourt family in the parish of Stanton Harcourt, enclosed in 1774, rose from £1 415 in 1773, to £1 702 in 1776 and to £2 444 in 1777.[1] The degree to which landlords regarded the opportunity to increase rent as the major benefit of enclosure and the level to which they raised rentals beyond the new productive capacity of the land in order to achieve it are difficult to determine. Advertisements for the sale of land undergoing or about to undergo enclosure frequently stressed the soundness of this type of investment. For example, an open field farm of 161 acres at Milton under Wychwood, advertised for sale in 1830, was 'recommended to those desirous of investing capital in a manner decidedly advantageous' as it was 'very likely that an Inclosure will soon take place'.[2] In fact, the land was not enclosed until 1849. Similarly, it is difficult to decide whether increased landlord income resulted from a genuine rise in land value, or from the opportunity presented to landlords to reduce the margins between tenants' incomes and rental payments. There are strong indications that, in some cases at least, landlords were seizing the opportunity to raise rents, which had long been static and no longer represented the full rental value of the land even under the old system of production, to a level more fully attuned to post-enclosure values. For example, the land survey, taken at an early stage in the enclosure of Stanton St John, was suggested as justification for an immediate 10 per cent increase in rents even before the award had been made and post-enclosure increases of rental applied.[3]

If landlords were guilty of opportunism in increasing rents from unrealistically low levels before enclosure to extremely high levels afterwards, estate papers and associated material offer little indication of how far this affected the farmer. Thomas Smith of Stadhampton, in reply to a questionnaire circulated on behalf of the Board of Agriculture in 1793, stated that increased rents resulting from enclosure had 'stirred up the spirit of improvement'.[4] He makes no mention of the replacement of tenants as an integral part of this process. Among evidence which suggests that such may have been the case is the statement of the tithe owner of Oddington who, in accepting a corn rent £20 below his original demand, admitted that such a loss 'will be greatly compensated by the satisfaction I shall experience in being able to divide my estate into two or three farms, and to procure good and substantial tenants upon each of them, which I shall much despair of being able to do while the parish remains in its present unenclosed situation'.[5] This statement implies that enclosure could and did result in tenant turnover, but it cannot be assumed that such an attitude was widespread among landlords or that the threat of dispossession was invariably implemented.

The land tax assessments, on the other hand, offer opportunities, which have been hitherto neglected, for a numerical analysis of the rate of residential turnover of both tenants and owner occupiers immediately after enclosure, and a means of comparing post-enclosure rates with those occurring in periods when parliamentary enclosure was not taking place. A further advantage of this approach is its freedom from many of the alleged difficulties associated with the more usual forms of land tax analysis. Acreages do not have to be calculated accurately: it is adequate merely to ensure that the sample includes only those who were farmers. For this purpose, exclusion of all occupiers with aggregate taxable holdings in any one parish assessed at less than £1 was found to give an acceptable basis of operation, as suggested by Gray.

Mingay's strictures about voluntary redemption of the tax also proved inappropriate, for in all such cases the names of occupiers and owners were listed along with those who still paid the tax. Furthermore, in Oxfordshire, tax payments on large non-agricultural uses of land were virtually non-existent and taxable assessments in respect of payments on tithe or collection of excise were often listed as such. In situations where a landlord paid the tax for his tenant, the collector occasionally listed the name of the landlord, but not the names of the tenants, separately. These cases were relatively infrequent and such parishes could be excluded from the sample without incurring any bias. Above all, independent evidence proved that the assessments were sensitive in their response to changes of occupier or owner. Normally, assessments were made between the end of June and the middle of August for the tax year April to April. The assessors were usually two of the more substantial, if not necessarily the largest, occupiers of land in each parish, though in one Oxfordshire example, Upton, there appears to have been a rotation of office among all the farmers whatever their importance (Ward, 1953). With their detailed local knowledge, the assessors appear to have been alert to changes in proprietorship and tenancy.

To check their accuracy, known transfers of occupancy based on a random selection of auctioneers' advertisements published in *Jackson's Oxford Journal* were related to the last appearance of the occupier in the land tax assessment (Table 11.1).[6] In all cases it appears that the outgoing occupier's name was removed from the list to be replaced by that of his successor at the assessment immediately following his departure. Even Richard Smith's move from Beckley, early in May 1804, was noted in the assessment of that summer, while the Begbroke farm tenanted by

Table 11.1 Sample transfers of occupancy

Parish	*Farmer*	*Date farm advertised for sale or to let*	*Reasons for sale*	*Year of last land tax assessment in which farmer appears*
Aston Rowant	John Paine	10 April 1793	Leaving	1792
Beckley	Edward Ledwell	25 March 1794	Death	1793
Beckley	Richard Smith	4 May 1804	Leaving	1803
Begbroke	William Taylor	14 August 1797	Death	1796
Benson	Ann Burgiss	30 January 1802	About to decline business	1801
Kencot	Ann Cook	24 October 1795	Retirement	1795
Kidlington	Edward Nichols	15 October 1800	Leaving	1801
Kingham	Henry Smith	24 March 1804	Leaving	1803
Merton	John Grace	4 January 1798	–	1797
North Aston	William Bezley	4 April 1799	Leaving	1798
North Aston	Richard Weston	7 October 1799	Retirement	1799

Source: Jackson's Oxford Journal; ORO, QSD/L/1–318.

William Taylor was listed as lately occupied by him in the assessment of 25 July 1797, though not occupied by his successor, Charles Matthews, until the assessment of 1798. Taylor's death probably occurred well before the tenancy was advertised as vacant in the middle of August. In sum, the assessments may be considered to possess a high degree of accuracy in their representation of the turnover of farm occupiers.

To determine the rate of farmer turnover, a sample of fifty-one parishes, widely differing in size and location within the county, was chosen. The sample parishes were divided between those enclosing before, during, and after the period for which land tax assessment evidence is available, between 1785 and 1831 (Fig. 11.1). The assessment for each sample parish was examined for 1791, and thereafter at decennial intervals up to 1831, and compared with its predecessor of ten years earlier, except for the 1791 assessment, which was compared with that of 1785, the earliest surviving assessment for most Oxfordshire parishes. Occasionally it was necessary to select an assessment one year before or after the usual date, but in all cases it was possible to ensure that the sample period examined deviated by no more than one year from the standard interassessment period. Each change of occupier and of owner was noted, and rates of turnover between each assessment calculated as the percentage of owner occupiers, tenants and of all occupiers (measured both by number and by tax paid) listed in one assessment who had not appeared in the assessment ten years' earlier. For present purposes, transference of a holding within the same family was not treated as a change of occupancy. To compensate for the differing length of interassessment period, the percentage turnover for each class of occupier in each period was divided by the number of years in that period. This statistic is referred to here as the standardised turnover rate. It should not be regarded as a measure of annual turnover analogous to that which would be obtained if the assessments for each year were examined. The present procedure cannot allow for changes of occupancy occurring twice or more frequently within an interassessment period since these appear as only one change of occupier. For this reason, too, it would be expected that the standardised turnover rates for the initial six-year period would, other things being equal, be slightly higher than those for subsequent ten-year periods. In spite of these difficulties the standardised turnover rate does represent a measurable index of the residential stability of farm occupiers, providing an adequately sensitive comparative measure of variations in turnover.

Residential mobility of farmers in Oxfordshire

As a first stage of analysis the fifty-one sample parishes were divided into four groups: fourteen were enclosed completely before parliamentary

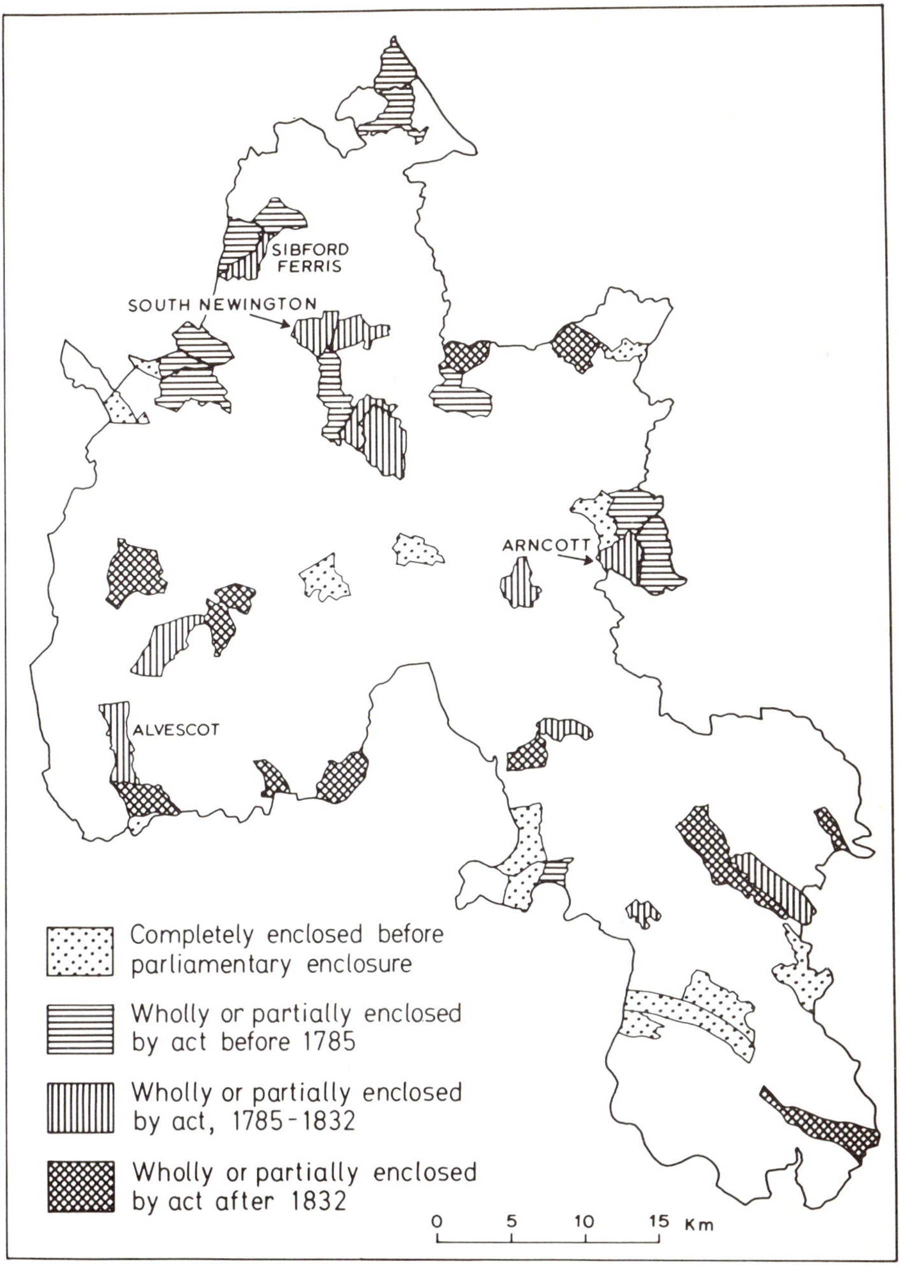

Fig. 11.1 The location of the sample parishes. Based on Gray, 1915

enclosure; of those enclosed by Act twelve were enclosed completely or partially before 1785, twelve were enclosed between 1785 and 1832, and thirteen after 1832. For each of these groups a standardised turnover rate was determined (Fig. 11.2 A–D), together with a composite rate for all the parishes in the sample (Fig. 11.2 E).

In general terms, the composite turnover rates (Fig. 11.2 E) appear to be explicable more or less by what is known about and what might be expected from variations in agricultural prosperity between 1785 and 1832. The interassessment periods 1785 to 1791 and 1791 to 1801 are marked by relatively low but divergent turnover trends. A period of advancing prices during the decade 1801 to 1810 is met by an increase in the rates of mobility of all farm occupiers, as compared with the two previous interassessment periods. The difficulties which beset all farmers, but especially grain producers, from 1815 onwards are reflected in an even higher turnover rate for tenants during the decade 1811 to 1820, a trend maintained in the following decade. On the other hand, there was a marked decrease in the turnover rate for owner occupiers during the 1811 to 1820 period, followed by an increase during the decade 1821 to 1830. This movement suggests either that the difficult years were not so serious for the owner occupiers or that the normal response of the owner occupier to a period of difficulty was decreased not increased mobility. In fact, these years are known to have been particularly severe for tenants as they took farms at the higher rentals prevailing during the Napoleonic Wars, a course leading undoubtedly to many failures, while the owner occupier was much less affected.[7]

The discrepancy between turnover measured by amount of assessment and by numbers of occupiers suggests that the mean size of tenanted holdings changing hands was consistently lower than the mean size of all tenanted farms throughout the period. This relationship appears not to hold true in the case of owner occupier turnover, where there is a much closer correspondence between the two curves (Fig. 11.2 E). However, owner occupied holdings were, in any case, of much smaller average size than those of the tenantry and, as the close proximity of the curves for tenant turnover and the turnover for all occupiers suggests, of considerably less importance as a part of the total ownership structure.

Initially there appears to be a low level of correspondence between high rates of turnover and the incidence of enclosure. Parishes with enclosure Acts before 1832 (Fig. 11.2 B and C) have turnover rates which, though high, are not uniformly higher than those of other groups of parish in the sample (Fig. 11.2 A and D). There was, for example, a particularly high rate of turnover of tenants in parishes with no enclosure Act during the interassessment periods from 1801 onwards, which was barely matched by turnover rates in parishes with enclosure Acts before 1832. An analysis of variance carried out on the turnover rates, measured by amount of assessment, for the decade 1801 to 1810 suggested that the differences between the groups were only marginally significant. However, given that the amount enclosed within each of the enclosing

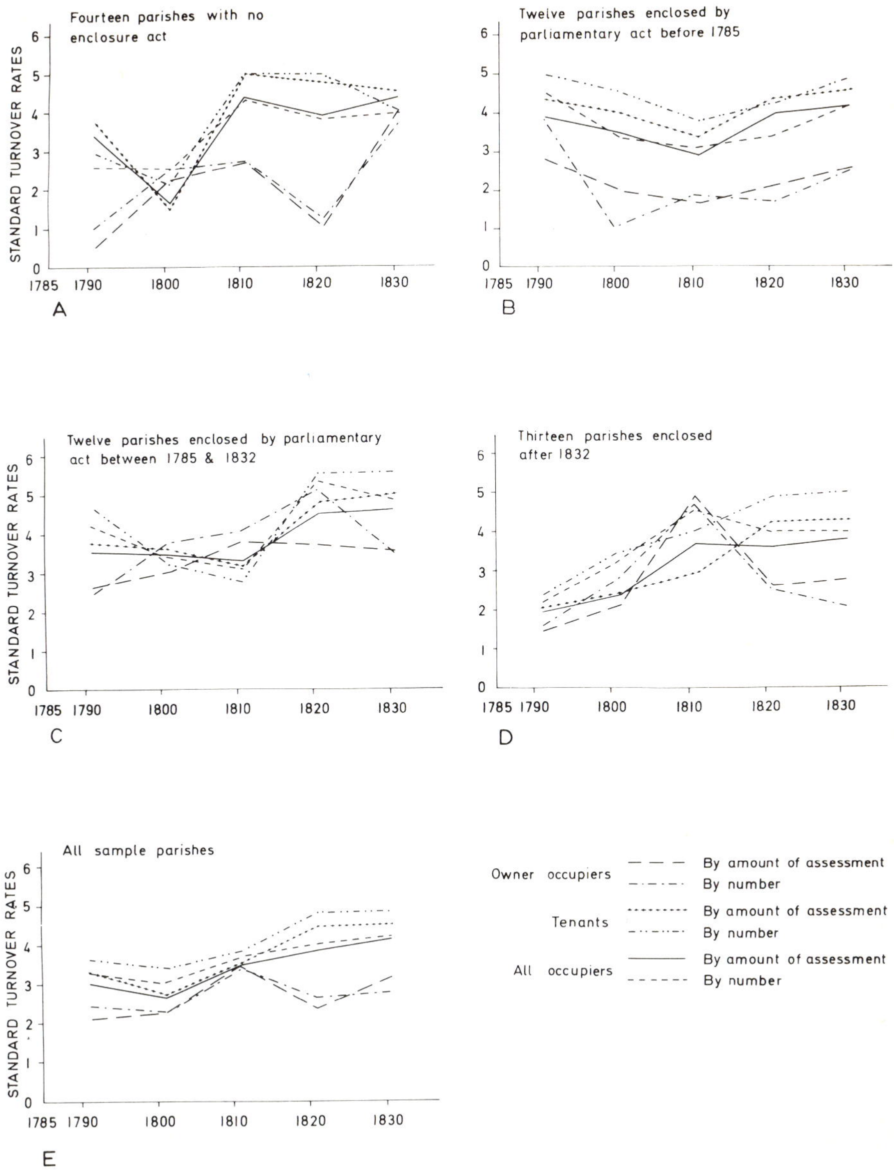

Fig. 11.2 Standardised turnover rates of farmers:

A. In fourteen parishes with no enclosure Act;

B. In twelve parishes enclosed by parliamentary Act before 1785;

C. In twelve parishes enclosed by parliamentary Act between 1785 and 1832;

D. In thirteen parishes enclosed after 1832;

E. In all sample parishes

Based on ORO, QSD/L/1–318; Gray, 1915

parishes was liable to vary considerably, in itself this is no proof that enclosure was not important as a major determinant of turnover variation. Furthermore, if high post-enclosure turnover rates were followed by those lower than normal in the period immediately after the post-enclosure phase, this would largely cancel out the high turnover rates directly associated with enclosure.

If the situation in parishes enclosing before 1785 is examined (Fig. 11.2 B), tenants' turnover rates for the interassessment periods from 1785 to 1791 and from 1791 to 1801 were higher both by number and amount than tenant turnover rates in any other group. Turnover for owner occupiers was also higher in this group between 1785 and 1791 and only matched in the 1791 to 1801 period by turnover rates in those parishes enclosing between 1785 and 1832. These features could well reflect the effects of enclosure occurring before 1785.

Parishes enclosing before that date are unique in showing a steady decline in turnover rates for both categories of occupier (measured by amount) for each successive interassessment period to 1811. This may be an indication of a trend towards increasing residential stability after the rapid turnover of the immediate period after enclosure. Indeed, the fact that this movement is not evident for other groups of parishes strongly suggests this possibility. It is, however, impossible to offer any firm conclusions concerning the relationship between enclosure and turnover without more detailed examination of individual cases.

Standardised turnover rates for four parishes enclosed between 1785 and 1832 have been plotted as deviations from average standardised turnover rates for all fifty-one sample parishes (Fig. 11.3 A–D). The date of enclosure and amount enclosed, as a percentage of the total parish acreage, are also marked, a high proportion of the total acreage being affected by enclosure in each of the four parishes. Though there was a tendency for turnover in the four parishes to display throughout a higher level of positive than of negative deviation from the sample means, the degree of correspondence between high positive deviations in turnover rate and the date of enclosure is striking. In three of the four parishes the enclosure award was made during the second or later interassessment period, enabling direct comparisons between turnover rates before and after enclosure. The level of turnover in the period before the award tended to be considerably closer to the sample average than that in the interassessment period in which the award fell, when large positive deviations from it were recorded.

Alvescot was exceptional in that the level of positive deviation declined in the case of owner occupiers, measured by amount of assessment, whereas an increase occurred in the level of turnover of owner occupiers measured by number. In the cases of South Newington and Sibford Ferris, the greatest degree of positive deviation from the mean

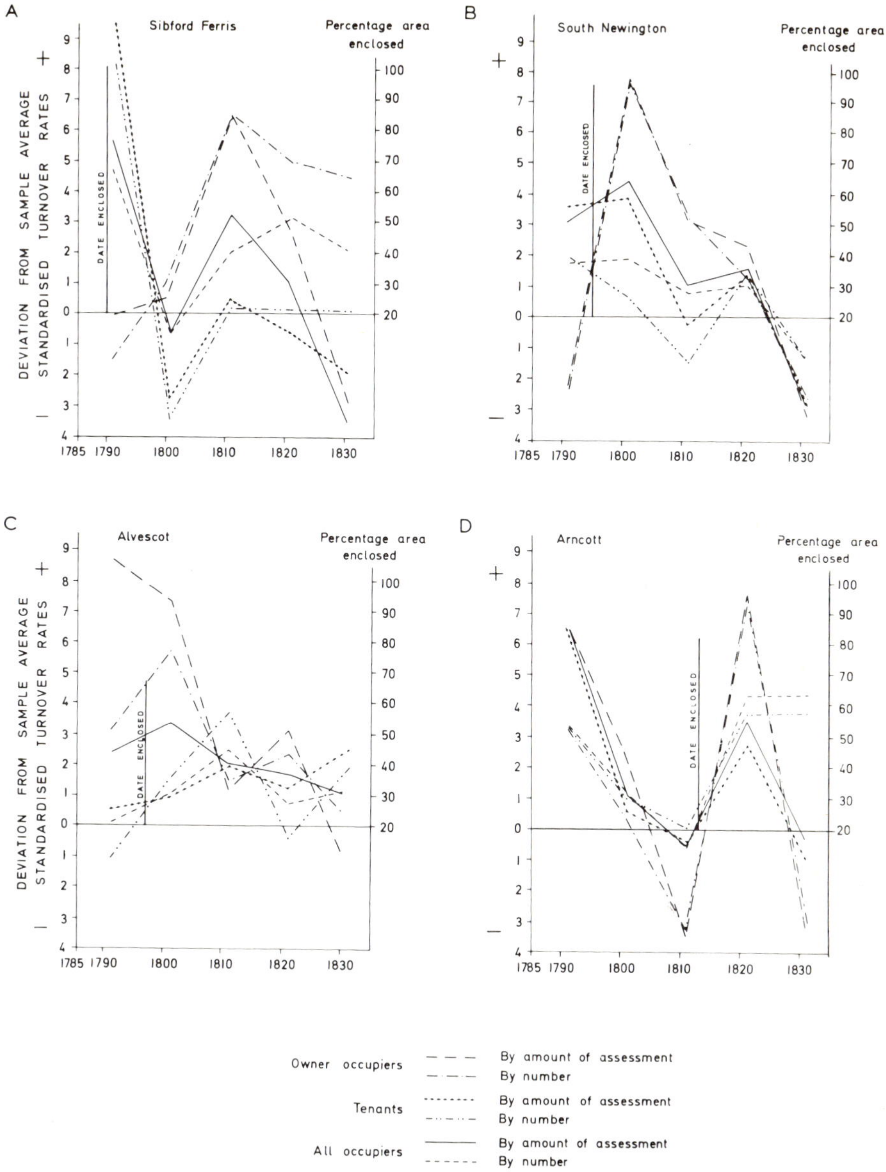

Fig. 11.3 Deviations from average standardised turnover rates for four parishes enclosed between 1785 and 1832:

A. Sibford Ferris;

B. South Newington;

C. Alvescot;

D. Arncott

Sources as for Fig. 11.2

turnover rates for all occupiers between 1785 and 1832, measured both by number and by amount of assessment, occurred during the inter-assessment period of the award. At Alvescot, this period also revealed the greatest turnover of all occupiers as measured by amount, though not by number, the peak for which took place at the end of the following decade. The weaker relationship between enclosure and increased turnover in the case of Alvescot may be explicable in terms of the relatively smaller proportion of the acreage of the parish enclosed by parliamentary Act, as compared with the other parishes examined. It is worth noting, however, that even in the case of Sibford Ferris, with almost 94 per cent of the parish acreage enclosed by Act, turnover rates for an admittedly small number of owner occupiers were no greater than the sample average during the enclosure period.

In general it would seem that enclosure in these four parishes imposed levels of turnover on farmers far in excess of average levels. The case of Arncott is particularly illustrative. In this parish, enclosure, in 1813, was preceded by average or lower than average rates of turnover in the decade 1801 to 1810, to be followed by rates in excess of average in the succeeding decade. This pattern contrasts with the situation in the other three parishes, where turnover rates were declining relative to the sample averages during the same period. However, high rates of turnover, which appear to be associated with enclosure, were not excessive in comparison with peaks in turnover rate which could occur at other periods. Arncott underwent a period of exceptionally high turnover between 1785 and 1791 which was more severe than that which occurred after enclosure, while Sibford Ferris and Alvescot experienced even higher turnover after the immediate post-enclosure peak. In the case of tenants in the last two parishes, a cyclical pattern of turnover may be involved, resulting from the introduction of leases for seven years or longer at enclosure, and, to that extent, may be explicable terms of enclosure.

Conclusions

Any conclusions about the causes of turnover in the absence of information about farming type, agricultural prices, differing policies of landowners and estates or of any other factor which might influence the rate of mobility of farm occupiers can only be speculative. However, more definite conclusions may be reached about the relationship between enclosure and turnover on the basis of the present evidence. Though the turnover rates of parishes enclosing between 1785 and 1832 were not considerably greater than those of other groups, those parishes which had been enclosed before 1785 appear to have had higher than average turnover rates during the period immediately after 1785. If parishes are examined individually, further evidence exists of a connection between the date of enclosure and turnover rates markedly higher than usual. These high turnover rates, however, were not in any respect unique and could occur independently of enclosure, even though such peaks may be related to enclosure through the operation of leases introduced at the time. In general, there appear to be good grounds for suggesting that although enclosure may not have placed pressure on owner occupiers and tenants to the extent that it alone resulted in the dispossession of small owner occupiers or the extinction of small tenancies, it was accompanied by an increase in the rate at which holdings of both these groups of farmers changed hands. Enclosure seems to have provided the opportunity for the appearance of new occupiers and tenants in Oxfordshire. Whether the new farmers were also newcomers to the area cannot be stated with certainty. It is almost as difficult to assess the exact nature of their contribution to agricultural change in the period after enclosure. Nevertheless the land tax evidence reveals that enclosure was associated with a considerable degree of residential mobility among farmers and this may have had important influences on methods of agricultural production and levels of output.

Notes

1. Stanton Harcourt MSS, Bodleian Library, Temporary box 13.
2. *Jackson's Oxford Journal*, 20 November 1830.
3. ORO, Misc/Bod IV/7f 5.
4. ORO, Wi IX/3.
5. ORO, Misc/Bod IV/2.
6. ORO, QSD/L/1–318, Oxfordshire land tax assessments.
7. Great Tew MSS, Miscellaneous leases and letters. On the Great Tew estate, nine of the ten tenants who took twelve-year leases in 1813 gave notice of leaving between October 1815 and March 1816.

References

CHAMBERS, J. D. (1940) 'Enclosure and the small landowner', *Econ. Hist. Rev.*, **10**, 118–27.

DAVIES, E. (1927) 'The small landowner in the light of the land tax assessments', *Econ. Hist. Rev.*, **1**, 88–98.

ERNLE, LORD (1912) *English Farming Past and Present*, London.

GRAY, H. L. (1910) 'Yeoman farming in Oxfordshire from the sixteenth century to the nineteenth century', *Q. J. Econ.*, **24**, 293–326.

GRAY, H. L. (1915) *English Field Systems*, Harvard University Press.

GRIGG, D. B. (1963) 'The land tax returns', *Agric. Hist. Rev.*, **11**, 82–94.

HAMMOND, J. L. and HAMMOND, B. (1911) *The Village Labourer*, London.

HAVINDEN, M. A. (1961) 'Agricultural progress in open-field Oxfordshire', *Agric. Hist. Rev.*, **9**, 73–83.

HUNT, H. G. (1958) 'Landownership and enclosure 1750–1830', *Econ. Hist. Rev.*, second series, **11**, 497–505.

JOHNSON, A. H. (1909) *The Disappearance of the Small Landowner*, London.

KERRIDGE, E. (1967) *The Agricultural Revolution*, Allen & Unwin.

MARTIN, J. M. (1964) 'The cost of parliamentary enclosure in Warwickshire', *Univ. Birmingham Hist. J.*, **9**, 144–62.

MARTIN, J. M. (1966) 'Landownership and the land tax returns', *Agric. Hist. Rev.*, **14**, 96–103.

MARTIN, J. M. (1967) 'The parliamentary enclosure movement and rural society in Warwickshire', *Agric. Hist. Rev.*, **15**, 19–39.

MINGAY, G. E. (1964) 'The land tax assessments and the small landowner', *Econ. Hist. Rev.*, second series, **17**, 381–8.

RAVENSTEIN, E. G. (1885) 'The laws of migration', *Jl R. Statist. Soc.*, **48**, 167–235.

RAVENSTEIN, E. G. (1889) 'The laws of migration', *Jl R. Statist. Soc.*, **52**, 242–305.

REDFORD, A. (1926) *Labour Migration in England*, Manchester.

THOMPSON, R. J. (1907) 'An inquiry into rent in the nineteenth century', *Jl R. Statist. Soc.*, **70**, 587–625.

WARD, W. R. (1953) *The English Land Tax in the Eighteenth Century*, Oxford Historical Series, Oxford University Press.

WHITE, L. J. (1969) 'Enclosures and population movements in England, 1700–1830', *Explor. Entrep. Hist.*, **6**, 175–85.

YOUNG, A. (1813) *General View of the Agriculture of Oxfordshire*, 2nd edn, London.

A. D. M. Phillips

12 Underdraining and agricultural investment in the Midlands in the mid-nineteenth century

With the introduction of cheap machine-made pipes and new systems of drain layout, underdraining in the 1840s was recognised as a viable, long-lasting agricultural improvement, enabling increased production from arable and pasture alike (Darby, 1964). At the same time it involved a considerable capital outlay, about £5 per acre, which rendered it essentially a landlord-financed improvement (Thompson, 1963). As a capital investment it augmented the value of land, thereby providing the opportunity of a return on landlords' expenditure in the form of increased rent. The level and security of investment in underdraining were recognised by contemporary agriculturalists as factors influencing the adoption of the improvement, especially in the mid-nineteenth century when agriculture was becoming increasingly capital intensive.[1]

Considerable variation in opinion existed, however, among nineteenth-century agriculturalists of the amount of return to be expected from draining investment. In the 1840s and 1850s draining was thought to enhance permanently the value of the inheritance to all future proprietors. Estimates of return on capital outlay in terms of increased rent varied but they were rarely less than 10 per cent and often as high as 25, averaging about 15 per cent.[2] The security of an ample return from investment in underdraining was accepted widely, Lord Stanley considering in 1841 that 'no bank, no commercial speculation, no new estimate of capital is so safe, sure and provident as borrowing the capital for the purpose of investing it under the ground of your soil' (Newman, 1845, p. 2).

The view that underdraining was not a speculative investment but one that produced a substantial and measurable return was claimed still in 1863 by G. Darby, one of the Enclosure Commissioners.[3] By 1873, however, the evidence given to a select committee of the House of Lords on the improvement of land revealed that opinions had begun to range widely on the value of underdraining as an investment. A small number of witnesses maintained that landlords could obtain an ample return from draining. G. Ryder of the Lands Improvement Company quoted seventeen examples of draining on estates over a ten-year period, and in ten cases increased rent expressed as a return on draining outlay exceeded 10 per cent.[4] The greater body of evidence did not endorse this view. In the report, underdraining was not regarded as a secure investment because it could break down. To compensate for this risk, landowners had to be assured of sufficient interest. In much of the evidence, draining was thought to provide a return of between 5 and 7 per cent, expressed as increased rent. As such, the opinion was voiced in the report that by 1873 underdraining as an improvement was not sufficiently lucrative to attract much capital.[5]

Presentday studies of the return obtained from investment in either draining or general agricultural improvement in this period have revealed little variation in their conclusions. Concentrating on the analysis of rent increase as a percentage of improvement expenditure, they have reinforced the views of the 1873 report that landowners' investment in draining was never an economic proposition (Thompson, 1959, 1960; Spring, 1963; Collins and Jones, 1967; Phillips, 1972). For example, on eight great estates, representing about 5 per cent of all land in England held in such estates and lying mainly in the eastern parts of the country, Thompson (1963) has shown that between 1847 and 1878 total improvement outlay produced a return of 2.36 per cent in the form of increased rent. Again, in a discussion of the adoption of mechanisation in reaping, David (1971) has argued that between 1851 and 1871 a rate of return of not higher than 4 per cent could be achieved in the corn growing areas of England from landlords' investment in 'mechanisation-cum-improvement projects', which ranged from underdraining to purchasing a manual sheaf-delivery reaping machine.

As presentday studies emphasise the uniformly low return on all agricultural improvement investment in the mid-nineteenth century, the diversity of contemporary estimates of the amount and security of return on underdraining warrants closer inspection, especially, as they were recorded as factors of importance in its adoption. This essay attempts to examine the amount of return that landlords and agriculturalists anticipated and were prepared to accept on capital outlay in draining in the mid-nineteenth century. To this end a study has been made of proposed draining schemes on 133 Midland estates, analysing the expected return in the light of the area intended to be drained and the reason for undertaking the improvement.

Background to source materials

The opportunity of examining these problems stems from the reports that Andrew Thompson made to the Enclosure Commissioners between

1857 and 1868 on loans for draining under the various land improvement Acts that came into existence in the second half of the nineteenth century. These Acts had their origins in attempts to alleviate the difficulties that contemporaries, considered tenants for life under strict family settlement, experienced in trying to improve their estates (Anon, 1845; Caird, 1878; Spring, 1963). Acts were introduced in 1840 and 1845, the aim of which was to enable owners of settled estates to defray the expense of draining by way of a mortgage. Although little used because of the lengthy and costly procedure involved in their administration, they established the principle that tenants for life might borrow for agricultural improvement on the security of their estates, repaying capital and interest in a rentcharge over a period of years.[6]

This principle was developed more effectively in the Public Money Draining Act, 1846, by which £2 million was made available to landowners to borrow for underdraining. By this Act, landowners wishing to borrow money to drain their estates applied to the Enclosure Commissioners, the administrative body, for an advance. The Enclosure Commissioners arranged for the land to be examined by an Assistant Enclosure Commissioner, or inspector as they became known, who reported on the proposed draining. If a favourable report was received, the Enclosure Commissioners issued a provisional certificate to the amount of the proposed loan, so that an applicant could obtain credit to carry out the improvement. When completed, an inspector examined the draining to ensure that it had been properly done. If satisfied, the Enclosure Commissioners ordered an absolute certificate to be issued by the Treasury to the amount spent by the landlord.[7] With the exception of the Landowners' Drainage and Enclosure Company's Act, 1847, and the West of England and South Wales Land Drainage Company's Act, 1848 (the companies established by which lent only a small amount for improvements to be made a charge on estates),[8] subsequent nineteenth-century land improvement Acts and Acts incorporating improvement companies in England retained the Enclosure Commissioners as the main administrative body for such loans: the Private Money Draining Act, 1849; the Public Money Draining Act, 1850; the Improvement of Land Act, 1864; the General Land Drainage and Improvement Company's Act, 1849; the Lands Improvement Company's Act, 1853; and the Land Loan and Enfranchisement Company's Act, 1860.[9]

Under all these Acts inspectors produced two reports on draining loans: the first commented on the proposed draining, detailing the land to be drained, the planned system of draining, the estimated cost and the probable increase in the value of land after draining; the second ensured that the draining had been carried out in the proposed manner. As the improvement legislation aimed at helping owners of settled estates, the Enclosure Commissioners were authorised to protect the reversionary interests of such estates, hence their concern with a system of draining that would be long-lasting and with the increased value that draining would bring to land. If draining brought no increased value or failed before the loan was amortised, the rentcharge became yet another burden on the estate and defeated the object of the legislation (Denton, 1868). Given these considerations of the Enclosure Commissioners, the reports of the inspectors should be of significance in examining draining as an agricultural investment.

The precise value of these reports depends on the efficiency of the inspectors. Although Spring (1963) has detailed the work of the Enclosure Commissioners, little is known about the ability of the inspectors to comment on the improvement's permanence and the increased land value that draining could produce. With regard to type of draining, the Enclosure Commissioners, believing the majority of landowners to be ignorant on the subject, authorised inspectors to institute a uniform system of pipe draining, 1.05 to 1.25 m deep, that was thought to be efficient and permanent in 99 per cent of cases.[10] This uniformity was not always appreciated and complaints were raised about the draining systems enforced by the Enclosure Commissioners and their inspectors. The ability of inspectors, however, to gauge the increased value of land after draining was widely recognised. In 1855, W. Napier of the Lands Improvement Company recorded that 'the inspectors . . . are all practical gentlemen, generally persons residing in the neighbourhood of the proposed improvements, who are engineers or surveyors or agriculturalists by education; and . . . their estimate as to the probable success of these improvements [draining] . . . is not speculative'.[11] Indeed Spring (1963) considered that the inspectors were the leading land agents of their time. By 1873 there were forty-six inspectors: at the beginning of the period under review they included in their number such agents and draining experts as J. Clutton, H. Davis, J. Dickson, J. H. Girdwood, A. MacCaw, S. Martin, R. Neilson, J. Parkes, L. H. Spooner and Andrew Thompson.[12]

The professional status of the inspectors would seem to give their reports an added importance in estimating the value of draining. At present, few of these reports have come to light, the main exception being those of Andrew Thompson. Thompson was appointed an inspector to the Enclosure Commissioners in 1847 on the recommendation of J. Parkes, the leading authority on underdraining at the time, and remained one until his death in 1869. Born in 1823, he possessed many of the professional qualities attributed to inspectors. He had been employed as bailiff to C. Arbuthnot's small Northamptonshire estate (700 acres). In December 1849, he came to Keele as agent to R. Sneyd's Staffordshire estate (9 200 acres). Arbuthnot recognised him as an excellent agent with the ability of upgrading farming systems in badly cultivated districts. In his twenty years as agent to Sneyd, he converted 'Keele into a model estate worked by good tenants, according to the most advanced principles and methods' (Spring, 1956, p. 553; Sturgess, 1965).

Thompson's reports to the Enclosure Commissioners under the land improvement Acts were collected in volumes, but only the last five covering the period between 1857 and 1868 are extant.[13] In these five volumes, Thompson records all inspections for all improvements sanctioned by these Acts. Only those relating to draining will be considered in this essay and they form the basis of the following discussion.

Thompson's reports

Thompson's report of his first visit to an estate to comment on the proposed draining usually detailed the estate's size, its location and soils, the area to be drained and in need of draining, the reasons for carrying out the improvement, the type of draining system and likely cost, and the increased yearly value of the land as a result of draining expenditure. His second visit to sanction the improvement produced a less detailed report, normally only noting the amount spent and whether the drains were functioning satisfactorily. Such reports provide the opportunity of examining the area that land owners intended to drain, the reasons for undertaking the improvement as seen by Thompson, and the anticipated return on such draining as estimated by a leading contemporary land agent.

Between 1857 and 1868 Thompson inspected 133 estates, which in this essay will be examined together. These estates may be regarded as forming case studies: for each there exist at least one initial report (for some estates there are several initial reports as owners continued to borrow money to extend their draining activities) and a report of a subsequent visit to sanction the draining, allowing a rentcharge to be established. Further information relates to draining on another fourteen estates, but, as only the sanctioning reports are available, the more useful initial report being confined to the missing pre-1857 volumes, these have been excluded from the analysis.

The exact size and location of some of the 133 estates are difficult to determine, especially when large estates spread over more than one county. In this analysis estate size has been calculated on a single county basis, while estate location has been taken to be that of its name. Where estate size has been omitted from Thompson's reports, it has been obtained by reference either to the *Return of Owners of Land in England and Wales, 1873* (BPP, 1874) or to Bateman (1883). The area covered by the 133 estates amounted to 367 474 acres or 1.3 per cent of

Table 12.1 Sample estates by size and acreage

Estate size (acres)	*Number of estates*	*Total acreage*	*Average acreage*	*Percentage of total estate acreage in England*
Less than 1 000	68	18 063	266	0.2
1 000–2 999	31	55 312	1 784	1.5
3 000–9 999	24	132 015	5 501	2.7
10 000 and over	10	162 084	16 208	2.3
Total	133	367 474	2 763	1.3

Sources: Sneyd MSS, Reports of Andrew Thompson, vols iv–viii; BPP, 1874; Bateman, 1883.

all land owned in England in 1873 (Table 12.1). Using the four major categories of estate size distinguished by Thompson (1963), more reports exist for smaller than for larger estates: sixty-eight estates of less than 1 000 acres are represented in the sample as against only ten over 10 000 acres. However, the area covered by large estates in the sample far exceeded that of the small. At the same time the larger estates represented a greater proportion of the total area of their respective categories of estate size than the smaller: the sixty-eight estates with less than 1 000 acres covered only 0.2 per cent of all land in England held in such estates, while the ten estates with over 10 000 acres occupied 2.3 per cent of their total estate area. Despite the small area covered in the sample, a range of examples of proposed draining schemes on estates of all sizes is provided.

The estates that Thompson inspected lay predominantly in the Midlands, in the area that Caird (1852) described as the principal grazing, green crop and dairy districts (Fig. 12.1). Only twelve estates (22 per cent of the total area in the sample) lay in Caird's corn district. This regional grouping might be expected given Thompson's agency at Keele and the Enclosure Commissioners' policy of providing each inspector with a district close to his home.[14] At the same time, the distribution of estates visited by Thompson demonstrates that inspectors were involved in covering considerable areas. The concentration of estates means that any conclusions reached in this essay will apply to the pastoral districts of the Midlands, a useful counterbalance to existing research, which has dealt in the main with the eastern corn counties of the country.

The data present several difficulties of analysis. The degree of bias in the sample cannot be determined. At this stage it is impossible to discover whether there was any inherent bias among owners who applied for draining loans under the improvement Acts. Moreover Thompson's inspections provide details of only 133 Midland estates. They are not a complete record of draining under the land improvement Acts in that area and the degree to which the sample is representative is difficult to judge. However, there is no evidence within the reports that the estates

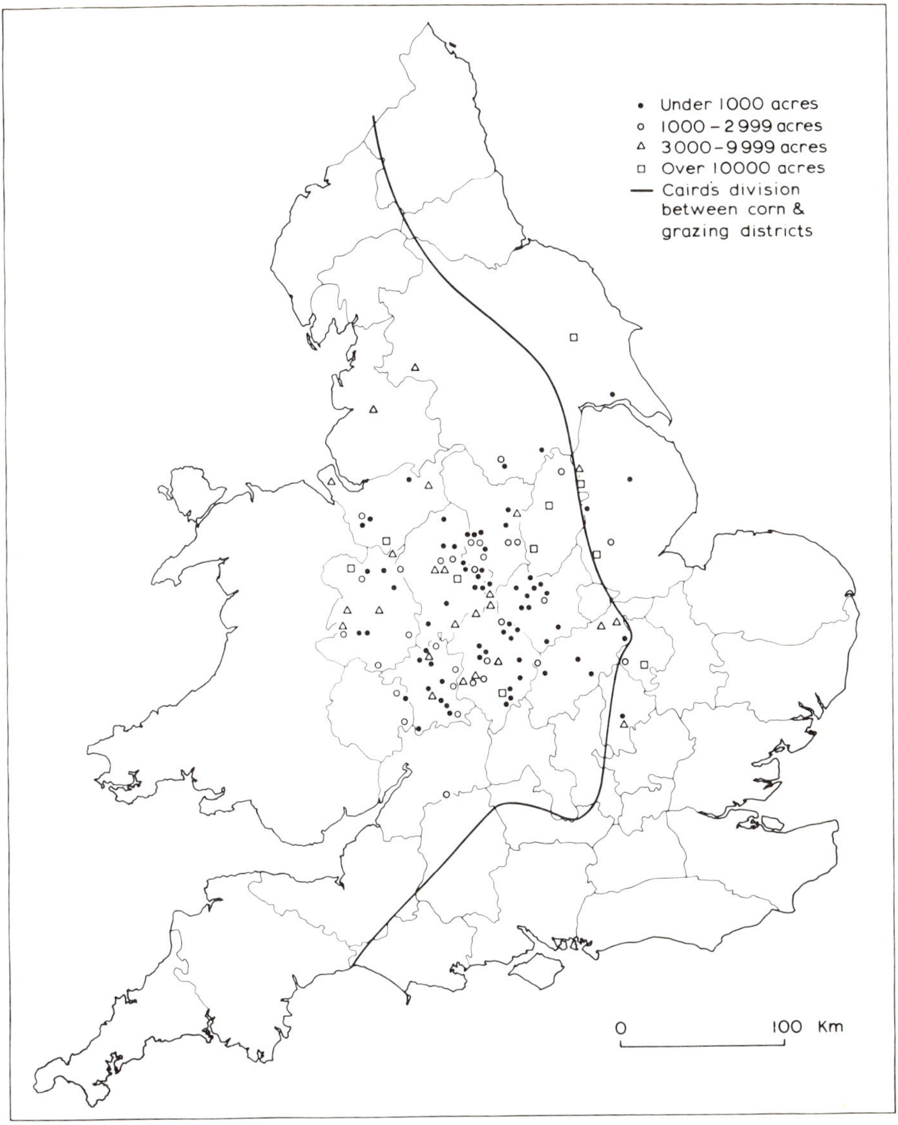

Fig. 12.1 Estates inspected by Andrew Thompson, 1857–68. Based on Sneyd MSS, Reports of Andrew Thompson to the Enclosure Commissioners, vols iv–viii; Caird, 1852

Thompson inspected were biased to any one size category, in general the inspectors being allocated districts rather than distinct groups of estates.

This analysis of anticipated return on draining is based solely on draining carried out under the land improvement Acts. The proportion that such draining formed of all draining is difficult to establish (Phillips,

1969). Some landowners, such as the Dukes of Bedford and Northumberland, financed all draining themselves (Spring, 1963; Thompson, 1963). Others relied on improvement Act loans to finance a high proportion of draining on their estates: on the Earl of Scarbrough's Yorkshire estate loans contracted under various improvement Acts accounted for 75 per cent of draining capital after 1863 (Phillips, 1972). Such a situation may have a bearing on the examination of the amount of land to be drained on the estates in the sample, but Thompson's estimate of the return on the improvement should not be affected.

Area proposed to be drained

On 99 of the 133 estates, Thompson noted the area in need of draining. He regarded this amount as wet land, leaving aside the question of whether or not draining would produce a return on the outlay. Acreages stated as needing draining tend to be estimates, being given in round sums. Although lacking precision, they provide an order of magnitude of the need for draining. Predictably, the absolute area increased with estate size: on average 159 acres of each estate under 1 000 acres required draining, as against 3 500 acres on estates in the largest category (Table 12.2). However, the amount of land needing draining, expressed as a proportion of either the total area of each estate category or average estate size demonstrated a reversed trend: as estate acreage increased, the proportion requiring draining declined. Whereas the average proportion of estates less than 1 000 acres in need of draining was 65 per cent, only 22 per cent of the area of estates over 10 000 acres required the improvement. Thompson's reports reveal that a higher proportion on small estates needed draining than on large ones.

The pattern of proposed draining on the 133 estates reflected the need for the improvement. Although the absolute area to be drained increased

Table 12.2 Land needing draining

Estate size (acres)	*Number of estates with data*	*Total acreage of estates with data*	*Acreage needing draining*	*Percentage of estates needing draining*	*Average estate size*	*Average acreage needing draining*
Less than 1 000	54	13 187	8 581	65	244	159
1 000–2 999	23	41 940	21 440	51	1 823	932
3 000–9 999	15	86 330	39 700	46	5 755	2 647
10 000 and over	7	112 324	24 550	22	16 046	3 507
Total	99	253 781	94 271	37	2 563	952

Sources: As for Table 12.1.

with estate size, proportionally a greater part of small estates was to be treated (Table 12.3). The average expenditure per acre by estate size increased from £0.84 for the largest to £2.50 for the smallest category, while the proportion to be drained rose from 13 to 43 per cent of the respective estate acreages. This proposed greater intensity of draining on the smaller estates was aided by a cost of draining which over this twelve-year period favoured them. For the ninety-nine estates for which data exist, it is possible to gauge how much wet land the draining schemes intended to treat (Table 12.4). For no estate category did the area proposed to be drained deal completely with all the wet land. With the exception of estates between 3 000 and 9 999 acres, the proportion of wet land to be drained was about 75 per cent on all categories of estate size. It would seem that a high proportion, if not all, of wet land was to be drained, the extent to which the problem was dealt with tending to be uniform among estates in the sample.

Three broad conclusions emerge from this analysis of the extent of proposed draining. In absolute terms, there was more wet land, more money to be spent on draining and more land to be drained on the larger estates. Proportionally, however, the amount of land in need of draining was greatest on the small estates, as was the area to be drained. Despite the varying amounts of wet land and land proposed to be drained, the intensity of draining the area of wet land tended to be constant, irrespective of estate size.

From the analysis of these 133 estates, the impression is given of greater draining activity on smaller estates. In reality, this may not have been the case. Small estates with less income had less capital to spend on agricultural improvement. The high proportion of wet land on such estates would suggest that draining had been neglected although needed. Such estates relied on loans under the land improvement Acts to remedy this situation by introducing large-scale draining schemes: hence their high level of proposed draining activity. Thus on T. K. Hall's Staffordshire estate of 1 620 acres, very little draining had been done despite the fact that fully two-thirds of it required draining. Similarly on L. Southwell's estate of 1 000 acres in Herefordshire, the amount of dry land measured about 100 acres: virtually no draining of any description had been carried out, the late Lord Southwell not spending anything on the improvement. Again on Rev. H. W. Sitwell's 1 800-acre estate in Derbyshire, 1 100 acres needed draining, as the improvement had been neglected.[15]

The number of estates in the sample and the absolute amount of land to be drained must also be considered in the context of greater draining activity on smaller estates. The average area proposed to be drained on estates of less than 1 000 acres was only 5 per cent of that on estates over 10 000 acres. Despite the greater number of the smallest estates, the

Table 12.3 Proposed draining on all estates

Estate size (acres)	*Total proposed expenditure* (£)	*Expenditure per acre* (£)	*Expenditure per estate* (£)	*Total acreage to be drained*	*Acreage to be drained as percentage of total estate area*	*Average acreage to be drained per estate*	*Average cost of draining per acre* (£)
Less than 1 000	45 137	2.50	664	7 686	43	113	5.87
1 000–2 999	115 811	2.09	3 736	18 830	34	607	6.15
3 000–9 999	176 327	1.34	7 347	28 343	22	1 181	6.22
10 000 and over	136 106	0.84	13 611	21 552	13	2 155	6.32
Total	473 381	1.29	3 559	76 411	21	575	6.20

Sources: As for Table 12.1.

Table 12.4 Proposed draining on ninety-nine estates where area in need of draining stated

Estate size (acres)	*Total proposed expenditure* (£)	*Expenditure per estate* (£)	*Total acreage to be drained*	*Acreage to be drained as percentage of estate area*	*Average acreage to be drained per estate*	*Acreage to be drained as percentage of area needing draining*
Less than 1 000	38 497	713	6 514	50	121	76
1 000–2 999	98 681	4 290	15 924	38	692	74
3 000–9 999	150 935	10 062	23 876	28	1 592	60
10 000 and over	119 150	17 021	18 487	17	2 641	75
Total	407 269	4 114	64 801	26	655	69

Sources: As for Table 12.1.

total acreage that was to be drained on these represented only 36 per cent of that on the largest estates. These figures are important, for in the country as a whole small estates were greater than the large both in number and in total area. The possibility exists that the small estates that Thompson inspected were exceptional. For example, Thompson made more inspections (twenty-two) in Staffordshire than in any other county. Of these, five were on estates of more than, and seventeen on estates of less than 3 000 acres. In Staffordshire the number of estates over 3 000 acres was thirty-three of which Thompson inspected 15 per cent, as against 9 205 estates between 1 and 3 000 acres, the number Thompson visited being 0.2 per cent of the total (Bateman, 1883). Not only were the small estate owners who proposed to drain few among their number, but they belonged to a distinct social group. Of the sixty-eight estates under 1 000 acres with proposed draining schemes, forty-one belonged to rectors who were attempting to improve their glebeland.

These considerations should modify the conclusions about draining intensity on different-sized estates in the sample. It would seem that the proportion of owners of small estates interested in draining was limited and restricted mainly to a specific social group. Not only was draining neglected on smaller estates, but only a few tried to remedy the situation. On those small estates where draining was adopted, the intensity tended to be greater than on others.

Reasons for draining

Draining was needed in varying degrees on all estates in the sample. In his reports, Thompson usually stated the immediate reasons for landowners contracting a draining loan. An analysis of these has relevance to landowners' anticipated return on draining. On estates less than 1 000 acres, two factors stand out. The more common was physical: land was excessively wet. Of the sixty-eight estates in this group, thirty-three were classed as very wet and on these Thompson regarded draining as of vital importance. On the majority of such estates there was no record of the land having been drained before: ridge and furrow existed on at least twelve and served as surface drains. Thompson noted the existence of tenant-financed and shallow draining systems on only six of the sixty-eight estates: these had failed, providing the need for new draining schemes. The less common reason for draining, recorded for only eight estates, was to reinvigorate land that had been neglected and required to be put in order to attract tenants and maintain rent. Thus H. Benton's 300-acre estate in Worcestershire had been rented at 15 shillings an acre, a product of neglect, and a considerable outlay was needed to improve its state of cultivation, while the draining of 35 acres on Rev. H. J. Peache's glebeland of 70 acres at Tutbury in Staffordshire was the result of land being long out of order.[16]

Excessive wetness of land remained a significant motivating force for draining on larger estates, being given as a reason on ten estates between 1 000 and 2 999 acres, thirteen between 3 000 and 9 999 acres and on three over 10 000 acres. Thus Thompson noted that he had never inspected as large an estate as that of Sir. H. B. Houghton in Lancashire more in want of draining, where out of 4 300 acres he had no difficulty in finding 2 100 acres in need of close draining.[17] However, with increasing size the neglect of estate improvement became more important in encouraging landowners to drain. Neglect could arise from various factors. Excessive financial burdens could restrict the capital available for improvement: on C. T. S. Kynnersley's estate of 2 570 acres in Staffordshire, Thompson recorded that little draining of any description had been done for the last fifteen years.

> The debt, family annuities and the burden including the keeping up of the estate in fair order have hitherto absorbed the whole of the income . . . and the present owner who in [1858] . . . succeeded to the property . . . will not be able to derive much advantage from it till the decease of the annuitants or until the rental can be safely increased through judicious improvements.[18]

On the other hand, previous owners could have ignored agricultural improvement. Lord Middleton in 1861 and 1863 proposed to spend £45 000 on draining his 39 900-acre estates in Lincolnshire, Nottinghamshire, Warwickshire and Yorkshire. Thompson noted that

> for more than half a century very little was done upon these large estates in the shape of permanent improvements . . . and though the late Lord Middleton left upwards of £300 000 to be laid out in the purchase of land, no portion of that large sum could be applied to the improvement of the settled estates. The late agent, a Mr Chowler . . . was also against modern improvements. Under such circumstances the time was sure to arrive when either a considerable loss in the revenue would have to be submitted to, or a heavy outlay incurred to meet the necessary requirements of the land in accordance with modern improvements.[19]

A similar situation was recorded on part of the Duke of Manchester's estate in Huntingdonshire, formerly held by Lady Olivia Sparrow. She had carried out little permanent improvement and 'like all other neglected estates there is a great deal to be done by the present possessor, otherwise he must submit to low rents and indifferent farmers'.[20] It is possible that improvement investment may have been less on detached portions of estates. Thus the Marquis of Bath's 3 510-acre estate in Shropshire had been allowed to get 'much out of order'.[21] In all neglect was recorded as a factor for draining on twelve estates between 1 000 and 2 999 acres, on seven between 3 000 and 9 999 acres and on five over 10 000 acres.

As a direct result of the decision to drain, Thompson recognised that on all estates the state of cultivation would be improved. Thus on estates under 1 000 acres where the aim was to drain pasture, as recorded for twelve such estates, the absence of the improvement had rendered pasture coarse and full of rushes, enforcing low stock densities. Of the eighteen estates under 1 000 acres where draining of arable was specified, Thompson considered that on seven which possessed light soils turnips and barley could be grown after draining, but only on two would draining allow the introduction of turnips to strong soils. On the remaining estates where arable was to be drained the achievement of heavier and more regular yields of wheat and beans was sufficient reward.

Thompson's reports suggest that the main reasons for landowners contracting a draining loan were attempts to remove excessive wetness from land and to improve land long neglected by estate management. Increased agricultural production was expected to result, but rarely was it conceived that new agrarian regimes would be adopted. Thompson considered that turnips and barley could be introduced after draining but mainly on light soils. There is little evidence of conversion of arable to pasture or vice-versa after draining, despite much contemporary advocacy for the latter (Trimmer, 1847; Sturgess, 1966). Draining was seen more as a means of securing the yield of crops already grown. Thompson emphasised underdraining as a measure to render manageable land in poor condition, which in its existing condition would have provided little profit for tenant and landlord alike.

Return on draining

For draining to provide a return, the improvement had to be efficient. Thompson enforced the adoption of what he considered a durable and satisfactory system of draining. Disliking tenant draining as unreliable, and shallow and turf drains because they were shortlived, he applied the Enclosure Commissioners' system of deep, pipe drains: 'if the plan proposed were in the least degree a departure from deep and permanent drainage, I should be the very last person to consent to it'.[22] Generally, he demanded a depth of 1.25 m for drains, regardless of soil type. If, occasionally, drains were sanctioned at lesser depths, they had to be closer together than those at 1.25 m: drains at 1.05 m were to be laid out not wider than 7.3 m apart.[23] Where land to be drained lay in ridge and furrow, drains were to be placed in the furrows. On such land he allowed drains 1.05 m deep but only on the basis that the ridges were to be levelled after draining, by which process the depth of drains in the furrows would be increased to 1.25 m.[24] The distance between drains depended much on the nature of the soil, varying from 7.3 to 11 m apart in clays and marls, and being much wider in lighter soils, 16.5 to 22.8 m apart or just occasional drains.

In spite of his enthusiasm for deep and permanent draining, he would not sanction systems that involved unnecessary complexities and additional costs. Thus on C. H. Cust's 15 000-acre estate in Lincolnshire, S. Hutchinson, the agent and author of a draining treatise, although in favour of 1.25 m drains, wanted to admit a current of air into the works, as propounded in his paper. This plan involved the expense of providing not only air gratings but an additional drain in each field along the top of the minor drains. Thompson rejected the plan: although 'admitting air into the drains can do no harm . . . it is generally admitted to do very little good [and] it is a question for consideration whether the expense attendant upon it should be charged upon the estate'.[25] As an inspector, Thompson tried to obtain the most efficient draining system at the lowest cost and on several occasions he pointed out instances where the proposed draining expenses were excessive. Thus the average cost of draining 2 000 acres of the Earl of Dartmouth's Staffordshire estate was calculated at £6 per acre, which Thompson considered was in excess of the real cost by 10*s*. Again, on C. D. Hudson's 1 500-acre Shropshire estate, 400 acres required draining. Of this amount, 200 acres were to be drained with occasional drains at an average cost of £7 per acre. To Thompson this sum was excessive and should have been reduced by £1 an acre.[26] The advantages of efficient draining were apparent to Thompson who saw it as the 'foundation for other improvements which ought to be made by the landlord or his occupier in the shape of improved cultivation and manures, of improved buildings and a fresh arrangement of fields'.[27]

In his reports Thompson expressed the increased value of land after draining in terms of shillings per acre. As he recorded the proposed average cost per acre of draining on each estate, the estimated increased value may be expressed as a percentage of the outlay. In every case Thompson considered that the proposed draining would augment the value of the land. The greatest absolute increases resulted from draining impoverished meadowland. Thus the draining of meadowland on R. Corbet's and C. D. Hudson's estates in Shropshire would produce an increased yearly value of 15*s* and 25*s* respectively.[28] The Houghton estate in Lancashire provided the example of the greatest absolute return. Thompson noted that this was the wettest estate he had seen and in parts the increased value per acre after draining would be 20*s* averaging 13*s*.[29] On the majority of estates, the increased value ranged from 7*s* 6*d* to 10*s* per acre, irrespective of size, location and land use.

Variations in the level of return depended on existing rents, soil quality, the degree of wetness of the land and the cost of draining. On those estates which possessed good quality land but were excessively wet and experienced low rents, the return on draining was likely to be high. The Houghton estate exemplified this situation, where the expected return on draining outlay averaged 12 per cent. Similarly the land of Rev. S. Smith's 160-acre glebe at Whitwick in Leicestershire was described as of fair quality for grain or pasture and the draining of 45 very wet acres was anticipated to produce a return of 10 per cent on the outlay, as was the draining of the wetter parts of Sir C. F. Smythe's Shropshire estate.[30] There was, however, little chance of obtaining a high return on those estates which, although in need of draining, already levied high rents. This was the case on Sir R. Peel's estates in Staffordshire and Warwickshire where, because the land was already highly rented, some of the tenantry objected to the improvement.[31] On soil of poor quality the anticipated return on draining was also low. Thus although draining would allow the claylands of the Warwickshire estate of the Earl of Abergavenny to produce good crops of wheat and beans, they were not worth more than 20*s* an acre in rent. Thompson was of the opinion that an increased yearly value of 8*s* per acre was as much as such land could return on draining alone. As the cost of draining these claylands was high (£6 an acre), the return was low, around 6.5 per cent.[32]

Partially reflecting soil conditions, the cost of draining had relevance to the anticipated return. Where the cost was low, there was a greater opportunity for the increased value to represent a high return. The draining on T. Tyron's Northamptonshire estate of 4 300 acres cost about £4 per acre. As the increased yearly value was estimated at 8*s* an acre, draining would produce a return of 10 per cent on the outlay.[33] However, on those estates where draining costs were high, the provision of a return was in doubt. The Wigglesworth estate of Countess Cowper in Yorkshire covered 3 700 acres. About 165 m in altitude, it was almost all pasture, and being wet was rented at 15*s* an acre. Lying on a clay and stone soil, the cost of draining was high at £7 per acre. According to Thompson, the land when drained would meet an increased yearly value of about 9*s*, providing a return of just over 6 per cent. However, Thompson considered that if the cost exceeded £7, the draining was unlikely to yield a return.[34]

The average anticipated return per acre on draining expressed as a percentage of capital outlay can be recorded for 124 estates in the

Table 12.5 Anticipated return on cost of draining on estates

Estate size (acres)	Number of estates with data	Level of percentage return						
		6	7	8	9	10	11	12
Less than 1 000	65	6	30	17	9	2	1	0
1 000–2 999	30	1	19	9	1	0	0	0
3 000–9 999	21	2	9	5	2	1	1	1
10 000 and over	8	0	1	7	0	0	0	0
Total	124	9	59	38	12	3	2	1

Sources: As for Table 12.1.

sample (Table 12.5). On these estates there is considerable uniformity in the level of return. For no estate did Thompson note a return that was less than 6 or greater than 12 per cent on the cost of draining. Moreover, the level of return varied little between different categories of estate size. The majority of estates belonged to a group where a return in the form of increased value of land in the order of 7 or 8 per cent was expected on the cost of the improvement. An analysis of Thompson's reports on proposed draining schemes reveals that a return could be expected on investment in underdraining, that it could be measured and that it was not speculative. However, the anticipated return was not as great as many contemporary agriculturalists suggested.

Such low returns may be related to the land improvement Acts legislation. As an inspector, Thompson had in his initial reports to comment not only on whether draining would increase the value of the land but also on whether the return on draining would exceed the rentcharge that the loans incurred. The rentcharge under the land improvement Acts was to repay both principal and interest on the loan over a number of years and was expressed as a percentage of the amount borrowed. The level of the rentcharge varied, dependent on the sum borrowed, being reduced with larger sums, the period taken to repay and the current interest rates on money. Loans under the Public Money Draining Acts were charged for twenty-two years at 3.5 per cent and involved a standard rentcharge of 6.5 per cent on the outlay. None of the other means of borrowing under the land improvement Acts could offer funds at so low a rate and rentcharges were higher and more varied. Assuming a basic interest rate on money of 4.5 per cent, the rentcharge incurred under the Private Money Draining Act 1849 was for twenty-two years at 7.25 per cent on the outlay. The improvement companies had the opportunity of extending the period of repayment and of thereby reducing the rentcharge level: thus over twenty-five years the rentcharge became 6.7 per cent, over thirty-one years 6 per cent and over fifty years 5 per cent on the outlay. It was against these levels that inspectors estimated the likely return on draining (Denton, 1868).

By examining the registers of absolute certificates issued under the improvement Acts, the percentage that the rentcharge formed of the outlay on those estates that Thompson inspected can be calculated.[35] In no case was the anticipated return on draining less than the rentcharge. The return estimated by Thompson in the majority of cases was the same as the rentcharge or only exceeded it by one or two percentage points (Table 12.6). It may be argued that Thompson's estimated low returns on draining outlay were a product of his concern in merely demonstrating that the increased yearly value would exceed the rentcharge.

Table 12.6 Relationship of draining return to rentcharge

Estate size (acres)	*Number of estates with data*	*Draining return and rentcharge the same*	*Excess of draining return over rentcharge by percentage points*					
			1	2	3	4	5	>5
Less than 1 000	65	14	27	18	4	1	1	0
1 000–2 999	30	3	15	12	0	0	0	0
3 000–9 999	21	3	7	6	2	1	0	2
10 000 and over	8	0	5	3	0	0	0	0
Total	124	20	54	39	6	2	1	2

Sources: As for Table 12.1; PRO, 1R3/6–38; Lands Improvement Company MSS, Registers of absolute certificates.

Such a possibility is unlikely. Thompson was of the opinion that draining did not produce the large returns that many contemporaries claimed. The 20 per cent return that Rev. C. Neville expected from draining part of his Nottinghamshire estate was thought by Thompson to be a considerable overestimate, which he reduced to 8 per cent.[36] He expressed the view that care should be taken 'to avoid in our calculations extreme notions of increased value', adding that 'it frequently happens that when a moderately wet farm of 200 acres and rented at £300 is drained and raised to £400 (which is 10*s* an acre) the occupier is in a somewhat worse position than he was before the farm was drained'.[37] Indeed, Thompson considered that the extraction of high returns on investment in draining was not common among landowners. Thus when Rev. A. Oakley required his tenants on his 2 800-acre estate in Shropshire to pay more than the rentcharge, he noted that there were 'fortunately few landowners who in carrying out the improvement of an estate committed to their keeping [took] this much'.[38]

Although returns were lower than many contemporaries estimated, Thompson made it clear that landlords were intent on obtaining some form of return on underdraining. On estates of all sizes, the adoption of draining depended on the tenantry paying interest on the outlay, whether that was the full rentcharge, which could range from 5 to 7.5 per cent or some standard rate, normally 5 per cent (Moore, 1965). Thus

Thompson noted that Rev. J. L. C. Cooper did not want the income from his 580-acre estate in Cheshire 'to be lessened in any way through any outlay in improvements and [that] the tenants would have to pay the full charge', as they did on the estates of C. T. S. Kynnersley in Staffordshire and of C. R. B. Legh in Cheshire.[39] Sir H. B. Houghton levied 6.5 per cent interest from tenants for draining his Lancashire estate, while on the estates of the Duke of Manchester in Huntingdonshire and of C. Turnor in Lincolnshire a high yearly charge for the outlay was expected.[40]

The payment of the rentcharge or interest on the outlay lessened tenant enthusiasm for the improvement. Tenants applied to have only 120 acres drained out of the 500 needing it on Rev. J. L. C. Cooper's Cheshire estate: 'working tenants are afraid of having interest to pay and I [Thompson] am not surprised that they did not apply for more draining'. Similarly on P. S. Wilkinson's Lincolnshire estate of 1 070 acres, the tenants admitted to only 275 acres as requiring draining, despite Thompson's belief that a much greater part of the estate was in need of the improvement.[41] Thompson's reports suggest that in most cases the tenant was the vital variable, for the interest he was prepared to pay on draining outlay dictated the area that the landlord was prepared to drain. Thus on the Northamptonshire estate of the Earl of Westmorland, as the occupiers were to pay the whole rentcharge, land was not to be drained against their wish, while Thompson recorded that on Lord Bagot's Staffordshire estate 'preference will be given to the occupiers who are most eager to have their land drained and are willing to pay a fair percentage on the outlay'.[42]

Conclusions

The average return on outlay that Thompson estimated draining would provide on these 133 Midland estates was much lower than that suggested by agriculturalists in the 1840s and 1850s and it corresponded closely to that put forward in the 1873 report on the improvement of land. The estimates of the 1840s and 1850s were likely to be excessive: a new system of draining had been introduced and the claims made for the improvement were based on enthusiasm rather than experience. Thus it was often reported that drains could last for ever or at least 200 years.[43] H. W. Keary could note in 1873 that twenty years before the return on draining had been exaggerated: 'we have learned that land is not altogether so expansive as some people thought it was'.[44] The view expressed in the 1873 report that draining provided insufficient return to attract capital was very much a reaction to the failure of landlords to obtain the levels of return imagined in the 1840s.

Thompson demonstrated that high returns on investment in underdraining could not be expected between 1857 and 1868. His reports reveal that the majority of landowners, whose estates he inspected, were aware of this situation and did not carry out the improvement in the anticipation of high returns. That landlords wanted some return was undeniable. According to Thompson's figures, the increased yearly value of land after draining represented on average 7 or 8 per cent of the outlay in this period, irrespective of estate size. The amount of return received, however, depended on how much the tenantry were prepared to pay. The evidence in Thompson's reports suggests that tenants were unwilling to pay large amounts, certainly not in excess of 7 per cent on the cost of draining. Such reduced expectations make the low returns actually obtained from draining, that presentday studies have revealed, less unusual. Moreover, it would be wrong to regard draining solely as a financial speculation, the view expressed in the 1873 report. Draining had a restorative function for land: Thompson cited as the two main reasons for draining on these 133 estates the need to render cultivable land that was excessively wet and to put in order land long neglected by estate management. The return was known not to be high but the improvement was carried out to maintain rental levels and to prevent loss. Whether or not the lack of high return from investment in draining made the improvement less acceptable to owners of small estates, which on the evidence of Thompson's reports had the greatest need of draining, remains a subject for further research.

Notes

1. BPP (1873) *Report of the Select Committee of the House of Lords on the improvement of land*, **16**, iii–iv.
2. BPP (1845) *Report of the Select Committee of the House of Lords . . . to enable possessors of entailed estates . . . to charge . . . for the purpose of draining* (Lords Sessional Papers), **18**, i, Q.590 and 657; BPP (1847–48) *Report of the Select Committee on Agricultural Customs*, **7**, Q.1450–51 and 4119–22; BPP (1849) *Report of the Select Committee of the House of Lords . . . to enable possessors of entailed estates . . . to charge . . . for the purpose of draining* (Lords Sessional Papers), **30**, Q.4–7; BPP (1852) *General Board of Health: minutes of information collected in respect to draining of the land*, **19**, Appendix 7, 69; BPP (1854–55) *Report of the Select Committee of the House of Lords . . . on the powers vested in the companies for the improvement of land*, **7**, Q.322; J. H. Charnock, (1844) *On Thorough Draining and its Immediate Results to the Agricultural Interest*, 2nd edn, London, pp. 6–7.
3. BPP (1863) *Report of the Select Committee of the House of Lords on the charging of entailed estates for railways*, **7**, Q.306.
4. BPP (1873), **16**, Q.976–80 and Appendix E, 383.
5. *Ibid*, iii–v and Q.1207, 2018–21 and 3025–7.

6. *An Act to enable the owners of settled estates to defray the expense of draining the same by way of mortgage*, 3 & 4 Vict., c.55; *An Act to alter and amend . . .; An Act to enable the owners of settled estates . . . by way of mortgage*, 8 & 9 Vict., c.56; Denton (1868), pp. 123–5.
7. *An Act to authorize the advance of public money to a limited amount to promote the improvement of land . . . by works of drainage*, 9 & 10 Vict., c.101; Spring (1963), pp. 161–5.
8. *An Act for incorporating the Landowners Drainage and Inclosure Company*, 10 & 11 Vict., c.ccxii; *An Act for incorporating the West of England and South Wales Land Draining Company*, 11 & 12 Vict., c.cxlii; BPP (1854–55) 7, Q.2; BPP (1873) 16, Q.303–4.
9. *An Act to promote the advance of private money for drainage of lands in Great Britain and Ireland*, 12 & 13 Vict., c.100; *An Act to authorize further advances of money for drainage and the improvement of landed property in the United Kingdom*, 13 & 14 Vict., c.31; *The improvement of land Act, 1864*, 27 & 28 Vict., c.114; *An Act for the incorporation of the General Land Drainage and Improvement Company*, 12 & 13 Vict., c.xci; *An Act for incorporating the Lands Improvement Company*, 16 & 17 Vict., c.cliv; *An Act for incorporating and granting other powers to the Land Loan and Enfranchisement Company*, 23 & 24 Vict., c.clxix.
10. BPP (1854–55), 7, Q.71–82.
11. *Ibid*, Q.321; BPP (1865) *Rules and Practices of the Enclosure Commissioners*, 47, 429.
12. BPP (1849), 30; BPP (1852), 19; BPP (1854–55), 7, Q.113.
13. Sneyd MSS., Keele University, Reports of Andrew Thompson to the Enclosure Commissioners, vols iv–viii.
14. BPP (1854–55), 7, Q.32, 114.
15. Sneyd MSS, Reports of Andrew Thompson, v, pp. 290–11; vi, pp. 80–2 and 272–4.
16. *Ibid*, iv, pp. 85–8 and 163–5.
17. *Ibid*, iv, pp. 277–80.
18. *Ibid*, iv, pp. 174–7.
19. *Ibid*, vi, pp. 214–22.
20. *Ibid*, vii, pp. 227–34.
21. *Ibid*, vii, pp. 171–6.
22. *Ibid*, vi, p. 61.
23. *Ibid*, viii, pp. 147–9 and 200–1.
24. *Ibid*, v, pp. 96–8.
25. *Ibid*, v, pp. 157–61; vii, pp. 71–4 and 123; S. Hutchinson, (1851) *Practical Instructions on the Drainage of Land*, Grantham.
26. Sneyd MSS, Reports of Andrew Thompson, iv, p. 201; v, p. 83; viii, p. 111.
27. *Ibid*, v, p. 215.
28. *Ibid*, vii, p. 65; viii, p. 111.
29. *Ibid*, iv, pp. 277–80.
30. *Ibid*, viii, p. 2; v, pp. 300–3.
31. *Ibid*, vi, p. 87.
32. *Ibid*, vi, p. 64.
33. *Ibid*, v, p. 262; vi, p. 68.

34. *Ibid*, vi, p. 266.
35. PRO, 1R3/6–38, Certificates of Draining Advances under the Public Money Draining Acts; Lands Improvement Company MSS, London, Registers of absolute certificates.
36. Sneyd MSS, Reports of Andrew Thompson, v, pp. 352–3.
37. *Ibid*, v, pp. 43–8 and 215.
38. *Ibid*, vii, p. 136.
39. *Ibid*, vii, p. 179; iv, pp. 174 and 265.
40. *Ibid*, iv, pp. 277–80, vii, p. 227; vi, p. 158.
41. *Ibid*, vii, p. 179; iv, p. 47.
42. *Ibid*, v, pp. 217 and 176.
43. BPP (1845) **18**, Q.561–63; BPP (1847–48) **7**, Q.3908–23, 4195–201; BPP (1849) **30**, Q.116.
44. BPP (1873) **16**, Q.3025–27.

References

ANON (1845) *Entailed and Settled Estates Improvement Bill: speeches of James Adam Gordon . . .; James Smith . . .; and John Charnock*, London.

BATEMAN, J. (1883) *The Great Landowners of Great Britain and Ireland*, London.

BPP (1874) *Return of Owners of Land in England and Wales, 1873*, **72**, parts 1 and 2.

CAIRD, J. (1852) *English Agriculture in 1850–51*, London.

CAIRD, J. (1878) *The Landed Interest and the Supply of Food*, London.

COLLINS, E. J. T. and JONES, E. L. (1967) 'Sectoral advance in English agriculture, 1850–80', *Agric. Hist. Rev.*, **15**, 65–81.

DARBY, H. C. (1964) 'The draining of the English clay-lands', *Geogr. Z.*, **52**, 190–201.

DAVID, P. A. (1971) 'The landscape and the machine: technical interrelatedness, land tenure and the mechanization of the corn harvest in Victorian Britain', in D. N. McCloskey, ed., *Essays on a Mature Economy: Britain after 1840*, Methuen, pp. 145–205.

DENTON, J. B. (1868) 'On land drainage and improvement by loans from government or public companies', *J. R. Agric. Soc.*, second series, **4**, 123–43.

MOORE, D. C. (1965) 'The corn laws and high farming', *Econ. Hist. Rev.*, second series, **18**, 544–61.

NEWMAN, C. (1845) *Practical Hints on Land Draining*, London.

PHILLIPS, A. D. M. (1969) 'Underdraining and the English clay-lands, 1850–80: a review', *Agric. Hist. Rev.*, **17**, 44–55.

PHILLIPS, A. D. M. (1972) 'The development of underdraining on a Yorkshire estate during the nineteenth century', *Yorks. Archaeol. J.*, **44**, 195–206.

SPRING, D. (1956) 'Ralph Sneyd: Tory country gentleman', *Bull. John Rylands Libr.*, **38**, 535–55.

SPRING, D. (1963) *The English Landed Estate in the Nineteenth Century: its administration*, Johns Hopkins Press.

STURGESS, R. W. (1965) 'The response of agriculture in Staffordshire to the price changes of the nineteenth century', unpublished PhD thesis, University of Manchester.

STURGESS, R. W. (1966) 'The agricultural revolution of the English clays', *Agric. Hist. Rev.*, 14, 104–21.

THOMPSON, F. M. L. (1959) 'Agriculture since 1870', in E. Crittall, ed., *Victoria History of the County of Wiltshire*, 4, 92–114.

THOMPSON, F. M. L. (1960) 'English great estates in the nineteenth century (1790–1914)', *Contributions to the First International Conference of Economic History*, Paris, pp. 385–97.

THOMPSON, F. M. L. (1963) *English Landed Society in the Nineteenth Century*, Routledge & Kegan Paul.

TRIMMER, J. (1847) *On the Improvement of Land as an Investment for Capital*, London.

P. D. Wilde

13 Locational problems of the English silk industry in the mid-nineteenth century

Regional specialisation in manufacturing was a distinctive feature of the industrial geography of nineteenth-century Great Britain. Most industries which supplied more than the immediate local market had a distribution in which a single manufacturing region was dominant, with few outliers. The textile industry provides some classic examples of this localisation of production. In the cotton industry in 1850 over 250 000 factory operatives were found in Lancashire and the neighbouring parts of Cheshire, and this region contained over 86 per cent of the industry's factory workforce (BPP, 1850). Again the hosiery trade, though still essentially domestic, was highly localised and 90 per cent of the knitting frames recorded in 1844 were in the east Midlands, a region containing a similar proportion of the total workforce of 100 000 (Smith, 1963). This trend was reflected also in the distribution of the pottery industry, the manufacturing of fine steel and cutlery, and footwear, as well as industries more restricted in their locations, such as shipbuilding.

These compact distribution patterns of individual industries are usually explained in terms of the benefits deriving from the close locational association of related firms. External economies arose in particular from the ready availability of skilled labour, the close contact with ancillary activities and specialised commercial institutions, and the resultant lower transport costs. Moreover, in an era of restricted communication, localisation of industry was vital for the diffusion of techniques and innovations (Pred, 1971). A local coal supply was of paramount importance in many industries as a power source, and many industrial concentrations were found on the coalfields, whose locational attractions for manufacturers were at their peak during the nineteenth century. However, the compact distributions assumed by industries such as footwear and hosiery, in which steam power was of little importance until the later nineteenth century, indicate the more general significance of external economies of association.

Regional specialisation had another dimension. As an industry tended to concentrate in one region, in turn an individual region's economy was usually dominated by a single basic activity. An industry, once established, appears to have been able to deprive any competing activity in the region of new capital, labour and entrepreneurial skill. As long as an established industry remained profitable, alternative activities were either eliminated or excluded from the region (Rawstron, 1964). This situation had distinct implications for the establishment of new activities. An existing regional specialism could not be replaced, even by a vigorously expanding industry, unless its profitability and employment were declining. Once a competitive situation arose between two industries, the weaker would be unable to recover resources, especially labour, lost during slumps, while the stronger would increase its control over the region's resources during booms (Rawstron, 1958). Only after a long period of such competition would the regional specialism change.

Given the strength of this competition among industries for resources and the intensity of regional specialisation which resulted, the distribution of industry at the larger national scale is explicable in terms of comparative advantage. The reasons for any particular pattern of localisation cannot be found solely in the relationships between the industry concerned and its factors of production, but must involve the industry's relationships with other activities both within the region and elsewhere (Chisholm, 1970). Rawstron (1958, p. 28) suggests that in the nineteenth century 'comparative advantage for industry in general declined away from the coastal coalfields, and those industries with fewer demands on location were established at ports away from the coalfields or took interior positions, mainly on or near coalfields'. Heavy export industries with their demand for the cheapest available transport consequently pre-empted the coastal coalfields. The major textile industries, pottery and fine steel, whose high value products were better able to absorb transport costs, adopted locations on the interior coalfields. Industries such as knitwear and footwear, with little need for a coalfield situation, took up marginal locations providing maximum protection from the competition of locationally more demanding industries (Rawstron, 1958; Mounfield, 1965).

Within this broad framework the process by which specific industries were established on the various coalfields was a function of chance, individual enterprise, economic growth and the availability of labour and capital, factors which could be more important than details of site and location. Each region acquired its own specialised industry and no inland activity, not even cotton manufacturing, could survive for long in the more competitive coastal locations (Chisholm, 1970). Even the locationally least demanding industry was able to gain considerable external economies by concentration and in general its distribution pattern also was dominated by a single region of production.

Although a new and expanding industry could replace gradually an existing regional specialism whose profitability was in doubt, such changes were rare during the nineteenth century. Only in transitional periods, when comparative advantages were in a state of flux, were national and regional patterns of activity most susceptible to modification. The Industrial Revolution brought a major upheaval in location patterns as pre-industrial technology and organisation were replaced or substantially modified (Rawstron, 1964). Since locational values (based chiefly on the high cost of transport and accessibility to sources of power) were relatively stable throughout the nineteenth century, so were regional comparative advantage and the broad features of industrial distribution. This stability was reinforced by two characteristics stemming from regional specialisation itself. The first was the increase in external economies with the expansion of a localised industry to the point when the agglomeration itself provided the strongest reason for its continued existence. At this stage it became virtually impossible for a manufacturer isolated from the benefits of agglomeration to remain competitive. Second, the size of the industry's workforce and its domination of the region's industrial structure were sufficient to provide a built-in stability and to make any change a slow process.

The silk industry

In contrast to these general features of nineteenth-century manufacturing, the silk industry in England displayed a scattered distribution in which no single region was dominant. In addition its distribution was far from stable, for its development involved alternating phases of spatial expansion and contraction. The reasons for these unusual characteristics lay in the industry's economics, especially in the extreme fluctuations in profitability experienced.

Economic instability

The industry's economic instability stemmed from unpredictable changes in production costs and in product demand and prices. The vagaries of fashion, on which the industry increasingly depended as fine cottons replaced plain silks in many goods, frequently affected the economic fortunes of thousands of looms and throwing mills (Hertz, 1909; Daniels, 1920; Unwin *et al.*, 1924; Prest, 1960; Davies, 1961). Competition from abroad, especially France, constituted a still more significant source of uncertainty. High tariffs or prohibition of imports reduced the level of competition from continental manufacturers with their advantages of lower labour costs (Badnall, 1828; Tariff Commission, 1905). Nevertheless,

French silks were so highly prized that there was a well organised smuggling trade to cater for the demand (Warner, 1921). As a result the weaker English industry displayed an inverse relationship with that of the French, thriving when the French were at war and slumping during peace.

Raw material costs were critical, being highly variable and accounting for between 50 and 80 per cent of the final selling price of a yarn or cloth (Tariff Commission, 1905; Warner, 1921; Coleman, 1969). The variability in costs was due to the susceptibility of the silk worm to disease, reducing raw silk production to low levels in some years, and to the inability to adapt output to meet fluctuating demand from silk yarn producers. Manufacturers' profits were consequently erratic depending on the margin between variable raw material costs and the uncertain price for their products in a volatile market (Hughes, 1960; Coleman, 1969).

In the nineteenth century boom and slump alternated rapidly. A long period of prosperity and expansion ended with the conclusion of the Napoleonic Wars in 1815. A depression followed but was reversed by a boom some ten years later, especially in the throwing branch when the reduction in the duty payable on raw silk released considerable capital for the industry. Profits fell again in 1826 when foreign goods could be legally imported for the first time (Badnall, 1828; Prout, 1829). The high duty levied on imports, at a rate in excess of 30 per cent *ad valorem*, was sufficient to protect even relatively inefficient producers and after the initial shock, the industry enjoyed some twenty years of relative stability. Even halving the rate of duty on manufactured imports in 1845 had little impact. A renewed boom marked the 1850s. Employment in the powered branch rose by one-third in six years to over 55 000 in 1856, mainly as a result of a 52 per cent increase in the number of powered looms (BPP, 1850; 1857).

The removal of the last vestiges of tariff protection in 1860 produced a prolonged slump throughout the industry, intensified by a world shortage of raw silk between 1863 and 1865 (BPP, 1866). The domestic section of the industry entered a permanent decline from this time, though the powered branch was still able to benefit from lucrative orders during the Franco–Prussian War (Davies, 1961). As a result employment in 1870 was only 15 per cent below its recorded peak in 1856 (BPP, 1871). After 1872 the industry entered a phase of continuous contraction for the remainder of the century.

Distribution

The distribution of the industry about 1840 reflects the period of relative economic and locational stability. At this time the pattern was

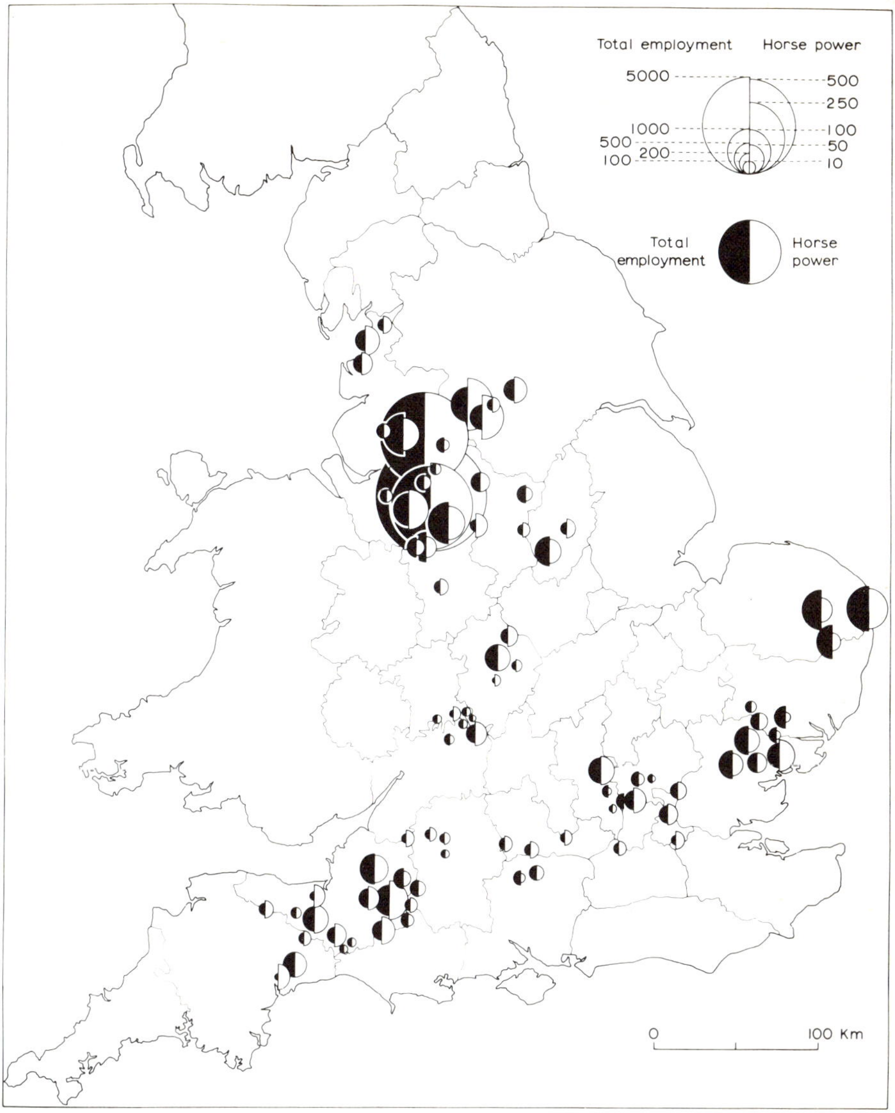

Fig. 13.1 The distribution of the powered silk industry in 1838. Based on BPP, 1839

unaffected either by intense speculation pressures, encouraging territorial expansion into marginal locations with inefficient power or an inadequate labour supply, or by the effects of a slump, when many normally profitable mills would have been idle or producing other textiles.

From the returns of the factory inspectors for 1838 (BPP, 1839) there

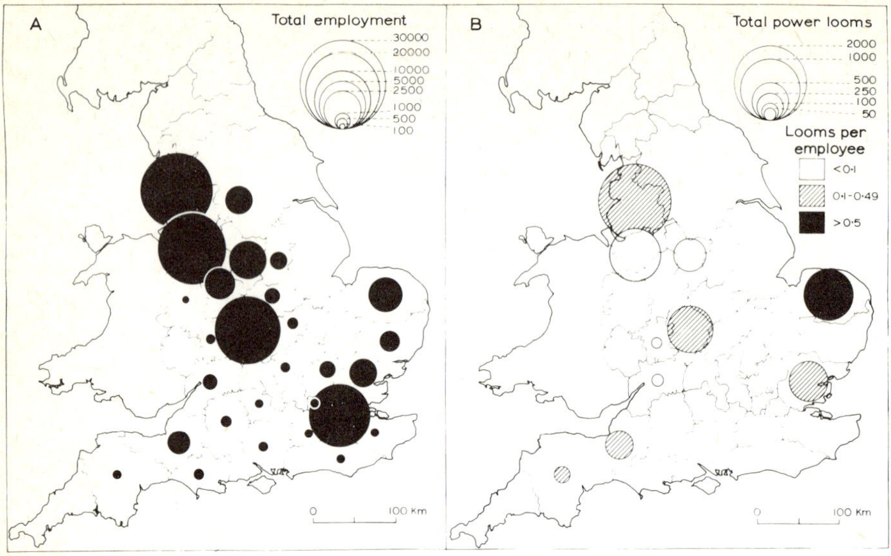

Fig. 13.2 A. The distribution of employment in the silk industry in 1851; B. The distribution of power looms in the silk industry in 1850. Based on BPP, 1852–53 and 1850

were in England 263 mills producing silk, scattered over eighty-eight parishes and employing 33 533 persons. The most important region was the southwest Pennines where parishes in eastern Cheshire and north Staffordshire accounted for 40 per cent of the national employment and 36 per cent of the power consumed. The remainder of the Pennine province accounted for a further 30 per cent of the total of both labour and power to give northern England an overwhelmingly dominant share of the industry. Within the Pennine province silk manufacturing was considerably more scattered than other textile industries and in southern England the 10 000 factory employees were distributed throughout sixty parishes in fifteen counties (Fig. 13.1).

These returns give an incomplete picture for, unlike the major textile industries, the bulk of silk production was not mechanised by the middle of the nineteenth century. The throwing mills, where silk yarns were produced, were almost always steam or water powered, but most types of cloth were still woven by hand. This relative delay in applying power to the looms was attributed chiefly to the nature and delicacy of the fibre, though the low quality of the imported reeled silk and the poor techniques of silk throwing also contributed (Fairbairn, 1865; Mason, 1910). In fact, in the early 1850s factory employment accounted for barely 30 per cent of the 135 569 silk workers in England (BPP, 1850; 1852–53) and the distribution of the total workforce (Fig. 13.2 A) was very different from that of factory workers.

Table 13.1 Regional employment in the silk industry, 1850–51

District	*Total employment* (000s)	*Factory employment* (000s)	*Factory employment as percentage of total*
Southwest Pennines	27.1	13.9	51
Lancashire	31.7	8.2	26
Yorkshire	3.5	1.7	48
East Midlands	8.2	5.3	65
Total for the Pennine province	70.5	29.1	41
Warwickshire and Cotswolds	24.2	3.5	14
Norfolk	5.8	1.4	23
Essex and Suffolk	5.7	2.9	50
London and Chilterns	23.1	1.7	7
Berkshire and Hampshire	0.7	0.1	17
Southwest England	3.9	3.1	78
Other districts	1.6	0.0	0
Total	135.5	41.8	31

Source: BPP, 1850; 1852–53.

The distribution of the industry's domestic workers must be taken into account. Although not distinguished, the division of workers between the manual and the powered branches of the industry can be estimated, at a county level at least, by comparing the 1851 Census figures with the factory returns of the previous year (BPP, 1850). The difference between these two figures can be taken to represent the number of hand loom weavers, since virtually all throwing was by power, and non-productive workers such as merchants and equipment makers are shown by local directories to form a small proportion of the total workforce.

The relative importance of the domestic industry varied considerably between regions. Warwickshire and London, which were of negligible significance in terms of the powered industry, were major centres of handloom weaving. Lancashire was even more important in the total industry than in the factory-based section, and in the southwest Pennine region there were as many domestic weavers as factory workers. Compared with these four regions of production, which together accounted for 78 per cent of total employment, other centres were small but in these, too, distinct variations in the relative importance of the powered and manual branches occurred. In particular the industry in southwest England and the east Midlands was dominated by the powered section, and Norfolk, with its high proportion of domestic workers, contrasted sharply with the rest of East Anglia.

The considerable regional variation in the industry's structure is also revealed in data from the factory returns for 1850 which record the place of weaving within the powered section. On a regional basis, weaving varied in importance both relatively and in absolute terms. More significant, however, is the fact that those regions in which powered weaving was most developed (Lancashire, Warwickshire and Norfolk) were also regions with proportionately large domestic industries (Fig. 13.2 B).

Regional specialisms

The comprehensive information provided by these two sources, used in conjunction with a wide variety of other evidence, allows the various regional specialisms to be identified (Wilde, 1971). The major long-established centres of handloom weaving, Coventry and the Spitalfields district of London, showed considerable specialisation in ribbon weaving and broad silks respectively. Powered ribbon looms were being introduced in Coventry by 1850, but the Spitalfields trade remained entirely manual. There was virtually no throwing carried on in either area and their yarn was supplied by small mills, chiefly powered by water, located in the Cotswolds and Chilterns. Although Coventry remained the only significant centre of ribbon weaving in England, with almost 95 per cent of all narrow looms (BPP, 1840) broad silks were woven in many places other than Spitalfields by the mid-nineteenth century. This was largely because of the industry's spread from its London nucleus in the early eighteenth century when, as in other trades, the restrictive practices and legislative constraints hindered the development of a new and larger scale organisational framework (BPP 1821, 1823; Hertz, 1909; Clapham, 1916; Spate, 1938).

It was in Lancashire rather than in Coventry or Spitalfields that the biggest concentration of handloom weavers existed, and where factory weaving was most important, with almost one-third of the powered silk looms at work in England. Although throwing was an important activity, Lancashire being second only to the southwest Pennines in the number of spindles installed (BPP, 1850), weaving dominated.

Throwing was especially important in the southwest Pennines, where 30 per cent of the nation's silk spindles were found in 1850. The quantity of yarn produced exceeded the needs of local weavers and manufacturers served the widely scattered national market, in particular meeting much of the demand from Lancashire and Spitalfields (Davies, 1961). Macclesfield was the organising centre of the southwest Pennine trade, which was also strongly established in Congleton and Leek. In Leek the production of weaving yarns was limited and throwsters concentrated on sewing silks, braids, bindings and other specialised yarns.

In the east Midlands and Yorkshire, specialised items also dominated output. In the former region the industry, centred in Derby, was largely engaged in supplying yarn to the local hosiery manufacturers, though plain ribbons and braid goods were produced in limited quantities. In Yorkshire virtually all mill workers were employed in *spinning* silk, that is making yarn from short lengths of 'waste' silk fibre, rather than in the more usual process of *throwing* the continuous filament (Singer *et al.*, 1957). The domestic weavers took their place in Yorkshire's mixed weaving industry which integrated a variety of yarns and textile trades.

The more scattered industry of southern England was organised on a different scale. Typically a small powered throwing mill provided yarn for domestic weavers in a few surrounding villages, though by 1850 the industry's structure was changing as handloom weaving became less profitable. Norfolk retained a large weaving trade (both powered and manual), but in Suffolk and Essex, and in the southwest of the country the domestic weaving trade declined and a surplus of yarn was produced which was sold in London. Some power looms had been installed even in the smaller centres, chiefly by a few progressive firms (such as Courtaulds in Essex), and it was on these firms that the future of the industry in these regions depended (Fig. 13.2 B).

Although the silk industry had a dispersed distribution, individual areas usually concentrated on a particular branch of the trade. Broad-weaving and the throwing of simpler yarns were duplicated in a number of regions, but other activities, for example ribbon weaving, waste silk spinning, and specialised yarn production, were localised in individual regions. At this more local scale of individual towns and districts, specialisation occurred within the industry, reminiscent of that in Lancashire in particular grades of cotton spinning or weaving. At this level the silk industry appears to have conformed to the general tendency for specialisation. On the other hand it was distinctive at the macroscale in that the various branches failed to become concentrated into a single dominant region of manufacture and a scattered distribution was retained.

Comparative advantage and the silk industry

This distribution pattern, though unusual, conforms with Rawstron's concept of comparative advantage, for the location of the silk industry was a function of its competitive abilities with regard to other activities rather than of its own factors of production. The slumps to which the industry was prone resulted in repeated competition from stronger, more stable industries for the resources it used. The silk industry's resources were easily lost in periods of recession and could be regained only with difficulty when prosperity returned. Examples of retreat from regions where more consistently profitable and technically progressive industries challenged the position of the silk industry are frequent. The dispersed distribution pattern of the industry reflects the need to adopt locations where competition was at a minimum.

This distribution pattern was thus affected by external competition because its internal production costs and requirements did not restrict the industry to regions with well defined characteristics. In particular, transport and power costs were of little importance in comparison with their role in most other forms of nineteenth-century manufacturing. Transport costs could be absorbed more easily by the high value product and in this respect the scattered locations, involving considerable movement of raw, semimanufactured and finished goods, were no great drawback.

Power requirements were perhaps more important in that they gave the industry a flexibility which helped to determine those regions in which its comparative advantage lay. Even in the powered section, only a small part of the total industry, power supplies were not as critical as in many other industries (Rodgers, 1960; Wallwork, 1968). This situation arose partly because the product could absorb additional costs of fuel transport or of water engineering, but more significantly because the mechanised silk industry used proportionately less power than any other textile. On average, silk mills had smaller engines installed (whether using steam or water power) but each unit of power involved the employment of over twice as many workers as in any other textile industry (Table 13.2). Moreover the use of inferior power sources does not appear to have caused serious harm to the competitive position of firms within the industry, so there were few incentives to improve efficiency. Even in the Pennine province, water power was used in silk far more extensively and for a longer period than in other textile industries. In some locations, mills were sited with little regard for water power supplies but were built to use horses and even man power, as at Nottingham, Winchester, Taunton and Sherborne (Lowe, 1798; BPP, 1831–32; Warner, 1921). Power supplies would seem to be relatively unimportant in silk manufacturing and the industry was likely to be ousted from the most attractive power sites, and from the coalfields in general, by more demanding activities. On the other hand, those locations adopted by silk were unlikely to be disadvantageous and, indeed, were to the industry's overall benefit in that competition was reduced.

Table 13.2 Average power requirements of textile industries, 1838

	Silk	*Cotton*	*Wool*	*Worsted*	*Flax*
Horsepower per mill	12.3	31.4	15.4	17.3	25.2
Employees per horsepower	10.4	4.4	2.7	4.4	3.9

Source: BPP, 1839.

Regional analysis

The effects of the competition for resources between silk and industries which were both stronger and more locationally demanding is vividly illustrated in Lancashire and the east Midlands. In Lancashire only those resources which were surplus to the needs of the cotton industry were available to silk. As the labour force in silk was less than one-tenth of that in cotton, these resources were never used significantly. In the powered branch, silk goods were produced in mills which changed their yarn according to the varying fortunes of the different branches of the textile trade and it was only infrequently that silk could match the profitability of cotton. As a result the number of mills returned as working silk, and the associated employment, could fluctuate violently, as it did in the twenty years after 1850 (BPP, 1850; 1857; 1862; 1867–68; 1871).

The domestic branch of Lancashire's silk industry was even more dependent on factors beyond its control and its apparent strength in 1851 was illusory. Large numbers of skilled weavers had been made available to silk by the introduction of power looms in the cotton industry, especially in the 1820s when silk's fortunes were booming. In the 1850s the cotton industry again entered a period of prosperity, and labour and other productive resources were once more reclaimed by the region's dominant activity. The 1851 employment figures represent a peak for the silk industry in Lancashire. Of the estimated 23 000 handloom workers, 5 000 were lost over the next decade and after 1860 the rate of decline accelerated. Although Lancashire was one of the most important centres of silk manufacturing, the industry was never firmly established nor did it have a significant impact on the region's economy.

In the east Midlands the importance of silk manufacturing in Derby and its insignificance in Nottingham also resulted from silk's competitive weakness. The hosiery industry was expanding rapidly in the region with Nottingham as the major centre of control and innovation. Employment in Nottinghamshire increased in the first half of the nineteenth century from 7 000 in 1812 to over 16 000 in 1844 (BPP, 1845; Smith, 1963). In such an economic climate, the silk industry was unable to gain a foothold. Around Derby, in contrast, the hosiery industry was small and declining. The local specialism was in the traditional silk branch of the trade but, as fine yarns from cheaper materials, such as cotton, were developed, this was being superseded and the Derby hosiers seemed to have lacked any innovative impetus to redress the situation (Smith, 1962). Moreover, it was in the countryside rather than the town that the industry was predominantly located (BPP, 1845), leaving Derby itself relatively free for alternative manufactures. As a result of these external factors, the silk industry was able to develop in Derby and rival the hosiery trade for labour and other resources. By the mid-nineteenth century, the silk industry was well-established in Derby and there was a marked contrast in employment between Derbyshire and Nottinghamshire (Fig. 13.2 A). Although the resources of Derby were dominated for a time by silk manufacturing, the industry continued to depend for a major part of its market on the hosiery trade, which remained the region's dominant textile.

In such regions, where another activity was established which could use labour and other resources more profitably, the silk industry was always insecure. In most other parts of England, the industry gained a measure of security by locating in regions whose resources were comparatively unattractive to other industries and where, in consequence, competition could be avoided. The existence of the industry, especially the powered branch, in the southwest Pennines illustrates this situation. This region was too far from cheap and plentiful power sources and from the control centre of Manchester to attract the cotton industry, and its resources appear never to have encouraged any other competing activity, even in periods of severe depression in silk. The power available in the southwest Pennines was adequate for silk's needs, and this area, on the fringes of the Lancashire textile province, with links to the latter's ancillary industries and commercial organisation, was probably the most advantageous location open to the industry. Silk's long tradition in the southwest Pennines gave its organisational framework a strength which made competition difficult, but it was chiefly the region's comparative disadvantage which enabled the silk industry to dominate its resources for so long.

The scattered locations of the silk industry through southern England were also protected from competition. The industry tapped the child and female labour of small rural communities where employment on the land was contracting and alternative occupations were few. In addition, as in Derby, silk manufacturing was established in towns where a major textile trade was declining. This latter process was the more significant, accounting for many of the centres of silk manufacturing in East Anglia and southwest England. The woollen industry in southwest England had become more concentrated in the early nineteenth century, with compact areas of production developing on the Mendip flanks and in the Stroud valley. Domestic manufacturers and mill operatives in the rest of the region were reported to be turning from other textiles to silk (BPP, 1821; 1840). Though the silk industry was able to establish itself on the retreating margins of the wool producing areas (as was flax manufacturing), wool retained its control over the resources of the core (Fig. 13.3).

In Coventry and Spitalfields, the traditional centres of domestic weaving, the industry's relationship with other activities may have been slightly different. In these centres, at least, the industry appears to have

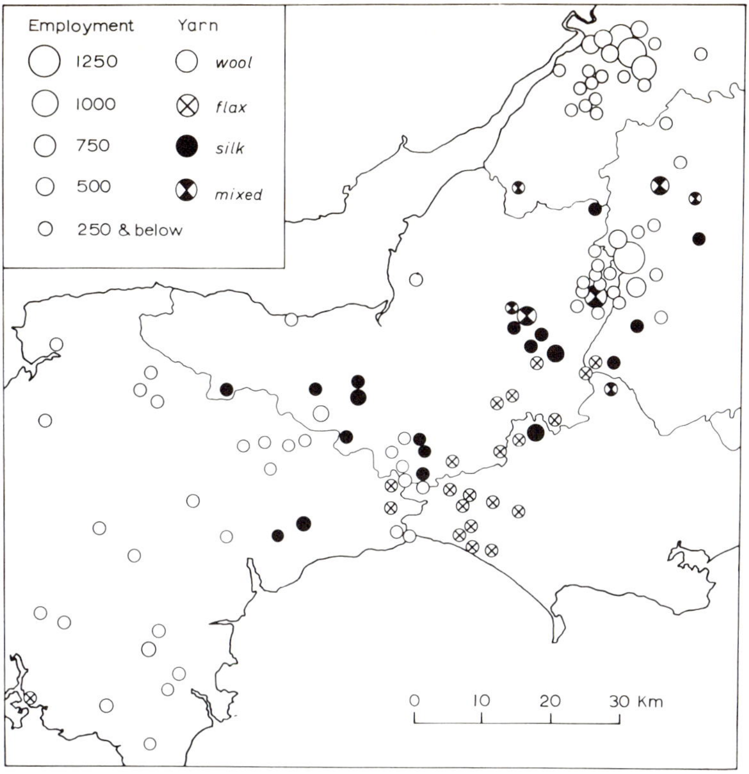

Fig. 13.3 Employment in textile mills of southwest England in 1838. Based on BPP, 1839

been capable of dominating employment and staving off competition to a considerable degree. In Coventry the only attempt to establish another trade was when watchmaking was introduced during one of the ribbon industry's slumps in the 1830s (Prest, 1960). Watchmaking was never a viable competitor and in 1851 employed only one-tenth the numbers found in silk (BPP, 1852–53). In Spitalfields, the silk industry continued as the only important source of employment, even though the migration of much of the industry had left it weak and unprogressive. The reasons for this apparent capacity for survival are debatable. They lay either in the inherent resilience and organisational monopoly of a long-established industry, or in the resources available in these places, which were so unattractive relative to those of other regions that competing activities were never encouraged. As in the southwest Pennines, it would appear that the comparative disadvantage of these two centres was of considerable significance in the industry's location.

In Yorkshire, the silk industry was unusual in that it was able to avoid competition by integrating with the region's major textiles. Silk spinning, but not throwing, had strong technical links with the worsted, and, more particularly, the flax industries of Yorkshire, in that all three handled long fibres. In addition, silk weaving in Yorkshire formed part, along with a variety of other fibres, of a mixed industry unique among the major textile regions (Forbes, 1852). Not only were individual towns concerned with more than one fibre, in itself an unusual feature of the period (BPP, 1839), but individual firms and mills manufactured a variety of products with different or mixed fibres. This close integration with a stronger textile trade removed the element of competition and made the Yorkshire silk industry far more resilient than elsewhere in England. The rest of the national silk industry, and particularly the throwing and pure silk weaving branches, did not have such close ties with any section of the textile trade and so comparable technological and economic support was not available.

The relationship of the silk industry to other activities would seem crucial in determining its distribution. Where competition was fierce, silk was likely to be excluded from a region's resources or its development shaped largely by external influences. Only where competition was almost non-existent, as in the southwest Pennines, Coventry, and many of the small towns of southern England, could the industry dominate the local workforce and associated resources. In most regions the mutual exclusiveness of manufacturing industry operated to the detriment of silk, which in its periods of depression could be replaced by a more stable industry. Silk's comparative advantage lay in locations where competition was minimal.

Reasons for dispersal

A consideration of the failure of the silk industry to assume the concentrated distribution typical of nineteenth-century industry is necessary. In part this anomalous situation was due to the development of dispersal early in the industry's history and to the resistance to change which the strong regional specialism engendered. In addition, lower levels of dependence on fuel and transport costs did not encourage spatial concentration within a coalfield region, as occurred in most other industries. Of most significance, however, was the persistent competitive weakness of the industry, which prevented the development of regional concentrations. Only in the southwest Pennines did silk's sphere of influence extend to encompass a region within which a number of sizeable towns were committed to silk. Even here the industry was not consistently strong enough either to counter the long-term attraction of Lancashire towns for the local surplus of rural labour (Smith, 1951), or to grow to a sufficient size to dominate all branches of the national industry.

This failure to become concentrated into a single region inevitably led to the limited growth of external economies of agglomeration within the industry. The scale of manufacturing was only large enough for associated activities to be strongly developed in the southwest Pennines. Despite the existence of about fifty ancillary firms in 1834 (Pigot, 1834), Macclesfield's powered silk industry was unable to support its own specialised engineering trade. Most of the machinery used in Cheshire was developed and manufactured by firms, such as Fairbairns of Manchester, which regarded silk's requirements as a sideline to their principal engineering business (Fairbairn, 1865).

The needs of the ribbon trade effectively concentrated at least the narrow weaving branch of the silk industry into a single centre. Because of the volatile nature of the consumer market, an immediate awareness of fashion trends and ready access to a supply of competent labour at short notice were vital requirements met by the concentration of the trade. The industry's organisation reflected and reinforced this need for concentration, for throughout most of its history some twelve merchant manufacturers maintained a monopolistic control over trade channels (Prest, 1960). For other branches of production, and for the silk industry as a whole, the external economies created by agglomeration were never sufficient to attract new enterprises to a single region and curtail development elsewhere. Lacking the competitive strength to dominate a major region and without a concentration large enough to create powerful external economies, the silk industry remained economically weak and locationally scattered.

Dynamic aspects

Although the broad outlines of the distribution of silk manufacturing remained stable, details were being modified constantly with changing economic conditions. Silk's frequent slumps were a major factor in determining locations where, in the long term, the industry could compete for resources; its booms provided brief periods in which any location could become highly profitable. The timing rather than the location of entrepreneurial ventures was the crucial consideration in their success and manufacturers were willing to operate widely dispersed concerns to benefit fully from rapid increases in demand or price. In these booms silk manufacturers could not encroach on established and prosperous industrial regions, but they were able to profit by taking up unused productive capacity in almost any location. Disused mills, pockets of under-employed agricultural labour and any other resources which could be brought rapidly into use were exploited. The mills adopted were usually small, old and water powered, and techniques of production were normally labour intensive (Badnall, 1828; BPP, 1857). While profits were high, inefficiency and high costs were of little consequence and in a

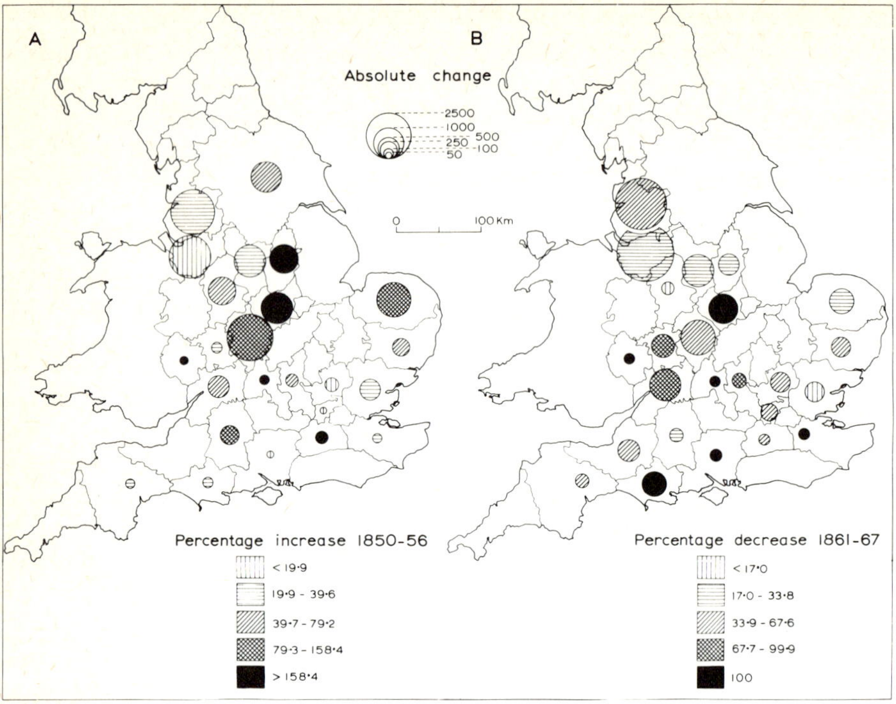

Fig. 13.4 The response of employment in the powered industry to economic change:
A. Expansion in the 1850s;
B. Depression in the 1860s
Based on BPP, 1850, 1857, 1862 and 1867–68

subsequent slump the enterprise could be abandoned with little loss. Such uncommitted resources were available mainly in southern England and on the less developed fringes of the Pennine province. In these areas, marked spatial expansion of the industry occurred in boom periods, while contraction into established cores characterised its depressions.

Major centres of production, such as in the southwest Pennines, tended to react less violently to economic fluctuations as they possessed a large proportion of locally available resources, particularly labour, which were already committed to silk: in Macclesfield, for instance, 63 per cent of all employment was in the silk industry in 1851 (BPP, 1852–53). However in the more highly capitalised silk industry of the Pennine province increased output could be achieved without increasing employment, and therefore employment data may understate the situation (BPP, 1833; 1857). The effects of fluctuating economic conditions may be exemplified by the changing distribution of the powered industry in the twenty years after 1850 (Fig. 13.4). The effects of prosperity are clear: the industry in Cheshire, Lancashire and Derbyshire declined in relative importance, though accounting for one-third of the absolute increase in employment, while on the margins of this province and in much of southern England employment expanded at a greater rate. In the depression that followed, the industry contracted more rapidly in southern England than in the Pennines, and by 1867 it had almost been eliminated from eleven of the eighteen southern counties in which silk mills had been working a decade earlier.

Conclusions

In terms of comparative advantage, factors of production in the silk industry, and the least-cost location based on these, had little relevance to its distribution. In periods of boom profits were high and location of little importance; in depression production was located, not where conventional costs were at a minimum, but where silk's comparative advantage was greatest in terms of reduced competition from other industries. These locations, which were least attractive to other activities, formed the long-term cores of the industry.

Chance may play a large part in determining an industry's location but this study has emphasised that there can be an economic rationale for its survival and for the nature of its distribution pattern. It has also been suggested that the extension of an industry's domination of labour supplies from an individual town to a region is a complex process. The alternation of prosperity and depression that was responsible for locational instability in the silk industry differed only in frequency and intensity from the economic changes experienced by all industry. Though less easily discernible, the same factors that promoted change in the silk industry must have influenced the locational patterns of all nineteenth-century manufacturing, since competition for resources and locations was a basic aspect of economic development.

References

BADNALL, R. (1828) *A View of the Silk Trade: with remarks on the recent measures of government in regard to that branch of manufacture*, London.

BPP (1821) *Second Report from the Select Committee of the House of Lords . . . into foreign trade (silk and wine)*, **7**, 421–64.

BPP (1823) *Lords Committee on the Bill to Repeal Certain Acts relating to the wages of persons employed in the manufacture of silk . . .* (Lords Sessional Papers), **156**, 57–178.

BPP (1831–32) *Report of the Select Committee to Enquire into the Present State of the Silk Trade and the effects of a change in tariffs*, **19**, 1–1050.

BPP (1833) *Report from the Select Committee inquiring into Manufactures, Commerce and Shipping*, **6**, 1–801.

BPP (1839) *Return of all Mills and Factories . . . June 1838*, **42**, 1–351.

BPP (1840) *Report of the Commissioners . . . into the Condition of Unemployed Hand Loom Weavers*, **23** and **24**, 49–800 and 17–717.

BPP (1845) *Report of the Commissioners appointed to inquire into the Condition of the Framework Knitters*, **15**, 1–1053.

BPP (1850) *Number of . . . Cotton, Woollen, Worsted, Flax and Silk Factories in Each County*, **42**, 455–75.

BPP (1852–53) *Census of 1851 population tables II*, **88**, parts 1 and 2.

BPP (1857) *Number of . . . Factories subject to the Factories Acts in Each County . . . (1856)*, **14**, 173–95.

BPP (1862) *Number of . . . Factories . . . in Each County . . . (1861)*, **55**, 629–59.

BPP (1866) *Factory Inspectors' Report for 1865*, **24**, 239–51.

BPP (1867–68) *Number of . . . Factories . . . in Each County . . . (1867)*, **64**, 811–45.

BPP (1871) *Return of Factories . . . showing the counties in which they are carried on*, **62**, 109–383.

CHISHOLM, M. (1970) *Geography and Economics*, 2nd edn, G. Bell.

CLAPHAM, J. H. (1916) 'The Spitalfields Acts, 1733–1824', *Econ. J.*, **26**, 459–71.

COLEMAN, D. C. (1969) *Courtaulds: an economic and social history*, 2 vols, Oxford University Press.

DANIELS, G. W. (1920) *The Early English Cotton Industry*, Manchester.

DAVIES, C. S. (1961) *A History of Macclesfield*, Manchester University Press.

FAIRBAIRN, W. (1865) *Treatise on Mills and Millwork*, 2 parts, London.

FORBES, H. (1852) *Rise, Progress and Present State of the worsted, alpacca and mohair manufacturers of England*, London.

HERTZ, G. B. (1909) 'The English silk industry in the eighteenth century', *Engl. Hist. Rev.*, **24**, 710–27.

HUGHES, J. R. T. (1960) *Fluctuations in Trade, Industry and Finance: a study in British economic development, 1850–1860*, Oxford University Press.

LOWE, R. (1798) *A General View of the Agriculture of the County of Nottingham*, London.

MASON, F. R. (1910) 'The American silk industry and the tariff', *Am. Econ. Ass.*, third series, **11**, 1–182.

MOUNFIELD, P. R. (1965) 'The footwear industry of the east Midlands (III) Northamptonshire 1700–1911', *E. Midld Geogr.*, 3, 434–53.

PIGOT, J. (1834) *Pigot and Co's National Commercial Directory . . . of Chester*, London.

PRED, A. R. (1971) 'Large city interdependence and the pre-electric diffusion of innovations in the U.S.', *Geogr. Analysis*, 3, 165–81.

PREST, J. (1960) *The Industrial Revolution in Coventry*, Oxford University Press.

PROUT, J. (1829) *Practical View of the Silk Trade embracing a faithful account of the result of the measures enacted in 1824 for the encouragement of that manufacture*, Macclesfield.

RAWSTRON, E. M. (1958) 'Some aspects of the location of hosiery and lace manufacture in Great Britain', *E. Midld Geogr.*, 2, 16–28.

RAWSTRON, E. M. (1964) 'Industry', in J. W. Watson and J. B. Sissons, eds, *The British Isles: a systematic geography*, Nelson, pp. 297–318.

RODGERS, H. B. (1960) 'The Lancashire cotton industry in 1840', *Trans. Inst. Br. Geogr.*, **28**, 135–53.

SINGER, C. *et al.* (1957–58) *History of Technology*, 5 vols, Oxford University Press.

SMITH, C. T. (1951) 'The movement of population in England and Wales in 1851 and 1861', *Geogr. J.*, 117, 200–10.

SMITH, D. M. (1962) 'The silk industry of the east Midlands', *E. Midld Geogr.*, 3, 20–31.

SMITH, D. M. (1963) 'The British hosiery industry at the middle of the nineteenth century: an historical study in economic geography', *Trans. Inst. Br. Geogr.*, 32, 125–42.

SPATE, O. H. K. (1938) 'Geographical aspects of the industrial development of London till 1850', *Geogr. J.*, 92, 422–32.

TARIFF COMMISSION (1905) *Evidence on the Silk Trade*, London.

UNWIN, G., HULME, A. and TAYLOR, G. (1924) *Samuel Oldknow and the Arkwrights*, Manchester University Press.

WALLWORK, K. L. (1968) 'The calico printing industry of Lancashire in the 1840s', *Trans. Inst. Br. Geogr.*, **45**, 143–56.

WARNER, F. (1921) *The Silk Industry of the United Kingdom: its origin and development*, London.

WILDE, P. D. (1971) 'Growth, decline and locational change in the English silk industry of the nineteenth century: a study in historical geography', unpublished PhD thesis, University of Keele.

B. J. Turton

14 British minor railways

Amalgamations and mergers are recurrent themes in British railway growth and have exercised considerable influence on the evolving patterns of routes. Few pioneer companies of the late 1830s and the 1840s survived unchanged in name or in the areal extent of their operations into the later nineteenth century, most of them being incorporated within much larger undertakings which in turn were reorganised under the provisions of the 1921 railway company regrouping legislation. Progressive amalgamations during the nineteenth century resulted in the existence of twelve major networks by 1921, each with a route length of over 1 000 km, accounting for 84 per cent of the standard gauge system in Britain and carrying 78 per cent of all freight and 52 per cent of all passenger traffic in the country. The remainder of the national network was shared between 108 smaller companies, of which only thirteen owned over 100 km of railway and forty-five less than 32 km. Thus, many minor lines had survived as independent companies by the time of the regrouping, although several were adjuncts of much larger undertakings, being dependent on them for their commercial livelihood. A number of these minor companies also retained a separate identity after the implementation of the regrouping legislation in 1923, being designated as 'principal subsidiary' companies within the four newly formed national railway groups. These minor companies finally disappeared with the railway nationalisation of 1948 and, in common with many sections of their parent undertakings, the majority are now closed to traffic.

Before the 1923 regrouping, all minor undertakings owning less than 32 km of track were recorded in a separate section of the official Railway Returns. Of these minor companies, forty-five were listed in the 1921 Returns but comprehensive traffic statistics are available only for twenty-three systems with a route length of between 8 and 32 km. Studies of the history of several of these minor companies are already available but no comparative analysis of the twenty-three systems as a whole has been made, a gap in railway literature that this account seeks to remedy. Traffic composition and density have been selected as the principal items for analysis as these reflect the overall character of each railway in terms of its commercial activities and of its relationship with the region in which it is located, although the traffic generated by each system was affected also by neighbouring, and in many cases, competitive railways.

Location, growth and network character

These twenty-three minor systems illustrate effectively many of the principal factors that played a part in shaping the national railway network. Carriage of coal and other minerals was the prime motive behind the construction of several undertakings in South Wales, south Yorkshire and west Cumberland, where small companies operated alongside more powerful and extensive systems built with the same objective. Other minor systems were established primarily as passenger-carrying lines in the growing urban areas of Merseyside and Manchester. These heavily used freight and passenger systems contrasted with railways built in rural areas, where traffic potential was low. Many of these rural ventures were never commercial successes, being supported financially by larger and more prosperous companies which acquired controlling interests in them.

Although any railway with a limited extent of track is restricted in the complexity of its network, the minor undertakings displayed considerable variation in their patterns and in the manner in which they linked with neighbouring systems. The simplest form that may be recognised is the rudimentary 'dead-end' branch line, although subdivisions on the basis of traffic may be made. Such branch lines are distinguished as the lowest-ranking elements in the hierarchy of railway types devised by Wallace, and seven of the minor railways came within this category (Wallace, 1958). A second group, comprising ten systems, displayed the same linear form but were joined at two or more points with other railway networks. In most cases the links were with pre-existing lines but connections were made also with railways built subsequently, as in the case of the Manchester and Altrincham railway and its links with the Cheshire Lines Committee system. Six systems developed slightly more complex networks, with branches from their original linear core and with several termini. These companies often made agreements with neighbouring undertakings for the operation of through services, this arrangement being particularly characteristic of systems in urban areas.

The earliest example of a minor line built as a rudimentary branch was the Bishop's Castle railway, opened in 1866, although originally it was intended as a cross-country link. A proposal to link Craven Arms, on the Shrewsbury–Hereford line, with Montgomery, on the Cambrian Railway's Welshpool–Machynlleth route, did not materialise and the line was not continued beyond Lydham Heath, where a short branch to Bishop's Castle was begun (Fig. 14.1 D). The undertaking in its truncated form was never a financial success and closed in 1935 (Christiansen and Miller, 1967).

The Mid-Suffolk line was built under the provisions of the Light Railways Act 1896, which sanctioned the construction and operation of railways, not expected to carry large volumes of traffic, to less stringent standards than those applied to main lines (Oxley, 1901). As in the case of the Bishop's Castle line, the Mid-Suffolk railway in its final form represented only a part of a more ambitious project, which was to connect Haughley, on the Great Eastern (GE) Ipswich–Norwich route, with Halesworth, on the GE Ipswich–Yarmouth line, penetrating a part of Suffolk that was served poorly by the GE. Opened in 1904, the line was built only as far as Laxfield and like its Shropshire counterpart was a commercial failure, closing in 1952 after many years of dwindling traffic (Fig. 14.1 B).

Construction of the Cleobury Mortimer and Nidd Valley lines also followed the passing of the 1896 Act, the latter having the distinction of being the first railway to be owned by a local authority. It was opened in 1907 by Bradford Corporation to convey building materials to the Gouthwaite reservoir in the upper Nidd, extending from the terminus of the North Eastern (NE) Pateley Bridge branch up-valley to Lofthouse

---

Fig. 14.1 Rural branch lines

A. Freshwater, Yarmouth and Newport (FYN); Isle of Wight (IW);
B. Mid-Suffolk Light;
C. Knott End;
D. Bishop's Castle

Abbreviations of railway companies adopted in Figs. 14.1–6

BAR	**Barry**
BIR	**Birkenhead**
B and M	**Brecon and Merthyr**
BP and GV	**Burry Port and Gwendraeth Valley**
FUR	**Furness**
IWC	**Isle of Wight Central**
M and C	**Maryport and Carlisle**
MID	**Midland**
NS	**North Staffordshire**
RHYM	**Rhymney**
SE and C	**South Eastern and Chatham**
TV	**Taff Vale**
WCE	**Whitehaven, Cleator and Egremont**

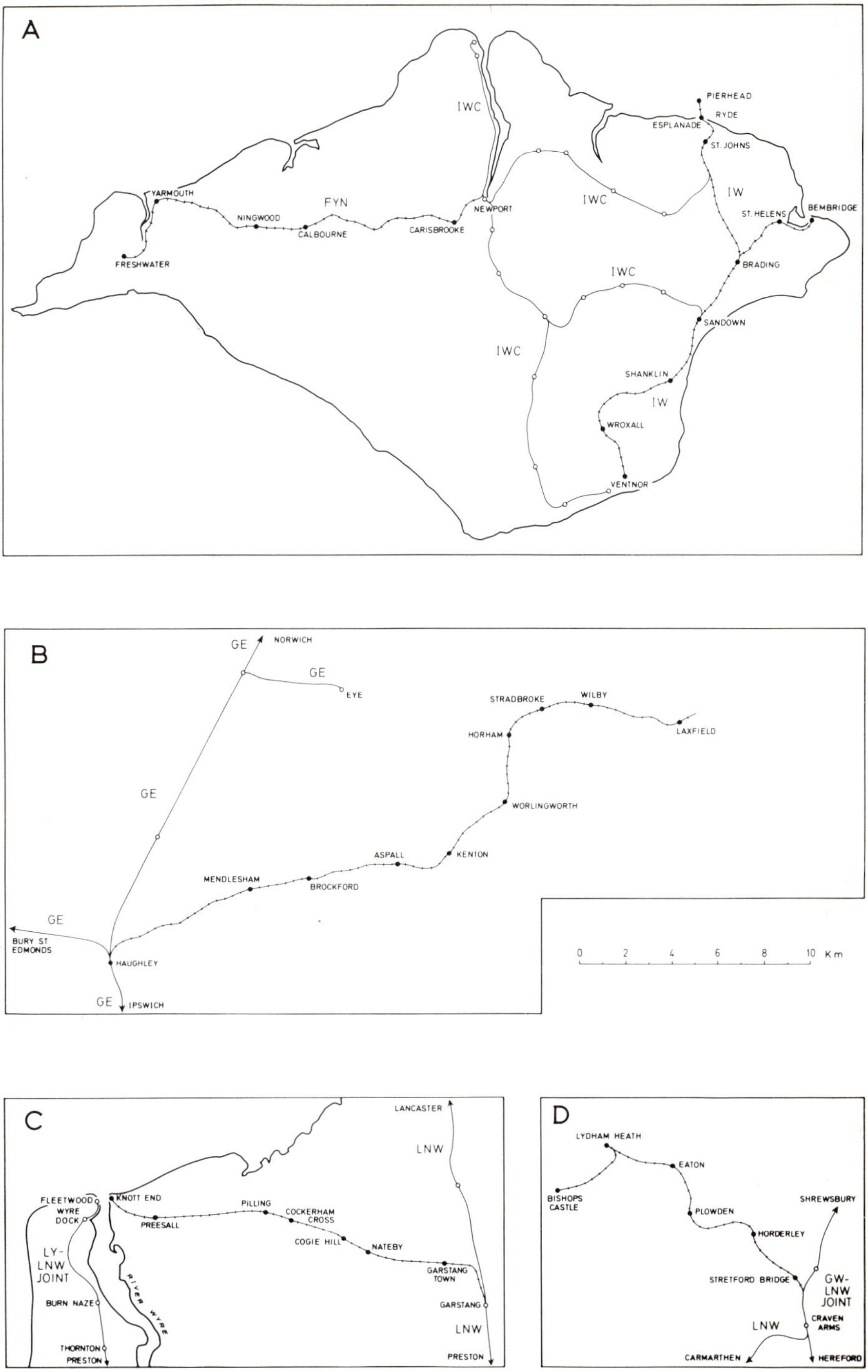
A
PIERHEAD
RYDE
ESPLANADE
ST. JOHNS
IWC
IW
YARMOUTH
FYN
NINGWOOD
CALBOURNE
CARISBROOKE
NEWPORT
ST. HELENS
BEMBRIDGE
FRESHWATER
BRADING
SANDOWN
SHANKLIN
WROXALL
VENTNOR
B
GE
NORWICH
EYE
STRADBROKE
WILBY
LAXFIELD
HORHAM
WORLINGWORTH
KENTON
ASPALL
MENDLESHAM
BROCKFORD
BURY ST EDMONDS
HAUGHLEY
IPSWICH
0 2 4 6 8 10 Km
C
LANCASTER
LNW
FLEETWOOD
WYRE DOCK
KNOTT END
PILLING
COCKERHAM CROSS
PREESALL
COGIE HILL
NATEBY
GARSTANG TOWN
GARSTANG
LY-LNW JOINT
BURN NAZE
RIVER WYRE
THORNTON
PRESTON
D
LYDHAM HEATH
EATON
BISHOPS CASTLE
PLOWDEN
SHREWSBURY
HORDERLEY
STRETFORD BRIDGE
GW-LNW JOINT
CRAVEN ARMS
CARMARTHEN
HEREFORD

(Fig. 14.2 B). A passenger service was introduced in 1908 but withdrawn in 1929, the line being then submerged beneath the Scar House reservoir (Jennings, 1967). In southern Shropshire the Cleobury Mortimer light railway was opened in 1908 to link the Clee Hill dolerite quarries with the Great Western (GW) system at Cleobury (Fig. 14.3 D), a passenger service operating between Cleobury and Burwarton until 1938.

Despite the restricted opportunities for traffic generation the Isle of Wight supported three independent railway companies until the 1923 regrouping (Robbins, 1963). The network in the eastern part of the island was complete at the time when the Freshwater, Yarmouth and Newport line was projected in 1880, the line being opened in 1889. Unlike its two eastern counterparts the Freshwater line served no large resorts and possessed no regular passenger ferry service with the mainland (Fig. 14.1 A).

In the coastal area of Lancashire between the Ribble and Wyre estuaries railway building was directed first towards linking the fishing port of Fleetwood and the growing resort of Blackpool to the main London and North Western (LNW) west coast line at Preston. A more northerly branch was opened in 1870 between Garstang on the LNW and Pilling. The original goal was Knott End, on the lower Wyre estuary opposite Fleetwood, but this was not achieved until 1908. As both freight and passenger traffic were light the line was closed to the latter in 1930 (Fig. 14.1 C).

The Llanelly and Mynydd Mawr was one of the smallest representatives of a family of systems in South Wales whose common interest was the transport of coal from collieries to ports on the Bristol Channel. In the late eighteenth and early nineteenth centuries anthracite from the western section of the coalfield was conveyed seawards by tramways, which were replaced gradually by standard railways, the Llanelly line being built in 1882 along the course of an 1806 tramroad (Fig. 14.4 B). This linked the Gwendraeth Fawr collieries with the Llanelly docks, to some extent duplicating the facilities offered by the Burry Port and Gwendraeth Valley system.

The ten minor railways connecting with adjacent systems at two or more points included coal-carrying lines in South Wales and south Yorkshire, three urban or suburban systems and several rural lines similar to those already described. In Glamorgan the Cardiff railway formed part of the remarkable concentration of four lines in the middle Taff valley, each line being owned independently before 1923 but all being concerned primarily with the down-valley coal trade to docks at Cardiff, Barry and Penarth. At its northern end the Cardiff line joined the Taff Vale system near Treforest but unlike the Taff Vale it possessed no independent route to the sea, relying on the Rhymney railway for access to Cardiff docks (Fig. 14.4 A). It was opened in 1911 as the last element in

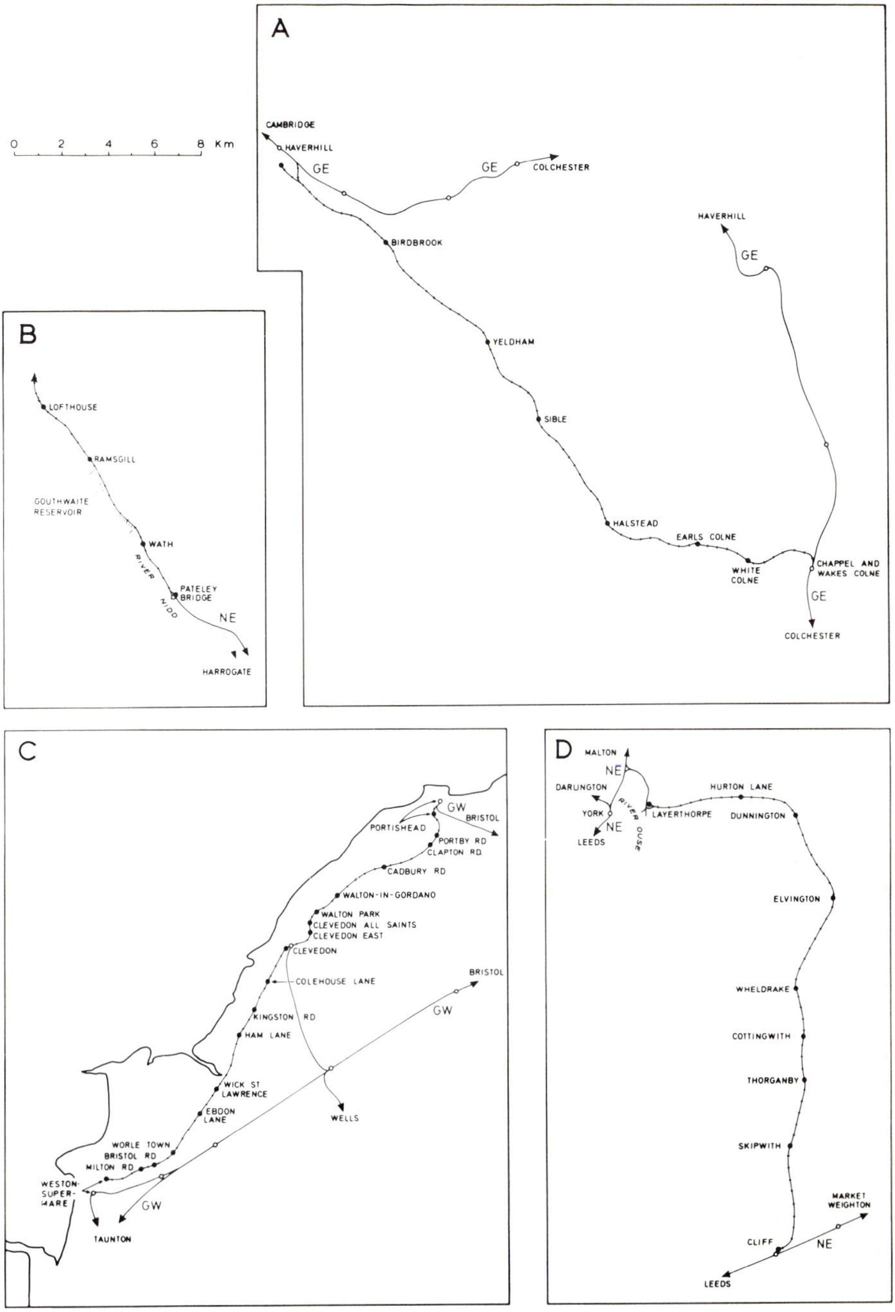

Fig. 14.2 Rural branch lines

A. Colne Valley and Halstead;
B. Nidd Valley Light;
C. Weston, Clevedon and Portishead;
D. Derwent Valley Light

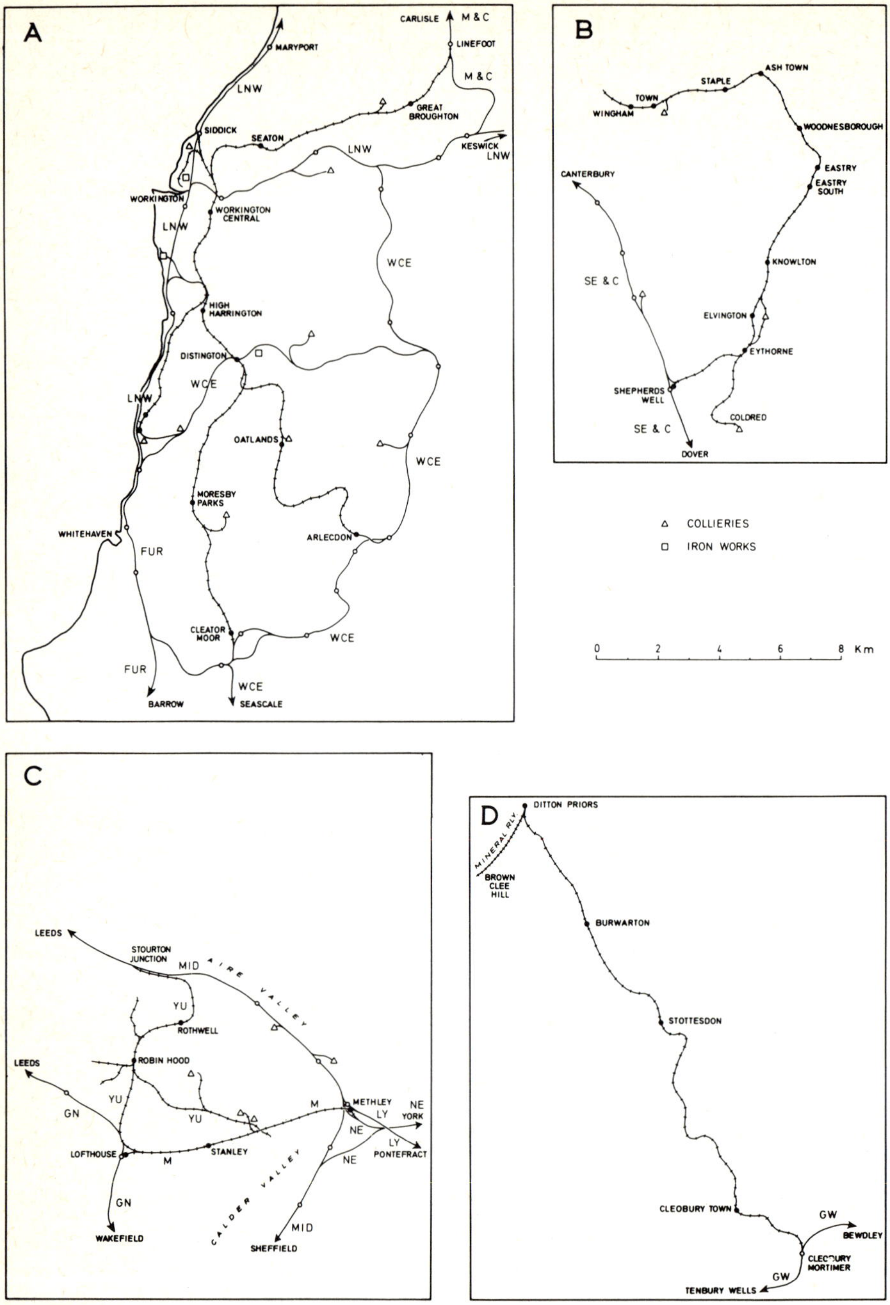

Fig. 14.3 Mineral systems

A. Cleator and Workington Junction; B. East Kent Light; C. Methley (M); East and West Yorkshire Union (YU); D. Cleobury Mortimer

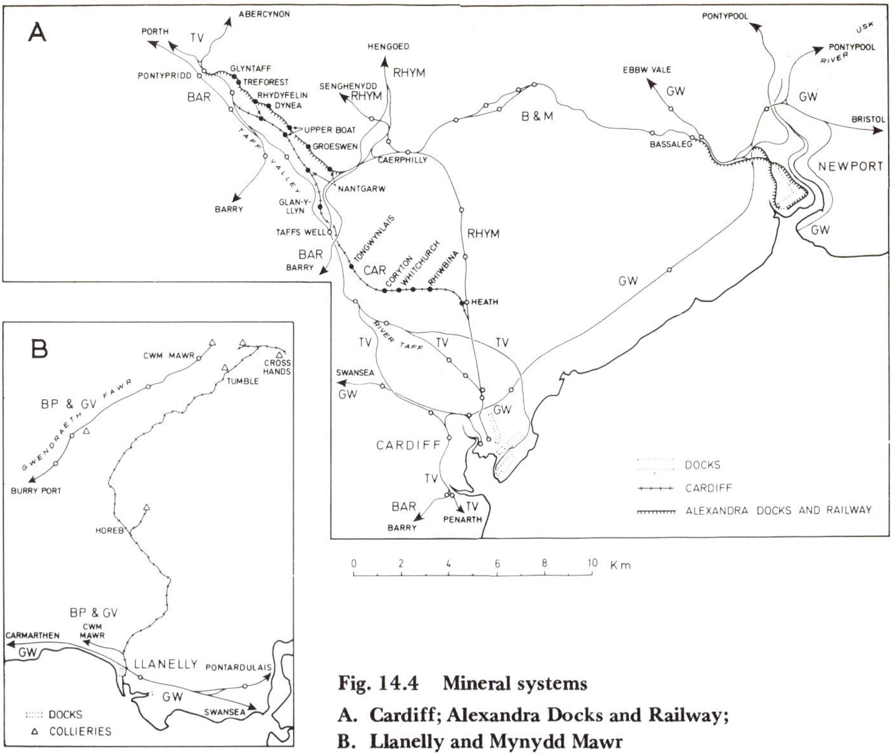

Fig. 14.4 Mineral systems
A. Cardiff; Alexandra Docks and Railway;
B. Llanelly and Mynydd Mawr

the complex of railways focusing on Cardiff, but its full potential as a coal route was never realised as the junction with the Taff Vale system was not used (Barrie, 1950). The middle Taff valley was occupied also by a second minor railway, the Alexandra Docks system, which, like the Cardiff line, linked with the Taff Vale (near Pontypridd) and then joined the Rhymney railway west of Caerphilly. The second section of the Alexandra Docks system extended from Newport Docks to Bassaleg, whence running powers over the Brecon and Merthyr railway gave access to the Taff valley section (Fig. 14.4 A). Despite its fragmented nature, however, the Alexandra railway was a heavily used freight line, enjoying the particular advantage of its own dock facilities at Newport.

Although established as an independent company the Methley railway was constructed as a joint venture by the Great Northern (GN), Lancashire and Yorkshire (LY) and NE companies between Lofthouse, on the GN Wakefield–Leeds line and Methley, on the LY Pontefract branch (Fig. 14.3 C). It served collieries on the Aire–Calder ridge and provided a direct link between the GN West Riding network and the Humber ports. The Methley line was thus similar to the minor railways of South Wales in that it carried a heavy mineral traffic, but it also enjoyed a strategic importance as a link between adjoining and much larger systems. The desire to establish a more direct route was also the motive behind the Great Central (GC) and North Staffordshire line, connecting Macclesfield on the latter system with Marple, on the joint Midland and GC route from Hayfield to Manchester (Fig. 14.5). Opened in 1869, the line was intended originally to provide North Staffordshire traffic with access to Manchester independently of the LNW. However a subsequent agreement between the LNW and the North Staffordshire companies involving use

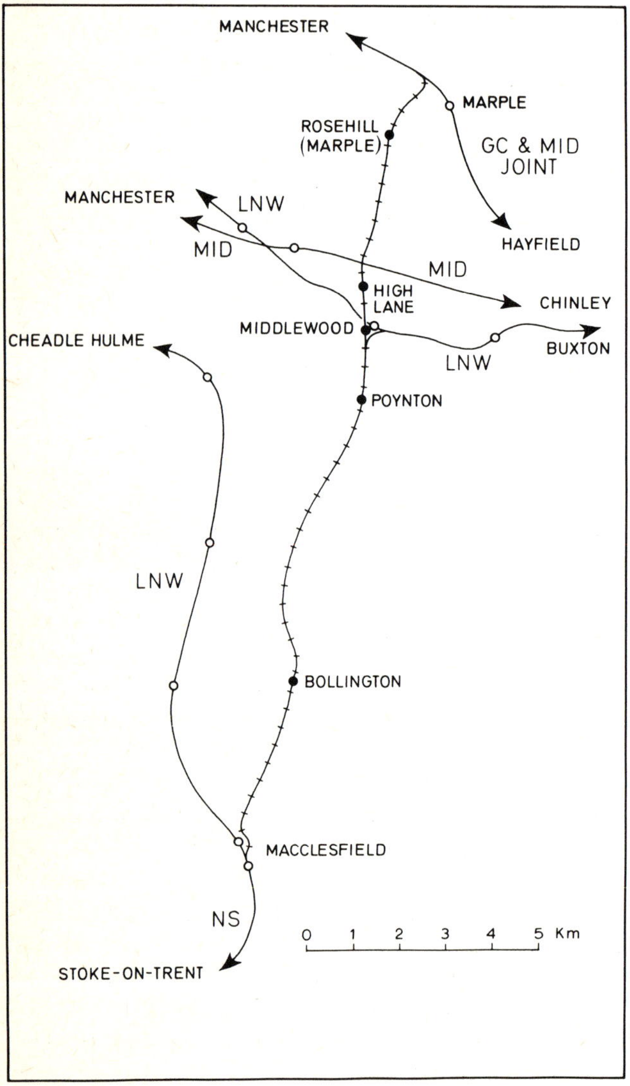

Fig. 14.5 Great Central and North Staffordshire Joint

of the Macclesfield–Cheadle–Stockport route robbed the GC and North Staffordshire line of much of its significance, although it did generate some suburban passenger traffic in later years (Christiansen and Miller, 1971).

The GE company had developed to exercise a near-monopoly over rail facilities in East Anglia, and the Colne Valley and Halstead railway was effectively a GE line, although operated as an independent concern. It was opened in 1863 to link the GE line between Colchester and Bury St Edmunds with that between Shelford and Long Melford and proved a little more successful than its neighbour the Mid-Suffolk, although neither line carried more than light agricultural traffic (Fig. 14.2 A). The Derwent Valley line, opened as a light railway in 1913, catered for the agricultural traffic in the areas east and south of York and was one of several light railways constructed for this purpose in the Humber–Ouse basin in the early twentieth century (Appleton, 1956). Like the Colne Valley line, the Derwent railway lay entirely within the sphere of operations of a larger system, the NE, connecting that company's Selby–Market Weighton line with York (Fig. 14.2 D).

The name of the Weston, Clevedon and Portishead light railway was derived from three Somerset coastal towns, each of which also possessed GW stations (Fig. 14.2 C). Connections between the two companies were made at the three towns although through train facilities were not available. The Weston line was built between 1897 and 1907 and its freight consisted almost entirely of limestone.

The three remaining examples of linear systems with two or more connections were primarily urban in character, the Manchester, South Junction and Altrincham line being the first of the group of minor railways under review to be completed. The company was formed in 1845 during a period when several plans were being considered to connect the four separate rail terminals then established in Manchester (Patmore, 1964). In 1844 the termini of the routes to Bolton and to Leeds were linked and in 1849 the Manchester South Junction was completed to connect the LNW London Road station with the terminus of the Liverpool line near Liverpool Road station (Fig. 14.6 A). An extension was opened in the same year to Altrincham, adopting a route parallel to the Bridgewater Canal and, with the growth of Sale, Brooklands and Timperley, it became a heavily trafficked suburban line. The catchment of the Altrincham line was extended with the opening of the Cheshire Lines Committee (CLC) Altrincham–Chester route and at Timperley a link was made also with the LNW line to Warrington and Widnes and the CLC railway to Cheadle and Stockport.

Several intersections with other lines focusing on Manchester were made by the Oldham, Ashton-under-Lyne and Guide Bridge Junction system, opened in 1861 as part of the 'outer belt line' around the city.

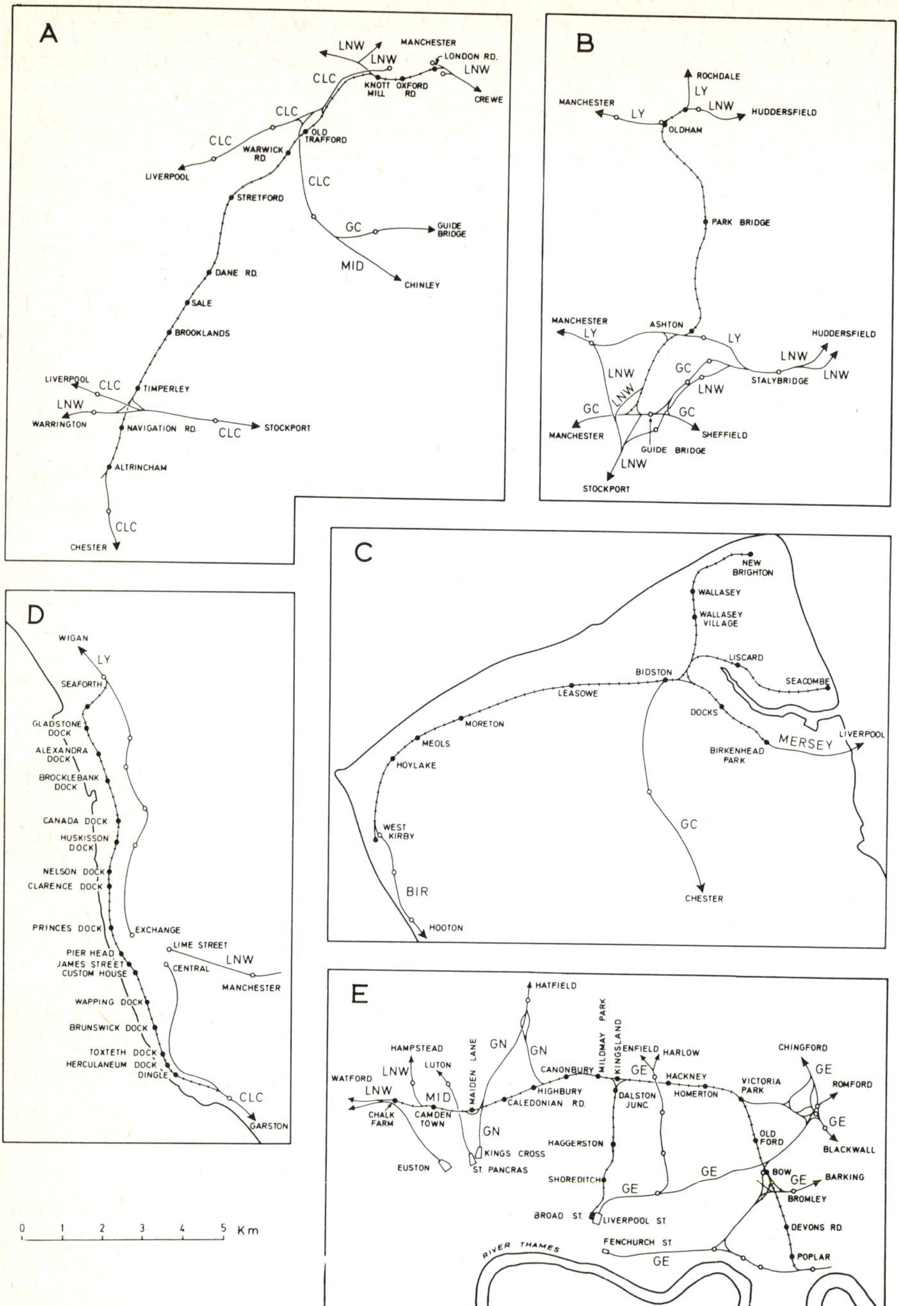

Fig. 14.6 Urban systems

A. Manchester, South Junction and Altrincham;
B. Ashton-under-Lyne and Guide Bridge Junction;
C. Wirral;
D. Liverpool Overhead;
E. North London

The Oldham line linked the LNW at Oldham with the LY (Manchester–Huddersfield) at Ashton and the GC (Manchester–Sheffield) at Guide Bridge, with a southern connection to the LNW Stalybridge–Stockport route (Fig. 14.6 B). The LNW and GC companies became joint operators of the Oldham line which carried cross-country traffic between Stockport and Rochdale.

The unique Liverpool Overhead railway was opened in 1903 to serve the chain of docks between Dingle and Gladstone, being extended to Seaforth two years later (Fig. 14.6 D). Only passenger services were operated and the line survived until 1957 as Britain's only example of an elevated urban railway (Box, 1959). It was the only line to provide through services between the northern and southern outskirts of Liverpool and although it was linked with the LY at Seaforth and the CLC near Dingle it remained a purely urban railway.

Six minor railways added branch lines to their original linear form and this group includes two of the more important urban systems, the North London and the Wirral. The North London was planned originally as a link between the London and Birmingham railway and the London docks and was opened in 1852 from Chalk Farm to Poplar, a junction with the Eastern Counties (later GE) line into Fenchurch Street being made at Bow (Fig. 14.6 E). Thus freight could be transferred more easily from the London and Birmingham railway to the docks but passenger traffic also grew in importance, using Fenchurch Street as the city terminus. With the expansion of the suburbs of north London, it was decided to obtain a more direct route into the city and in 1865 the Dalston–Broad Street branch was opened (Course, 1962). Services from both western and eastern arms of the system were operated into Broad Street and North London railway trains also ran over the GN, LNW and North and South Western Junction (Willesden–Kew) lines.

The provision of through services between a minor railway and adjacent larger systems also characterised the Wirral undertaking, which continued the lines of the Birkenhead and GC companies across the head of the Wirral peninsula and in turn linked with the Mersey railway (Fig. 14.6 C). The first section of the Wirral network was opened in stages between 1866 and 1872 and linked Hoylake with Birkenhead docks, whence ferries to Liverpool were available. In 1878 the western terminus was extended to West Kirby and in 1888 the line was continued eastwards to Birkenhead Park, connecting here with the recently-completed Mersey tunnel railway to Liverpool Central station. In the same year a branch to the growing suburbs of Wallasey and New Brighton was built, followed in 1895 by the Seacombe branch. Completion of the Birkenhead company's (joint LNW and GW) West Kirby–Parkgate line along the west coast of the Wirral peninsula in 1886 and of the GC Chester–Bidston line (1896) linked the Wirral system with the rest of the

Cheshire network but through services were not developed (Patmore, 1961; Highet, 1961).

During the nineteenth century Cumberland and Westmorland remained as strongholds of minor railways, the only representative of the principal networks being the LNW with its isolated lines in the Workington and Whitehaven districts. The west Cumberland coalfield was connected with the trunk routes of the LNW and Midland companies by three independent railways and, within the coalfield, two minor companies competed with several larger undertakings for the coal and iron ore traffic. Of these two the Whitehaven, Cleator and Egremont company was owned jointly by the Furness and LNW undertakings, and its traffic data were not recorded in the Railway Returns but included within the totals of the two parent companies. The second small coalfield system, the Cleator and Workington Junction, operated independently although 30 km of its total length were leased to other companies and the available traffic records only relate to the remaining 18 km. Built between 1882 and 1886 the Cleator system tapped the Cumberland coalfield and the haematite ironstone at Cleator Moor, and was the third railway established to compete for this traffic, the Furness and Whitehaven–Egremont systems having been opened earlier in the century. Between Siddick and Whitehaven the Cleator system duplicated the LNW coastal line and, like the latter, connected the Maryport and Carlisle railway in the north with the Furness system in the south (Fig. 14.3 A). Branches were built to Arlecdon, Siddick and Lowca and the Cleator was one of the most complex of the minor railways.

In south Yorkshire the East and West Yorkshire Union railway served the same mining district as the Methley undertaking, linking the Aire valley section of the Midland line at Stourton with the GN trunk between Wakefield and Leeds (Fig. 14.3 C). Several colliery branches were provided and a link also with the Methley line near Stanley. Unlike the Methley railway, no passenger services were operated on the East and West Yorkshire. In southeast England the East Kent light railway was opened in 1916 to link the concealed coalfield, discovered in 1890, with the South Eastern and Chatham system at Shepherds Well, north of Dover. A branch was built subsequently to Sandwich, on the South Eastern Dover–Ramsgate line (Fig. 14.3 B).

The Isle of Wight railway was the most prosperous of the three island systems, serving the three leading resorts of Sandown, Shanklin and Ventnor and controlling the connection at Ryde with the ferry to the Portsmouth terminus of the London and South Western and of the London, Brighton and South Coast networks (Fig. 14.1 A). Junctions with the Isle of Wight Central were made at Sandown and south of Ryde and through services to Newport and the west of the island were developed (Robbins, 1963). The independent Bembridge Harbour branch, opened in 1892, was intended to operate in conjunction with a train-ferry service from Hayling Island, on the Hampshire coast, but the venture failed and the branch became a part of the Isle of Wight system in 1898.

Traffic

Lack of data in sufficient detail makes it impossible to construct traffic flow maps for the British railway network at any stage in its development to 1923. The annual records of freight and passenger traffic related to each system as a whole and there is no reliable means of discovering the amount of traffic carried by particular sections of the network. The great variation in traffic generation potential throughout Britain was reflected in the differing densities carried by the various railways but no accurate way of assessing these differences exists. For the systems with less than 32 route-km, however, the relationship between traffic levels and the economic character of the region served may be defined more clearly in view of the limited area covered by each of these minor railways. Analyses of railway traffic using annual freight and passenger records have been made for several North American networks, traffic density being expressed in terms of the amount of freight or numbers of passengers per unit length of railway, and this technique has been adopted for comparing the prosperity of each of the minor railways that survived in Britain until the 1923 regrouping (Wallace, 1958; Thomas, 1962).

The total annual tonnages carried in 1921 ranged from over 250 000 on the mineral systems of the Alexandra, Cleator, East and West Yorkshire and Llanelly companies to less than 20 000 on the more remote rural branches such as the Bishop's Castle or Mid-Suffolk light railways (Table 14.1). Coal was responsible for at least 80 per cent of the freight originating on five companies (Cardiff, East and West Yorkshire, East Kent, Llanelly, Methley), a proportion similar to that recorded by many other larger mineral systems in South Wales and west Yorkshire (Turton, 1969). Freight on four systems (Colne Valley, Derwent, Mid-Suffolk, Wirral) was dominated by the 'general merchandise' category, agricultural produce being the leading commodity in each case. On several systems coal was replaced by another mineral as the principal component of freight traffic, the leading resources concerned being limestone (Alexandra, Weston), dolerite (Cleobury Mortimer), and building materials (Freshwater, Knott End).

The four leading companies in terms of freight density (Table 14.2) owed their position to their function as intensively used mineral systems conveying coal, ironstone and limestone either to export points or to centres of heavy industry within coalfields. A distinction can be drawn, however, between the two Welsh lines, which acted as self-contained

Table 14.1 Traffic data for minor railways

Company	*Route length* (km)	*Passengers* (000s)	*Freight originating* (000 tons)	*Coal class traffic as percentage of total freight*	*Other minerals as percentage of total freight*
Alexandra Docks and Railway	14.8	180.4	338.1	0	63
Bishop's Castle	15.6	25.0	10.8	31	21
Cardiff	18.8	134.9	88.8	84	13
Cleator and Workington Junction	18.0	446.3	436.8	73	26
Cleobury Mortimer	19.2	8.0	89.8	4	92
Colne Valley and Halstead	30.4	134.5	32.7	0	17
Derwent Valley Light	25.6	36.8	28.0	10	0
East and West Yorkshire Union	11.2	0	386.6	88	10
East Kent Light	26.8	7.4	98.6	93	0
Freshwater, Yarmouth and Newport	18.8	142.2	4.1	5	69
Great Central and North Staffordshire Committee	17.6	630.6	22.0	15	56
Isle of Wight	24.0	887.0	29.1	32	32
Knott End	18.0	82.0	19.6	5	60
Liverpool Overhead	14.4	13 100.3	0	0	
Llanelly and Mynydd Mawr	20.8	129.5	297.5	95	4
Manchester, South Junction and Altrincham	15.2	3 170.5	17.6	43	31
Methley Joint Committee	9.2	134.4	40.5	99	0
Mid-Suffolk Light	31.2	30.0	12.0	0	0
Nidd Valley Light	9.6	65.8	10.9	12	41
North London	22.8	11 409.4	0	0	0
Oldham, Ashton-under-Lyne and Guide Bridge Junction	10.0	360.1	6.7	1	49
Weston, Clevedon and Portishead	23.2	89.6	42.7	0	97
Wirral	22.0	2 498.2	65.2	10	9

Source: Railway Returns, Section C, 1921.

Table 14.2 Freight traffic density, 1921

Company	*Tons per route km originating* (000s)
East and West Yorkshire Union	34.51
Cleator and Workington Junction	24.26
Alexandra Docks and Railway	22.84
Llanelly and Mynydd Mawr	14.30
Cardiff	4.72
Cleobury Mortimer	4.67
Methley Joint Committee	4.40
East Kent Light	3.67
Wirral	2.96
Weston, Clevedon and Portishead	1.84
Great Central and North Staffordshire Committee	1.25
Isle of Wight	1.21
Manchester, South Junction and Altrincham	1.15
Nidd Valley Light	1.13
Derwent Valley Light	1.09
Knott End	1.08
Colne Valley and Halstead	1.07
Bishop's Castle	0.69
Oldham, Ashton-under-Lyne and Guide Bridge Junction	0.67
Mid-Suffolk Light	0.38
Freshwater, Yarmouth and Newport	0.21
Average – all minor companies	4.75
Average – all railways in Great Britain	7.25

Source: Railway Returns, Section C, 1921.

systems linking colliery with port, and the East and West Yorkshire and Cleator railways, which formed part of more extensive networks covering their respective coalfields. The density of freight traffic on these four minor lines was comparable with that on much larger systems such as the CLC of south Lancashire and Cheshire or the Hull and Barnsley in Yorkshire.

All the other minor systems carried densities below that for the British railway network as a whole, although many similar light densities existed on lines within the networks of larger railway undertakings. In some instances the poor receipts from freight were compensated by a flourishing passenger traffic, the Manchester and Altrincham, Wirral and Isle of Wight companies in particular deriving the bulk of their revenue from such a source. Of this group the most successful in terms of freight conveyed was the Weston, Clevedon and Portishead, with its traffic in limestone between the Clevedon and Weston-in-Gordano quarries, and

Portishead docks. Although coal transport was the motive behind the construction of the East Kent light railway the total coal traffic was very much less than that carried by its northern counterparts but amounts did increase with later development of the field. The light railways in rural East Anglia and Yorkshire were intended only to cater for a modest agricultural traffic in areas where provision of main line facilities would not have been justified and this is borne out by their very low freight densities. The great difference between the success of the four leading companies as freight carriers and the remaining companies is emphasised by the fact that these four accounted for 75 per cent of all freight originating on the minor railways in 1921 (Table 14.2).

A similar pattern emerges from the analysis of passenger traffic. Four urban systems carried 89 per cent of all passengers originating on the minor systems (Table 14.3). Only these four undertakings achieved a

Table 14.3 Passenger traffic density, 1921

Company	*Passengers per route km originating* (000s)
Liverpool Overhead	909.70
North London	500.41
Manchester, South Junction and Altrincham	208.58
Wirral	113.55
Isle of Wight	36.95
Oldham, Ashton-under-Lyne and Guide Bridge Junction	36.01
Great Central and North Staffordshire Committee	35.82
Cleator and Workington Junction	24.79
Methley Joint Committee	14.60
Alexandra Docks and Railway	12.18
Freshwater, Yarmouth and Newport	7.56
Cardiff	7.17
Nidd Valley Light	6.85
Llanelly and Mynydd Mawr	6.22
Knott End	4.55
Colne Valley and Halstead	4.40
Weston, Clevedon and Portishead	3.86
Bishop's Castle	1.60
Derwent Valley Light	1.43
Mid-Suffolk Light	0.99
Cleobury Mortimer	0.41
East Kent Light	0.28
Average – all minor companies	77.08
Average – all railways in Great Britain	46.10

Source: Railway Returns, Section C, 1921.

passenger traffic density above that for all railways in Britain, relying on commuter services for much of their revenue. Freight traffic was of little importance on the Manchester and Altrincham but the Wirral and North London systems generated some freight and acted as links within a more extensive network of freight services. The Liverpool and North London systems had a more balanced daily traffic than the Manchester and Wirral lines, which were more suburban in character and carried fewer passengers during off-peak hours.

The high ranking of the Isle of Wight railway is due almost entirely to its function as a continuation of the Portsmouth–Ryde ferry link, carrying passengers to the three large resorts in the southeast of the island. A large proportion of its annual traffic was carried during the summer season and the traffic levels in winter were much lower and comparable to those recorded by railways in the lower ranks of the table, such as the Freshwater and Yarmouth line.

Although their primary business was coal transport many of the mineral railways developed passenger services between mining communities and larger towns that also functioned as centres for the coal traffic. The Cleator, Alexandra, Cardiff and Llanelly systems were noteworthy in this respect, their passenger densities being higher than those of light railways serving rural districts of the country.

Minor railways in the period following regrouping

Although many of the minor railways were associated closely with larger systems before the regrouping, the Railways Act of 1921 brought them all within the orbit of the four new undertakings. Under the title of 'principal subsidiaries' the minor railways became subject to the operating policies of the new groups and many of the changes made were due primarily to management decisions. After 1923 both minor and major systems were affected by changing economic conditions and the growing challenge of road transport in urban and rural areas. The reasons behind the closure of many of the minor systems that took place in the 1920s and 1930s were associated not only with railway operating policy but with more general changes occurring within the field of inland transport as a whole. In his analysis of passenger service withdrawals, Patmore (1966) discusses five main reasons for closures of sections of the railway network and two of these, the rationalisation of duplicated facilities inherited by the post-1923 companies and remedying the over-provision of facilities, were largely responsible for the demise of several minor lines in the interwar period.

The incorporation of the Weston, Cardiff, Alexandra and Llanelly lines into the enlarged GW created several clear cases of duplicated services. This situation, coupled with declining receipts as road transport competition intensified, led to the abandonment of passenger services on several of these lines before 1939. In Somerset the direct service provided by the Weston line was withdrawn in 1940, although the towns of Weston, Clevedon and Portishead retained their GW facilities. In South Wales the GW inherited a collection of undertakings which had competed vigorously for mineral traffic and had led to many outstanding instances of route duplication. Cross Hands, a mining district at the head of the Gwendraeth valley, was linked to Llanelly by the Llanelly and Mynydd Mawr and by the GW, while further east the Alexandra and Cardiff companies became a part, together with the Rhymney, Taff Vale and Brecon systems, of a complex network of duplicated routes focusing on Newport and Cardiff. The London, Midland and Scottish company acquired all the undertakings on the west Cumberland coalfield, and services on the Cleator system, which duplicated many of the former LNW and Whitehaven and Egremont routes, were withdrawn before the Second World War.

In the more remote rural areas the light railways and several of the 'dead-end' minor lines were unable to match the flexibility of road passenger transport and the Nidd Valley, Knott End, Derwent Valley, Cleobury Mortimer and Bishop's Castle undertakings all lost their passenger services in the interwar period. In East Anglia the Mid-Suffolk and Colne Valley lines survived this period of rationalisation but the former lost its passenger facilities in the third year of state railway ownership. In 1953 the Freshwater branch was closed, together with other sections of the Isle of Wight system, and in 1962 the Colne Valley line saw the cessation of passenger services after almost a century's existence, the last of the rural minor lines to succumb to road competition. The passenger services on the Oldham line were withdrawn between Oldham and Guide Bridge in 1959 and in the following year the Seacombe branch of the former Wirral system was closed. All that now remains of the former GC and North Staffordshire Joint line is a short spur between Romiley, on the Hayfield to Manchester line, and Rose Hill (Marple), the service to Macclesfield having been discontinued in 1970. In Glamorgan the lower section of the Cardiff line has survived as part of the suburban network, passenger services being retained between Cardiff and Coryton. In the London area Broad Street still acts as the terminus of services operated over the former North London railway to Camden Town and thence to Richmond, and of occasional services from Watford using the former LNW main line as far as Chalk Farm.

The Manchester and Altrincham line has survived successive regroupings and rationalisation policies and remains an important element in the network of suburban services south of Manchester. Electrification to Altrincham has increased the capacity of the line, which also carries traffic originating on the former CLC route to Knutsford and Chester. Thus the minor system that developed, as Patmore (1964) has noted, as almost an afterthought to the link between two early Manchester termini has survived by virtue of its importance as a commuter route.

Conclusions

The minor railways provide a useful medium through which many of the factors that have influenced the changing patterns of the British railway network may be investigated. They form small-scale examples of rail developments displaying trends that have much in common with more extensive networks, encountering the same problems that faced larger systems during the twentieth century. Several minor lines were promoted as single traffic railways, with particular emphasis on either passengers or minerals, and their fortunes were governed largely by the continued existence of that traffic. Other lines were built to challenge already established undertakings and added to the complexity of the networks in many coalfield areas, ultimately being removed from the railway map during the post-1923 phases of rationalisation. Many ventures of the late nineteenth and early twentieth centuries were less ambitious in terms of the traffic expected, and experienced a precarious existence in rural districts, often surviving only because of financial support from more prosperous railway companies.

The widespread withdrawals of passenger services that followed the 1923 regrouping and the 1948 nationalisation of railways produced a steady decrease in the number of minor railways. Today the only examples that survive are those forming part of the suburban networks of Greater London, Manchester, Merseyside and Cardiff. Their continued existence is dependent on government attitudes towards the subsidising of rail services in these areas where public transport by rail is still an important part of the travel pattern.

References

APPLETON, J. H. (1956) 'The railway network of southern Yorkshire', *Trans. Inst. Br. Geogr.*, 22, 166–7.

BARRIE, D. S. (1950) *The Taff Vale Railway*, 2nd edn, Lingfield, Oakwood Press.

BOX, C. E. (1959) *The Liverpool Overhead Railway, 1893–1956*, Liverpool.

CHRISTIANSEN, R. and MILLER, R. W. (1967) *The Cambrian Railways*, vol. 1, David & Charles.

CHRISTIANSEN, R. and MILLER, R. W. (1971) *The North Staffordshire Railway*, David & Charles.

COURSE, E. A. (1962) *London Railways*, Batsford.

HIGHET, C. (1961) *The Wirral Railway*, Lingfield, Oakwood Press.

JENNINGS, B. (1967) *A History of Nidderdale*, Huddersfield, Advertiser Press.

OXLEY, J. S. (1901) *Light Railways*, London.

PATMORE, J. A. (1961) 'The railway network of Merseyside', *Trans. Inst. Br. Geogr.*, **29**, 231–43.

PATMORE, J. A. (1964) 'The railway network of the Manchester conurbation', *Trans. Inst. Br. Geogr.*, **34**, 166–7.

PATMORE, J. A. (1966) 'The contraction of the network of railway passenger services in England and Wales 1836–1962', *Trans. Inst. Br. Geogr.*, **38**, 105–17.

ROBBINS, R. M. (1963) *The Isle of Wight Railways*, 4th edn, Lingfield, Oakwood Press.

THOMAS, F. H. (1962) 'Some relationships between a railroad route and its region', *Tijdschr. Econ. Soc. Geogr.*, **53**, 155–61.

TURTON, B. J. (1969) 'British railway traffic in 1921', *Trans. Inst. Br. Geogr.*, **48**, 155–71.

WALLACE, W. H. (1958) 'Railroad traffic densities and patterns', *Ann. Ass. Am. Geogr.*, **48**, 352–74.

A. Moyes

15 Second World War manufacturing industry and its significance for north Staffordshire

For many years, location theorists have sought to interpret spatial patterns of industrial activity as the outcome of locational forces. For simplicity's sake, these forces are often regarded as operating on some isotropic surface at constant rates through time. In reality, locational forces cannot operate so straightforwardly. First, they interact on a surface which is littered with the spatial patterns generated by now vanished forces; they rework the stock of fixed and floating capital into new forms. Second, they act erratically rather than continuously. Periods of locational stability are separated by what has been termed 'parametric shock' (Pred, 1967). These locational shocks may buffet an industrial activity which, under previous forces, was stable at a location. Dziewonski (1966) has suggested that when the new forces have sufficient power to uproot the activity, it has 'vanishing stability' at that location. These dynamic concepts seem appropriate to the study of industrial location and relocation, but, among geographers, only Lewis (1970) has attempted to measure locational forces. Although difficulties arise in isolating and quantifying all the real and perceived locational forces which might have been at work in a particular case, the concept forms a framework for analysis which has been adopted in this essay.

This account has two principal aims, which stem from its focus on the period during and immediately before and after the Second World War. The first is to describe the locational significance of one parametric shock on manufacturing industry at the national scale, namely the shock of setting up a command economy mobilised mainly to produce war materials. Second, it attempts to reconstruct the wartime industrial geography of the north Staffordshire conurbation, both to illustrate the implications of this controlled economy at a local level and to assess the significance of these developments for postwar industrial diversification of its economy.

The Second World War economy: distinctive characteristics

Defence requirements

Armaments manufacture by commercial companies has always been strongly influenced by government control. Their prosperity has depended often on erratic bursts of government expenditure, and thus such an industry generally has been insufficiently equipped for sudden demands. To meet these deficiencies governments have therefore become involved in arms manufacture on their own account.

A nucleus of government-run armament works had been maintained since the early nineteenth century in the form of three Royal Ordnance Factories (ROFs) in the London area at Enfield, Waltham Abbey and Woolwich Arsenal. During the First World War, additional works were built at Birtley (County Durham), Irvine in Ayrshire, Gretna near Carlisle, Holton Heath near Poole in Dorset and Hereford. Shipbuilding, commercial gun and cartridge manufacturing, the emerging vehicle firms, and the chemical industry were mobilised; some firms had government sponsored extensions. However, these shortlived additions to the country's manufacturing capacity barely altered the regional balance of employment after 1918.

The Second World War's technology was different from that of the First. Although the amount of small arms ammunition and firepower employed in each was roughly similar, there was a vast increase in production of more sophisticated weapons – automatic weapons, tanks, wheeled vehicles and particularly aircraft (Hornby, 1958). To make this complex precision weaponry in quantity required new techniques of production and industrial organisation. It needed widespread interchangeability of parts, which could best be achieved by setting up highly mechanised production lines devoted to the mass manufacture of one or two components only. Numerous similar production lines would feed large-scale subassembly and final assembly plants. Rearmament, which began in a small way in the spring of 1934 (Postan, 1952), placed quite different demands on the country's economy from those made in 1914. Some areas of the country had industrial capacity immediately adaptable to war needs; in others, no such capacity existed. The whole economy required restructuring and heavy investment for rearmament.

Apart from these technological changes since 1914, new strategic and political requirements became apparent in the interwar years. The possibility of aerial bombardment made industrial concentrations highly vulnerable. Much of southern and eastern England lay within bombing range of mainland Europe. Accordingly in 1934 an 'unsafe' zone was defined by the Ministry of War Production, inside which industrial development was regarded as unwise (Fig. 15.1). The definition remained

Fig. 15.1 Changing definitions of 'unsafe areas', 1934–41. Based on Hornby, 1958

as one of the criteria taken into account in location decisions until 1941, by which time most of the largest war factories had been planned. In political terms, rearmament was seen to be as powerful a means of altering the regional balance of industrial prosperity as the more widely publicised 'distressed area' measures taken after 1934. Since many of the war factories were anticipated to be large employers of labour, many of the areas of high interwar unemployment (mostly lying outside the unsafe zone) were strong contenders for government investment in rearmament.

Coastal bombardment was viewed as less relevant to planning the strategic location of factories at this stage. Inevitably, much rearmament for the Navy had to be conducted in coastal areas, but this represented a final assembly stage in production. The manufacture of components and naval ammunition was possible in inland safe areas. After the fall of France, however, the vulnerability criterion was altered to curb new industrial development within 35 km of the coast for fear of invasion, but this had been reduced to a band from Land's End to the Wash by November 1941 (Fig. 15.1).

The organisation and location of the enlarged armaments industry

Industry was adapted for war production in five ways, each of which had different locational implications. First, certain (mostly consumer) industries not adaptable to war production were concentrated in the hands of a few firms, liberating surplus manpower and premises for re-use *in situ* or elsewhere. Second, the Ministry of Supply set up a Directorate of Industrial Planning in January 1934. By 1939 it had almost completed an inventory of the potentialities of each manufacturing plant for war work. At the least, premises were earmarked for storage purposes; at the most, government-owned machine tools were installed in spare floor-space. The proportion of space allocated was greatest in engineering works, particularly in car and aircraft assembly and component firms, even though most of them were sited in the unsafe area.

Third, the directorate identified firms which could act as major contractors for the supply of finished articles (ranging from small arms ammunition to aircraft) and produced a supporting list of potential subcontractors to undertake work for the major contractors. Car manufacturers were the first major contractors selected, mainly for aero-engine production. The importance of this development lay in establishing information on subcontracting facilities and available space. Elaborate chains of subcontracting and dispersed manufacturing were set up, using small reserves of productive capacity at varying distances from the parent works (Mensforth, 1947). Since subcontracting firms were usually smaller than their parents, they were countenanced more easily

Table 15.1 Regional distribution of agency factories and ROFs handed over to civilian use by December 1946

Standard region	*Number*	*Floorspace* (000 m^2)	*Largest factory's floorspace* (000 m^2)	*Average factory's floorspace* (000 m^2)
Northern	23	408.7	195.7	17.8
Yorkshire and Humberside	22	199.6	51.7	9.1
East Midlands	17	137.8	39.1	8.1
South Eastern	31	253.2	23.0	8.3
South Western	31	340.3	67.5	10.9
Wales	30	667.9	202.9	22.3
West Midlands	40	867.1	204.4	21.7
North Western	47	1 554.2	142.9	33.1
Scotland	23	364.3	175.2	15.8
Northern Ireland	14	118.5	23.4	8.5

Source: Board of Trade data, 1947.

in the unsafe zone where scattered small engineering resources might already be available (Nixon, 1946).

Fourth, the Air Ministry began an 'agency' factory scheme in January 1936 in which major contractors (not necessarily engineering firms) were given the task of managing government built and equipped factories or extensions to be used for component manufacture, subassembly or final assembly. The scheme was slow to mature partly because most contractors were unwilling to manage plants at great distances from their parent works (Ward, 1946). After September 1939 more rapid progress was made. By 1945, 206 agency schemes had been instituted, involving some 330 factories, at a cost of £286 million (Hornby, 1958). The regional distribution of those agency factories (and the small number of ROFs) which had been handed over to civilian manufacturing by the end of 1946 is shown in Table 15.1. Although no standard region was without agency factories, those regions which lay mainly or wholly within the unsafe zone received little of the new floorspace created. The major exception was in the west Midlands, where agency factory building was permitted in the Birmingham conurbation and in Coventry because the Ministry of Supply feared disruption of component flows between related firms if they were forced to disperse unduly. In the safe zone, a sizeable concentration of medium-sized agency factories was in southeast Lancashire, reinforcing the existing engineering traditions of Greater Manchester. In comparison, Merseyside and the Severnside portion of the South Western standard region received fewer large plants. The depressed areas of peripheral Britain received little agency factory building. Wales and the Northern England standard region had a few very large ROFs, which were handed over to civilian use at the same time as their agency factories. Similarly in Scotland, the largest factory (for Rolls-Royce aero-engines near Glasgow) accounted for almost half the new factory floorspace created and the remaining twenty-two factories were small. Even in the context of a command economy it was difficult to stimulate much industrial interest in the depressed areas.

The final way in which the country's peacetime economy was mobilised was by expanding the scale of operation, but not the scope, of the ROFs. The three old-established ROFs, vulnerably located near London, employed barely 9 000 persons in 1937 (Hay, 1949). By 1939 twenty-three additional ROFs had been approved, of which seven, including the re-opened Hereford and Birtley, were in operation at the start of the war. At their peak there were forty-four ROFs employing nearly 350 000 men and women, out of 1 million engaged directly in munitions manufacture. The ROFs were of three generally single-purpose types (Fig. 15.2). Least numerous but most expensive were the new explosives plants which were mainly located on the western seaboard or adjacent to substantial inland water supplies. They all required extensive areas of low-lying, level land (Hornby, 1958) amounting to 1 012 hectares at Bishopton, the first and largest of these new plants. The three factories making propellant fuels were major employers (12 000 workers each at Bishopton and at Wrexham, and 4 000 at the later-established Ranskill). The others, making TNT and other explosives, rarely employed more than 2 000.

The second type was termed 'Engineering ROF', at most of which the main activity was mass production of ammunition cases. These were less expensive works to build; the contract for the largest, at Radway Green on the northern edge of the north Staffordshire conurbation, was valued at £944 000 (Kohan, 1952). They needed less land and were generally smaller in employment capacity. Other factories in this group produced gun parts. Some engineering ROFs were run as agency factories by commercial firms, and five assembled finished guns. The location of these relatively smaller plants was not heavily influenced by vulnerability considerations, but the ammunition case-making units were generally near the filling factories which they were to supply (Fig. 15.2).

The ammunition filling factories formed perhaps the most distinctive group of ROFs. The first dispersal Royal Filling Factory (RFF) was announced in May 1935, to be built at Chorley in Lancashire at a cost of £11 million. Construction of two more, in South Wales, began in the following year. On the declaration of war work began on another RFF at Swynnerton in north Staffordshire and one at Risley in south Lancashire. All these first-generation RFFs were designed as vast single-purpose workshops intended to employ up to 30 000 people (mostly women) on

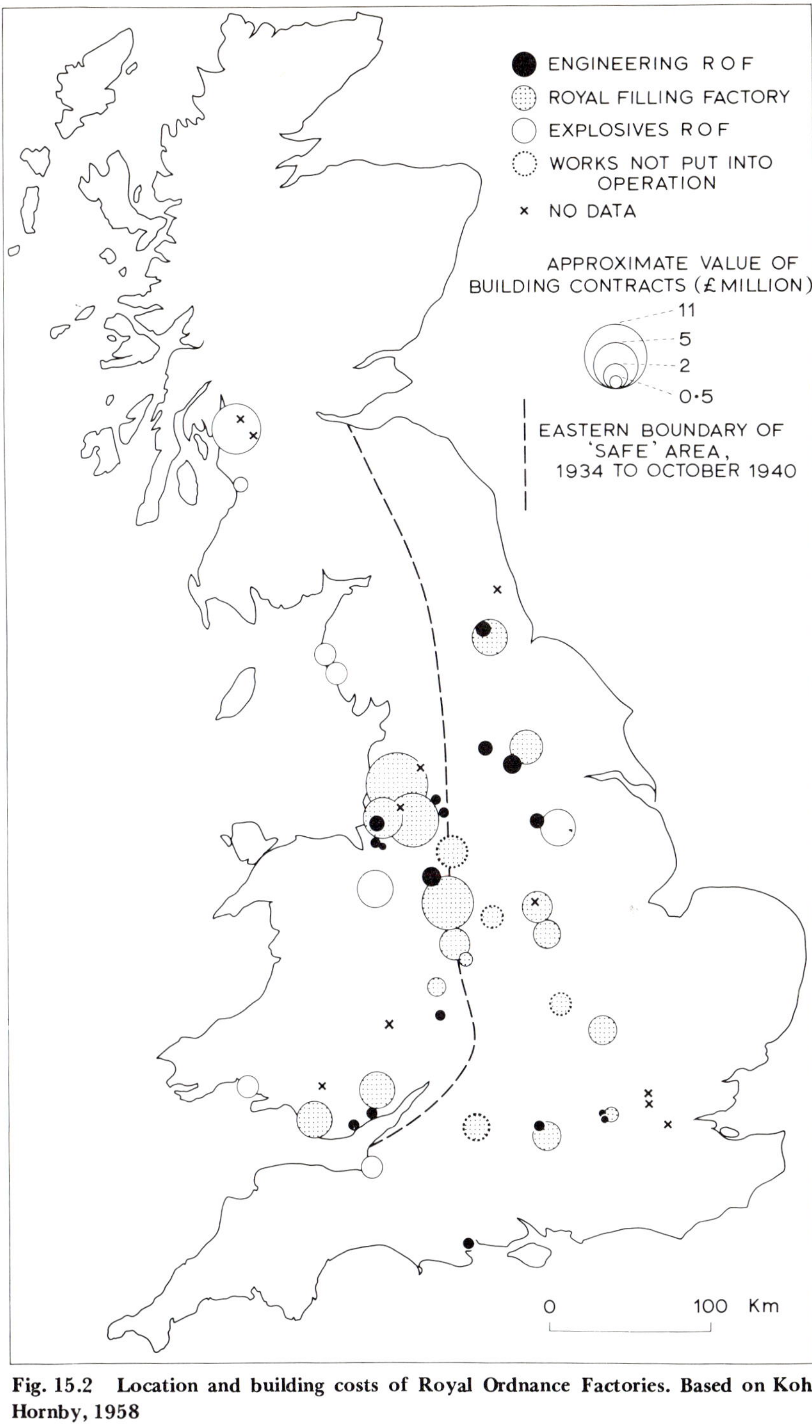

Fig. 15.2 Location and building costs of Royal Ordnance Factories. Based on Kohan, 1952; Hornby, 1958

three shifts and each required sites of roughly 400 hectares. Chorley proved to be too large to operate successfully, so subsequent RFFs were reduced in size, allowing them to be sited at points where smaller pockets of labour were available. As the anticipated need for ammunition rose, previously examined sites in vulnerable west Yorkshire (Thorp Arch) and County Durham (Aycliffe) were developed. Not until after the redefinition of the unsafe area in October 1940 were smaller filling factories built in the east Midlands, the west Midland conurbation, and near Bedford (Elstow) and Reading (Burghfield). Rising productivity at earlier RFFs rendered some of these surplus soon after completion and the smaller RFFs planned at Swindon, Northampton, Tutbury and Macclesfield never materialised. Thus the aggregate distribution of ROF capacity showed a broad adherence to the 1934–40 vulnerability principle even after it had been altered.

These measures to mobilise the economy for war production were contingency plans. Their investment schemes took time to create new manufacturing employment. Once completed, direct and indirect munitions employment rose to a maximum of 5.2 million by mid-1943 (Postan, 1952). In May 1941, with the new pattern of war industry taking shape, the Board of Trade collated estimates of the amount of labour needed to man fully the plants then coming into production. A considerable demand arose for labour in work of national importance to the west of the vulnerability line and in existing manufacturing converted to war production. The million males unemployed in June 1939 were absorbed rapidly by the armed forces and manufacturing. The only substantial increase in the total labour force came from increasing the proportion of females at work. The female labour force rose from 5.1 million to 7.3 million in the first four years of war. By May 1941 small reserves of immobile female labour were available, but mainly in coastal areas ('Green' Areas) at one time or another in the vulnerable zone (Fig. 15.3). Even at the peak war effort larger numbers were available in central Scotland and on the Northumberland coalfield, but most main contractors would not or could not subcontract work over such distances, and there was already severe overloading of the main east coast railway line serving these areas (Bell, 1946). Direction of labour to jobs was the only alternative.

Wartime industrial activity in north Staffordshire

One subregion with a substantial need for manpower in war production was north Staffordshire (Fig. 15.3). The 48 000 vacancies in war industry in this conurbation amounted to almost one-third of its total peacetime labour force, possibly the most extreme example of temporary restructuring of any subregion's economy. It is in this light that

Fig. 15.3 Vacancies for unskilled labour on war work, 1941. Based on PRO BT 96/77X/5540

the wartime industrial geography of this conurbation will be reconstructed.

For present purposes, the north Staffordshire conurbation is defined as the city of Stoke-on-Trent with Newcastle-under-Lyme MB and Kidsgrove and Biddulph UDs. In this conurbation 40 per cent of the total workforce of 162 000 was employed in pottery manufacture in 1939, and 16 per cent in coalmining. Although the former had increased its nominal workforce in the interwar years and the area's insured total population had risen at the national rate, some 20 000 pottery workers were either unemployed or underemployed. Female activity rates in Stoke were 25 per cent higher than the national rate, whereas in the towns around it relatively few women were employed. Two much smaller manufacturing sectors each employed about 3 000 persons in 1939: metal manufacture, with declining employment, and engineering, which largely supplied the needs of the two staple industries, with stagnant employment. The only major new interwar activity was the Michelin tyre factory in Stoke, which offered 1 600 jobs in 1939. Service industries had grown in the 1930s to comprise 27 per cent of the conurbation's employment by 1939, a low proportion even by west Midland standards. The remaining employment sectors absorbed only 10 per cent of the workforce.

The conurbation was typical of many subregions which developed a localised specialisation in the Industrial Revolution. By 1939 it still contained 80 per cent of the country's pottery employment (Fogarty, 1945). Its largest single activity was a consumer goods industry with a distinct craft technology barely amenable to wartime use. It was the nearest large population cluster in a safe area to the vulnerable west Midlands conurbation, in which manufacturing capacity for war purposes was large. A reserve of labour displaced from a concentrated pottery industry would inevitably arise, and the area's potential for new industrial activity was soon harnessed, in each of the five ways already described (Fig. 15.4).

Contractions of sectors of the peacetime economy

Pottery manufacture was one of the industries most severely reduced in the Second World War. It employed only 37 per cent of its 1939 total in the conurbation in 1942 (Forsyth and Morris, 1943), whereas jobs in other consumer goods industries like cotton and woollen textiles, and boot, shoe and clothing manufacture only fell to between 55 and 61 per cent of their prewar norms. Since Stoke contained three-quarters of the country's electrical porcelain industry (an activity which expanded during the war), the fall in employment in domestic pottery, sanitary ware and glazed tile manufacture was even sharper. The pottery industry

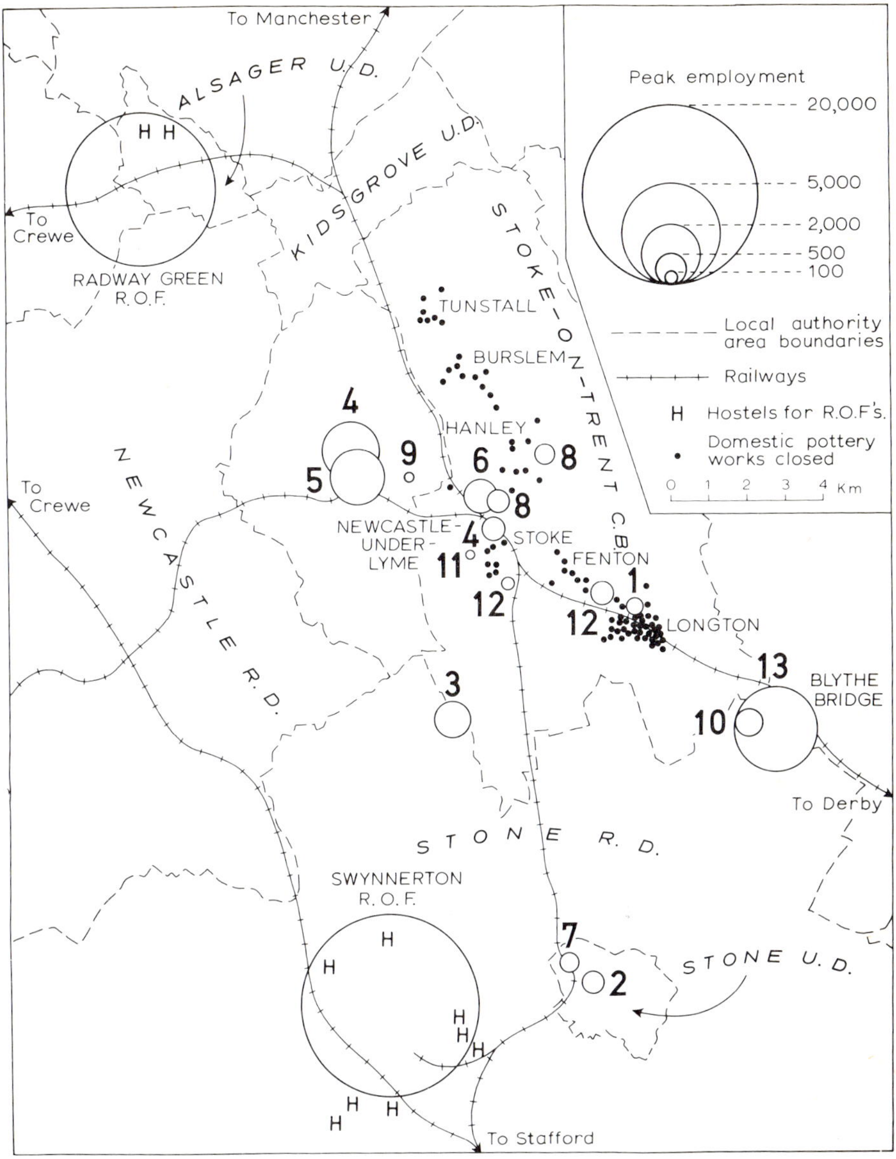

Fig. 15.4 Major wartime amendments to the industrial structure of north Staffordshire

Key to firms:

1. **Aero Tanks**
2. **Alvis**
3. **Bank of England**
4. **BSA**
5. **British Thomson Houston**
6. **Dunlop**
7. **English Electric**
8. **N. C. Joseph**
9. **Martin Colour**
10. **R. B. Page**
11. **Perry Chains**
12. **Rolls-Royce**
13. **Rootes**

labour force began to decline in 1940 as men were gradually conscripted; then the decline accelerated after November when the government ordered a cutback of production for the home market. By then, large new factories in the area sought manpower at a rate of 1 450 per month, and unless the pottery industry contracted further it was anticipated that local voluntary labour supplies would be exhausted by the spring of 1941 (Watkinson, 1941). Contraction came after May when the industry reluctantly submitted its concentration scheme; of 575 000 m^2 of industrial floorspace requisitioned in the conurbation, 304 000 were in pottery factories alone.

Of the 272 pottery factories in Stoke, eighty-eight were retained as a nucleus to supply a narrow range of undecorated utility ware, in addition to a small export of decorated pottery; forty-two other firms moved the vestiges of their activity into these nucleus premises; thirty-five firms closed completely; and over a hundred dormant firms' premises were allocated to storage, though little of their equipment was removed. By 1943, some self-explanatory shifts had occurred in the use of many premises in the city (Table 15.2).

Table 15.2 Number of industrial and commercial premises in selected uses in Stoke

Uses	*1939*	*1943*
Earthenware, china and porcelain works	237	153
Glazed tile works	46	26
Pottery ancillary trades	97	91
Brick and tile works	37	21
Engineering works	41	47
Government property	48	286
Factories for storage only	0	120
Total of all uses	78 149	77 399

Source: Reeves, 1945.

The adaptability of existing industry for war purposes

Only in 1937 were a limited number of north Staffordshire firms placed on priority lists, eight years after the first assessments of the war potential of industry had been authorised. They were six electrical porcelain works and two non-ferrous foundries in Stoke, the Michelin tyre factory, and a uniform manufacturer in Newcastle. The only subsequent additions were the Shelton iron and steel works, an already heavily committed aluminium rolling mill, eight makers of sanitary ware in the city, fifteen potters' millers (supplying raw material to electrical porcelain firms), and one Biddulph and one Stoke engineering works, each of which had capacity for machining shell and bomb cases.

Although all these were adaptable to war demands (and manpower at the foundries doubled) there was otherwise little potential in the local economy to produce war materials directly or by subcontracting. The limits of usefulness of the bulk of the pottery industry were the use of two of its continuous kilns for heat treating metal components.

Movement of manufacturing firms into north Staffordshire

Superficially, the conurbation seemed well placed to receive the dispersed industrial activity of major contractors vulnerably sited in the Birmingham area. However, labour shed by the contracting pottery industry was intended for ROF work, and little of its discarded floorspace was usable, because the premises were often old and difficult to adapt, and their owners were reluctant to allow machinery to be moved. The BSA gun company, bombed out of Birmingham in August 1940 before its agency factory at Newcastle was ready, was the first seeker of space in the area for dispersal units. After acquiring premises locally BSA concentrated its later searches for floorspace in textile mills in Mansfield, Leicester and Kidderminster (Ward, 1946). Later Rolls-Royce used a pottery at Fenton and a railway carriage works at Stoke in association with its taking over the BSA factory in 1944. The only large industrial re-use of a pottery was by Dunlop, which took over Wedgwood's Etruria works in 1940 and had converted it into Europe's largest tyre remoulding unit by 1945 (Storrs, 1946). Though several dispersing firms favoured north Staffordshire they preferred non-pottery premises – Daimler in textile mills at Leek (Nixon, 1946) and Alvis in a shoe factory at Stone (Light, 1967). Subcontractors needing little floorspace showed similar preferences; the Stratford-on-Avon hollow-ware maker N. C. Joseph used a Hanley furniture warehouse and a canalside flour mill for welding bomb cases. Manufacturing dispersals generally brought only a handful of key workers. In summary, north Staffordshire's favourable location for dispersals was largely offset by its shortage of adaptable premises.

Agency factories

Government decisions endowed the conurbation with almost 100 000 m^2 of new single-storey light engineering works and 9 000 jobs in new trades, neither of which would have materialised in peacetime. The four factories involved remained government-owned throughout the war, but were offered selectively on lease from 1945, a process which was to be an important part of postwar regional development policy. These factories occupied level unconfined sites at the then edges of the

Table 15.3 The four north Staffordshire agency factories

<table>
<tr><th rowspan="2">Location</th><th rowspan="2">Agent</th><th rowspan="2">Floorspace (m^2)</th><th rowspan="2">Date approved</th><th colspan="2">Peak employment</th></tr>
<tr><th>male</th><th>female</th></tr>
<tr><td>Newcastle</td><td>BTH Co, Rugby</td><td>7 875</td><td>September 1939</td><td>340</td><td>1 360</td></tr>
<tr><td>Newcastle</td><td>BSA Ltd, Birmingham</td><td>36 248</td><td>October 1939</td><td>800</td><td>1 200</td></tr>
<tr><td>Blythe Bridge</td><td>Rootes Securities Ltd, Birmingham</td><td>6 440</td><td>January 1940</td><td rowspan="2">2 000</td><td rowspan="2">3 000</td></tr>
<tr><td>Blythe Bridge</td><td>Rootes Securities Ltd, Birmingham</td><td>46 670</td><td>January 1940</td></tr>
</table>

Source: Hornby, 1958.

conurbation, free from mining subsidence and near expanding residential areas (Table 15.3).

The first agency factory in Newcastle was taken over by British Thomson Houston (BTH) to augment their existing shell-producing capacity in London (BTH, 1946). The second factory was brought into production by BSA in April 1941, making parts for, and assembling, aircraft cannons and using components from previously established local dispersal units (Ward, 1946). By April 1944 this factory was transferred to Rolls-Royce, which required additional production capacity for a new jet engine. The last two factories were built at Blythe Bridge in Stone RD adjacent to Stoke's municipal aerodrome. They were operated by Rootes to modify and repair aircraft. By January 1944 the smaller factory was being used for making rifle parts for ROF Maltby.

The Royal Ordnance Factories

The two ROFs were planned as complementary industrial units, sufficiently far apart at opposite ends of the conurbation to tap fully its labour reserves (Fig. 15.4). Both these ammunition plants were authorised at the outbreak of war, and were to become major employers. Radway Green works employed a maximum of 9 736 women and 4 820 men in February 1943, and the Swynnerton plant engaged 15 313 women and 5 086 men at its peak in August 1942 (House of Commons Reports, 1945).

The Radway Green plant occupied a site with direct access to the Crewe–Stoke railway. It resembled the agency engineering factories in the conurbation in its heavy endowment of machine tools and its large main building, but it conducted more processes. The labour force was drawn from a 20 km radius but some key staff from Woolwich were housed in a 400 house estate in Alsager, where a hostel was also built.

Swynnerton RFF lay adjacent to the main London–Manchester railway, 14 km southwest of Newcastle in a thinly populated area. It was the most expensive RFF (£13.6 million) though not the largest. The 523 hectare site was remodelled to create underground storage for explosives. Because of its isolation seven hostels were built, providing 4 409 beds – by far the largest hostel scheme at any ROF (Forbes-Smith, 1943). As with other RFFs, the various production stages within the site were divided among innumerable small workshops to minimise the danger of explosions. The labour force reached 1 000 in November 1940 and rose steadily to 13 000 nine months later, when local voluntary labour supplies were considered exhausted. Peak employment was achieved only after the Control of Employment (Directed Persons) Order of January 1942 and by the end of 1943 the factory had 8 553 workers whose homes were outside commuting range (Ministry of Supply, 1944). This peak was shortlived, and rising productivity in RFFs and decreasing ammunition demands cut Swynnerton's total workforce to 8 500 by 1945, a rate of decline paralleled by other ROFs outside the development areas.

Wartime employment in the conurbation: a summary

A locational shift in job opportunities occurred as Stoke's manufacturing and service activities contracted and new manufacturing arose outside it (Fig. 15.4). From 1939 to 1943, Stoke probably lost 22 000 jobs for men and 1 000 for women, while Alsager, Biddulph, Newcastle and Stone RD gained 1 000 jobs for men and 15 000 for women. The number of transfers between industries and the number acquiring new skills are difficult to determine. Ministry of Labour statistics, on which postwar reconstruction teams had to rely, were inaccurate and misleading. Local workers were drafted to other areas where their skills were more valuable and replaced by labour from elsewhere (including 'Bevin boys'). Although the Potteries had only 0.8 of the country's male and 1.1 per cent of its female working population in 1939, they contributed 1.7 per cent of Coventry's male immigrant labour force from 1940 to 1941, and 3.9 per cent of its female (Shenfield and Florence, 1944). Independent estimates suggest that two-thirds of the pottery industry's labour moved into engineering and explosives manufacture, contributing most to their growth from 3 000 jobs in 1939 to 60 000 at their wartime peak (Forsyth and Morris, 1943). The reverse transfer of labour after 1945 was to be almost as large.

Reconstruction of the conurbation's economy in the postwar period

As a measure of the extent of wartime disruption, the shifts of employment in the conurbation's industries from 1938 to 1943, and thence to

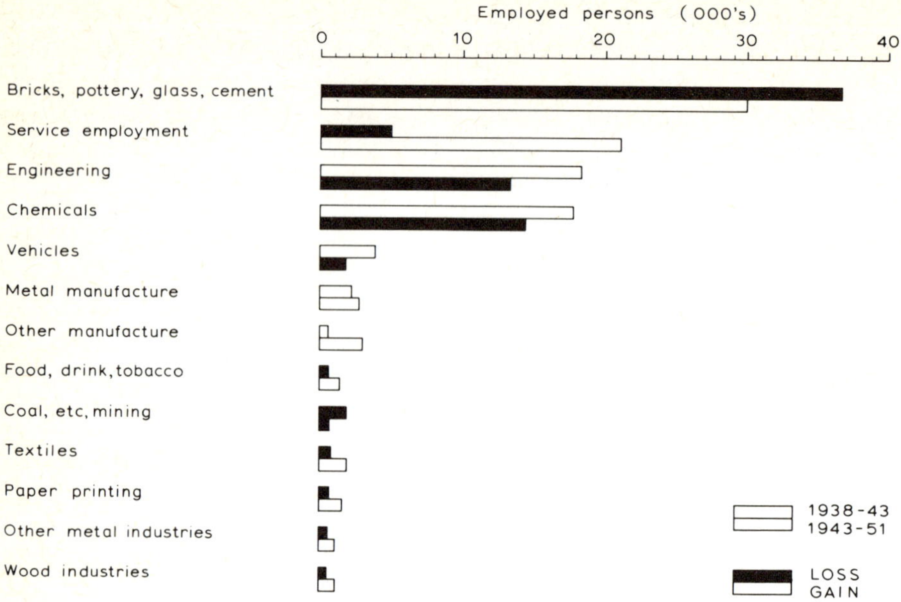

Fig. 15.5 Estimated shifts in employment in selected industries in north Staffordshire, 1938–43 and 1943–51. Based on Ministry of Labour returns

1951 have been recorded (Fig. 15.5). Despite heavy gross changes within some industries, net employment change between 1938 and 1951 was generally small. Ceramics regained all but 6 700 of their employment, and by 1951 were again the dominant sources of manufacturing jobs in the conurbation. Four other types of shift can be distinguished. The mining industry alone lost men during the war and continued to lose employment after 1943, despite strenuous postwar recruitment drives. Food and drink, textiles, paper and wood industries, 'other metal industries' and basic services like distribution more than recovered wartime losses. Only two industrial groups (metal manufacture, including foundries, and 'other manufacturing industries', mainly the rubber industry) sustained their wartime growth. A final group of activities, engineering, chemicals (explosives) and vehicles, which had mushroomed during the war, lost employment almost as quickly afterwards, leaving little net growth. However, some underlying changes had occurred which have continued to influence the conurbation's economic performance.

The forces shaping readjustment were a mixture of local and national political and economic interests. The early postwar years saw the Board of Trade in a strong position to influence the location of new industries through its control of building materials, its issue of building licences and its power to allocate surplus government facilities to new tenants. Its attitude to north Staffordshire was influenced strongly by the pottery industry's wish to rebuild its export trade, and was stated explicitly in the Distribution of Industry Act 1948: 'In view of the shortage of labour in the pottery industry, especially female, the time (was) inappropriate for the introduction of more industries other than factories to provide work for the disabled and partially-fit men who constitute the majority of those presently registered as unemployed.' The Board made available as many building licences as it could to modernise pottery works *in situ*, and supported schemes to build three new suburban potteries outside Stoke, but these foundered because of local political squabbles.

Since north Staffordshire was not given development area status (because its future prospects were officially considered good), its hopes for industrial diversification lay in adapting the surplus war capacity it had acquired. However, the dispersal firms had additional manufacturing space at their parent plants, and were in premises which their previous users now required. Only the Dunlop rubber remoulding unit was to remain, until the company ceased remoulding in 1960. The agency firms were different in that they normally had first option on their factory's lease. Rootes did not wish to remain at Blythe Bridge, partly because the factory was unsuitable for car assembly, and Rolls-Royce, with surplus space in older-established works, vacated their plant in February 1946. Only BTH immediately took up their lease, adapting the factory to make small electric motors produced previously at Coventry. A period of uncertainty followed; eventually the Rolls-Royce factory was occupied by Rists Wires and Cables, a firm making wiring sets for cars, which had been forced out of its Lowestoft premises in 1939 and had since outgrown them. The larger of the former Rootes factories was allocated to the Birmingham-based Simplex company for domestic appliance manufacture in March 1946, but the smaller remained unlet for another year until taken by a local bus-body builder. The slow rate at which these factories were let indicates the difficulty of finding tenants who were not under pressure to relocate in development areas.

Before 1945 the peacetime industrial possibilities for the two ROFs had been investigated, along with others throughout the country. Local Board of Trade officials suggested that Radway Green could serve as a cooperatively run engineering shop to cater for the growing needs of a more mechanised pottery industry. Swynnerton had little use save as a munitions factory, the local chamber of trade seeing little scope for even small-scale firms using its scattered, tiny buildings; yet a long-term plan using the site as a trading estate and housing area for population overspill was proposed. However, as both these ROFs were among the twenty-three to be retained as standby armaments plants in postwar defence plans, development schemes were shelved. In December 1945, Radway Green employed 2 062 and Swynnerton 5 314; of the total, 3 900 jobs were for local men, and these were used as an argument by the Board of

Trade against the stimulation of further industry to move to north Staffordshire from other areas. Reviews of defence expenditure in the mid-1950s brought a further reduction in ROF capacity; as part of this scheme, Swynnerton, by then employing 2 500, closed in June 1958. Radway Green was retained, to produce 90 per cent of the country's needs for small ammunition, but with a labour force reduced to about 2 000 by 1972.

The function of the north Staffordshire ROFs as employers of local labour influenced Board of Trade attitudes towards making the region more attractive to footloose industry requiring new locations. Without the reductions in activity at both ROFs in 1958 two industrial developments on the northwest edge of the conurbation would certainly not have been countenanced, and the growth potential of the larger would probably have passed north Staffordshire by. These developments were a branch factory at Alsager for the Stoke sanitary ware manufacturer Twyfords, and an expansion of a small branch factory of English Electric at Kidsgrove, from which the present International Computers factories have grown to employ 4 000 people.

After its closure, the possibilities of the Swynnerton site were again appraised, but this time as part of west Midlands, rather than north Staffordshire, regional planning strategy. Grandiose schemes for a new town to accommodate some of Birmingham's overspill population and industry had collapsed by 1967, Swynnerton's inappropriateness for such a role stemming from its location outside commuting range of Stoke, its unsuitable premises and small-scale site and the lack of local development area status (Ministry of Housing and Local Government, 1967; Moyes, 1971). As no other government-sponsored manufacturing body had use for the site, part was returned to agriculture, and part reserved for army training.

Conclusions

The nature of the external environment is but one of the forces specified by Dziewonski as acting on an activity at a location. The inertial forces (such as the varying tie of fixed equipment as it proceeds towards amortisation) and frictional forces (such as a labour force *in situ* which could, with varying effort, be transferred to some new site) influence the permanence of an activity. Forces tending to disrupt stability are external tensions, diseconomies which could arise as the industry's technology or the demand and supply situation in which it operates alter. The balance between the inertial and frictional forces and the external tensions resolves itself in four states. Of these the first, complete stability, occurs only at some so-called optimal location, where external tensions in the form of a change in external environment since the plant was established are absent. The second, partial stability, exists where inertial and frictional forces are larger than external tensions; stability can be restored by strengthening internally the economics of the activity. Thirdly, vanishing stability results from external tensions sufficiently strong to overcome inertial and frictional forces. The fourth state, full mobility, is the condition at which the question of relocation arises.

These concepts fit the wartime economy of north Staffordshire most closely if an additional term 'conditional stability' is added, in which 'the location is optimal only through the intervention of a potentially unstable factor' (Tarrant, 1968, p. 442). This state applies readily to the incoming dispersal firms occupying requisitioned property; these firms' locational stability depended wholly on the maintenance of sufficient contract work to keep their dispersal units at full capacity. Wholesale cancellation of war contracts began in September 1945 and immediately put these units into a state of vanishing stability. Their temporary occupants had no rights over the requisitioned premises which thus exerted no inertial forces. In an external environment which had reverted to a peacetime character, there was no compelling logic for them to retain these marginal production units. The few local firms which had adapted their activities for war production had not changed their fixed equipment sufficiently during the war to point them towards new technologies or commercial possibilities in peacetime. They had not assembled inertial or frictional forces powerful enough to fit them to different peacetime markets; moreover the external environment of the early postwar period was extremely favourable to these local firms because of the strong demand for their normal products. The probabilities of either of these wartime adjustments to the area's economy being permanent were therefore weak.

The agency factories, as purpose-built premises which had been engaged in a type of flow-line engineering similar to that of peacetime, presented strong inertial forces for the retention of this activity. These forces were non-existent in the agency firms themselves; they were also conditionally stable, depending on contracts and only expressing interest in agency factories if the external environment of the early postwar years was broadly similar to that experienced during the Second World War. Only BTH was able to contemplate moving peacetime production immediately into their agency factory in north Staffordshire.

The Swynnerton site represents conditional stability at its most pronounced. The weak, inertial forces which allowed it to retain ammunition filling until 1958 were neutralised by the forces of a rapidly changing external environment (changing defence requirements) leaving the site with vanishing stability for its designed activity. The forces necessary to mount normal production would have had to counter the disincentives which the site in its abandoned condition afforded. Swynnerton's failure to commend itself as a potential industrial estate may be compared with the development area RFFs. As at Swynnerton, the chaos of small buildings at these sites has acted more as a deterrent than an incentive to orderly development of industrial estates; buildings had no inertial force. The other RFFs differed from Swynnerton in the strong frictional force which their manpower presented. If no attempt to provide replacement employment on their closure had been made, their labour forces would have been released into local economies which were unable to absorb them. In the context of the development areas, the barely suitable buildings of vacated RFFs were viewed as a potential industrial asset in 1945. In the context of north Staffordshire after 1958, the Swynnerton site has proved no more than an embarrassing non-conforming land use.

This study has emphasised the subtlety of the forces acting to shape industrial location and relocation. In north Staffordshire, the repercussions on the economy originating in the immediate prewar and wartime period are still working themselves out in the 1970s, though the re-armament decisions and strategic concepts of the 1930s that first created them have now been largely forgotten.

References

BELL, R. (1946) *History of the British Railways during the War 1939–45*, London.

BTH COMPANY (1946) *War Achievements of the BTH Company*, Rugby.

DZIEWONSKI, K. (1966) 'A new approach to the theory and empirical analysis of location', *Pap. Reg. Sci. Ass.*, **16**, 17–25.

FOGARTY, M. P. (1945) *Prospects of the Industrial Areas of Great Britain*, London.

FORBES-SMITH, A. (1943) *Estimate of saving in bus-miles to be secured by transferring ROF workers to existing hostel accommodation*, PRO MT50/123, paper CTC 188.

FORSYTH, M. and MORRIS, M. (1943) *Stoke-on-Trent: a review of the industries, services and development of the area*, London, Nuffield College Social Reconstruction Survey; mimeo.

HAY, I. (1949) *ROF: the story of the Royal Ordnance Factories 1939 to 1948*, London.

HORNBY, W. (1958) *Factories and Plant: official history of Second World War*, London.

HOUSE OF COMMONS REPORTS (1945) *Official Report, Parliamentary Debates, Commons*, 5th series, **417**, col. 1070.

HOUSING AND LOCAL GOVERNMENT, MINISTRY OF (1967) Circular letter to local authorities concerning the Swynnerton site.

KOHAN, C. M. (1952) *Works and Buildings: official history of Second World War*, London.

LEWIS, P. W. (1970) 'Measuring spatial interaction', *Geogr. Annalr.*, **52B**, 22–39.
LIGHT, B. (1967) *The Alvis Story*, Coventry.
MENSFORTH, E. (1947) 'Air frame production', *J. Instn Mech. Engrs*, **156**, 27.
MOYES, A. (1971) 'Post-war changes in the distribution of employment and population in north Staffordshire', unpublished MA thesis, University of Keele.
NIXON, St J. C. (1946) *Daimler 1896–1946: fifty years of the Daimler Company*, London.
POSTAN, M. (1952) *British War Production: official history of Second World War*, London.
PRED, A. (1967) *Behaviour and Location: foundations for a geographic and dynamic location theory, Part 1*, Lund studies in geography, Series B.
REEVES, A. J. (1945) *Postwar Employment in Stoke-on-Trent*, Stoke-on-Trent.
SHENFIELD, A. and FLORENCE, P. S. (1944) 'Labour for the war industries: the experience of Coventry', *Rev. Econ. Stud.*, 12, 45.
STORRS, SIR R. (1946) *Dunlop in War and Peace*, Hutchinson.
SUPPLY, MINISTRY OF (1944) *History of ROF Swynnerton*, mimeo.
TARRANT, J. R. (1968) 'Industry in south-east Ireland: a study in industrial location', *Ir. Geogr.*, 5, 440–58.
WARD, D. M. (1946) *The Other Battle*, Birmingham.
WATKINSON, G. L. (1941) *Letter to Minister of War Production, 7 January*, PRO BT 96/62.

G. Coleman

16 Regional development policy: the Merseyside experience, 1949–1966

Regional development policy in Britain was initiated as a response to the massive unemployment which affected particularly those areas of Britain heavily dependent on basic heavy industries. While the impact of policy in the prewar period was negligible (Odber, 1965), it has had important implications for postwar policy, first in the emphasis given to the relief of unemployment and second in the way in which imbalances in the employment structure were seen as the main cause of unemployment. Much postwar regional policy has been viewed as a social service designed to relieve unemployment (McCrone, 1969; McCulloch, 1956) and 'an implicit objective has been to diversify or restructure the economy in such a way as to strike at the roots of the unemployment problem' (Lloyd, 1970, p. 374).

While the views that high unemployment is indicative of general economic distress and that it results from structural problems are questionable (Smith, 1968; Cunningham, 1970; Coleman, 1972), it is clear that the reduction of local unemployment and the restructuring of the employment bases have been the major aims of postwar regional policy in Britain. Any attempt to assess the impact of development area policy on Merseyside between 1949 and 1966 must emphasise these objectives.

The methods of sponsoring regional development in Britain have changed relatively little during the postwar period. Inducements offered to industry to move to the development areas have increased considerably both in size and scope and by 1968 annual expenditure on regional policy amounted to about £270 million. The major control over industrial expansion has been the requirement for an Industrial Development Certificate (IDC) for any new factory or factory extension of 465 m^2 or more. By judicious use of both the 'stick' and the 'carrot', the Board of Trade (the executive body for regional policy) has attempted to channel jobs from the more prosperous regions to the development areas.

Within the context of this paper Merseyside refers to the development area scheduled in 1949 (Fig. 16.1), but other definitions of Merseyside have been used where necessary. In an appendix to *The North West: a regional study* (Department of Economic Affairs, 1965) a rather larger definition of Merseyside was adopted, though not so large as that used by the Board of Trade in their examination of the movement of manufacturing industry in the northwest (Board of Trade, 1969). Because of the aggregated nature of the statistics available from these sources it has been often impossible to break them down to provide standard areal units and it has been necessary to compare indices for different definitions of Merseyside.

The data problem is not limited to the definitions of Merseyside. Lloyd and Dicken (1968) have noted the technical problems which result from the generally unsatisfactory nature of the areal units for which official figures are compiled and the unreliability of the employment data except in their most aggregate form. These problems can be reduced by working with aggregated data (at the Standard Industrial Classification (SIC) level) and by noting only the stronger trends. It must be recognised that the data are subject to error and that this problem, associated with that of different regional definitions, means that the various indices must be interpreted with care. Given these constraints, this paper attempts to examine some of the changes wrought on Merseyside by the application of regional policy between 1949 and 1966. Since unemployment and employment structure were clearly the main targets of the policy these will be examined in some detail, and the various factors which have contributed to their change will be analysed to see how the policy has been applied and the extent to which it has been successful.

Merseyside: the development area

Though Merseyside was not declared a development area until 1949 the problems of the subregion date from the interwar period when the overdependence on port based activities (Baines, 1970) and the failure to develop an alternative manufacturing sector led to high levels of unemployment. Despite this, Merseyside was never scheduled as a 'special area' but, through a special Act (1936) Liverpool Corporation acquired the kind of powers made available by prewar legislation. Little was achieved before 1939 but these powers were significant in the immediate postwar period. As in the other problem areas of Britain, the outbreak of war effectively solved the unemployment problems of Merseyside but the war also perpetuated its outdated employment structure and the return to peacetime conditions saw unemployment rise sharply (Lloyd,

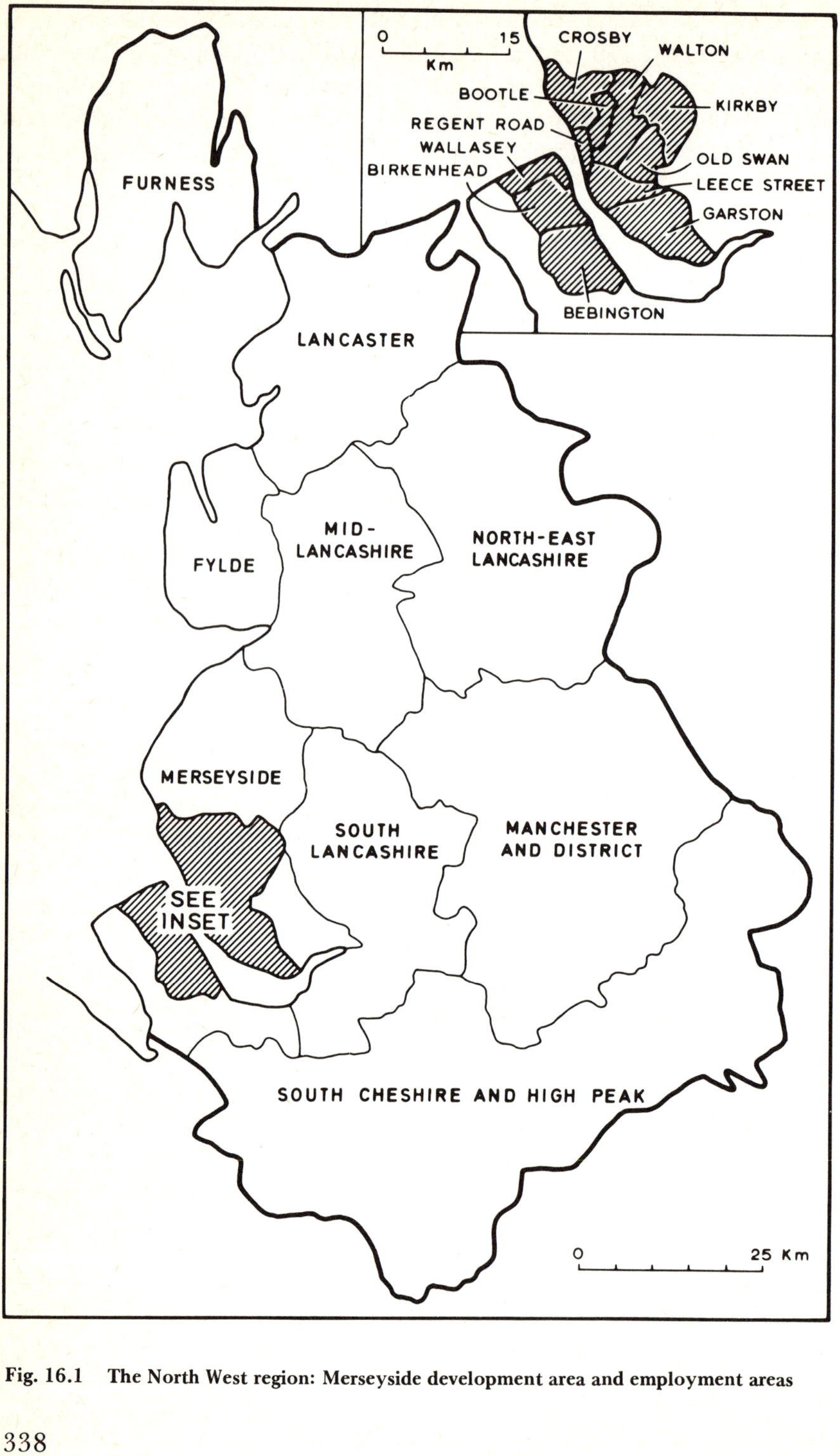

Fig. 16.1 **The North West region: Merseyside development area and employment areas**

1965). In the immediate postwar years Liverpool used its powers to attract new industry, postponing the need for development area status (Board of Trade, 1948). This increased the economic disparity between Liverpool and the surrounding local authorities (Lloyd, 1970) and by 1948 male unemployment on Merseyside was three times the number of male jobs promised in the whole of the postwar factory-building programme.

Merseyside was scheduled a development area at an unfortunate time in terms of policy application. In the immediate postwar period a combination of circumstances, some of them temporary, had made the development areas particularly attractive to industry (Odber, 1965), gaining between 1945 and 1947 more than half the new industrial building in the country but with only one-sixth of Britain's labour force in manufacturing. As a result of this apparent success the impression was created that 'the momentum of development in the outer regions of Britain would . . . be taken over by private enterprise so that government spending could be allowed to taper off' (Odber, 1965, p. 339).

The active implementation of regional policy was halted temporarily by the 1947 balance of payments crisis and in the following years expenditure on the policy gradually declined with a reduction in the pressure on businessmen to build new factories in the development areas. The favourable circumstances which had helped the postwar development areas were also fading and after the scheduling of Merseyside *The Economist* (2 April 1949) noted that the advantages of development area status were 'likely to be much less tangible' than they had been twelve months earlier. The decline in the carrying out of regional policy is illustrated clearly in the case of Merseyside where, in its first year as a development area, the amount of industrial building completed fell below the figure for 1948 and fell further in 1950 to only 25 500 m^2 (Table 16.1). Not only did the absolute total fall but the share of the northwestern total going to Merseyside fell from 20 per cent in 1948 to 7.6 per cent in 1950, while Merseyside had almost 15 per cent of the northwestern total of manufacturing employment.

A more satisfactory position may be indicated by the fact that a larger Merseyside gained 61 per cent of the employment resulting from inward moves of manufacturing industry to the northwest during the 1945–51 period, although of these 27 000 jobs some 20 000 were in existing premises (Fig. 16.2). This suggests that many of the new jobs were in the area's large ex-government wartime factories which had been converted to peacetime production. Only 7 000 jobs were created by 1951 in new premises, including those jobs which had been attracted by Liverpool Corporation. The initial impact of regional policy on Merseyside was slight and the attraction of the region was the presence of old wartime factories which generally influenced industrial relocation in the immediate postwar years (Odber, 1965). Unemployment on Merseyside had

Table 16.1 Industrial building completed in the Merseyside development area, 1947–65

Year	*Total completions* (000 m²)	*Percentage of northwestern total*
1947	10.6	9.6
1948	56.7	20.0
1949	43.6	14.7
1950	25.5	7.6
1951	155.9	31.8
1952	80.8	16.0
1953	191.6	28.9
1954	73.5	15.6
1955	67.0	11.3
1956	108.1	15.9
1957	104.7	18.1
1958	114.3	15.9
1959	79.0	18.8
1960	119.8	20.0
1961	236.6	31.2
1962	86.7	14.8
1963	296.7	57.2
1964	114.0	27.9
1965	140.5	32.9
Total	2 105.6	21.5

Source: Board of Trade, North West Regional Office.

been reduced to 2.3 per cent by 1951 against a national figure of 1.2 per cent (Table 16.2); it followed the national trend rather than reacting to any major input of new employment opportunities.

Some delay is likely between the granting of development area status and the appearance of tangible results. The increase in 1951 in both the level of industrial building and the share of the northwestern total going to Merseyside probably represents the first fruits of regional policy. In the three years 1951–53 there were 428 300 m² of industrial building completed in the area, representing almost 29 per cent of the northwestern total for the period and 40 per cent of the 1947–59 total for Merseyside. During the former period Merseyside received almost double its 'share' of industrial development compared with the northwest as a whole.

The main developments during the 1947–53 period indicate a concentration in those fields in which Merseyside traditionally had specialised. More than 20 per cent of the factory completions were for the food, drink and tobacco industries, while engineering and chemicals accounted for a further 25 per cent. There were also large developments in the metal and the vehicle industries, which together accounted for 31 per

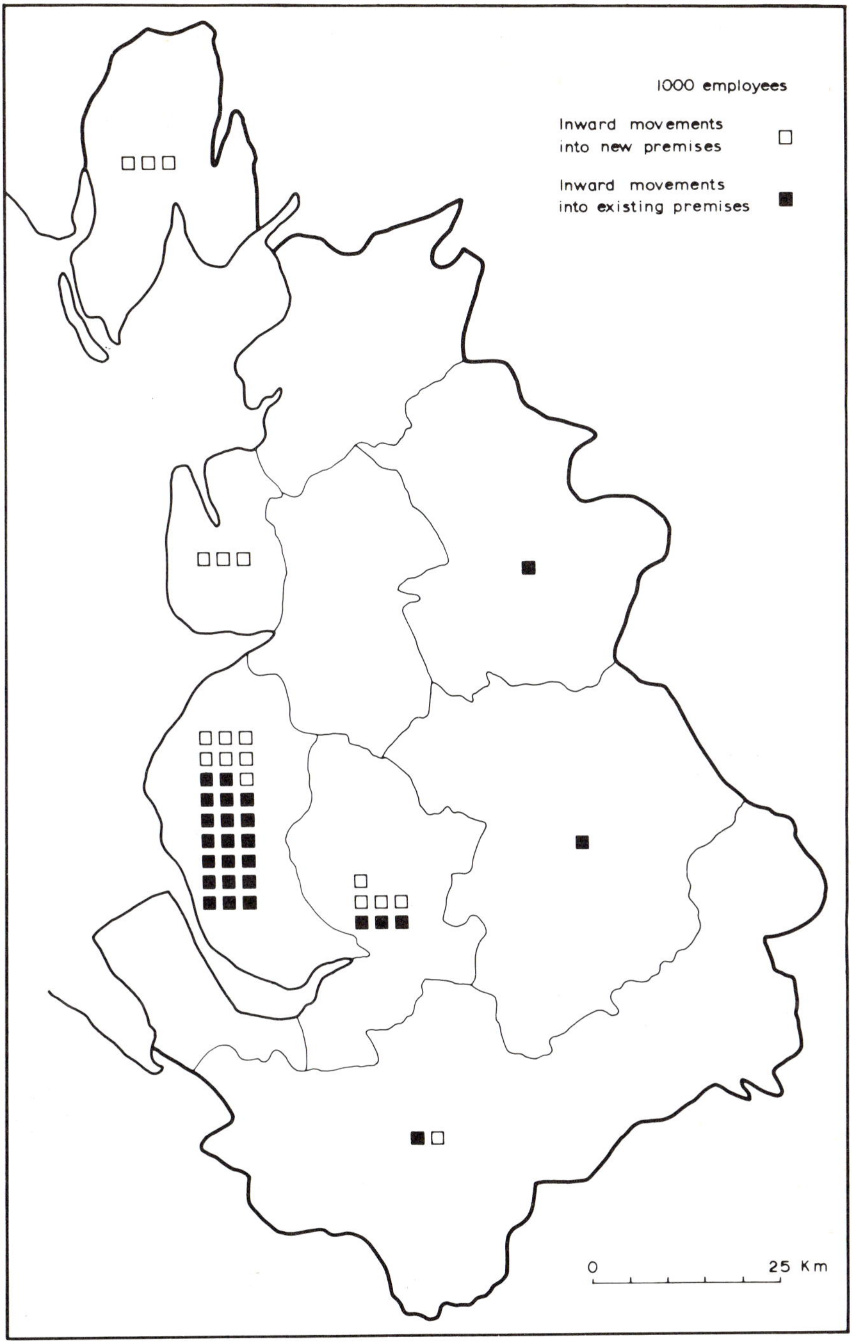

Fig. 16.2 Movement of manufacturing industry in the North West region, 1945–51. Based on Department of Employment and Productivity statistics

Table 16.2 Merseyside unemployment, 1951–67

Year	*Number unemployed* (000s)	*Percentage unemployed*	*UK percentage unemployed*
1951	15.0	2.3	1.2
1952	21.4	3.3	2.0
1953	23.3	3.6	1.6
1954	18.9	2.9	1.3
1955	14.2	2.1	1.1
1956	14.1	2.1	1.2
1957	19.0	2.8	1.4
1958	25.8	3.9	2.1
1959	28.3	4.2	2.2
1960	22.1	3.2	1.6
1961	19.6	2.9	1.5
1962	28.6	4.1	2.0
1963	34.2	5.0	2.5
1964	25.7	3.6	1.6
1965	15.4	2.2	1.4
1966	13.7	1.9	1.5
1967	21.7	3.0	2.4

Source: Department of Employment and Productivity statistics.

cent of the industrial building. Apart from these latter groups there is little suggestion of any significant attempt to diversify the employment structure and the employment structure in 1953 showed few changes from that of 1938. Thirlwall (1967) has examined the performance of the postwar development areas with a version of the 'shift-and-share' technique which attempts to explain the difference between regional and national rates of employment growth. The technique attempts to differentiate employment change resulting from either the employment structure of the area (proportionality shifts) or the individual performance of industries within the area (differential shifts). The terminology used to describe the components of shift-and-share analysis has varied but since the technique has been extensively examined (Zelinsky, 1958; Fuchs, 1959 and 1962; Dunn, 1960; Hemmings, 1963; Thirlwall, 1967; Stilwell, 1969), no further attempts will be made to describe its intricacies. Thirwall's analysis of the Merseyside development area for the 1949–52 period indicated that employment grew at a faster rate than in the country as a whole. The increase on Merseyside was 5.5 per cent while that for Britain was 2.3 per cent and the analysis indicated that the 'excess' of 3.2 per cent was due entirely to the differential component. The proportionality component was negative, indicating that in 1949 Merseyside had an employment structure unfavourable to growth. The area managed to overcome this handicap and produce a high rate of growth due to the buoyancy of its individual industries.

The situation by 1953 reflects the mixed fortunes of the subregion. The employment structure was poorly balanced, receiving little diversification, but the area fared well in terms of jobs created during the immediate postwar years. Industry in general was growing faster than the national rate, unemployment was falling and from 1951 the level of industrial building was high. While the natural attractions of available factory space and the activities of Liverpool Corporation were responsible for development to 1948, development area status appears to have been significant in some respects from 1951. Following the recession of 1952–53 and the loosening of IDC restrictions in the more prosperous regions, the fortunes of Merseyside declined. In 1953 unemployment in the development area rose to 3.6 per cent and in the following years total industrial building fell from the 1953 peak. The proportion of the northwestern total industrial building going to Merseyside declined to an annual average only slightly above the area's share of manufacturing employment (Table 16.1). The reduction in the application of the policy is marked by the sharp fall in the number of operatives in plants 'entering' the development area (Fig. 16.3). In 1954, the number of operatives in plants leaving Merseyside exceeded the number entering and, apart from a brief respite in 1955, this trend continued to 1959. A shift-and-share analysis for the 1953–59 period for Merseyside shows a complete

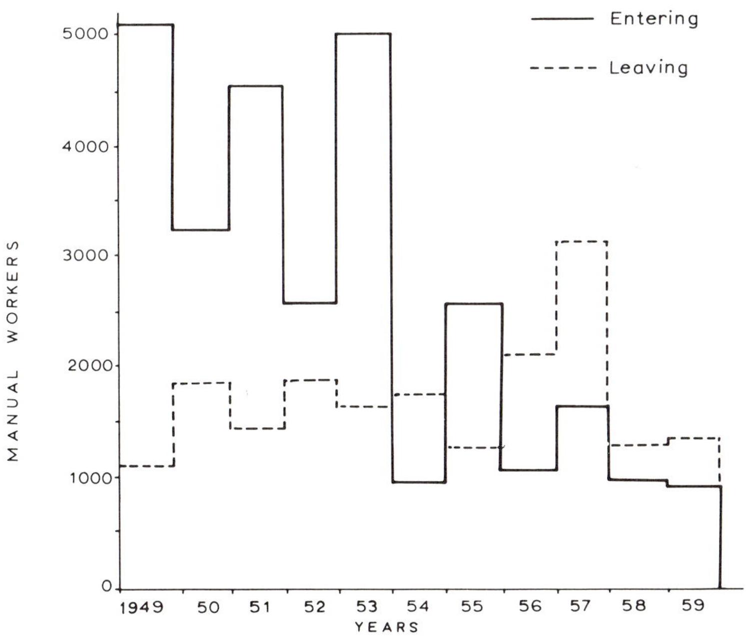

Fig. 16.3 Manual workers entering and leaving Merseyside, 1949–59. Based on Lloyd, 1965

Table 16.3 Merseyside shift and share analysis, 1953–59

Total growth	5 562
Regional share	27 580
Total shift	–21 928
Proportionality shift	7 404
Differential shift	–29 332
Proportionality modification shift	–2 755

reversal of previous trends (Table 16.3). Total employment grew by only 5 652 employees, or 1 per cent, compared with an expected (regional share) growth of 27 580 employees, thus indicating a 'loss' of 21 928 jobs. The total shift loss comprised two diverging elements – a positive proportionality shift and a negative differential shift. The positive proportionality shift indicated that the employment structure of Merseyside had been favourable to growth to the extent of 7 404 jobs in 1953. The replacement of an unfavourable structure in 1949 by a favourable one in 1953 is difficult to explain, since the employment structure apparently had changed little during this period. Changes among the growth industries at the national level between 1953 and 1959 may have provided Merseyside with a favourable structure. The component reflects the extent to which employment on Merseyside was dominated by fast-growing, slow-growing or declining industries at the national level. Large proportionality losses were recorded in the nationally declining food, drink and tobacco, shipbuilding and transport and communications industries while large gains came from the nationally fast growing engineering, insurance, banking and finance, professional services and distribution industries.

The second element of the total shift was the differential component, indicating the extent to which individual industries on Merseyside fared better or worse than the same industries nationally. The negative score of 29 332 jobs demonstrates that Merseyside fared badly. The potential for growth indicated by the positive proportionality shift did not come to fruition in the ensuing period. Only the food, drink and tobacco, shipbuilding, paper, printing and publishing and transport and communications industries did markedly better than their national counterparts, while the chemicals, engineering, vehicles, other metals, bricks, pottery and glass, timber and furniture and other manufacturing industries, construction, insurance, banking and finance, professional services and miscellaneous services all fared worse.

The distribution of the differential shift is important for it shows that Merseyside increased employment in three nationally declining industries: food, drink and tobacco, shipbuilding and transport and communications. This could lead eventually to a worsening of the employment structure and the prospect of a negative proportionality shift at some time in the future. This feature is emphasised by the proportionality modification shift (Stilwell, 1969) which measures the extent to which an area has improved its employment structure during the period under study. Between 1953 and 1959 this component was negative for Merseyside, indicating that the employment structure of the subregion had deteriorated to the extent of 2 755 prospective jobs (Table 16.3).

This deterioration is reflected in the distribution of new building among the various industries of Merseyside. In the 1953–59 period 42 per cent of all completions by area was for the food, drink and tobacco industry while a further 7.3 per cent was for the nationally declining shipbuilding, textiles, clothing and timber and furniture industries. Only the 28.2 per cent devoted to the fast growing engineering industry ameliorated this situation.

Between 1952 and 1959 there was an inward movement of 20 000 jobs in manufacturing industries to a region rather larger than the development area, representing 62 per cent of the northwestern total for this period (Fig. 16.4). At the same time manufacturing industry in the development area lost 5 628 employees. The non-coincidence of both areal units and dates makes exact comparisons difficult but it does appear that there was a loss by closure or contraction of about 25 000 jobs in manufacturing industries in the eight-year period. This fact partially explains the general failure of individual industries on Merseyside to expand at their national norms, indicated by the large differential shift loss.

The ineffectiveness of regional policy on Merseyside in this period is not surprising since development area policy was in abeyance from 1951 to 1958 (Odber, 1965). The policy never recovered the initiative after the sharp break in 1947, and the general prosperity of the 1950s, when unemployment rates averaged only 1.5 per cent for the country (but 2.9 per cent for Merseyside), brought about a further reduction in the application of policy by the 1955–56 Parliamentary Estimates Committee (McCulloch, 1956). During this period of *laissez-faire* Merseyside gained a 'fair share' of the northwestern total only in terms of employment resulting from inward moves of manufacturing industry. While unemployment remained persistently above the national level (Table 16.2), manufacturing activity underwent a considerable contraction in employment and the employment structure deteriorated.

The worsening economic difficulties, which began to appear in 1955, culminated in 1958 when the end of the long postwar boom, which had staved off the long-term decline of several of the heavier industries, coincided with a trough in the trade cycle (Brechling, 1967). Unemployment throughout the country rose sharply and the development areas

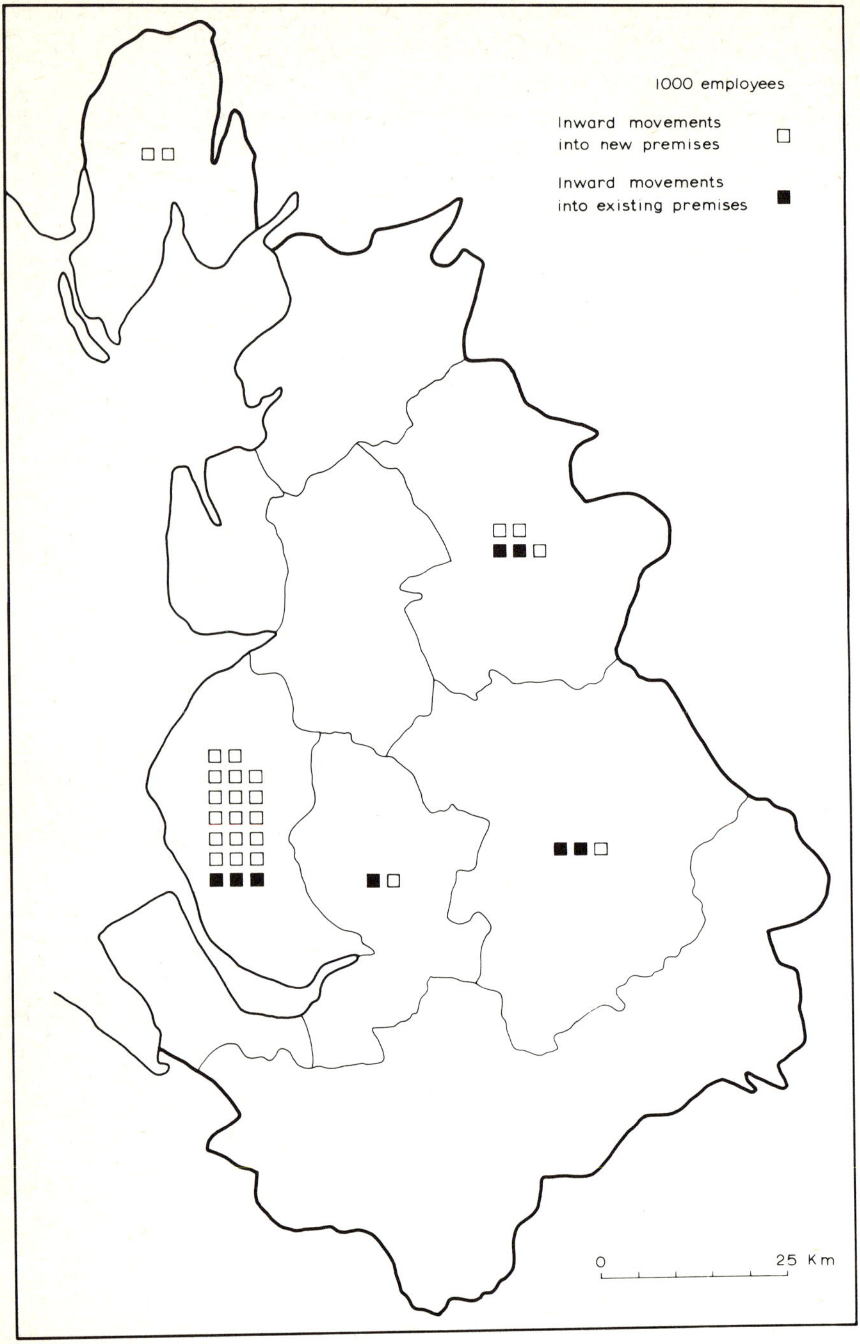

Fig. 16.4 Movement of manufacturing industry in the North West region, 1952–59. Source as for Fig. 16.2

were particularly affected since their dependence on heavier industries had remained substantial. These economic difficulties resulted in the renewal of legislation and, more important, a renewed vigour in the application of policy. Thus, on Merseyside the period after 1959 saw a further dramatic reversal of previous trends. While the rate of growth of employment for 1953–59 had been only 1 per cent (5 652 employees) and the manufacturing sector had seen an absolute decline in employment of almost 6 000, the period 1959–66 saw a growth of 5.6 per cent, an increase in total employment of more than 30 000 and an increase in manufacturing employment of almost 17 000.

Table 16.4 Merseyside shift and share analysis, 1959–66

Total growth	33 072
Regional share	42 444
Total shift	–9 372
Proportionality shift	1 883
Differential shift	–11 255
Proportionality modification shift	4 075

The major elements of a shift-and-share analysis for the period reveal that although the employment structure as a whole was less favourable to growth than it had been in 1953, the total growth was much higher because the total shift, although still negative, had been more than halved from the 1953–59 figure (Table 16.4). This reduction in the total shift loss was attributable to a far smaller differential loss for the period 1959–66, a reduction from –29 332 (1953–59) to –11 255 (1959–66). Although the individual industries on Merseyside continued to lag behind their national counterparts in terms of growth in employment, their deficit had been considerably reduced. A notable feature is that while there had been a worsening of the total structure (foretold by the negative proportionality modification shift for 1953–59) during the initial part of the 1959–66 period, the trend was later reversed because a positive proportionality modification score of 4 075 indicates an improvement of the structure by 1966.

The change in the trend of the differential loss can be attributed in the main to two industries. The engineering industry turned a 1953–59 loss of 3 439 into a 1959–66 gain of 12 879 with an increase in employment of almost 22 000, representing a greater than 50 per cent increase between 1959 and 1966. The same industry also accounted for 20.7 per cent of the total area of industrial building completed for the 1960–65 period. The vehicles industry converted 1953–59 loss of 1 210 into a 1959–66 gain of 9 946 with an increase in employment of almost 10 000 employees, an increase of 127 per cent over the 1959 total. Again, the figures for industrial building completed during 1960–65 show that 28.5 per cent of the total was devoted to the vehicles industry, which represented only 3.5 per cent of the manufacturing employment on Merseyside in 1959. Moreover, the figures for industrial building completed do not include the 278 700 m^2 development for the Vauxhall Motor Company which was built outside the development area proper at Ellesmere Port but which employed labour drawn largely from Merseyside.

In the services sector the professional and scientific services reversed a 1953–59 loss of 8 063 on the differential component to record a gain of 4 047 for the 1959–66 period. Thus, after almost a decade during which individual industries had failed consistently to match national rates of growth in employment, the development area outpaced Britain in the growth of two of the fastest growing manufacturing industries and of the fastest growing service industry.

The moderate proportionality gain was the result of large gains in the service industries, especially professional and scientific services, and large losses from the main nationally declining manufacturing industries, notably shipbuilding, textiles, clothing and food, drink and tobacco. The largest loss, however, was made on the nationally declining transport and communications industry, with a figure of –12 148. These proportionality losses should be viewed in their proper context for the shipbuilding, textiles and transport and communications industries contributed a positive total of 5 473 to the proportionality modification score, that is, the proportionality losses contributed to an improvement in the employment structure of Merseyside by reducing employment in nationally declining industries.

The reversal of fortunes after 1960 was seen not only in an increase in the area of new industrial building on Merseyside but also in a marked increase in the percentage of the northwestern total (Table 16.1). During the six years 1960–65 there was a total of 994 200 m^2 of such building in the development area compared with only 546 600 for the previous six years and only 1 111 300 m^2 for the 1947–59 period.

After 1959 statistics are available relating to planning approvals of building for both manufacturing and service industries. The data are of limited use since about 25 per cent of all approvals are never completed, but they have a great advantage in that they show the immediate impact of any change in development area policy or in the degree of its application. In the year of the Local Employment Act, 1960, approvals on Merseyside more than doubled, though much of this increase was due to the single large development by Ford at Halewood (Table 16.5). Merseyside was descheduled under this Act in July 1960 because of the number of jobs generated from this and other developments, and the result can be seen in the drastic drop in approvals in the area for 1961 and 1962 and the halving of the share of the northwestern total going to

Table 16.5 Building approved in the Merseyside development area, 1959–66

Year	*Building approved* (000 m^2)	*Percentage of northwestern total*
1959	278.2	33.5
1960	569.2	33.2
1961	120.1	16.6
1962	70.1	17.0
1963	178.8	26.6
1964	121.2	20.4
1965	215.4	26.7
1966	261.8	27.9

Source: Board of Trade, North West Regional Office.

Merseyside. Because the high rate of unemployment continued after the prospective jobs had been filled, Merseyside was rescheduled in February 1963, the effect of which can be seen in the total approvals in that year, an almost threefold increase over 1962 and a return to a greater than proportionate share of approvals going to the development area (which contained 20.6 per cent of the employed population of the northwest in 1966).

An accurate analysis of the approvals data is not possible for the whole period 1959–66 for, after 1963, the SIC was withheld from a growing number of developments, amounting to almost three-quarters of the total by 1966. For the years 1959–63 the structural breakdown of the approvals data shows a pattern similar to that of the completions data. While approvals for such traditional Merseyside industries as food, drink and tobacco and chemicals were important, there were high levels of approvals in industries such as engineering, vehicles and paper, printing and publishing and generally the data suggest a significant movement towards diversification.

The increased tempo of industrial activity after 1960 is reflected also in an inflow of some 30 000 jobs in manufacturing industries for the larger Merseyside area, representing 66 per cent of the northwestern total for the 1960–65 period (Fig. 16.5). This compares with an inflow of only 20 000 such jobs for the preceding 1952–59 period.

The inadvisability of using crude unemployment totals and of equating these with the number of jobs to be created by induced industrial expansion is shown clearly by the experience of Merseyside during the early 1960s. In 1960 there were more than 20 000 unemployed on Merseyside and 30 000 jobs 'in the pipeline', half of them in the motor industry. This was the major cause of the descheduling of Merseyside later that year, for it was thought that the unemployment problems would be solved once the projected expansions were in full production.

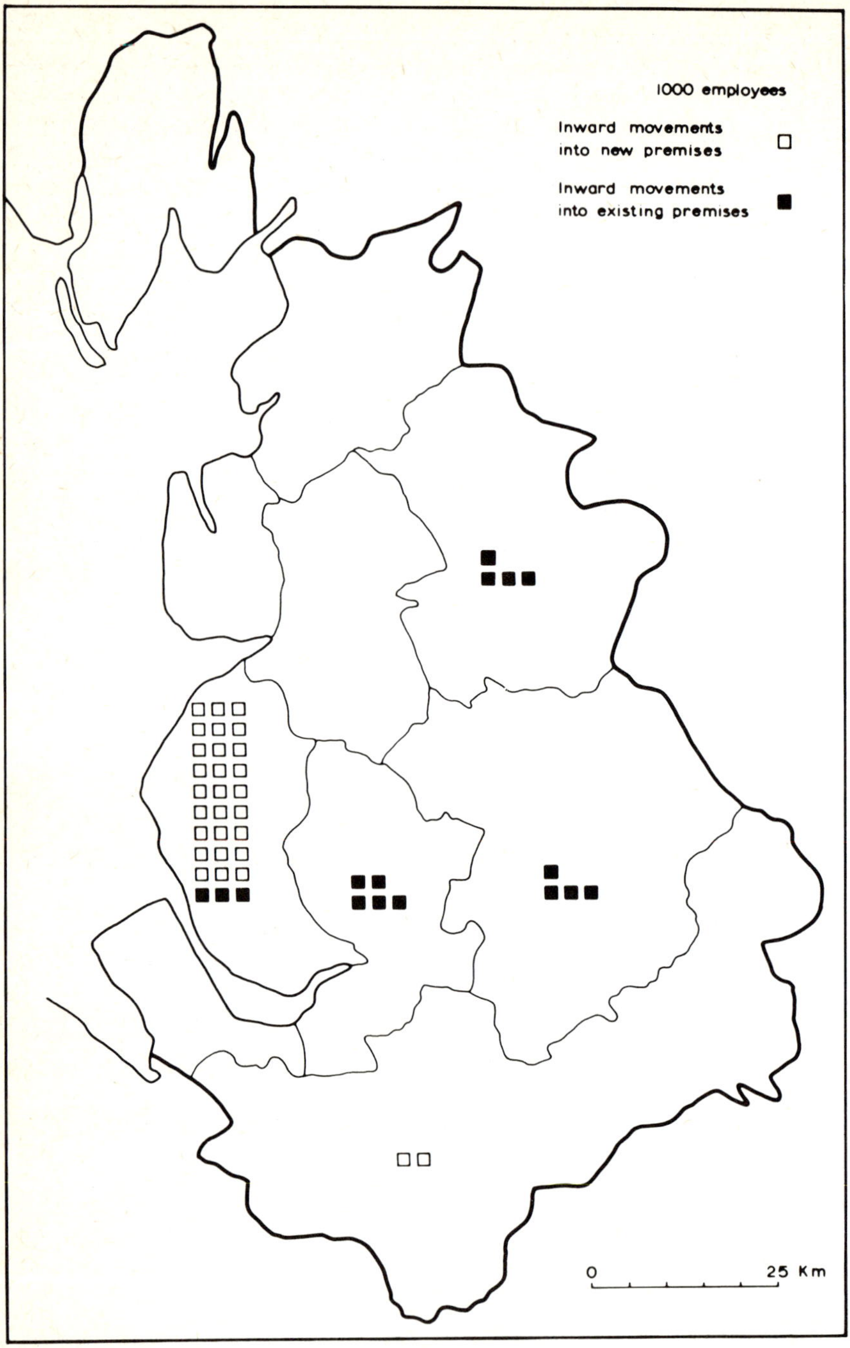

Fig. 16.5 Movement of manufacturing industry in the North West region, 1960–65. Source as for Fig. 16.2

Salt (1967) has shown, however, that of the total labour force (15 000) of both the Ford and Vauxhall works in 1965 only about 1 100 had been unemployed immediately before joining them and of these about 75 per cent had been out of work for less than six weeks; they were not among the 'hard core' of long-term unemployed. The labour policy of the two firms was partly responsible for this situation, but a short-term reduction in the level of unemployment was 'not feasible in an area like Merseyside where so many of the unemployed were unaccustomed to factory discipline and work or were virtually unemployable except in certain unskilled jobs' (Salt, 1967, p. 255). Instead of absorbing the unemployed these developments caused a large redeployment of men from other industries in the area, notably engineering, transport and communications and chemicals. Such industrial expansions are likely to have a favourable effect on unemployment rates but this is more likely to occur in the long run when the new industries have been fully integrated into the employment structure.

During the 1960–65 period development area policy was therefore extremely effective in terms of the total amount of building approved and completed, in the employment resulting from inward moves of manufacturing activity, in the increase in total and manufacturing employment, and in the improvement of the employment structure and in the share of the northwestern total of such items going to Merseyside. There can be little doubt, however, that the policy did not achieve the short-term objective of eradicating large-scale unemployment. In June 1967 there were still more than 21 000 unemployed on Merseyside, 3 per cent of the employed population. Since 1945 there has been an inflow of about 80 000 jobs in the manufacturing sector to the larger Merseyside area but during this period manufacturing employment has grown in the development area by only 20 000 (Fig. 16.6). This imbalance indicates the extent of closures and contractions which have taken place in existing manufacturing premises since the war and reinforces the statement that 'development has thus had to run fast to stand still, let alone get ahead quickly enough to absorb the existing thousands of unemployed' (Salt, 1967, p. 255). It also provides evidence to support those development measures (such as the Regional Employment Premium) aimed at helping industries already in the development areas since, at least in the case of Merseyside, the attraction of new industry has in itself been insufficient.

Conclusions

While the policy can be said to have had periods of both success and failure in the allocation of the means by which unemployment was to be

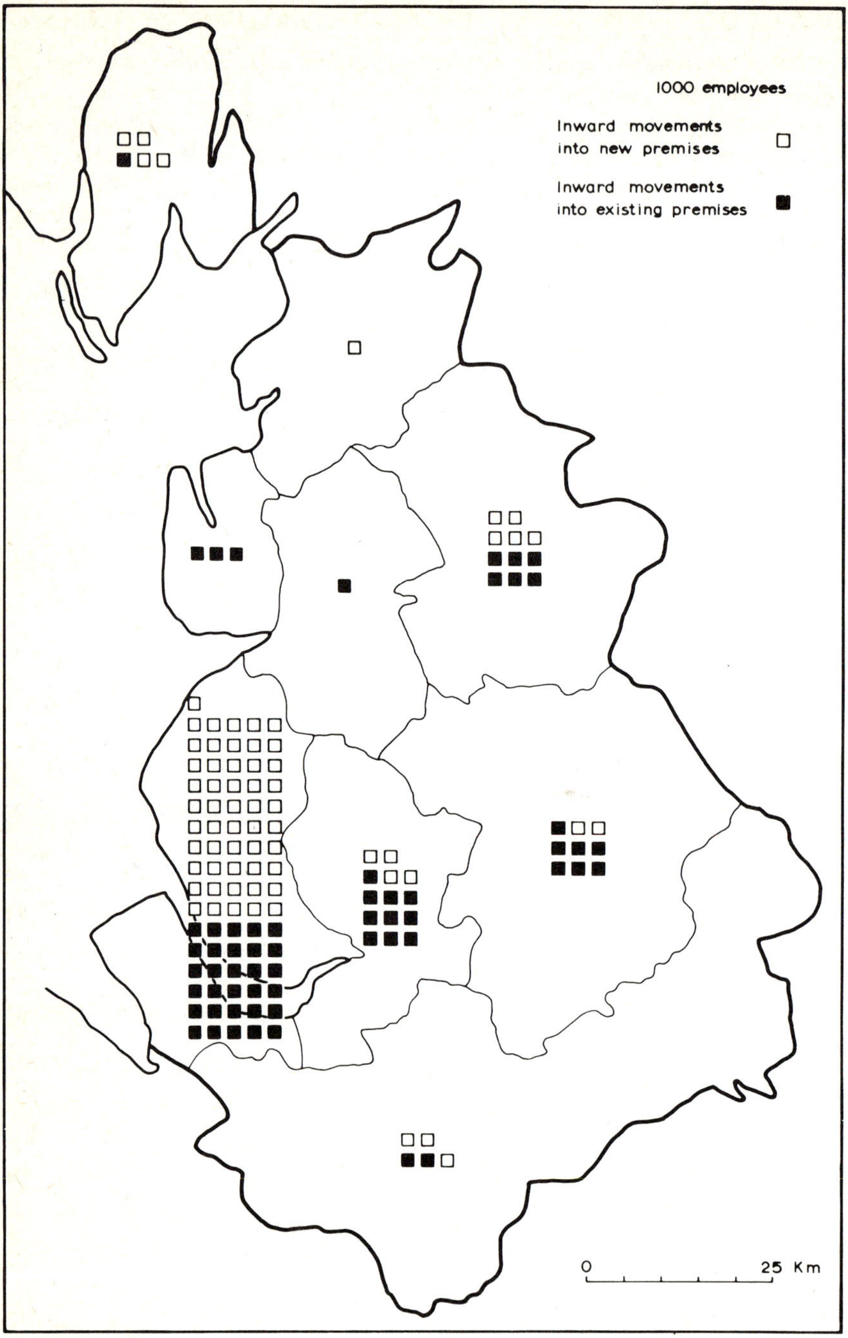

Fig. 16.6 Movement of manufacturing industry in the North West region, 1945–65. Source as for Fig. 16.2

reduced, the ultimate objectives have not been attained. There has been no consistent downward trend in the numbers unemployed during the 1951–67 period (Table 16.2). The figures for Merseyside closely correlate with those for Britain as a whole ($r = 0.79$, significant at 99 per cent) and indicate that the unemployment fortunes of Merseyside have been linked to the overall economic wellbeing of the country rather than reflecting the activities of regional policy. In terms of changing the employment structure, however, the policy has contributed a great deal when applied vigorously. Declining industries have been replaced gradually by modern, fast-growing industries and the total employment structure has been improved to an extent that the area was no longer dependent on activities associated with the port (Allen, 1968). By 1966 the application of regional development policy had achieved part of its purpose by bringing about a substantial transformation of the employment base of Merseyside. From the shift-and-share analysis for 1959–66 it can be seen that Merseyside had not completely solved its structural problems. The transformation had been significant but only partial and further assistance would have been necessary to consolidate the gains made by 1966.

References

ALLEN, E. G. W. (1968) 'The new industrial revolution in the North West', *Sunday Times*, 28 January.

BAINES, D. E. (1970) 'Merseyside in the British economy: the 1930s and the Second World War', in R. Lawton and C. M. Cunningham, eds, *Merseyside: social and economic studies*, Longman, pp. 58–71.

BOARD OF TRADE (1948) *Distribution of Industry*, London.

BOARD OF TRADE (1969) *The Movement of Manufacturing Industry in the United Kingdom 1945–1965. Study of movement affecting the North-West region*, London.

BRECHLING, F. (1967) 'Trends and cycles in British regional unemployment', *Oxf. Econ. Pap.*, 19, 1–21.

COLEMAN, G. (1972) 'The impact of regional development policy on the North West since 1945', unpublished MA thesis, University of Manchester.

CUNNINGHAM, N. J. (1970) 'Unemployment on Mersyside 1957–67', in Lawton and Cunningham, *Merseyside: social and economic studies*, pp. 202–34.

DEPARTMENT OF ECONOMIC AFFAIRS (1965) *The Problems of Merseyside*, London.

DEPARTMENT OF ECONOMIC AFFAIRS (1965) *The North West: a regional study*, London.

DUNN, E. S. (1960) 'A statistical and analytical technique for regional analysis', *Pap. Proc. Reg. Sci. Ass.*, 6, 97–112.

FUCHS, V. R. (1959) 'Changes in the location of U.S. manufacturing since 1929', *J. Reg. Sci.*, 1, 1–17.

FUCHS, V. R. (1962) 'Statistical explanations of the relative shift of manufacturing among regions of the United States', *Pap. Proc. Reg. Sci. Ass.*, **8**, 105–26.

HEMMINGS, M. F. W. (1963) 'The regional problem', *Natn. Inst. Econ. Rev.*, 25, 40–57.

LIVERPOOL CORPORATION (1936) *Liverpool Corporation Special Act 1936*, London.

LLOYD, P. E. (1965) 'Industrial change in the Merseyside Development Area 1949–59', *Tn. Plann. Rev.*, 35, 285–98.

LLOYD, P. E. (1970) 'The impact of development area policies on Merseyside 1949–1967', in Lawton and Cunningham, *Merseyside: social and economic studies*, pp. 374–410.

LLOYD, P. E. and DICKEN, P. (1968) 'The data bank in regional studies of industry', *Tn. Plann. Rev.*, 38, 304–16.

McCRONE, G. (1969) *Regional Policy in Britain*, Allen & Unwin.

McCULLOCH, F. J. (1956) 'Government and industry: a new policy?', *Tn. Plann. Rev.*, 27, 164–85.

ODBER, A. J. (1965) 'Regional policy in Great Britain', in US Department of Commerce, *Area Redevelopment Policies in Britain and the Countries of the Common Market*, Washington, pp. 327–433.

SALT, J. (1967) 'The impact of the Ford and Vauxhall plants on the employment structure of Merseyside, 1962–65', *Tijdschr. Econ. Soc. Geogr.*, **58**, 255–64.

SMITH, D. M. (1968) 'Identifying the grey areas: a multivariate approach', *Reg. Stud.*, 2, 183–93.

STILWELL, F. J. B. (1969) 'Regional growth and structural adaptation', *Urban Stud.*, **6**, 162–78.

THIRLWALL, A. P. (1967) 'A measure of the proper distribution of industry', *Oxf. Econ. Pap.*, **19**, 46–58.

ZELINSKY, W. (1958) 'A method for measuring changes in the distribution of manufacturing activity: the United States, 1939–1947', *Econ. Geogr.*, 34, 95–126.

M. E. C. Sant

17 Interregional industrial movement: the case of the non-survivors

The great majority of empirical studies of industrial location have dealt with enterprises which have been successful in so far as they had survived, at least until the date at which the researcher enumerated them. In addition most studies have assumed static economic conditions or, as in the case of historical, economic studies, have taken cognisance of little more than long-term trends. In this essay the objective is reversed; it is a systematic analysis of the closure of manufacturing establishments following their movement from other locations during the period 1945 to 1971. The prime concern is with time series of annual and cohort closure rates and with their broad regional variations. By themselves, these parameters provide important insights into the penumbra of locational changes; in addition they may be used to advance theories of industrial location and to refine regional policies.

According to the definitions used by Howard (1968), over 4 000 establishments were involved in some form of interregional movement in the United Kingdom between 1945 and 1965. By the end of 1966 almost one in four of these had closed. By standard region the proportion varied from less than 15 per cent to about 35 per cent. Over the whole country about 2 per cent of the plants which moved in any one year closed in the following year. Subsequent enumeration by the Department of Trade and Industry (DTI) has used different definitions for establishments but the picture between 1966 and 1971 remained much the same.

These figures demonstrate that much effort and investment have been expended on industrial movement, only to see the effects nullified by closure. Closure also is more likely to occur in some regions than in others, and movers are more vulnerable at certain times than at others. However, these observations must be kept in perspective, for closure occurs not only among recent movers, but also in the much larger population of stationary plants. There are no comparable figures for the latter group and it is impossible to tell whether they are more or less vulnerable than the plants which had moved. To make proper comparisons information is required on non-mobile firms with the same characteristics (size, product, organisation) as the mobile group. The British evidence so far examined has tended to be based on samples of small, specialised businesses. In a study of the 'birth' and 'death' of establishments in the west Midland metal trades from 1923 to 1938 it was found that annual average closure rates were about 5 per cent (Beesley, 1955). More recently, the 1971 Committee of Inquiry on Small Firms, dealing with those employing 200 or less, reported that in their sample of 1 251 manufacturing firms existing in 1963, 23 per cent had gone into liquidation, ceased trading or been taken over by 1970: of the sample about 20 per cent were classified as 'failures'. In the distributive trades the figures for total mortality were substantially higher — 32 per cent in wholesaling and 28 per cent in retailing. In comparison the mobile manufacturing firms enumerated by the DTI tend to have been more viable. Nevertheless, the fact that a large proportion closed within a few years of their opening (over 10 per cent within the first five years) must raise questions about the nature of the movement process, the choice of location and the need for better financial provision in the early years. These closures also have implications for the type of regional incentives favoured by governments. Above all it is worth while to consider whether businesses put themselves at greater risk by moving (an action for which they may have little precedent or experience) than by staying at their original location. The picture drawn by Beesley of mobile plants 'marching stolidly into the ambush' is, though exaggerated when applied to all movers, probably an apt description of some.

Ideally, to answer questions of this kind case histories of every closure are needed, together with comparable information for control groups of surviving movers and non-movers. It is essential to know the probable fate involved in not moving, the manner and circumstances in which the move was carried out, post-move experience and performance, and the reasons for the closure. None of this information exists, and it would be an insuperable task to assemble it. However, data of a fairly simple kind on closure is available, collected as part of the DTI's analysis of industrial movement (Howard, 1968). These tabulations record the year of opening and of closure for each establishment which moved (*a*) between 1945 and 1965 and closed before the end of 1966, and (*b*) between 1966 and 1971 and closed before the end of 1971. The two series are based on dissimilar definitions and must be treated separately. Moreover they are straightforward 'head counts', simply recording that certain numbers of establishments failed to survive. Changes of ownership are excluded, except where these were followed by a total change in the nature of production at the establishments, in which case both a closure and a new opening were recorded. Thus, the scope of this paper is restricted to providing a description of national and regional patterns of closure and to drawing some general, but (for reasons already given) only partly substantiated conclusions about their implications for regional policy.

Industrial movement and closure

The reasons for closure within a few years of establishments moved to new locations are complex. In carrying out the investment associated with a move the great majority of firms are confident, at the time of the decision to relocate, that their new plants will have a long life. Given the alternatives to movement (Townroe, 1968), few plants would be transferred or branches opened if it were known that they would close within a short time. As there is evidence that plants which move have, on average, much faster growth rates than those which are stationary (South East Joint Planning Team (SEJPT), 1971; Howard, 1968), the closures, especially the 'early' ones (within the first five years), seem to be major anomalies in an otherwise highly successful category of industry.

Closure is not only associated with failure. The first distinction to be made among the possible causes of closure is between those plants on which closure is forced by poor economic performance and inability to compete, and those for which closure is part of a process of growth or of a strategy for rationalisation. Even this distinction is not necessarily perfect, for it is possible to conceive of large multiplant companies whose overall performance is weak but some of whose branches are successful (Rolls-Royce, for example). An additional factor in closure is the effect of exogenous events such as a compulsory purchase order on the industrial premises into which the plant has moved or an 'act of God'. These three general, albeit partly indistinct, causes of closure form the basis of a classification (Fig. 17.1) to which examples of more specific factors can be allocated.

Closure associated with a successful operation is likely to be related to plans for expansion or to a need for rationalisation. Among the former group there may be cases where insufficient forward planning had gone into the original move, leading to an unsatisfactory choice of location followed by unnecessary disruption of normal operations to carry out a further move. On the other hand, many expanding firms deliberately pursue a stepped process through a series of rented factories. By taking short leases rather than buying freehold sites their capital requirements are held down and production and capacity kept in reasonable harmony. In Luttrell's (1962) study the intended performance of branch plants revealed wide variations. Small subsidiaries were primarily intended to be temporary, although intentions tended to understate reality. In contrast,

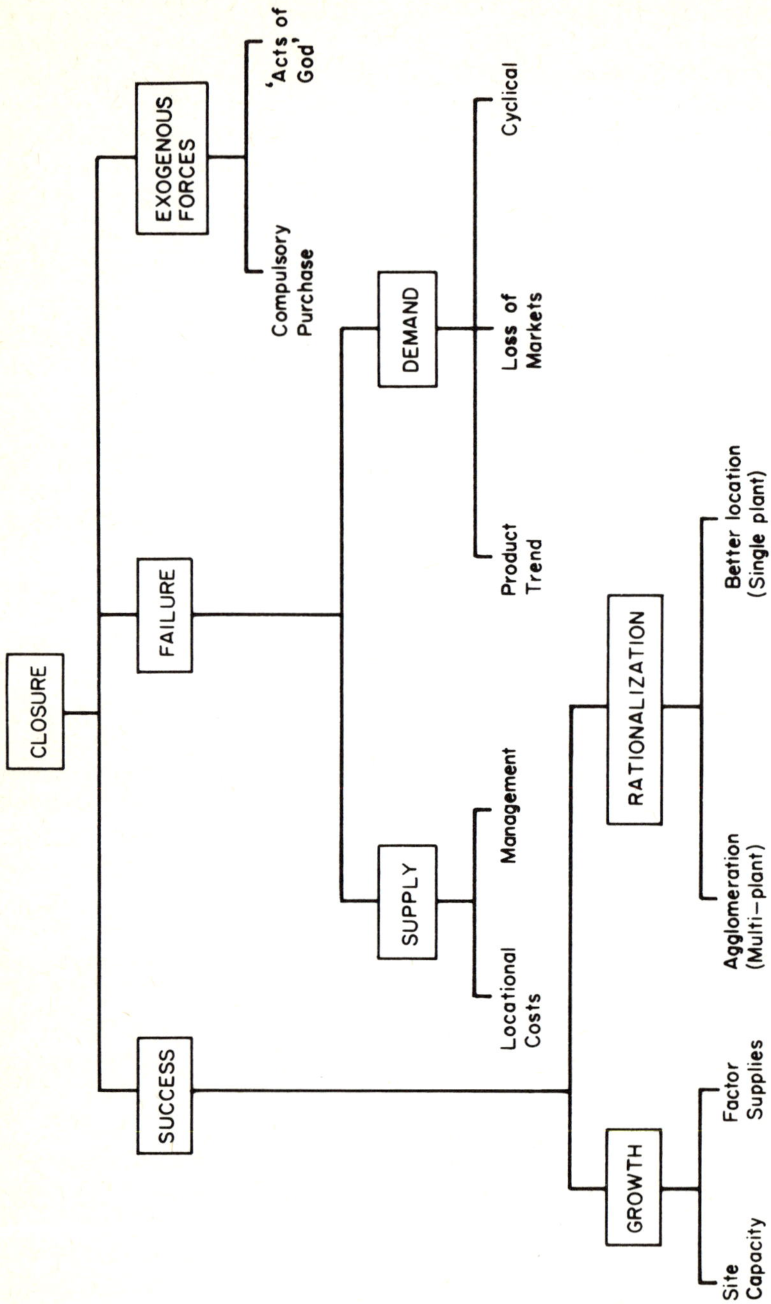

Fig. 17.1 Schematic classification for causes of industrial closure in re-located establishments

self-contained branches were mainly planned as permanent additions to the firm's capacity. However, site capacity is not the only constraint on continued growth and a second move may be necessitated, for example, by labour shortages or by changes in the nature and location of suppliers. Thus in parts of southeast England firms moving from London to areas where labour appears to be readily available have found that their presence, together with that of later arrivals, has transformed the labour market into one of scarcity and higher costs (Sant, 1970; SEJPT, 1971).

Normally, rationalisation implies an attempt to reduce costs by improved organisation which, in turn, may require a major change in the number and location of branches in a multiplant company. In the present context this is likely to take the form of agglomeration economies, with branches being closed and production concentrated in larger plants, a situation conforming closely to the classical Weberian cost-saving solution based on his graphical analysis of intersecting isodapanes (Weber, 1929). Although not usually considered as rationalisation, the closure and relocation of a single plant firm or of an individual branch in response to changing markets, techniques or sources of materials may also be included. Such migration appears to have been common during critical periods in the development of some industries in the nineteenth century and there is no reason to suppose that it does not still occur.

Closure, when associated with failure, is likewise attributable to a variety of causes. These are divisible into two groups: those related to supply problems and those resulting from changes in demand for the plant's product. On the supply side, industrial movement has been shown to incur certain costs. In the short term costs can be attributed directly to the move (loss of production or the purchase of new plant and machinery) but these ought to have been foreseen and absorbed by the firm. Luttrell (1962) has given evidence of the size of the differential and the period taken to reduce costs to the level at the former location. In his sample of branch factories, total costs in the first year were, on average, double those of the parent, while in the third year their costs were only about 15 per cent higher. Cases may have occurred where firms have never achieved a significant reduction in locational costs and such firms have lost the competitive strength previously held. Instead of being strengthened by the move, they have become weaker and eventually closed. On the supply side, there are problems associated with management policy. Many of these are not the sole preserve of establishments which have moved, but in the case of branch plants controlled from a remote head office the managerial skill within the branch may be inadequate to deal with more than day-to-day decisions and insufficient attention may be paid to policies on which success depends (Loasby, 1967).

Besides incurring locational costs the transferred plant is in danger of losing markets served from its former location. To ensure success these losses must be minimised and new markets developed. Demand is never static and one of the risks faced by the recent mover is that the investment and effort involved may be nullified by new trends in demand for his product or by cyclical shifts in income. Some firms, though themselves healthy, operate in a declining industry. Despite confidence in their future at the time of movement, they may later find themselves caught in the industry's downward spiral unless able to diversify. Cyclical effects are much more vicious and less discriminating in their impact, and expanding firms in growth industries are not immune. If the income elasticity of demand for the product is high, output will suffer drastically during a cyclical downturn, and even a firm with good growth prospects may be unable to maintain enough liquidity to withstand financial pressures. This list does not exhaust the possible causes of closure and many which have been discussed could be further subdivided. It indicates that, from the industrial viewpoint, some closures are a mark of success, to be welcomed.

Industrial closure in the United Kingdom, 1945–66

Closures can be examined under two broad headings: (*a*) the year-by-year balance between movement and closure, and (*b*) the likely rate of closure or survival for a set (cohort) of relocated plants opened in a given year. Neither is an entirely straightforward value for three major reasons, each of which must be considered before adequate conclusions can be drawn. First, the data from which the figures for movement and closure are derived are truncated, and any statement about potential performance must be based on extrapolation of trends over a relatively short period. Second, in calculating closure rates, care must be taken to account for cumulative movements and survivals from previous years. Finally, the survival of particular cohorts may be above or below average, as a possible result either of a certain industry-mix of movers in a given year, or of cyclical fluctuations in demand affecting the new plant in its earliest years.

On the basis of Howard's definition of movement, the numbers of establishments involved between 1945 and 1966 were 4 087 total moves (1945–65), of which 953 were closures and 3 134 survivors.

The time series of total movement has been discussed elsewhere (Sant, 1975), but the main findings from a lagged regression analysis are significant relationships between movement, on the one hand, and structural (policy) and cyclical (economic) variables on the other (Fig. 17.2). The

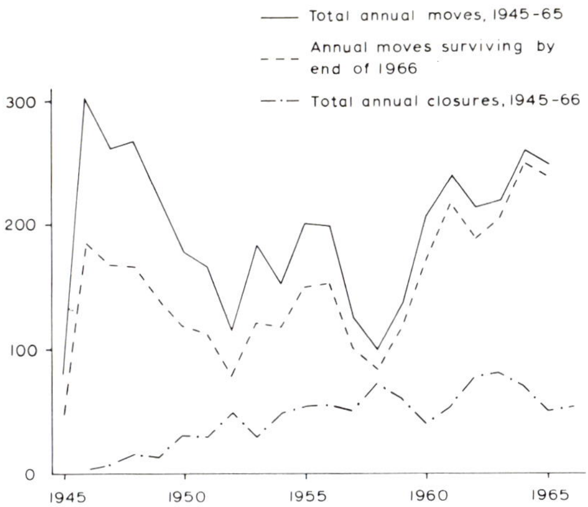

Fig. 17.2 Time series of industrial movement and closure in the United Kingdom. Based on DTI unpublished statistics.

gap between the two graphs of total movements and survivors from each year still in existence at the end of 1966 represents the non-survivors (Fig. 17.2). For clear reasons this gap diminished towards 1966, as insufficient time had elapsed for the number of closures to equal that in earlier cohorts. Conversely, annual numbers of closures tended to increase as the number of cohorts grew, with each adding its closures to those from cohorts already in existence. This pattern dominated the first ten years of the series, after which closures fluctuated widely around a more slowly rising trend.

The 'closure ratio', the proportion of each cohort which closed between the year of opening and the end of 1966, fell from more than 40 per cent in the 1945 cohort, to about 25 per cent in the 1955 cohort, and less than 5 per cent in 1964 (Fig. 17.3). The fall is approximately linear, though fluctuations occur around the best-fit line. The 'annual closure rate' is calculated by expressing closures in a given year as a percentage of the movers in previous years still in existence at the start of that year. To give some insight into the relationship between this

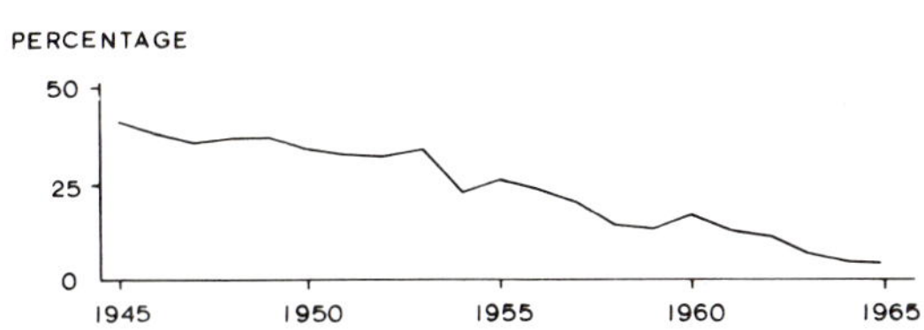

Fig. 17.3 Cohort closure ratios in the United Kingdom, 1945–65: percentage of moves per annum closed by the end of 1966. Source as for Fig. 17.2

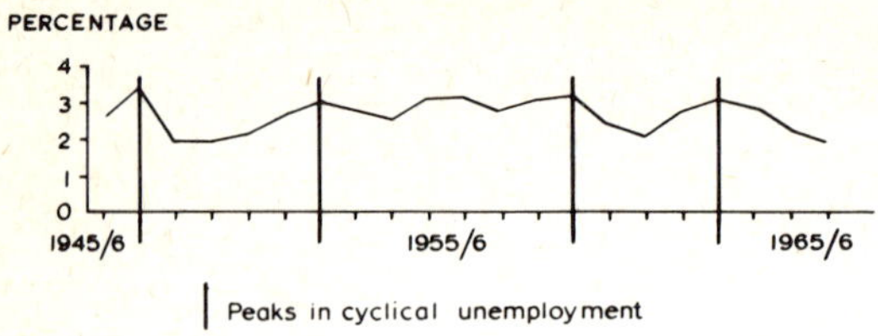

Fig. 17.4 Annual closure rates in the United Kingdom, 1945–66. Source as for Fig. 17.2

curve and other cyclical changes the peak years in national unemployment have been included (Fig. 17.4). The series has been smoothed by applying a two-year running mean and, in consequence, the extreme values have been suppressed. Even so, quite wide variations occurred, between 1.8 per cent in 1965–66 and 3.4 per cent in 1946–47.

The 'cohort closure rates' are derived from the annual average rate of closure of the openings carried out in each year (Table 17.1). Three sets

Table 17.1 Cohort closure rates to 1966; over the first three years; over the fourth and fifth years of the cohort's life

Year of opening (cohort)	*Annual closure rates*		
	to 1966	*years 0–3*	*years 4–5*
1945	1.9	0.5	3.3
1946	1.8	1.9	3.0
1947	1.8	1.1	2.3
1948	2.0	1.4	4.0
1949	2.1	2.0	2.6
1950	2.0	2.5	1.7
1951	2.0	1.2	4.2
1952	2.2	1.4	3.0
1953	2.4	2.4	3.5
1954	1.7	3.2	0.4
1955	2.2	2.6	3.2
1956	2.1	2.2	3.4
1957	2.0	1.8	1.7
1958	1.6	2.0	1.5
1959	1.7	1.8	0.8
1960	2.4	1.8	4.0
1961	2.1	2.6	2.4
1962	2.3	2.0	3.0
1963	1.6	1.8	–
1964	1.4	1.4	–
1965	2.0	–	–
Mean (1945–65)	1.9	1.8	2.6
Standard deviation	0.27	0.61	1.16

Source: DTI unpublished statistics.

of rates have been calculated relating to closures: (1) between the year of opening and the final year of the series (1966); (2) from birth to the third year of each cohort's life; and (3) in the fourth and fifth years of each cohort's life. These rates provide further insight into the scale and variability of closures during the earlier years of each cohort. A more detailed examination of individual years in the life of each cohort is provided by the use of 'age specific closure rates' (Fig. 17.5). Each point on the graph describes the mean closure rate in the cohorts' first, second . . . 'n'th year. Because the series is truncated at the end of 1966, each year has a smaller number of observations than preceding years (the first year has twenty-one observations, the twenty-second year has only one). To avoid too small a sample the closure rates in the last four years (the nineteenth to twenty-second) are ignored. The values on which the graph is based are two-year running means to reduce sharp changes in the curve.

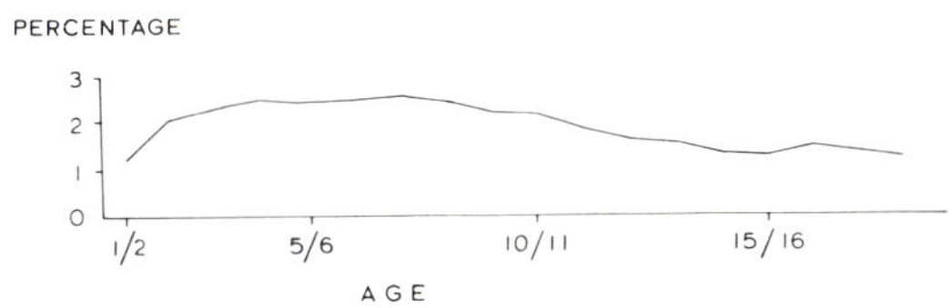

Fig. 17.5 Age specific closure rates in the United Kingdom, 1945–65. Source as for Fig. 17.2

Together Table 17.1 and Fig. 17.5 make it possible to reconstruct an idealised history of the life of a cohort of relocated establishments. Before doing this, however, it is worth considering the conditions under which no variation of closure rates through time might be expected. This will occur if all mobile establishments represent an entirely homogeneous body, having the same annual performance levels, and if all the components of the national economy are perfectly stable. None of these conditions is realistic, but they do provide a yardstick for comparison. If they are discarded it becomes reasonable to expect cohort closure rates to respond (*a*) to economic conditions at, and shortly after, the time of establishment, and (*b*) to the stage in the life of a cohort.

Over the period from the year of establishment to the end of 1966, the average closure rate in each cohort (Table 17.1, col. 1) was 1.9 per cent per annum, with a standard deviation of only 0.27 per cent. The two major features of these values are the great similarity of cohorts and that, despite changes in the economy since 1945, there was no discernible trend in closure rates – one cohort was much like any other. This long-term stability, implied in rates over a twenty-year sequence of cohorts, has considerable importance for forecasting the impact of planned industrial movement.

Contrasts may be made between the experiences of cohorts during the early years of their existence. It is assumed that the date of establishment occurs at mid-year, in order to calculate closure rates over the first thirty months of existence (Table 17.1, col. 2). By the end of this period the annual average rate was 1.8 per cent, slightly below the whole-life mean, but the standard deviation was, at 0.61 per cent, considerably higher. This could be expected if, as was hypothesised above, early closure rates responded to economic cycles, or if cohorts differed from each other in their viability, but it is not possible to conduct an adequately detailed analysis of either cause. Without discounting the latter, there does appear to have been some significant correlation between the survival of the first two and a half years and the state of the national economy, measured in terms of changes in the index of production and average unemployment over corresponding periods.

In comparison, average closures over the next two years (Table 17.1, col. 3) were somewhat higher, at 2.6 per cent per annum. This indicates the possible combination of two likely causes; an increased chance of closure following failure to solve the problems, such as high costs or lack of liquidity, attendant on the early development of a plant; and the emergence of forces to relocate yet again when the establishment has had a highly successful early period. The balance between these alternatives cannot be assessed, except to suggest that the former is probably of greater importance. In addition to possessing a higher closure rate, this period was more variable, with a standard deviation of 1.16 per cent. It is difficult to explain this increase without knowledge of the cases involved, but it is possibly associated with the same cyclical factors which affected earlier closures.

Volume and variability of closure increased during the early life of a cohort. Subsequently, the high closure rate reached at about the fifth year was not a peak but the edge of a plateau of high rates (Fig. 17.5). These began to turn down only from about the eighth year, until by the eighteenth year they had fallen almost to 1 per cent per annum on average. The point should be emphasised that this trend cuts across all cyclical fluctuations. Notwithstanding these, it can be inferred that the longer an establishment remains in its new location, the greater becomes its life-expectancy.

The effect of regional policy, 1945–71

The creation of higher and more stable levels of employment in the assisted areas implies that closures should be minimised, or at least brought into line with the rates in non-assisted areas. Regional variations in closures stem from two causes: differences in the kinds of establishments and industries attracted by the regions, and disparities in locational costs. The first tends to lie outside the direct control of regional policy, in that little can be done to determine the character of mobile industry or the types of industry going to different regions, except in a negative way. It is not possible, for example, to ensure that only firms in growth industries should relocate in the development areas, or that the same areas should receive a lower proportion of branch plants. On the other hand, the second factor has been the primary concern of regional policy, and a variety of measures, giving different levels of subsidy, has been enacted since 1945 (Brown, 1972).

Despite the need for a cautious interpretation imposed by imperfections in the data, it is perhaps one of the most striking features of industrial movement that regional variations in cohort closure rates (adjusted for size of cohorts) have altered markedly. Between 1945 and 1960 the peripheral areas experienced rates well above the national average, with an index of 117 (UK = 100) compared with only 83 in the rest of the country. In the 1960–65 period the rates in the two regions were equal, and between 1966 and 1971 the trend continued with the peripheral areas becoming less vulnerable to closure.

In so far as it is possible to generalise about comparative locational advantages in different regions since 1945, it appears that the balance, initially, did not favour the development areas. Taking into account the then existing scale of subsidies, Hague and Dunning (1954) concluded that, compared with the London area, the development areas had slightly higher costs and the new towns slightly lower, and that the differences were greatest in the early years in the life of a new establishment. These costs related only to the static position and ignored the effects of dislocation on the firm's links with suppliers of materials and services and with former markets. On each of these counts it is reasonable to conclude that while all moves were likely to incur some disadvantages, those firms going to development areas, and thereby usually having the greatest spatial dislocation, suffered most and found it most difficult to adjust to the problems.

The 1950s were years of low regional incentives; in particular there was little, if any, help to firms in overcoming the initial costs of movement. In comparison, the 1960s saw a progressive strengthening of inducements, with the total outlay on various policy instruments increasing from around £5 million per annum in the late 1950s, to £50 million in 1966–67 and about £350 million in 1970–71 (at current values). These increases reflected the demands of new inducements, most of which were designed to improve the liquidity of firms, especially in the initial years of their investment. Under the Local Employment and Finance Acts of 1963 these included investment grants for plant and machinery, and 'free depreciation' which allowed firms, if they wished, to defer tax payments until their investment had been written off. In 1966 the free depreciation incentive was abolished but investment grants were increased, as they were again in 1967, giving more help to firms which might not see a profit on their investments and hence not benefit from free depreciation. Another, more massive, change occurred in 1967, namely the introduction of the Regional Employment Premium which, although intended as a subsidy on the employment of labour, also had the side effect of increasing the cash flow to firms in development areas. These, together with other forms of assistance, have been estimated (Wilson, 1973) to contribute a cost subsidy to the development areas of about 2.5 to 5 per cent of net value added. At this level these ought to do much to remove interregional cost disparities and probably account for most of the change in closure ratios in the peripheral areas.

Table 17.2 Adjusted closure indices 1945–71 in peripheral and non-peripheral areas (UK = 100)

Areas	*1945–60*	*1961–65*	*1966–71*
Peripheral*	117	100	91
Non-peripheral	83	100	107

* Peripheral areas are defined as Northern Ireland, Scotland, Wales, Northern England, the North West and South West standard regions.

Source: DTI unpublished statistics.

In the process of this turnabout in the viability of firms moving to different regions there appear to have been two important side effects: (1) the distance involved in industrial movement has a less direct effect on closure; (2) new differentiation is created between branch plants and transferred establishments. Evidence before 1960 indicates that longer moves sustained a higher incidence of closure (Table 17.2). Moves to peripheral areas, with a median distance from the original location of about 280 km, had closure rates about 50 per cent higher than moves to other regions, which had a median distance of less than 120 km. This is supported by Keeble's (1968) findings: of firms moving from northwest London between 1940 and 1964, those locating in the provincial zone (more than 160 km from London) had closure rates twice as high as those staying within the metropolitan zone – 20 per cent compared with 9 per cent.

In the 1966–71 period the pattern was substantially different (Table 17.3). Classifying moves according to whether the destination lay in the same subdivision as the origin, in the same standard region (but a different subdivision) or in a different region, and differentiating between peripheral and other areas, only one group was markedly different. This was the non-assisted middle-distance movers, relocating within the same region, with a closure index of over 150 compared with between 88 and

Table 17.3 Closure ratios by categories* of origins and destinations, 1966–71

Categories of origin and destination	*Destinations*					
	peripheral areas		*other areas* (total moves in brackets)		*total UK*	
Within same subdivision of a standard region	90	(746)	88	(1 123)	89	(1 869)
Within same standard region but in different subdivisions	88	(188)	150	(431)	132	(619)
In different standard regions	102	(687)	102	(287)	102	(974)
Total	91	(1 621)	107	(1 841)	100	(3 462)

* Data for 1966–71 relate to the eleven standard regions and the sixty-two statistical subdivisions of the United Kingdom (Central Statistical Office, 1971)

Source: DTI unpublished statistics

102 in the other five groups. In practice, it appears that this category places itself at greatest risk. By moving as far as they do, the firms in this group (predominantly single plant transfers rather than branch plants) lose many of the advantages of a wide range of beneficial linkages with their former locales. At the same time they have not moved far enough to gain the assistance available in the peripheral areas. In comparison, the equivalent group within the peripheral areas, nurtured by greater financial assistance and, almost certainly, by careful advice from government agencies, had a low rate of closure.

From the data before 1960 it may be inferred that branch plants were more vulnerable than transferred establishments, the former accounting for over 80 per cent of moves to peripheral areas but only for 55 per cent of moves to the rest of the country. Keeble also showed that branch closure rates were almost one and a half times as high as transfer closure rates (15 per cent and 11 per cent respectively), with the worst performance being recorded by branches in the provincial zone and the best by transfers in the metropolitan zone. The evidence for 1966–71 shows major differences (Table 17.4). Branches generally had become more

Table 17.4 Closure indices for branch plants and transferred establishments, 1966–71

Destination	*Branch plants*	*Relocated establishments*	*All moves*
Peripheral areas	80	100	91
Other areas	121	102	107
Total UK	94	102	100

Source: DTI unpublished statistics.

viable in the peripheral areas but had lost ground, relatively, in other areas.

In spite of insufficient data it appears safe to conclude that each of these changes (closures by region, by distance moved, and by industrial organisation) is largely the result of alterations in the balance of regional advantages brought about by the greatly increased level of incentives and financial support available in the assisted areas. The only other major factors which, in theory, could have been effective are the characteristics of industrial movers. However, these tend to change slowly and, compared with the position before 1960, there had been little change in size of firm or ratio of branch plants to transfers going to different regions. There may have been some change in the industrial composition of movers to different regions. Before 1960, the peripheral areas tended to take a significantly poorer mixture than the rest of the country, with over a third of their immigrant firms coming from declining industries. In comparison the proportion in the other regions was less than one-fifth. Recent incentives, with considerable benefits to profitable, capital intensive firms, may have altered the balance, but as yet there is no empirical support for this view.

Conclusions: closures and regional policy

Industrial closure is not inherently undesirable; to a large degree it is necessary in order to release resources for the growth of more profitable sectors. Undesirability arises when the rate of closure is too high, either in industry as a whole, or in specific sectors or regions. Moreover, there is no easy definition of rates which are 'too high'. The monopolist might take a very different view of closure from that of a worker rendered unemployed by closure, and what might be desirable to a specialised region, in danger of losing a traditional industry, might not be the optimal strategy for the national economy. However, when establishments close shortly after investing considerable resources in relocating their operations, there is cause for concern. No doubt some fall into the category of desirable closures, but many (probably the great majority) raise questions about the nature of industrial movement which have implications for regional policy.

The main findings in this analysis of closures can be summarised in the following manner.

1. Nationally, a quarter of the establishments involved in interregional movement between 1945 and 1965 had closed by the end of 1966. This was associated with an annual cohort closure rate of about 2 per cent.
2. National age specific closure rates showed the greatest vulnerability between the fifth and eighth years.

3. Annual closure rates tended to be quite closely associated with cyclical fluctuations in business conditions.
4. Regional variations occurred in closure rates. Initially the peripheral areas were at a marked disadvantage compared with the rest of the country, but as locational subsidies were increased through the 1960s the disparity was reduced, and eventually the peripheral areas had lower rates of closure.

Although not hitherto discussed in the context of industrial location theory, it is clearly useful to place these findings in such a framework, in order to lend coherence to arguments about their implications for regional policy. For this purpose the most appropriate starting point is the neoclassical theory developed by Smith (1971). In this, Smith proposes a model of 'spatial margins to profitability', into which it is possible to build not only variable costs and revenues, but also the effects of areally selective subsidies and a number of other complicating factors. It can be inferred that, in a 'policy free' situation, the closer a mobile plant moves to the optimal location, the greater its chance of success and survival. Subsidies under regional policy widen these spatial margins and closure rates can bear eventually little relation to the distance of movement. A novel concept, that of 'temporal margins to profitability', can be added to Smith's model. Although conclusive evidence is lacking, it is not unreasonable to suggest that, in the absence of selective subsidies, the well located plant will overcome its relocation costs faster than the poorly located one. This concept of temporal margins needs considerable research before verification, but if it is a valid notion it augments the spatial margins model as an explanation of variations in the incidence of closure within the United Kingdom. In particular, it provides a possible reason for the different closure rates found in the peripheral areas and in the rest of the country, and their apparent relationship with subsidy levels.

References

BEESLEY, M. (1955) 'The birth and death of industrial establishments: experience in the west Midlands conurbation', *J. Ind. Econ.*, 4 (1955–56), 45–61.

BROWN, A. J. (1972) *The Framework of Regional Economics in the United Kingdom*, Cambridge University Press.

CENTRAL STATISTICAL OFFICE (1971) *Abstract of Regional Statistics*, London.

COMMITTEE OF INQUIRY ON SMALL FIRMS (1971) *Dynamics of Small Firms*, Research Report No. 12, London.

HAGUE, D. C. and DUNNING, J. H. (1954) 'Costs in alternative locations: the radio industry', *Rev. Econ. Stud.*, 22, 203–13.

HOWARD, R. S. (1968) *The Movement of Manufacturing Industry in the United Kingdom, 1945–65*, Board of Trade, London.

KEEBLE, D. E. (1968) 'Industrial decentralization and the metropolis: the north-west London case', *Trans. Inst. Br. Geogr.*, **44**, 1–54.

LOASBY, B. (1967) 'Making location policy work', *Lloyds Bank Rev.*, **90**, 34–47.

LUTTRELL, W. F. (1962) *Factory Location and Industrial Movement*, National Institute of Economic Research, London.

SANT, M. E. C. (1970) 'Age and area in industrial location', *Reg. Stud.*, **4**, 349–58.

SANT, M. E. C. (1975) *Industrial Movement and Regional Development*, Pergamon.

SMITH, D. M. (1971) *Industrial Location*, Wiley.

SOUTH EAST JOINT PLANNING TEAM, DEPARTMENT OF THE ENVIRONMENT (1971) *Strategic Plan for the South East: report of Economic Consultants Ltd*, London.

TOWNROE, P. M. (1968) 'Locational choice and the individual firm', *Reg. Stud.*, 3, 15–24.

WEBER, A. (1929) *Alfred Weber's Theory of the Location of Industries*, trans. C. J. Friedrich, Chicago.

WILSON, T. (1973) 'British regional policy in the European context', *The Banker*, 123, 168.

M. J. Pullen

18 Objectives, constraints and spatial behaviour

Regional scientists are concerned with analysing the operation of economic systems with specific space-time locational characteristics. This essay discusses a theoretical foundation for the analysis of such spatial economic systems. An economic system, in the broad sense used here, can be either closed, where its behaviour is independent of any external influences, or open, where its behaviour is potentially partly determined by outside forces and where the system itself also potentially influences the operation of the external world. This distinction is well illustrated by the various input-output models in use in regional analysis (Richardson, 1972) and by the application of gravity models to the analysis of shopping behaviour (Black, 1966; Gibson and Pullen, 1972). This essay is concerned specifically with closed economic systems. Any system whose behaviour is heavily dependent on exogenous variables is an incorrectly specified model of the unit under study, in that the model could be improved by enlarging the system under scrutiny to include the exogenous variables.

There is a strong interventionist tradition in regional science in that areas of study have tended to be those where policy recommendations would be welcomed by government. Such an approach implies the possibility of controlling the behaviour of the economic system. It is useful to contrast two extreme situations, which are obviously linked to the economist's models of a complete command economy and of a perfectly competitive free market economy. First, there is a system where the behaviour of its components is completely determined by centralised decision making. Second, there is a system where the individual units comprising the system act as independent agents making independent decisions (Marschak, 1959; Isard, 1969). The policies operated by an authority, whether government, government agency, board of directors or head of a large family, will raise many subtle problems in the second type of situation compared with the first.

This interventionist tradition has led to difficulties of maintaining a clear distinction between normative regional science, with a foundation of value judgments as to what are desirable objectives for the system, and regional science that is behaviourist in approach and attempts as far as possible to be 'value judgment neutral'. The theoretical model outlined here attempts to adopt the 'value judgment neutral' approach in that, although the existence of individual welfare functions is recognised, there is no system objective function built into the model.

A Markov process view of the economy

The economist distinguishes carefully the variables that are stock concepts from those that are flow concepts. Human and non-human resources have at any one time a space location, which can be observed to change over time (Hägerstrand, 1970). At a particular instant of time (τ_0) the stock of resources can be described by a $n \times m$ matrix where n represents the number of different types of goods in the economy (goods being defined to include human resources as well as inanimate objects) and m is the number of identifiable locations in the system. This matrix $[L(\tau_0)]$ is the stock position of τ_0. A second 'stock-taking' after an interval of time would give another matrix $[L(\tau_1)]$. The difference between $[L(\tau_0)]$ and $[L(\tau_1)]$ would indicate movement of human and non-human assets over space during the interval of time, $\tau_1 - \tau_0$.

If the goods in the closed economic system are differentiated spatially, the matrix $[F]$, describing the flow or movement of goods from one location to another between τ_0 and τ_1, will have the dimensions $(nm) \times (nm)$. This matrix can be partitioned into submatrices which indicate the change of location of one particular good, so that there are n $(m \times m)$ submatrices on the leading diagonal of the partitioned matrix $[F]$:

$$F = \begin{bmatrix} F_1 & 0 & \dots\dots\dots & 0 \\ 0 & F_2 & \dots\dots\dots & 0 \\ \cdot & \cdot & & \cdot \\ \cdot & \cdot & & \cdot \\ \cdot & \cdot & & \cdot \\ 0 & 0 & \dots\dots\dots & F_n \end{bmatrix}.$$

The $[F]$ matrix, recording flows of goods between geographical locations, can be transformed into a transition matrix $[T]$ in a Markov chain theory sense (Bishir and Drewes, 1970; Richardson, 1973) by dividing each element f_{ij} by $\Sigma_j f_{ij}$, where

$(i = 11, 12, \dots 1m, 21, \dots 2m, \dots, nl, \dots nm)$ and

$(j = 11, 12, \dots 1m, 21, \dots 2m, \dots, nl, \dots nm)$

to give t_{ij}, or the probability of a commodity moving to a new location during one period of time ($\Sigma_j t_{ij} = 1$).

This simple model assumes that economic activity consists merely of transactions which rearrange the location of an otherwise unaltered stock of resources; the only transformation that occurs is spatial and is costless in terms of transportation inputs.

Economic activity, however, consists also of the transformation, via a series of production processes (with fixed locations), of one collection of goods (inputs) into a new collection of goods (outputs) which in turn may then be subject to spatial transformation. Within the theoretical framework being outlined, when a good has reached a location where it is to be used in some production process, the good can be regarded as entering an absorbing state of the system and hence as reducing the volume of goods remaining in the system to be subject to further locational change in the next time period.

As the transformation of goods by a production process is treated as an absorbing state and leads to a vector of outputs leaving the system at the end of each time period, the goods that enter the absorbing state of the system must be processed and 'recycled' into the system if the Markov equilibrium situation is to be avoided where all goods after τ time periods (where τ approaches the limit infinity) end up in the absorbing state. The recycling is in fact the use of the goods as inputs at specific locations in the production of a new collection of goods.

The model now has the form of a partitioned probability matrix [P]

$$P = \begin{array}{c|c} T & Q \\ \hline 0 & I \end{array}$$

where [T] is the matrix of probabilities of the location of goods changing during a time period, [Q] the diagonal matrix of probabilities of goods being used during a time period in a production process whose outputs are other goods, and [I] an identity matrix of $(nm) \times (nm)$ dimension. The length of the time period used will determine the size of the probabilities in the [T] matrix and the [Q] matrix. If the time period is defined so that all goods are in the correct location for non-spatial processing then all the elements in the [T] matrix become zero.

Capital goods will appear in the [T] and [Q] matrices. Consumption or use of capital goods in production processes will appear as an element in the [Q] absorbing state matrix. Remaining unused capital (location fixed) will appear in elements on the leading diagonal of the [T] matrix. At the end of the life of the capital good, the output vector arising from current production will include scrap, which as a new commodity will be recycled into the system in the subsequent time period.

The ultimate absorbing state in the [Q] matrix for human resources is, of course, death. Training in new skills (transforming human capital) will

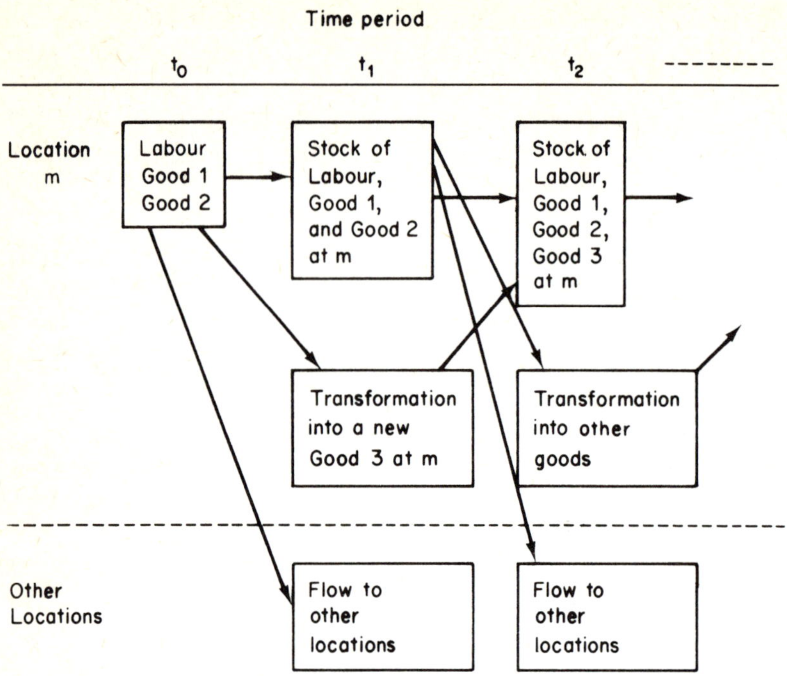

Fig. 18.1 The transformation process

also appear as an absorbing state for the flow system in the human resources account. The absorbing state 'death' for human resources in the system (unless reincarnation is accepted) is a final departure from the system – there will be no re-appearance as an input in the next time period. The human resource system input is live births (Stone, 1966).

In a closed system the outputs of one time period (the proportions of total resources ending up in the absorbing state matrix in the period) will be used in production processes to create the inputs for the next period (Fig. 18.1). The model of the economic system has now been split into two subsystems, one of flows or of spatial transformation, and another of non-spatial transformation or the use of production processes in the 'manufacture' of new goods. The problem arises of how and where the services of the transport industry (or 'movement industry') appear in the schema. The assets used by the transport industry in moving goods (people) through space are awkward to handle within the transition matrix framework. The asset location of the transport industry cannot be specified in the same terms as that of industries where 'production' is spatially fixed. The spatial transformation process is also not costless in terms of inputs required from the transport industries (Isard, 1956). To transform a male manual worker in Brighton into a male manual worker

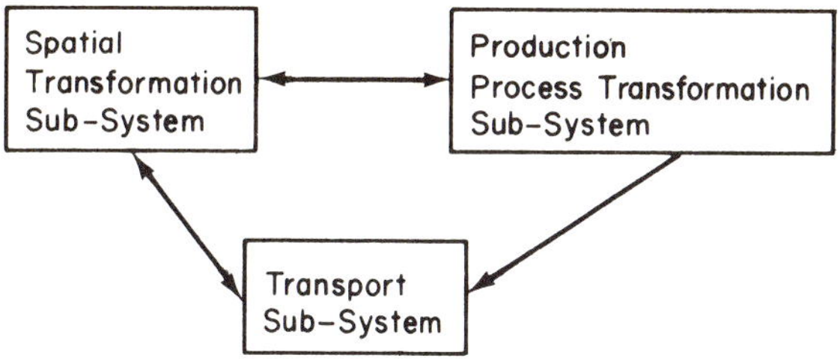

Fig. 18.2 Subsystem interrelationship

in central London will involve inputs of transport goods, such as rail services. This underlines the fact that the two subsystems are interdependent and are part of a global system.

Because of the peculiarities of the transport industry, its activities may be viewed as comprising another subsystem (Fig. 18.2). In effect, this comprises a sector of the economic system that combines the characteristics of the two other subsystems. A production process is undertaken using inputs of goods but is not undertaken at a fixed location being, more concisely, a production process that is mobile. It is not intended to examine the transport industry subsystem. In the short run it will be reasonable to assume that a proportion of the output of the production subsystem is syphoned off into the transport subsystem. The overall capacity of the transport sector therefore is fixed (although there is flexibility in the use of transport resources on particular journeys in the system) and hence provides a constraint on the movement of goods (people) possible in the spatial transformation subsystem.

The most probable state of the flow subsystem

If the time period used in the analysis does not exclude pure spatial movement, the spatially differentiated economic system will generate complex patterns of spatial flows in each time period. These flows represent the movement of goods from place of production to place of consumption (use), the movement of people from place of residence to place of work, and the movement of people for shopping or leisure purposes. A major research field of the economist, geographer and regional scientist is the analysis of the nature of these flows in a space–time dimension. The objective is usually in terms of what the aggregate flows will be between geographical locations and how they are likely to change over time. The interest is in the behaviour of the system, rather than in that of one individual who is deciding how to use the assets under his control.

The flows in the system could be described in minute detail, specifying where every good (person) moved to and from in each time period. This description would be what Wilson (1970) calls an *assignment*. The aggregate of individual trips between each origin-destination pair is the flow in the time period between the specified points in space. A description of the system in terms of these flows is termed a *distribution*. The assignment is a *microstate* description. There are many assignments which would result in a particular distribution or *macrostate* description of the system. If a macrostate description of the system is provided it is not possible to specify which of the many assignments of individual movements gave rise to that particular distribution. The specification of the correct assignment would involve knowledge of the welfare functions of each individual who operates in the economic system and the constraints operating on him. The theoretical approach is the one traditionally found in consumer choice theory in economics, where the individual maximises an independent welfare (utility) function subject to a budget constraint. In the absence of such impossibly detailed information the analyst interested in the behaviour of the system will wish to discover which is the most probable distribution: which distribution (or macrostate description) has the largest number of assignments (microstate descriptions) consistent with it.

As already suggested in the discussion of the probability matrices in the Markov chain view of the world, a major characteristic of knowledge about the behaviour of an economic system is uncertainty. First, there is what may be termed Shackle-type uncertainty due to the inability of mere mortals to possess *knowledge* about non-existent future events (Shackle, 1961). This type of uncertainty is experienced regardless of the structure of decision making in the system. Second, there is the uncertainty that arises from decentralised decision making within an interdependent system, a type obviously related to the nature of the control system in operation. As the completely centralised command economy situation is a figment of the theoretician's imagination, in any known economic system uncertainty can be expected to arise from decisions being made at various levels and places within the system. It is preferable then to build a theoretical framework that operates in probabilistic terms rather than in a deterministic fashion. For example, the probabilistic formulation of gravity models for trip generation, whether shopping trips, journey-to-work trips or leisure trips, seems preferable to deterministic formulations which imply omnipotent powers of knowledge on the part of persons who make decisions within the system (Huff, 1963).

A static one-period allocation model will be considered to illustrate the problem of predicting the system flow behaviour at macrostate level. The system examined is one concerned with shopping trip flows in a region. If there are K individuals making shopping trip decisions to J

different shopping centres then the number of possible origin-destination trips per shopping trip decision is $K \times J$. As it is usual in such models to use zones as origins and as destinations for shopping trips, the number of origin zones can be denoted as I, so that $I = J$ and aggregation of individual shoppers in the, ith zone is assumed where $(i = 1, \ldots . I)$ and $(j = 1, \ldots . J)$. This means that any ith origin zone can generate shopping trips to more than one shopping centre. If all the individuals are making independent decisions where to shop according to their particular welfare functions, the problem, in the absence of a full scale survey to determine where each shopper shops, is to extract as much information about aggregate shopping trip flows between the ith and jth zones as possible.

The analyst is seeking the most probable trip distribution in the system, which will be that associated with the largest number of individual trip assignments. If each shopper is making a single shopping trip decision the number of shopping trips in an assignment is K, which is equal to $\Sigma_j \Sigma_i K_{ij}$. The matrix $[K_{ij}]$ shows the aggregate flow between origin and destination pairs – it is a *distribution*. The number of assignments that give rise to the distribution $[K_{ij}]$ is the number of combinations of choosing the number of trips from zone 1 to zone 1 from the total number of trips K, multiplied by the number of combinations of the number of trips from zone 1 to zone 2 from the total number of trips $K - K_{11}$, and so on.

$$\text{Thus } w([K_{ij}]) = {}^{K}C_{K_{11}} \cdot {}^{K-K_{11}}C_{K_{12}} \cdot {}^{K-K_{11}-K_{12}}C_{K_{13}} \ldots\ldots\ldots \quad (18.1)$$

where $w([K_{ij}])$ is the total number of assignments that are associated with the distribution matrix $[K_{ij}]$. This may be written more fully as:

$$w([K_{ij}]) = \frac{K!}{K_{11}!(K - K_{11})!} \cdot \frac{(K - K_{11})!}{K_{12}!(K - K_{11} - K_{12})!} \cdot \frac{(K - K_{11} - K_{12})!}{K_{13}!(K - K_{11} - K_{12} - K_{13})!} \ldots\ldots$$

which simplifies to:

$$w([K_{ij}]) = \frac{K}{K_{11}!\, K_{12}!\, K_{13}! \ldots} = \frac{K!}{\Pi\, K_{ij}!} \quad (18.2)$$

As the $[K_{ij}]$ distribution matrix that has the largest number of assignments associated with it is to be found, it is necessary to

$$\textit{Maximise} \quad w([K_{ij}]) = \frac{K!}{\prod_{ij} K_{ij}!} \quad (18.3)$$

Statement (18.2) refers to the total number of assignments associated with one distribution. The total number of assignments for all possible distributions in the system (W) can be written:

$$W = \Sigma\, w([K_{ij}]) \tag{18.4}$$

The probability of any distribution in the system $[K_{ij}]$ may be defined as:

$$p_d = \frac{w\,([K_{ij}])}{\Sigma\, w([K_{ij}])} \tag{18.5}$$

where p_d is the probability of the particular macrostate distribution $[K_{ij}]$ occurring, and where $\Sigma_d\, p_d = 1$. This means, therefore, that the most probable distribution is the one which maximises $w([K_{ij}])$.

In recapitulation to find the most probable pattern of shopping trips in this simple model it is necessary to maximise $w([K_{ij}])$ subject to the constraints:

$$\sum_j K_{ij} = K_i \tag{18.6}$$

$$\sum_i K_{ij} = K_j \tag{18.7}$$

where K_i is the number of shopping trips originating in zone i,
and K_j is the number of shopping trips terminating in zone j.
A Lagrangian expression can then be formed where λ_1 and λ_2 are Lagrangian multipliers:

$$\text{Max. } \phi = \ln w + \sum_i \lambda_1 \left(K_i - \sum_j K_{ij}\right) + \sum_j \lambda_2 \left(K_j - \sum_i K_{ij}\right) \tag{18.8}$$

The logarithm of w is adopted to enable the use of Stirling's approximation,

$$\ln N! = N \ln N - N \quad \text{and hence} \quad \frac{\partial \ln N!}{\partial N} = \ln N - 1$$

so that:

$$\frac{\partial\ \phi}{\partial K_{ij}} = -\ln K_{ij} - 1 - \lambda_1 - \lambda_2 = 0$$

$$\frac{\partial\ \phi}{\partial \lambda_1} = K_i - \sum_j K_{ij} = 0$$

$$\frac{\partial\ \phi}{\partial \lambda_2} = K_j - \sum_i K_{ij} = 0$$

and if $(1 - \lambda_1 - \lambda_2) = -\lambda, \quad \ln K_{ij} = -1 - \lambda$ (18.9)

A similar result can be derived from information theory (Jaynes, 1957; Theil, 1967). If an event is defined as being a macrostate distribution in the shopping system, then probabilities (p_d) can be assigned to each event, where $d = I, \ldots\ldots D$ so that:

$$\sum_{d=1}^{D} p_d = 1 \quad \text{and} \quad p_d \geqslant 0$$

The information content of a message is defined as the logarithm of the reciprocal of the probability p_d:

$$h(p_d) = \ln \frac{1}{p_d} = -\ln p_d \tag{18.10}$$

Thus the higher the probability of a particular event occurring the lower is the information content of the message saying that the event has occurred. Before the message is received (before the event has occurred) the information content of the forthcoming message is not known, so it is necessary to compute the expected information content $H(p_d)$:

$$H(p_d) = \sum_d p_d\, h(p_d) = \sum_d p_d \ln \frac{1}{p_d} = -\sum_d p_d \ln p_d \tag{18.11}$$

To maximise the expected information content subject to $\Sigma_d\, p_d = 1$ the Lagrangian is formed:

$$\text{Max. } \theta = -\sum_d p_d \ln p_d - \lambda\left(\sum_d p_d - 1\right)$$

so that

$$\frac{\partial\ \theta}{\partial p_d} = -1 - \ln p_d - \lambda = 0$$

and

$$\ln p_d = -1 - \lambda \quad (d = 1, \ldots\ldots D) \tag{18.12}$$

The expected information content of the forthcoming message is, therefore, the same for all the macrostate distributions in the system, or alternatively the probability of each and every macrostate distribution is identical. With D possible macrostate distributions, therefore, $p_d = \dfrac{1}{D}$.

Substituting into equation (18.1):

$$H(p_d) = -\sum \frac{1}{D} \ln \frac{1}{D} = \frac{1}{D} \ln D \tag{18.13}$$

The maximum value that $H(p_d)$ can take is $\ln D$, when $D = 1$. Where $D > 1$, $H(p_d) < \ln D$. Accepting the customary definition that $\dfrac{1}{D} \ln D = 0$

if $D = 0$, and that the expected information $H(p_d)$ cannot be negative then:

$$0 \leqslant H(p_d) \leqslant \ln D \tag{18.14}$$

The expected information concept, $H(p_d)$, is the entropy or measure of disorder in the system. The maximum value ($\ln D$) increases as the number of macrostate distributions of shopping trips (D) in the example increases.

The vital result from using an entropy maximising approach is that, to decrease the entropy, or uncertainty, in the system, it is necessary to impose constraints on the operation of the system so that certain macrostates become more probable than others — the worst information situation of having to assume all $p_d = \frac{1}{D}$ does not exist. In the extreme, the imposition of constraints may make some macrostates impossible (their $p_d = 0$), but will make it possible to select one macrostate as being the most probable (Wilson, 1967).

To return to the simple shopping model, the result in equation (18.9) illustrated the maximum entropy situation where the only constraints were (18.6) and (18.7). It will be remembered that the flows in the spatial transformation subsystem are subject to the capacity constraints of the transport sector subsystem. Such a constraint may now be added to the simple shopping trip model:

$$\sum_j \sum_i g_{ij} K_{ij} = G \tag{18.15}$$

where G is the total shopping transport capacity of the system, and g_{ij} is the proportion of total system capacity used in making one shopping trip from origin i to destination j.
With the extra constraint the Lagrangian expression becomes:

$$\text{Max. } \phi = \ln w + \sum_i \lambda_1 \left(K_i - \sum_j K_{ij}\right) + \sum_j \lambda_2 \left(K_j - \sum_i K_{ij}\right) + \lambda_3 \left(G - \sum_j \sum_i g_{ij} K_{ij}\right) \tag{18.16}$$

where λ_1, λ_2, λ_3, are Lagrangian multipliers, and setting the partial derivative with respect to K_{ij} to zero, it is possible to obtain:

$$\frac{\partial\ \phi}{K_{ij}} = -\ln K_{ij} - 1 - \lambda_1 - \lambda_2 - \lambda_3 g_{ij} = 0$$

or $$K_{ij} = e^{-1-\lambda_1-\lambda_2-\lambda_3 g_{ij}} \tag{18.17}$$

By substituting equation (18.17) into the constraint equations (18.6) and (18.7), expressions enabling the evaluation of λ_1 and λ_2 can be derived, and by a process of calibration either λ_3 if the g_{ij} are known, or $\lambda_3 g_{ij}$ if the g_{ij} are unknown, can be estimated and the entropy maximising macrostate of the system distribution of shopping trips $[K_{ij}]$ can be found. Thus by adding an extra constraint to the knowledge about zonal totals for trip origins and destinations (embodied in constraints (18.6) and (18.7)), the most probable macrostate of the system has been found, and because of the extra constraint the system entropy has been reduced.

The technique applied here to a shopping trip example can be generalised to apply to any system involving spatial flows that result from a large number of individual decisions being made within it. The information content attached to a message that a specific event has occurred is related to the probability of it occurring, and this probability depends on the number of constraints on the operation of the system. In a completely closed command type of economic system with all behaviour determined centrally, the system state (S_1) has a probability (p_1) equal to unity. There is no information content in the message stating that S_1 has occurred. At the other extreme, in a decentralised market economy where the decision making of individuals is unconstrained, the regional scientist should assume all the system states S_i where $i = 1, \ldots . n$ are equally probable so that $p_i = \frac{1}{n}$.

The traditional economic analysis of consumer and producer choice in a space–time environment has concentrated on the optimisation of some welfare function for each individual usually subject to some total asset constraint (for example, a budget constraint in traditional consumer choice theory (Green, 1971)). These individual welfare functions are not known and cannot in practice be known for a complex spatial economic system; but it is feasible to specify the constraints operating on individual choice for small groups of spatially adjacent individuals. From a positive viewpoint it would seem preferable, in the absence of knowledge about individual welfare functions on the part of the regional scientist, to concentrate on the constraints operating on individual decision making within the system with a view to reducing the entropy or uncertainty about its possible state.

A system objective function?

When the economist is concerned with prescription it is not usually very long before the esoteric concept of a social welfare function is produced *deus ex machina*. To standardise the terminology in the remainder of this essay the social welfare function can be termed the system objective function (Arrow, 1951; Rothenberg, 1961). The idea of the system objective function being derived by some form of democratic voting procedure by all participants is fraught with theoretical difficulties as well as not being, even with the greatest stretch of the imagination, a description of the way in which business or government objectives are formulated. The alternative is that a system objective function is imposed by one or a committee of top decision makers who may or may not be subject to the 'lobbying' activities of other participants in the system. Ultimately it is the action of one *individual* or small group of *individuals* that *de facto* determine the system objective function as evidenced in system top level policy decisions. The individual making a decision about the spatial distribution of company plant is acting to maximise his own individual welfare function in which his activities as company director feature as arguments in addition to his activities as a consumer of food or as a member of a golf club.

From a positive or behavioural viewpoint more progress is likely to be made if the system objective function concept is abandoned. Instead, the behaviour of the system under analysis should be regarded as the outcome of individual action where each individual fulfils several roles between which there can be conflict, and where the individual is constrained not only by the rules attached to fulfilling a particular role (such as managing director, civil servant or father-in-law) but also by interaction between individuals especially if the decision making is within a committee framework. The analyst does not possess information about all the individual welfare functions. He can proceed, however, using an entropy maximising approach without entanglement in the social welfare function concept, which depends on the existence of 'corporate beings' independent of the individuals who make up the corporate organisation (Wilson, 1971).

If the regional scientist is interested in positive analysis, as opposed to prescriptive or normative analysis, he should be wary of specifying models that embody a system objective (Isard, 1958; Stevens, 1958; Beckmann and Marschak, 1965; Takayama, 1966). It would be more productive of insights into the operation of a spatial economic system if the attention is concentrated on specifying the constraints on the space-time behaviour of individuals in the system and then seeking its most likely macrostate.

Constraints

The specification of the constraints that are likely to be operating upon an individual is a major research task. It is activities rather than quantities of goods that form the arguments in the individual's welfare function (Lancaster, 1966 and 1971; Evans, 1972). The individual, therefore, selects a group of activities from the set of all possible activities open to him so that he cannot improve his overall level of satisfaction by choosing some other group of activities. Each activity that the individual undertakes requires inputs of goods; for example, the activity of freshwater fishing requires the input of bait, fishing rod, fishing rights on a stretch of water and transport to and from the fishing site. The input requirements for each activity emphasise the multiple use made of economic goods and involve the construction of a matrix which specifies a consumption technology – the required inputs for undertaking a particular activity. The consumption technology represents a set of constraints imposed on the activities of the individual; for example, if no ownership of fishing rights on a particular stretch of water then no (legal) fishing activity. It should be noted that transport inputs to overcome spatial separation feature in the consumption technology matrix. To revert to the fishing example, if the fisherman possesses only a bicycle to overcome spatial separation, then if he is constrained to not being away from his home base for more than sixteen hours and the nearest rights to fish are over eight bicycle-hours away, the fishing activity is outside the possible 'space-time prism' of the individual (Hägerstrand, 1970).

For the moment the other important constraint on the activities chosen by the individual is his income since this imposes a maximum limit on his expenditure on inputs to undertake consumption activities. Economists have concentrated on the budget constraint in their models of consumer choice. The activity of fishing would have been possible for the fisherman only if he had sufficient income to afford a car, all other activities being constant (Fig. 18.3).

The choice model outlined so far can be stated formally.

1. Maximise an individual welfare function

$$U = f(a)$$

where a is a vector of activities.

2. Subject to the consumption technology constraints

$$a = C\,x$$

where x is a vector of inputs of goods, and C is a matrix specifying the quantities of each good required for each activity. An element c_{ij} would represent the quantity of good j required per unit of activity i, and $\Sigma_j\, c_{ij}$ would represent the total inputs of j goods ($j = 1, \ldots . n$) required for a unit of activity i.

3. Subject to a budget constraint on the purchase of goods

$$\sum_j s_j\, x_j \leqslant Y$$

where s_j is the price of the jth good, and Y is the consumer's total income.

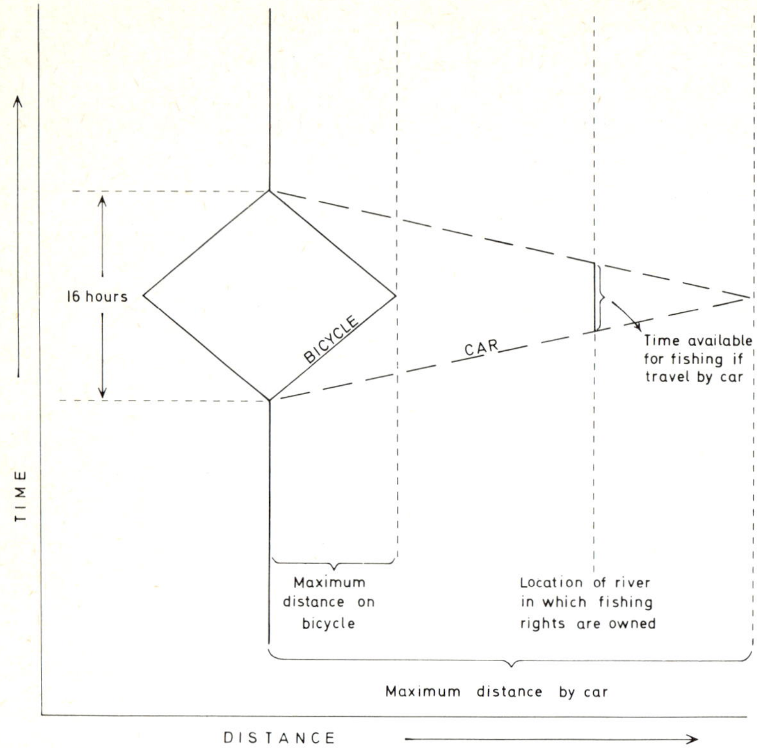

Fig. 18.3 Space-time constraints. Based on Hägerstrand, 1970

4. Subject to non-negativity requirements

$$a, x \geqslant 0$$

Given the price of goods vector (which, because labour is included as a good which is sold for income, has wage rates as negative prices), $s_j c_{ij}$ will represent the value of the jth good incorporated per unit of activity i, and $\Sigma_j s_j c_{ij}$ will represent the unit cost of operating the ith activity. The model can therefore be abbreviated to:

$$\text{Maximise } U = f(Cx)$$

$$\text{Subject to: } sC\,x \leqslant 0$$

$$x \geqslant 0$$

where the activity space is transformed into goods space. There are, however, good reasons for not performing this transformation (Lancaster, 1971) as the information embodied in the consumption technology matrix is hidden in the individual's unknowable welfare function, so that the $U = f(Cx)$ function is an amalgam of the characteristics of the non-linear $U = f(a)$ function and the linear consumption technology matrix. It is preferable, therefore, as Lancaster demonstrates, to map the consumption technology constraints from goods space into activity space.

Because the space-time constraints so far have been hidden in the consumption technology matrix, it is necessary to elaborate on the nature of these constraints (Becker, 1965; Evans, 1972). The simplest interpretation of a time constraint is that the total time inputs in performing activities should not exceed the total time available. The total time could range from twenty-four hours, if diurnal behaviour patterns are of interest, through to a Modigliani-type average expected life span (Modigliani and Brumberg, 1954). The second type of time constraint is the one that is stressed by Hägerstrand (1970) with his space-time prisms and by Evans (1972), namely that each activity will involve a minimum time input to perform. For example, the activity of consuming a performance of Berlioz's *Symphonie Fantastique* will involve a necessary time input of fifty minutes. The activity of selling one's labour at a place of work will involve a time input not less than the length of the minimum working day. Increasing the minimum necessary time input in an activity (for example the purchase of a joint of beef) is one way of rationing scarce goods. If it becomes necessary to queue to purchase a joint of beef the time taken to purchase beef successfully will increase, and beef will be distributed in favour of those persons to whom time is not a scarce resource (the unemployed and able-bodied senior citizens). The third type of constraint involving time is where two or more activities are linked together in a joint product situation. For example, to achieve a work activity of eight hours it may be necessary to link it to a travelling activity of two hours for the journey to and from the place of work so that:

$$b\,\Upsilon_w \leqslant \Upsilon_t$$

where b is a constant representing the necessary relationship between time spent on the two activities, work time Υ_w, and travel time Υ_t. In this example $\Upsilon_t \geqslant 0.25(8)$. To incorporate these aspects of time constraints Evans uses time expended on each activity as the variables in the individual's welfare function:

$$U = f(\Upsilon_w, \Upsilon_t, \Upsilon_i) \quad (i = 1, 2, \ldots . n)$$

where Υ_w is time at work, Υ_t is time spent travelling and Υ_i is the time spent on the ith activity, and where utility is assumed to be positively related to the time spent on each activity $\left(\frac{\partial\, v}{\partial \Upsilon_i} > 0\right)$.

The consumer choice model then becomes:

$$\begin{aligned} \text{Maximise} \quad & U = f(\Upsilon) \\ \text{Subject to:} \quad & sR\Upsilon \leqslant 0 \\ & B\Upsilon \geqslant 0 \\ & 1\Upsilon = T^{*} \\ & \Upsilon \geqslant 0 \end{aligned}$$

where R is the matrix describing the input of goods required for the ith activity, B is the matrix representing certain fixed relationships linking activities together, 1 a unit vector and T^{*} the total time available.

The individual decision maker operating in an economic system will be subject to income constraints and space-time constraints. He will also be subject to constraints that do not have a pure economic or geographic foundation, namely, discriminatory and institutional role-playing constraints. The discriminatory constraints may take many forms, from racial discrimination in the operation of the housing market, discrimination against non-union labour in taking up job opportunities to discrimination in awarding club membership rights to fish or play golf. All these constraints are likely to modify the spatial behaviour of individuals from what it would have been in the presence only of economic and geographic constraints. The institutional role-playing constraints are related to the multirole situations in which most individuals find themselves, and to the fact that there are social norms of behaviour for each role being fulfilled. One aspect of the norm for fulfilling the role of managing director of a company could be his presence in the managing director's office for a certain minimum period per week. This institutional constraint could well alter the allocation of time between work and golf, and so alter the location of the individual concerned in the space-time dimension. In a less frivolous example, the social norms expected in terms of time inputs for family life at home and for work situations will have important influences on the spatial relationship between work place and place of residence in terms of what is an acceptable period to spend commuting.

In the analysis of the behaviour of a regional economic system, in which the individuals operating the system are free to make decisions at one level or another, the individual welfare function is a useful theoretical construct but is in practice unobservable. Given that spatial models must have some geographical aggregation (km^2 or $100\ m^2$) it is suggested that a fruitful approach for the regional scientist to adopt is the investigation of the 'average' constraints to which the occupants of one spatial zone are subjected. By becoming aware of what spatial behaviour is possible within the bounds of the constraint set, the analyst will reduce the system entropy and increase the probability of a particular pattern of flows (of goods or people) in a particular time period.

Given these constraints, by maximising the system entropy the most probable flow pattern can be discovered and used as the basis for the probabilities in a Markov chain transition matrix of change in location.

The production subsystem

So far the behaviour of an economic system within one time period has been discussed. This concluding section indicates how a time dimension, in the sense of simulating the time path of aggregate variables in the model, can be introduced. Given an initial stock of goods including human resources, a Markov transition matrix (based on an entropy maximising submodel) has been used to indicate the flows or changes of location within the system in one time period, and the absorbing state matrix has been used to indicate the probability of goods entering a production process which transforms one vector of goods into a new vector of goods. The absorbing state matrix reveals nothing about the relationship between the goods entering the 'absorbing' production system and the outputs of the production system that form the next period's inputs into the Markov chain model. Technological relationships between inputs and outputs in the production system are required. One model which assumes constant and linear relationships between inputs and outputs is the well developed input-output model of modern economics (Leontief, 1966; Yan, 1969; Richardson, 1972). If an input-output table is constructed on a commodity basis, instead of an industrial sector basis, so that it conforms with the previous analysis, the basic transactions matrix $[X]$ records in each of its columns the inputs procured of other commodities for the manufacture of a particular commodity. If it is assumed that n is the number of goods and m is the number of locations, so that the input-output table is given a spatial dimension, the production system can then be represented (Fig. 18.4).

The first column represents the inputs required to produce commodity 1 in location 1 so that $x_{11.11}$ is the use in the production of commodity 1 at location 1 of commodity 1, $x_{21.11}$ is the use in the production of commodity 1 at location 1 of commodity 2, and so on. The important feature to note is that the partitioned matrix $[X]$ is diagonal:

$$X = \begin{bmatrix} X_1 & 0 \;\ldots\ldots\ldots & 0 \\ 0 & X_2 \ldots\ldots\ldots & 0 \\ \cdot & \cdot & \cdot \\ \cdot & \cdot & \cdot \\ \cdot & \cdot & \cdot \\ 0 & 0 \;\ldots\ldots\ldots & X_m \end{bmatrix}$$

		LOCATION 1			LOCATION 2			LOCATION m		FINAL OUTPUT (z)	GROSS OUTPUT (x)
		11	21	n1	12	22	n2	1m	nm		
LOCATION 1	11	$X_{11.11}$	$X_{11.21}$	$X_{11.n1}$						Z_{11}	X_{11}
	21	$X_{21.11}$	$X_{21.21}$	$X_{21.n1}$						Z_{21}	X_{21}
	n1	$X_{n1.11}$	$X_{n1.21}$	$X_{n1.n1}$						Z_{n1}	X_{n1}
LOCATION 2	12				$X_{12.12}$	$X_{12.22}$	$X_{12.n2}$			Z_{12}	X_{12}
	22				$X_{22.12}$	$X_{22.22}$	$X_{22.n2}$			Z_{22}	X_{22}
	n2				$X_{n2.12}$	$X_{n2.22}$	$X_{n2.n2}$			Z_{n2}	X_{n2}
LOCATION m	1m							$X_{1m.1m}$	$X_{1m.nm}$	Z_{1m}	X_{1m}
	nm							$X_{nm.1m}$	$X_{nm.nm}$	Z_{nm}	X_{nm}
GROSS INPUT		X_{11}	X_{21}	X_{n1}	X_{12}	X_{22}	X_{n2}	X_{1m}	X_{nm}	Z	X

Fig. 18.4 Production subsystem input-output table

This differs from other interregional input-output models (Isard, 1952; Stone, 1961; Leontief and Strout, 1963) in that the relationships recorded do not include any movement of goods through space – the relationships are those concerned solely with transformation by a production process at a specified location. The results of this production process are to produce a final output vector (Z). It is this output vector that becomes the next period's input vector for the Markov chain model. The gross output of each commodity n at location m (X_{nm}) consists of output that is 'consumed' as inputs in the production of other commodities and the final output of the commodity which will enter the next period's transition matrix and be transformed spatially.

A matrix of pure technical input coefficients $[A]$ can be derived from the $[X]$ matrix by dividing each element x_{ij} by X_j (or X_i since $X_i = X_j$):

$$a_{ij} = \frac{x_{ij}}{X_j} \tag{18.18}$$

where $i = 11, 12, \ldots .\ 1m, 21, \ldots .\ 2m, \ldots .\ nl, \ldots nm$
and $\quad j = 11, 12, \ldots .\ 1m, 21, \ldots .\ 2m, \ldots .\ nl, \ldots nm$

The model can now be summarised so that each row reads:

$$\sum_j x_{ij} + Z_i = X_i \tag{18.19}$$

Substituting (18.18) into (18.19):

$$\sum_j a_{ij} X_j + Z_i = X_j$$

$$\text{or} \quad Z_i = X_j - \sum_j a_{ij} X_j \tag{18.20}$$

There are $m \times n$ equations of this form, one for each row. These equations may be summarised using matrix notation:

$$Z = (I - A)\, X$$

Given X, which is the output of the Markov absorbing state process in time Υ_0, and given the technical input coefficient matrix $[A]$, it is possible to calculate the vector Z, the final outputs which form the stock of inputs entering the transition and absorbing states in time Υ_1.

The attention of economists has been focused on what in this essay has been called the production system. When regional problems became of interest in economics, the single region input-output framework was modified into an interregional input-output model in order to handle spatial movement. This led to problems of hybrid input coefficients in which it is known that the spatial element is not stable. The approach outlined here has the merit of separating the spatial transformation process from the production (location fixed) transformation process. It also places emphasis, long given by geographers, on flows of commodities and people in the space dimension, stressing the greater degree of uncertainty attached to the behaviour of the flow system as opposed to the behaviour of the production system.

References

ARROW, K. J. (1951) *Social Choice and Individual Values*, Yale University Press.

BECKER, G. S. (1965) 'A theory of the allocation of time', *Econ. J.*, **75**, 493–517.

BECKMAN, M. and MARSCHAK, T. (1965) 'An activity analysis approach to location theory', *Kyklos*, **8**, 128–41.

BISHIR, J. W. and DREWES, D. W. (1970) *Mathematics in the Behavioural and Social Sciences*, Harcourt Brace.

BLACK, J. (1966) 'Some retail sales models', unpublished paper presented to Urban Studies Conference.

EVANS, A. W. (1972) 'On the theory of the valuation and allocation of time', *Scott. J. Polit. Econ.*, **19**, 1–17.

GIBSON, M. and PULLEN, M. (1972) 'Retail turnover in the East Midlands: a regional application of a gravity model', *Reg. Stud.*, **6**, 183–96.

GREEN, H. A. J. (1971) *Consumer Theory*, Penguin Books.

HÄGERSTRAND, T. (1970) 'What about people in regional science?', *Pap. Reg. Sci. Ass.*, **24**, 7–21.

HUFF, D. L. (1963) 'A probabilistic analysis of shopping centre trade areas', *Land Econ.*, **53**, 81–90.

ISARD, W. (1952) 'Interregional and regional input-output analysis: a model of a space economy', *Rev. Econ. & Stat.*, **33**, 318–28.

ISARD, W. (1956) *Location and Space-economy*, MIT Press.

ISARD, W. (1958) 'Interregional linear programming: an elementary presentation and general model', *J. Reg. Sci.*, **1**, 1–59.

ISARD, W. (1969) *General Theory, Social, Economic and Regional*, MIT Press.

JAYNES, E. T. (1957) 'Information theory and statistical mechanics', *Phys. Rev.*, **106**, 620–30.

LANCASTER, K. (1966) 'A new approach to consumer theory', *J. Polit. Econ.*, **74**, 132–57.

LANCASTER, K. (1971) *Consumer Demand, a New Approach*, Columbia University Press.

LEONTIEF, W. W. (1966) *Input-Output Economics*, Oxford University Press, NY.

LEONTIEF, W. W. and STROUT, A. (1963) 'Multi-regional input-output analysis', in T. Barna, ed., *Structural Interdependence and Economic Development*, International Conference on Input-Output Techniques, 1961, Macmillan, pp. 125–89.

MARSCHAK, T. (1959) 'Centralization and decentralization in economic organizations', *Econometrica*, **27**, 399–430.

MODIGLIANI, F. and BRUMBERG, R. (1954) 'Utility analysis and the consumption function: an interpretation of cross section data', in K. K. Kurihara, ed., *Post*

Keynesian Economics, New Brunswick, Rutgers University Press, pp. 142–201.

RICHARDSON, H. W. (1972) *Input-Output and Regional Economics*, Weidenfeld & Nicolson.

RICHARDSON, H. W. (1973) *Regional Growth Theory*, Macmillan.

ROTHENBERG, J. (1961) *The Measurement of Social Welfare*, Prentice-Hall.

SHACKLE, G. L. S. (1961) *Decision, Order and Time in Human Affairs*, Cambridge University Press.

STEVENS, B. H. (1958) 'An interregional linear programming model', *J. Reg. Sci.*, 1, 60–98.

STONE, R. (1961) 'Social accounts at the regional level: a survey', in W. Isard and J. H. Cumberland, eds, *Regional Economic Planning*, Paris, pp. 37–84.

STONE, R. (1966) *Mathematics in the Social Sciences and other essays*, Chapman & Hall.

TAKAYAMA, T. (1966) *Spatial Economic Analysis*, Inaugural lecture, Armidale, New South Wales.

THEIL, H. (1967) *Economics and Information Theory*, Amsterdam.

WILSON, A. G. (1967) 'A statistical theory of spatial distribution models', *Transpn. Res.*, 1, 253–69.

WILSON, A. G. (1970) *Entropy in Urban and Regional Modelling*, Centre for Environmental Studies, London, Pion 1971.

WILSON, A. G. (1971) 'On some problems in urban and regional modelling', in M. Chisholm, A. E. Frey and P. Haggett, eds, *Regional Forecasting*, Butterworths, pp. 179–220.

YAN, C. (1969) *Introduction to Input-Output Economics*, Holt Rinehart & Winston.

R. P. Bradshaw

19 The development of regional planning in Spain

Rural poverty and regional underdevelopment were first recognised as serious problems in Spain during the sixteenth century, and the search for remedies has been a major preoccupation of Spanish intellectuals and reformers from Feijoo and Jovellanos in the eighteenth century to contemporary writers such as the economist Ramón Tamames. The basic causes of rural poverty have been the excessive pressure of population on the land, an arid climate, poor soils, and the inequitable and inefficient distribution of land that occurs with large estates (*latifundios*) and small fragmented farms (*minifundios*). Until the late 1950s all attempts to change this situation were concentrated on land improvement schemes, especially irrigation and colonisation projects, and to a lesser extent on land redistribution. In recent years the advent of industrialisation has lifted Spain out of its state of general poverty, but at the same time has accentuated the differences between the wealthy regions, which have manufacturing centres, and the poor regions which are still dependent on agriculture. The present situation is best summarised in the studies of the Banco de Bilbao (1973), which show that in 1971 the *per capita* net income in each of the five richest provinces was more than double that in each of the ten poorest. During recent years new policies have been applied, the most important being the attempt since 1964 to industrialise the poorer regions. This latest stage in regional planning is still in process of development.

Early attempts to remedy regional underdevelopment

Official concern about regional problems dates from the sixteenth century, with the discovery that large areas of central Spain were becoming depopulated. The government ordered a series of inquiries, the *Relaciones Topográficos* of 1575 and 1578, to investigate the social and economic causes of this depopulation. However, despite these inquiries no attempt was made to improve the harsh conditions under which the rural peasantry was forced to exist, and which contributed so much to the strong movements of population to the richer coastal areas of Spain, to the new American colonies and to the Spanish empire in Europe (Nadal, 1966). Not until the reign of Carlos III (1759–88) was any serious attempt made to solve these problems. Carlos III's leading minister, the Count of Campomanes, introduced the first schemes for promoting regional development in Spain, based on colonisation projects and the construction of large multipurpose canals. Work was begun on two canals, the Canal de Castilla which runs from Alar del Rey to Medina del Rioseco but not finally completed until 1849, and the Imperial Canal of Aragón, which runs alongside the River Ebro from Tudela to Zaragoza, and was completed by 1789. Both canals were used for transport as well as irrigation, and contemporary reports suggest that they did much to raise the standard of living in their respective regions (Banco de Aragón, 1960).

The colonisation policy was an attempt to solve the rural problems of Andalusia. During the eighteenth century agriculture in southern Spain was marked by three distinct features: large areas of uncultivated land, the *baldíos*; the predominance of large estates, the *latifundios*; and the desperate land hunger of the vast mass of the rural peasantry. The policy was to introduce new peoples to the region who would be unaccustomed to the bad practices and customs of the area, who could have their own rights and laws, and who would not fall under the influence of the *latifundistas*. The scheme was approved by royal charter in 1767, and catered for the installation of 6 000 Bavarians in a number of new settlements on uncultivated lands in the Sierra Morena. These settlements were to populate the zones along the Madrid–Cádiz and Valencia–Andalusia highways, in order to combat the local banditry and to create model settlements which, by their example, would lead to improvements throughout the rest of Andalusia.

In the event 7 764 colonists were accepted (Palacio Atard, 1964) and included French, Italians, Swiss and Flemings as well as Bavarians. Each colonist was given land, a house, livestock and agricultural implements in return for a rent paid to the crown. Special provision was made to prevent the growth of large landholdings, the subdivision of land and subletting, and except for the parish priest all religious orders, aristocrats and other non-productive classes were prohibited from entering the settlements. Eventually forty-four villages and eleven towns were established over an area of 2 500 km^2 (Herr, 1958). The English writer Joseph Townsend (1791), travelling through the area in 1786–87, described the new settlements as appearing like oases in the Spanish countryside.

News of the scheme quickly spread and many Spaniards flocked to the area and installed themselves on the edge of or within the new settlements. Soon the foreign-born colonists were in a minority and despite the apparent success of the scheme, they began to desert the settlements, especially after 1770. Many could not survive the difficult conditions of a harsh climate and poor soils, and many succumbed to disease. After 1770 Madrid insisted that the settlements should show a bigger financial return. Pablo Olavide, one of the originators of the scheme and its first superintendent, was forced to raise the colonists' rents, at which point many deserted to work in the mines in the area. Despite these difficulties the scheme continued to make some progress until 1776, when Olavide was arrested and condemned by the Inquisition, the last major figure in Spain to be so treated (Palacio Atard, 1964). After the fall of Olavide the settlements gradually declined until 1835, when they were absorbed into the rest of the administration. Although the experience gained in these early attempts at regional planning was soon lost in the political and military upheavals caused by the Napoleonic Wars, López de Sebastián (1970) has shown that many features of the Sierra Morena scheme were revived in the Badajoz Plan of 1953.

The development of irrigation policy

It was not until the end of the nineteenth century that any fresh attempt was made to develop the backward regions. Influential writers of the time such as Joaquín Costa and Macías Picavea pointed to the contrast in agricultural productivity between the *regadío*, Spain's rich irrigated lands, and the *secano*, the poor, dry farming areas. Irrigation, it was argued, could be the panacea for rural poverty, an argument which was accepted generally in Spain. The development of irrigation policy can be seen in the progress of legislation during the period from 1870 to 1952 (Gómez Ayau, 1961; Naylon, 1967). The earliest legislation referred only to major hydraulic works; however, experience was to show that for irrigation to be effective, projects would have to include all construction work, including levelling land, building new roads and farmhouses. The irrigation laws increasingly recognised the need for state intervention. The 1870 irrigation projects were left entirely in the hands of private companies, whereas by 1952 they were under complete state control.

The legislation also involved a progression in scale from small local projects to the planning of entire valley regions. The emphasis on river valley development was largely the result of a long campaign waged by a public works engineer, Laurenzo Pardo. His campaign was for a single authority that would manage an entire river basin and would be responsible for all aspects of a hydraulic scheme including irrigation, navigation, hydroelectric power generation and domestic water supply. Pardo emphasised that these works should be fully coordinated in order to gain maximum benefit. Although his principal aim was to increase the amount of irrigated land in Spain his proposals constituted a wholesale river basin development programme.

Pardo began campaigning for a river basin authority for the Ebro in 1906, but it was not until the dictatorship of Primo de Rivera in 1923 that he gained the attention of the government, especially the Minister of Public Works. On 5 March 1926 a law was passed enabling hydrographic confederations to be established in all the major river basins in Spain. These confederations were semi-autonomous bodies governed by a large assembly consisting of representatives from various government ministries and over a hundred delegates from interested groups such as municipal authorities, hydroelectric power companies and farmers. These assemblies established the major projects the confederations would undertake, while a smaller governing body organised their execution and the routine running of the river basin. The first of the confederations was the *Confederación Sindical Hidrográfica del Ebro* created later on the same day, 5 March 1926. This was the first water authority of its kind in the world and pre-dated the Tennessee Valley Authority by seven years. Subsequently other confederations were created – in 1927 for the Duero and Guadalquivir, and in 1928 for the Segura, the rivers of the eastern Pyrenees and the Tagus. The Ebro remained the most important scheme, having created 72 163 hectare of new irrigated land by 1930, although almost all of this was supplied from existing dams (Velarde Fuertes, 1968). The other confederations had hardly completed their preliminary studies when the government of Primo de Rivera fell on 26 January 1930.

Contrary to common belief, interest in irrigation policy did not diminish after the fall of Primo de Rivera; indeed the governments of the Republic spent annually up to four times more on irrigation projects than the governments of Primo (Malefakis, 1970). However, at the beginning of the Second Republic in 1931 everything connected with the dictatorship of Primo, including the hydrographic confederations and Pardo, was anathema. By 1933 the Minister of Public Works had revived the confederations, changing their name to *Mancomunidades* (unions), and appointing Pardo to take charge of the new Centre for Hydrographic Studies and prepare a national plan of hydraulic works (Ramos Oliveira, 1946). The main feature of Pardo's new plan was to divert some of the headwaters of the Tagus to the Segura valley in Murcia. The reason for this transference of water was the wide disparity in growing conditions between the two areas, the irrigated land of central Spain producing less than one-third of that in the Levante. The scheme, like the rest of the agrarian reforms of the Second Republic, had insufficient time to take effect before the Civil War began in 1936. However, the Segura scheme was revived in 1966, and in 1974 forms the largest irrigation project under construction in Spain.

During the Second Republic concern was expressed at the slow rate of transformation of land from *secano* into *regadío*. In some areas irrigable zones remained under *secano* twenty years after the completion of the main dams and distribution canals. In part this was due to the technical difficulties inherent in any major hydraulic scheme, but the major difficulties came from the *latifundistas* and the local political bosses, the *caciques*, who opposed any reform that threatened their way of life or would loosen their hold over the peasantry. Two solutions to these problems were proposed. First, a *Servicio de Obras de Puesta en Riego* was created in 1932 with the authority to construct all the secondary works necessary to bring water from the main distribution canals to the farmlands. Second, the idea of internal colonisation was revived. The Agrarian Reform Law of 15 September 1932 permitted the expropriation of 'surplus' *secano* land owned by *latifundistas* and its redistribution among colonists selected from landless peasants and farm labourers. The Fifth National Irrigation Congress held at Valladolid in 1934 proposed the extension of this practice to the irrigated lands, a proposal which was incorporated in all future irrigation projects. These ideas had little time to take effect before the start of the Civil War.

Despite the violent change in political ideology, the Franco government of 1939 continued the irrigation policies of the Second Republic. Large hydraulic works were to be constructed by the state authorities, and the irrigable lands were to be settled by private societies of colonists supervised by the *Instituto Nacional de Colonización* (INC), a semi-autonomous body created in October 1939. Little progress was made by these private societies and in 1949 the INC was given responsibility for all colonisation activities from the promotion of new colonisation schemes to research into methods for obtaining increased crop yields.

The Badajoz and Jaén plans

The development of an irrigation policy in the period from 1870 to 1949 made possible Spain's first attempts at large-scale integrated regional planning, the Badajoz and Jaén plans of 1952 and 1953. In the early 1950s Badajoz and Jaén were two of the most backward provinces in Spain. In terms of *per capita* income in 1955 Badajoz and Jaén were, respectively, the forty-fourth and forty-seventh poorest provinces in Spain (Table 19.1). They were overdependent on agriculture which, in 1950, employed over 70 per cent of Badajoz's and 69 per cent of Jaén's active population. Both provinces were characterised by *latifundios*, with large areas given over to the cultivation of olives, wheat and other

Table 19.1 Provinces ranked according to *per capita* incomes, 1955–71

Provinces	*1955*	*1957*	*1960*	*1962*	*1964*	*1967*	*1969*	*1971*
Vizcaya	2	1	2	1	1	2	2	1
Guipúzcoa	1	2	1	3	3	3	1	2
Alava	5	5	11	6	6	5	3	3
Madrid	4	3	4	2	2	1	4	4
Barcelona	3	4	3	4	4	4	5	5
Baleares	14	16	16	10	8	6	6	6
Gerona	16	13	10	5	5	7	8	7
Navarre	7	7	8	7	7	8	7	8
Santander	9	10	6	11	10	9	9	9
Lérida	21	20	15	12	12	18	11	10
Tarragona	15	15	7	9	11	13	10	11
Valladolid	11	12	18	17	17	12	14	12
Logroño	13	9	13	8	9	10	13	13
Zaragoza	10	11	12	13	14	11	12	14
Huesca	25	25	17	15	16	19	17	15
Oviedo	6	8	9	16	20	15	19	16
Burgos	19	18	21	19	18	16	16	17
Valencia	8	6	5	14	13	14	18	18
Castellón	20	17	14	18	15	17	15	19
Alicante	26	32	19	21	21	20	21	20
Guadalajara	30	24	30	28	27	27	23	21
Segovia	18	14	23	20	24	25	25	22
Soria	28	22	38	25	25	23	20	23
Palencia	12	19	22	23	19	21	22	24
Palmas (Las)	22	27	24	30	28	33	29	25
León	27	26	31	22	22	22	26	26
Seville	17	21	20	27	30	30	33	27
Teruel	32	30	36	24	23	28	24	28
Pontevedra	29	23	25	32	26	26	27	29
Cádiz	23	31	27	33	34	37	34	30
La Coruña	31	28	33	37	37	32	31	31
Salamanca	33	33	37	26	29	29	28	32
Santa Cruz de Tenerife	37	38	26	42	36	43	35	33
Toledo	41	40	40	39	38	39	32	34
Huelva	24	29	29	36	39	42	40	35
Murcia	36	39	35	38	31	35	30	36
Málaga	39	44	41	46	35	34	38	37
Córdoba	35	41	28	43	44	41	42	38
Albacete	45	43	45	34	33	38	41	39
Avila	43	45	44	45	43	45	43	40
Zamora	34	34	34	31	32	24	37	41
Cuenca	40	35	46	29	40	36	44	42
Ciudad Real	38	42	42	41	42	40	36	43
Badajoz	44	36	39	44	47	44	45	44
Almería	49	48	49	49	50	49	49	45
Lugo	42	37	43	35	41	31	39	46
Granada	50	50	48	48	46	46	46	47
Jaén	47	47	32	40	49	50	48	48
Cáceres	46	46	47	47	48	47	47	49
Orense	48	49	50	50	45	48	50	50

Source: Banco de Bilbao, 1973.

cereals, and to rough pasture lands. Large amounts of labour were required for small amounts of time, causing widespread seasonal unemployment. The social problems, such as poor housing conditions, low incomes and high illiteracy rates, were among the worst in Spain.

The plans for these areas included a vast programme of public works. In Badajoz province this involved the construction of five main dams and four smaller dams on the River Guadiana and its tributaries, and in Jaén province the construction of six dams on the upper reaches of the River Guadalquivir. Both plans included colonisation schemes for the new irrigation zones. By 1971 the main hydraulic works had been completed, representing a considerable technical achievement. In Badajoz 89 056 hectares of irrigated land and forty-one new INC villages had been created, and in Jaén 20 307 hectares of irrigated land and twenty-two new villages. The plans included reafforestation schemes, industrialisation and electrification projects (Naylon, 1966) and they can be considered as the direct descendants of Pardo's confederations, except in one important respect. For administrative convenience the plans apply only to the provinces of Badajoz and Jaén and not to the river basins of the Guadiana and the Guadalquivir. However, as the two provinces contain the major portions of these river basins, no great difficulties have occurred in practice.

One major criticism can be levied at these two plans. They were aimed at alleviating economic and social backwardness, yet, despite considerable technical achievements and vast expenditure, Badajoz and Jaén are still among the poorest provinces in Spain (Table 19.1). Indeed they have fared little better than other provinces suffering from similar problems but not endowed with large irrigation schemes. The ability of the two plans to promote regional development has depended on two factors: the ability to create additional jobs and to raise *per capita* output. In the early 1950s the huge army of unemployed caused overriding priority to be given to the first factor. This was reflected in article 10 of the colonisation law of 1949 which stated that it was the main aim of the INC to install the largest possible number of colonists on the irrigated lands. As a consequence of this policy the number of colonists on the *regadio* lands rose to over seven times the number of farm workers on *secano* land. However, this was only achieved by giving each colonist a reduced plot of land, so that net production per farm worker was no higher on the *regadio* than on the *secano* (Berger and Wienberg, 1957). Thus by pursuing the first factor, the second, raising *per capita* output, has been sacrificed. As a solution to the problem of rural unemployment the Badajoz and Jaén plans have been of minimal effect. The number of new colonists involved in each scheme remains very small. By 1971 there were only 5 419 colonists in the Badajoz plan, and 2 069 in the Jaén plan. These numbers could have been considerably higher but for the official policy of allowing existing private landowners in irrigated zones to continue holding a large proportion of their lands. Only land 'in excess' has been used for new colonisation. As a consequence INC colonists occupy only 32 per cent of the new irrigated areas (Comisaría del Plan de Desarollo, 1972). Including workers in the new industries, of which there were 6 628 in Badajoz and 9 087 in Jaén, the total employment created by the plans by 1971 was 12 047 in Badajoz and 11 156 in Jaén. These represent very small proportions of the total population of 818 000 in Badajoz and 781 000 in Jaén in 1950. In comparison, mass emigration has made a much greater impact on rural unemployment. Over the 1951–70 period, Badajoz lost 214 794 persons and Jaén 264 751 in this way (Bradshaw, 1972).

The Badajoz and Jaén plans have formed the standard pattern for the many subsequent irrigation and colonisation schemes which now exist in almost every region of Spain (Naylon, 1967). The engineers and agronomists who were given control of these plans have achieved a considerable degree of technical success, but have given little consideration to the costs. In general these plans have been expensive, slow in coming to fruition, limited in effectiveness, and have not provided an efficient means for developing Spain's backward regions, conclusions emphasised in the World Bank report of 1963 (International Bank for Reconstruction and Development, 1963).

New developments in regional planning during the 1950s

In 1946 the government created *Juntas Provinciales de Ordenación* to study the possibilities for growth and development in all provinces. This represented a new departure in policy, being the first regional development programme in Spain that was not based on land improvement. During the 1946–58 period over fifty *planes provinciales* were completed, but these were mainly studies and little of practical significance was achieved. In 1958 the *planes provinciales* were given responsibility for improving the socio-economic infrastructure of the Spanish countryside, and between 1958 and 1970 over 24 800 schemes under this policy were completed. The majority involved the improvement of roads, water and electricity supply, drainage and the telephone service. In the latter part of this period these schemes were concentrated in the main villages and centres of a region, so that at least in the selected villages there would be a level of public services adequate for the 1970s (Comisaría del Plan de Desarrollo, 1972). Economists have recognised the difficulty of assessing the contribution made by this type of scheme to the process of regional development (Hirschman, 1958). Although the *planes provinciales* have brought considerable benefits to rural areas, by themselves they have not been able to stimulate the rural economy and appear unable to prevent large areas of Spain from becoming depopulated. During the period of effectiveness of the *planes provinciales*, the 1960s, over two-thirds of Spain's parishes, the *municipios*, suffered an absolute loss in population.

In 1958 the economic council of the national trade union movement, the *Organización Sindical*, created a *Gabinete Técnico* composed of economists, statisticians and agronomists. This group carried out a series of studies of social and economic conditions in each of the fifty provinces, much of the information coming from the *Sindicatos*' own local experts. These studies were completed by 1963 and represent the first comprehensive survey of the problems of the various regions of Spain and the possibilities for their future development. They were used by the World Bank Mission and later in the preparation of the first national economic and social development plan. Because of their great value the *Sindicatos* were asked to repeat the survey in the early 1970s. In the mid-1960s the *Sindicatos* also began a series of regional studies, but based on natural regions rather than the administrative areas of the provinces. Several of these studies have led to plans, notably for the regions of the Tierra de Campos and the Campo de Gibraltar and for the southeast region of Spain. In sum the *Sindicatos*, although not involved in the organisation of regional planning in Spain, have by their researches and publications played an important role in its development.

Regional and national planning

In 1964 the Spanish government embarked on its first national economic and social development plan. This was the culmination of the economic policies of the new *Opus Dei* 'technocrat' ministers and represented the mechanism by which the Spanish economy was to be managed. Regional planning was introduced as an integral part of the first national plan, as it was thought that regional growth would make a significant contribution to national development. In the first 1964–67 national plan the objective of regional planning was stated as securing the balanced participation of all the geographic regions in the economic and social wellbeing of the country. This was to be achieved by four methods: the creation of industrial growth poles in areas of low income; assuring the better use of resources and promoting local initiatives by financial incentives and joint projects; intensifying the various agrarian reform programmes; and guiding internal and external migrations according to the needs of economic development (Comisaría del Plan de Desarrollo, 1963). However, in practice attention has been concentrated on the first method, the development of industrial growth poles.

The growth pole policy

Under this policy, industry was to be promoted in seven towns located in backward regions (Fig. 19.1). Of these towns five (La Coruña, Vigo, Seville, Zaragoza and Valladolid – the *Polos de Desarrollo*) had an existing industrial base, while two (Burgos and Huelva – the *Polos de Promoción*) had no important industry. Industrial expansion was to be achieved entirely through private enterprise, attracted to these growth poles by economic benefits granted by the state. The benefits have varied over time, changing with the size and importance of the firm attracted, and being generally higher in the *Polos de Promoción* than in the *Polos de Desarrollo*. They have included a 95 per cent reduction in company taxation, a 50 per cent reduction in import duties on essential equipment, preference in obtaining official credit, and a 10 per cent investment grant. This assistance has been granted to new firms, or for extensions to existing factories.

At first considerable success was attributed to these seven growth poles, but in retrospect it appears that their success has been limited. At the end of 1967 a total of 261 firms had received benefits under the scheme and had started production. From selected information given by 207 (80 per cent) of these firms it can be seen that the most successful

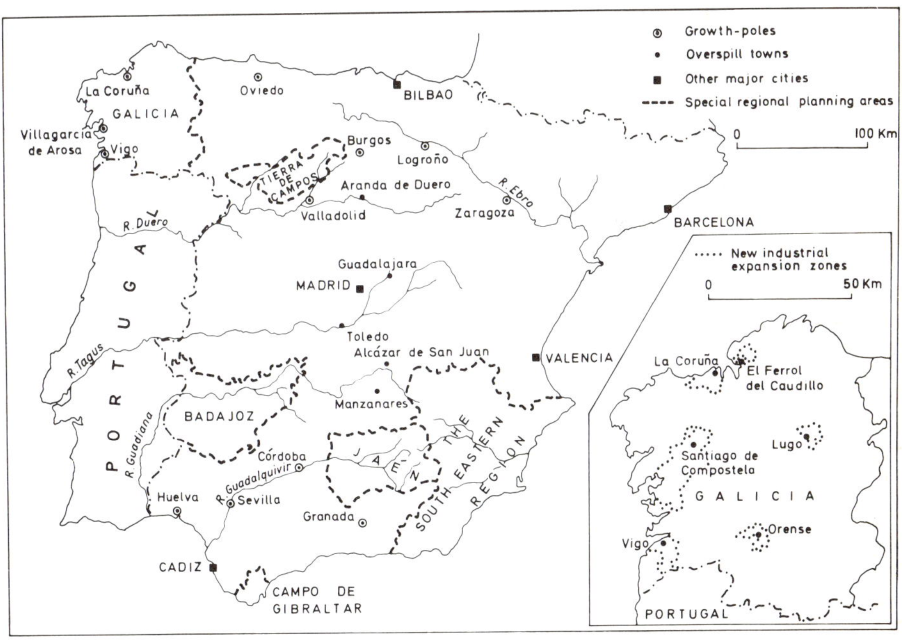

Fig. 19.1 Regional planning projects in the first, second and third national plans, 1964–75. Based on Comisaria del Plan de Desarollo Economico y Social, 1963, 1968, 1972

growth poles, in terms of number of approved schemes, have been Zaragoza, Burgos, Seville and Valladolid, while the generally poorer and smaller towns of Vigo, La Coruña and Huelva have not fared so well (Table 19.2). However, in terms of attracting completely new firms to the growth poles, Burgos possesses an outstanding record, Seville and Valladolid have been reasonably successful, while Huelva's twelve newcomers include big international oil and chemical companies.

The effect of government benefits in promoting this growth has been summarised in Table 19.2. Some firms would have come to the growth poles even if they had received no benefits. Others would have come anyway, but the benefits either allowed their plans to be brought forward or enabled larger projects to be undertaken. However, for certain firms the benefits have been vital and without them they would not have existed. Only in Burgos, and to a lesser extent Huelva, has the growth pole policy played a fundamental role in promoting the industrial growth of the city. The varied success of the policy is due largely to the different characteristics of each of the growth pole cities. Seville, Zaragoza and Valladolid were already important industrial towns able to generate a large number of firms, many of which were subsidiary or auxiliary to existing industries. To date the scheme has not generated new industries which would widen the economic base of these cities, and has had only a marginal influence on their development.

The lack of success of Vigo and La Coruña as growth poles can be ascribed to a variety of factors: their smaller size, the lack of a strong industrial base, the general poverty of the Galicia region, and distance from the rest of Spain. Clearly, additional incentives would be required to attract new firms to these growth points. Only in the cases of Burgos and Huelva has the growth pole policy made a fundamental impact on the city's economy, but since both towns were previously small and non-industrialised, any form of development would have had a similar effect. In Huelva, the nearby Río Tinto mines formed the basis of a large mineral and petrochemical complex, although the amount of new employment created has not been large. Burgos enjoys the locational advantages of being on the main road from Madrid to France, and acts as an overspill town from the Basque industrial region. As a result there has been a large inflow of medium-sized firms which have transformed Burgos into a thriving industrial town.

Several improvements could be made to the growth pole programme. The administration of the benefits is often slow and for many industries, where timing and speed of operation are important, this timelag has been costly. There is evidence that at times the level of benefits has been too low, indeed several firms claimed they had received no real assistance at all. More emphasis could be given to the preparation of factory sites, as a number of firms complained of delays in obtaining essential services such

Table 19.2 Results of the growth pole programme, 1964–67

	Firms within growth pole scheme		*Origins of firms*		*Effect of government benefits*		
Growth pole	*total approved*	*total investigated*	*new to growth pole*	*extensions to existing firms*	*no help*	*some help*	*vital to existence*
La Coruña	15	14	6	8	4	10	0
Vigo	23	20	10	10	6	13	1
Huelva	20	18	12	6	2	11	5
Seville	46	33	17	16	7	25	1
Valladolid	38	30	17	13	5	19	6
Burgos	59	45	37	8	6	18	21
Zaragoza	60	47	13	34	9	35	3
Total	261	207	112	95	39	131	37

Source: Field data, 1968.

Table 19.3 Duration of each growth pole scheme

Growth pole	*Year of inauguration*	*Duration in years*
Burgos	1964	11
Huelva	1964	11
La Coruña	1964	8
Seville	1964	7
Valladolid	1964	7
Vigo	1964	8
Zaragoza	1964	6
Granada	1970	10
Córdoba	1971	10
Oviedo	1971	10
Logroño	1972	10
Villagarcía de Arosa	1972	10

Source: Comisaría del Plan de Desarrollo, 1972.

as water supply, electricity and telephones. Under the second and third national plans of 1968–71 and 1972–75, some of the original growth poles are being phased out and new ones established, there being no more than seven growth poles at any one time (Table 19.3). The limited success of the scheme so far raises the question whether it is necessary to draw other government agencies into the programme. In this context it is surprising that Spain's National Industrial Institute does not participate, for it is government-controlled and has a wide range of industrial activities. It also raises the question whether it is possible to have a successful 'carrot' policy with no 'stick'; that is, in order to draw more industries into the backward regions it may be necessary to restrict their location in the developed regions, a practice adopted in other countries.

The specific regional plans

Although the national plan originally intended that the growth pole scheme should form the main element in its regional development programme, in recent years increasing attention has been given to a series of special regional plans. While the growth poles were regarded as a general panacea for underdevelopment, these regional plans were designed to meet the specific requirements of particular underdeveloped areas. Most of these plans have their origins in the *Sindicatos'* regional studies of the early 1960s; the first was for the Canary Islands, introduced as an integral part of the national plan in 1964. The Canaries have a number of specific problems and some unique advantages which give the islands a pattern of economic development distinct from the rest of Spain. Industrial growth is restricted by the small local market and the distance from the main Spanish domestic market; communications are difficult, both to the Spanish mainland and to the neighbouring islands; there is no adequate water supply. However, the islands enjoy a very benign climate suitable for tourism all the year round and for the cultivation of early crops. Taking account of these factors, the plan for the Canaries is essentially a general investment programme with special attention being given to irrigation schemes, construction of more hotels for the tourist trade, and improvements to the road network, port facilities and the water supply system.

The second of the regional plans applies to the Tierra de Campos region of the northern Meseta. The Tierra de Campos is a predominantly agricultural area which has a harsh climate with cold winters and hot summers, and is also an area of poor soils. The greatest difficulties arise from the *minifundios*, which are the predominant form of landholding in the area. The plan relies heavily on work done earlier for the *Sindicato* regional study. Inaugurated on 23 September 1965, it proposes to transform the entire economic structure of the region with special emphasis being given to agricultural improvements. Notable features are the proposals to irrigate 103 900 hectares of *secano* land, to apply land consolidation to all the *minifundios* of the region, to promote new types of farming enterprise, especially dairying, and to promote agricultural processing industries (Servicio de Acción Regional, 1968). The plan is not just an economic programme: it represents an attempt to break the social stagnation of the region and to give it a new and viable way of life. For this reason investment has been channelled into twenty-eight selected villages and small towns to improve their physical, cultural and urban services so that they attain at least a reasonable standard of living. Evidence so far suggests that these selected nuclei have not yet had any significant effect on their surrounding areas (Comisaría del Plan de Desarrollo, 1968; Pena Sanchez, 1973).

The third of the regional plans to be inaugurated during the first national plan was for the Campo de Gibraltar. Although this is one of the poorest and most depressed areas of Spain, the development plan was introduced not for economic reasons but as a new political factor in the quarrel with Britain over the future of the Rock of Gibraltar (Naylon, 1972). The Spanish government intends to create a prosperous conurbation round the bay of Algeciras so that the people of the Rock will wish to be linked permanently with the Spanish mainland. The Campo de Gibraltar plan, which also springs from an earlier *Sindicato* study, aims to improve the general economy of the area and includes new tourist zones, agricultural reform schemes, but with most emphasis being given to new industrial developments. Although the plan had a chequered history (the largest single employer, the textiles firm Confecciones Gibraltar SA went bankrupt in 1970), for overriding political reasons it will not be allowed to fail, despite the fact that many aims have still to be achieved. At the beginning of 1972, 3 946 jobs had been created, but this total was still far short of the 7 000 workers who had previously been employed on the Rock (Comisaría del Plan de Desarrollo, 1972).

The second national plan of 1968–71 saw a straightforward continuation of the regional planning programme. The third, 1972–75, has a similar policy, but with the addition of two more special regional plans. The first, for the southeast region of Spain, is essentially an irrigation plan based on the diversion of the headwaters of the Tagus to the Segura valley, but also includes the development of communications and urban services throughout the region. The other plan applies to the four provinces of Galicia, and includes an attempt to promote dairying and inshore fishing in various parts of the region, the construction of a motorway link with the rest of Spain and completion of the El Ferrol–Gijón railway. The most important feature is the construction of new industrial estates in two areas along the coastal strip from El Ferrol to Pontevedra, and around the interior towns of Lugo and Orense (Anon., 1973).

The management of territory policy

From a geographer's point of view the most interesting innovation of the third national plan has been the introduction of the policy called *vertebración del territorio* or management of territory. This starts with the assumption that cities are the principal proponents for political, social and economic change within a nation. The major metropolitan areas are considered the leading centres for the diffusion of innovations and the process of economic development is seen as intimately connected with the system of cities and their spatial structure. The policy of *vertebración del territorio* is an attempt to reach an optimum urban system that will maximise each city's contribution to the national income.

The concept of territory management is not entirely new in Spain. In the first national plan five *polígonos de descongestión* or overspill towns were designated around Madrid (Fig. 19.1), although except in the cases of Guadalajara and Aranda de Duero no real progress occurred. Under the new policy the first task has been to select and establish the hierarchical system of cities within Spain, together with their hinterlands, by such methods as the use of gravity models (Comisaría del Plan de Desarrollo, 1972). The objectives of *vertebración del territorio* include maximising national income, equalising standards of living throughout the country, and giving special assistance to the most dynamic metropolitan regions. Applying these objectives to Spanish cities, the planners have established a list of works including investment in transport, housing and urban developments for each major centre. This new policy is still in a stage of exploration rather than application, and it makes a large number of assumptions, especially about the link between urban systems and economic growth. Nevertheless, it represents a worthy attempt to attack the basic forces hampering economic growth in Spain.

Conclusions

In Spain regional planning has progressed from being a neglected feature of government policy to playing a prominent role in the national plans. At the same time the techniques of regional planning have been considerably refined, although they are still capable of further improvement. There has been a tendency towards introducing new plans rather than pursuing the execution of existing programmes. For example, if more attention has been given to the growth poles of La Coruña, Vigo and Villagarcía de Arosa, then the regional plan for Galicia would not have been so urgently required. To date there has been no attempt to assess the varying merits of each of the different types of regional plan currently in operation. Each plan is assessed on its own merits, rather than as one alternative among several possible plans. For example, the hydraulic plan has become an end in itself, rather than an alternative method for improving regional income. More harmful has been the power of vested interests to promote certain types of regional plan. The likelihood that large areas will be converted into irrigated land depends not so much on the profitability of such schemes – as Naylon (1973) has pointed out, these schemes are financially disastrous – but on the considerable ability of vested bodies such as the INC and the Ministry of Public Works to command a large share of the national budget.

Ultimately the justification for any regional development programme depends on its ability to raise *per capita* incomes. However, many of the provinces which have had development plans (Badajoz, Jaén, La Coruña, Pontevedra, Burgos, Valladolid, Zaragoza, Seville, Huelva) were relatively no better off in 1971 than they were in 1955, and in some cases they were poorer (Table 19.1). The need for effective regional planning, with the ability to raise incomes in the poorer regions to the national average, is as strong as it has ever been.

References

ANON (1973) *Desarrollo*, 28 October.

BANCO DE ARAGÓN (1960) *Aragón*, Zaragoza.

BANCO DE BILBAO (1973) *Renta nacional de España y su distribución provincial 1971*, Bilbao.

BERGER, J. and WIENBERG, D. (1957) 'Estudio económico de la evolución de la produción agrícola de las Vegas Bajas del Guadiana de 1953 a 1956', *Revta. Estud. Agro-soc.*, 21, 21–38.

BRADSHAW, R. P. (1972) 'Internal migration in Spain', *Iber. Stud.*, 1, 68–75.

COMISARÍA DEL PLAN DE DESARROLLO ECONÓMICO Y SOCIAL (1963) *Plan de desarrollo económico y social 1964–67*, Madrid.

COMISARÍA DEL PLAN DE DESARROLLO ECONÓMICO Y SOCIAL (1968) *II Plan de desarrollo económico y social 1968–1971*, Madrid.

COMISARÍA DEL PLAN DE DESARROLLO ECONÓMICO Y SOCIAL (1972) *III Plan de desarrollo económico y social 1972–1975*, Madrid.

GÓMEZ AYAU, E. (1961) *El estado y las grandes zonas regables*, Madrid.

HERR, R. (1958) *The Eighteenth-Century Revolution in Spain*, Princeton University Press.

HIRSCHMAN, A. O. (1958) *The Strategy of Economic Development*, Yale University Press.

INTERNATIONAL BANK FOR RECONSTRUCTION AND DEVELOPMENT (1963) *The Economic Development of Spain*, Johns Hopkins Press.

LÓPEZ DE SEBASTIÁN, J. (1970) *Política agraria en España 1920–1970*, Madrid.

MALEFAKIS, E. E. (1970) *Agrarian Reform and Peasant Revolution in Spain*, Yale University Press.

NADAL, J. (1966) *La población española*, Barcelona.

NAYLON, J. (1966) 'The Badajoz plan', *Erdkunde*, 20, 44–60.

NAYLON, J. (1967) 'Irrigation and internal colonization in Spain', *Geogr. J.*, 133, 178–91.

NAYLON, J. (1972) 'A challenge to the Rock – the Campo de Gibraltar development plan', *Geography*, 57, 1–9.

NAYLON, J. (1973) 'An appraisal of Spanish irrigation and land-settlement policies since 1939', *Iber. Stud.*, 2, 12–18.

PALACIO ATARD, V. (1964) *Los españoles de la ilustración*, Madrid.

PENA SÁNCHEZ, M. (1973) 'El Plan Tierra de Campos y su realización', *Estudios. Geogr.*, 33, 170–85.

RAMOS, OLIVEIRA, A. (1946) *Politics, Economics and Men of Modern Spain 1808–1946*, trans. T. Hall, Gollancz.

SERVICIO DE ACCIÓN REGIONAL (1968) *Tierra de Campos*, Madrid.

TOWNSEND, J. (1791) *A Journey through Spain in the years 1786 and 1787, with particular attention to the agriculture, manufactures, commerce, population, taxes and revenue of that country*, 3 vols, London.

VELARDE FUERTES, J. (1968) *Política económica de la dictadura*, Madrid.

J. Naylon

20 A new era for Spanish mining

The mineral wealth of the Iberian Peninsula is proverbial and has been known since earliest times, with Cretan, Phoenician, Greek and Carthaginian mining settlements along the south and east coasts being followed by six centuries of Roman activity. The Almadén mercury mines – the Argentum Vivum of the Romans – are still the richest in the world, although they are known to have been worked since the fourth century BC. Pliny, in his *Natural History*, mentions the plentiful iron deposits of Vizcaya. Nineteenth-century industrial expansion in western Europe created an almost unlimited market for Spanish minerals, and for a time Spain became the world's most important producer of low-cost lead from the Sierra Morena and the mountains of Cartagena, copper from Huelva, zinc from Cartagena, Asturias and Santander, tin from Galicia and, above all, iron ore from the Basque country. By 1877 Spain was the leading European producer of these minerals, holding this position until the outbreak of the First World War. Between 1871 and 1914, 80 per cent of the iron ore imported into Britain was Spanish. Even after the trade began to decline to but a shadow of its former self, between the years 1920 and 1970 Bilbao alone shipped out over 150 million tons of *rubios*.

The famous iron ores of Vizcaya are now nearly exhausted; in 1972 the eighteen workings still active produced only 1 127 489 tons, a drop of 16 per cent compared with the previous year. Whereas in 1900 the port of Bilbao sent 6 million tons of iron ore to Britain alone, now the mineral has almost ceased to figure in exports. On the contrary, iron ore is now imported to feed the rapidly growing Basque iron and steel industry. Unlike the beginning of the century, when Spain sold her iron ore to Britain and bought steel ships in return, Spain now imports foreign ore and sells ships to Britain, which she ousted in 1972 from third place in the world shipbuilding league. This is just one measure of Spain's economic transformation since 1959, from which time her economy has been the fastest-growing in western Europe. However, while such changes are gratifying evidence of progress, Spain illustrates the classic trade problem of developing economies, the development programme calling for increased imports of fuel and raw materials, and causing serious pressures on the balance of payments. In Spain this phase coincides with a period of scarcity and high costs of raw materials and instability in energy supplies, both of which threaten the development programme.

Minerals in the Spanish trade balance

The industrialised economies of the West have reached such rates of growth that their demands are threatening, not to exhaust, but to make scarce some basic resources. Consequently the availability, quality and price of metals for industry are subject to considerable fluctuations on world markets. During 1973, for instance, copper and tin prices reached new peaks and the price of zinc rose from £160.50 per ton in January to £742.50 per ton in November. Although Spain has great and varied mineral resources, home output has not increased enough to supply an industrial sector whose rate of growth in 1972 was the highest in the world including Japan. In a short space of time, therefore, Spain has changed from a mineral-exporting country to a mineral-importing country (Tables 20.1 and 20.2).

Whereas in 1964, at the beginning of the first national economic and social development plan, Spain's domestic output more than covered

Table 20.1 Spain's balance of trade in metallic and non-metallic minerals (£ million)

Year	*Imports*	*Exports*	*Deficit*
1970	56.7	28.8	27.9
1971	64.2	25.4	34.6
1972*	95.3	23.9	71.6

* 11 months only.
Source: Ministerio de Industria.

Table 20.2 Spain's balance of trade in metallic minerals (£ million)

Year	*Imports*	*Exports*	*Deficit*
1970	37.8	19.5	18.3
1971	45.0	14.4	30.6
1972	67.9	11.4	56.5

Source: Ministerio de Industria.

(122.1 per cent) her demands for metallic minerals, by 1972 she was covering only 16.8 per cent (Comisaría del Plan de Desarrollo, 1972). In non-metallic minerals the proportion fell from 59.6 per cent to 44.8 per cent. In 1958, Spain had a £30 million surplus on her exports of minerals and metals (not including liquid and solid fuels or iron and steel products). By 1965 this surplus had disappeared and by 1970 it had changed into a deficit of £75 million. Unless supply sources change, by 1975 the deficit on the mineral account will be of the order of £149 to £224 million. If fuel imports are added, the bill assumes alarming proportions: in 1972 Spain had a trading deficit of £61.3 million in metallic minerals, £16.2 million in non-metallic minerals, £48.2 million in imported metals and £365.5 million in imported solid, liquid and gas fuels. This total deficit of £491.2 million spent on minerals, metals (excluding steel) and fuels is an obvious handicap for a country in Spain's economic position.

The national mining plan of 1971

Given the circumstances of the world mineral and fuel markets, developing countries such as Spain must reduce their dependence on outside supplies as much as possible. Spain does not have to base her industrial growth solely on the massive import of basic raw materials as she can turn to her own mineral resources and re-evaluate them. Evidence suggests, however, that to do this she will have to permit foreign participation and spend more of her own money than has previously been the case.

Among the reasons for the decline of Spanish mining after the First World War were the discovery of larger and richer resources in Africa, Australasia and the Americas, strong export competition in iron, lead and pyrites, the introduction of new technical processes, the slowing down of developments in zinc, mercury and potash mining by monopoly interests, the sporadic and wasteful nature of much mining activity, and the working-out of the best deposits by foreign interests who obtained a stranglehold on Spanish mining after the passing of the *laissez-faire* 1869 mines law. Largely as a reaction to this situation, the nationalist government imposed restrictions on foreign capital investment after the Civil War of 1936–39, including the repatriation of capital and earnings. During the years of economic autarchy from 1939 to 1959 Spanish-owned mines made a living, but always with a good deal of protection and with inadequate investment. The liberalisation of the Spanish economy following the 1959 stabilisation plan caused a rundown of many mines – and mining settlements – which has continued to the present day. The decline has been most marked in metallic ores, especially iron and lead, with a decrease from 822 mines and over 50 000 miners in

Fig. 20.1 Mining activities under the Spanish National Mining Plan, 1971. Based on Instituto Geológico y Minero de España, 1961

1969 to 186 mines and 18 300 miners in 1970, and to only 142 mines and fewer than 16 000 miners in 1972.

Some half-hearted attempts at introducing fresh finance and technology from abroad were made after 1959 when the Spanish government lifted the ban on foreign investment and repatriation of earnings. Later foreigners were given the opportunity to control 50 per cent or more of a mining venture. These measures however constituted no more than the concessions which Spain was obliged to make in return for development finance and advice. Only since the government became aware of the 'crisis in minerals supply' and published the national mining plan in 1971 have tax rebates and official credits been announced on a scale which may make an appreciable impact. Besides setting aside almost £250 million of public money for new investment in mining, the plan aims to obtain rapid results by attracting international finance and technology. The administrative hurdles to obtaining large-scale prospecting concessions have been surmounted so that new and rapid survey techniques can be employed over extensive areas. The policy of 'first come, first served' in granting concessions and licences has given place to a more careful consideration of the applicants' financial solvency and technical equipment; this is one way in which the plan tries to reduce the speculation and concession-hunting which traditionally have characterised Spanish mining.

The return of foreign investment

These liberal and positive measures have injected new hope and activity into Spain's mining regions. In the first year and a half of the national mining plan private interests put £2.99 million into prospecting alone and a further £58.7 million into work already under way. Long-established foreign companies, which have survived the Civil War and subsequent monetary restrictions and restraints on investment, are expanding their operations rapidly in the new climate. The British-controlled Tharsis Sulphur and Copper Company has modernised its equipment and begun a long-term pyrites mining programme in Huelva. Peñarroya España is expanding its lead and zinc production and has inaugurated a 100 000 tons-per-year lead-silver blast furnace, the second largest of its kind in Europe. The Royale Compagnie Asturienne de Mines is also expanding lead and zinc output, and the Compañía Andaluza de Minas, another firm with French participation, has doubled its capital in order to increase iron ore output and modernise its ore-loading installations at Almería.

A number of newcomers from abroad have been induced to take a closer look at Spain's subsoil. International companies, concerned about dwindling resources and nationalistic attitudes elsewhere, are relieved to find the Spanish government welcoming them. The Canadian Cominco is sinking shafts into promising silver-lead deposits in northwest Spain. Chanter-Engelhard are prospecting in the province of Córdoba. Exploración Minera Internacional de España, jointly operated by United States and Canadian interests, has increased its capital to finance exploration work. Progemsa, with 40 per cent participation by the Bethlehem Steel Corporation, obtained prospecting concessions in the Ebro basin in 1971. Asarco (the American Smelting and Refining Company) has begun a prospecting programme in Murcia. On the Costa del Sol near Málaga the Royale Austrienne, the Compañía de Exploracione, Minero-Industriales, the Canadian companies Noranda and Placer Management, and Spain's National Industrial Institute have pooled their resources to search for nickel and industrial diamonds. United States' capital accounts for 99.8 per cent of the Compañía Fomentadora de Exploraciones Mineras, SA, while Panamanian interests hold 80 per cent of the Compañía Minera Metalmine Española, SA, prospecting and working mines and quarries at La Carolina (Jaén). In 1973 Laporte Industries Limited, of the United Kingdom, acquired 40 per cent of the capital of Minas de Gador, the most important Spanish company specialising in the mining and treatment of clay minerals.

The new mining and prospecting pattern

The national mining plan has investigated 106 minerals found in Spain and has drawn up a strategy for the selective development of some of them, according to reserves, economic working conditions, and domestic and overseas demands. These studies suggest that Spain comes first in western Europe in her proved reserves of eleven out of fourteen minerals essential to modern industry. The comprehensive national prospecting campaign, together with new concentrating and refining techniques, is increasing known reserves and allowing the utilisation of others which once would have been disregarded as too poor. An encouraging aspect of the scene is that much of the new mining and processing activity is taking place in Andalusia, Spain's most populous but poorest region.

Spain's huge pyrites and sulphur complexes, probably the largest in the world, have had their future transformed by new extractive techniques which permit integrated exploitation. Iron and copper pyrites mining, which is particularly associated with the Río Tinto area of Huelva, has been going through a long period of crisis. The essential problem is what to do with these large resources. Worldwide measures against pollution have included the elimination of sulphur from petroleum and natural gas; the consequent availability of large amounts of residual sulphur at low cost has caused a drastic fall in world sulphur prices which has particularly affected Spanish exports of crude pyrites. Competition from the Lacq natural gas field in France has been especially keen. In 1972 crude pyrites prices were only one-third of those in 1970. At the same time, and despite the nationalisation of the British-owned Río Tinto Company in 1955 and its substitution by the Compañía Española de Minas de Río Tinto, production methods have been traditional, relying on cheap labour.

The serious condition of the industry, particularly worrying because of its location in Spain's poorest region, induced the Ministries of Industry and Labour to carry out in 1964 a joint survey of the ways in which pyrites mining at Río Tinto could be restructured. Proposals included technical improvements in the extraction of sulphur and iron, wider prospecting, the deepening of the port of Huelva, and better marketing. An investment of over £44 million was thought necessary for modernising the exploitation of the pyrites and the other metals found with them, the only alternative being to close the mines completely. This amount of capital was not available from private sources and state assistance did not appear to be forthcoming until the minerals scarcity suddenly forced a re-evaluation. In October 1972 the Ministry of Industry and the Council of Ministers announced a general reorganisation programme for the pyrites industry. This involves the contraction of output to a level sufficient to meet local demand and export commitments to 1976; the expansion of efficient opencast mining; the reduction of natural sulphur imports to the minimum needed by agriculture; government aid for new sulphuric and phosphoric acid plants, which consume pyrites and are the bases of fertiliser production; and the building of a pyrites cinders upgrading plant in the growth pole at Huelva. Steel manufacturers have asked for the urgent installation of this plant to supply the fast-growing Spanish steel industry with at least 1 million tons of pelletised iron ore per year. The debit side of the programme, for an area of high unemployment, involves the phasing out of 1 641 jobs; 1 198 men are being prematurely pensioned off at the age of fifty-five, and the remainder can opt for cash severance pay or retraining for new jobs in the same industry.

While the government is trying to revive the Río Tinto mines, the combination of the minerals famine and of fresh extractive techniques has led to the opening-up of new pyrites deposits in the provinces of Huelva and Seville. The Auxini company, belonging to the National Industrial Institute, has discovered a deposit of 22 to 34 million tons in the abandoned Sotiel mine in Huelva, and is building there a concentration and roasting plant with an annual capacity of at least 1 million tons. The most encouraging event for the prospecting programme has been the discovery, by technicians of the German Metallgesellschaft, of a deposit of 50 million tons in the Apirsa mines at Aznalcóllar, 48 km from

Seville. The Aznalcóllar mines were operated in the first part of this century by the British-controlled Seville Sulphur and Copper Company, which sold out in 1942. The successor company Apirsa (Andaluza de Piritas, SA) was set up in 1960 but ceased mining in 1968 when part of the underground workings collapsed and were flooded. Since then the Banco Central (Spain's third largest private commercial bank), which holds majority control in Apirsa, and Metallgesellschaft have invested a modest £150 000 in prospecting and development, an outlay leading to the discovery of a reserve provisionally calculated at £579 million.

Test runs in a small pilot plant suggest that the Aznalcóllar pyrites have a high lead-zinc content. The 50 million tons of pyrites probably contain about 2.4 million tons of zinc, which is twice the total known zinc reserves in Spain, and a lead content, expressed in metal, of 1 million tons – about equal to Spain's total lead reserves. Besides some 23.5 million tons of sulphur and 21 million tons of iron (obtained by leaching and roasting the pyrites) there is some 250 000 tons of copper, 2 500 tons of silver and 50 tons of gold. In order to exploit the discovery the Banco Central has sold 40 per cent of Apirsa to Metallgesellschaft, which will provide the technical facilities. The initial investment has been variously quoted at between £23 and £43 million, for the provision of mining equipment, a milling and flotation plant for ore concentration, a sulphuric acid plant and a pelletising plant for reducing pyrites residues into material suitable for local steel manufacturing. The construction of a lead smelter at Aznalcóllar has also been suggested. Target annual production by 1975, from opencast mines, is 1.5 million tons, yielding 25 000 tons of copper concentrates, 115 to 125 000 tons of zinc, 45 000 tons of lead and 1.25 million tons of iron-sulphur pyrites.

The changed prospects of the pyrites industry illustrate the hopes which new techniques and demand conditions hold out for Spanish mining. According to the Ministry of Industry, the Aznalcóllar workings alone will save some £62.2 million which would have had to be spent on metal imports. From the known 400 million tons of pyrites modern techniques can extract 240 million tons of iron ore (equal to Spain's known iron ore reserves in 1971), 12 million tons of zinc, 6 million tons of lead, 2.5 million tons of copper and appreciable quantities of precious metals. The rational exploitation of Spain's pyrites alone could guarantee lead and zinc supplies for a long time. Furthermore, with the likelihood that the Huelva and Seville pyrites reserves may amount to 600 million tons, and with adequate finance to modernise and enlarge installations, an increased sulphur supply can be envisaged. This domestic source, coupled with phosphates from Bu-Craa in the Spanish Sahara (the world's largest reserves) and the new sulphuric and phosphoric acid plants under construction, may establish Huelva as a major fertiliser supplier to the North African countries.

Copper mining is closely associated with pyrites in its location and in view of the international supply situation, market demand is high. The price of cash copper wire bars on the London Metal Exchange rose from £426 per ton in late 1972 to £1 081 per ton in November 1973. Copper is becoming scarcer and stocks are dwindling as a result of setbacks in Chilean production, of difficulties in Zambia through the closure of the Rhodesian railways route, of a great increase in demand, and of uncertainty about 'paper' exchange rates for currencies causing consumers and speculators to invest in raw materials. Spain's copper needs, which are expected to double during the 1970s, are largely being met from indigenous sources in Andalusia and Galicia.

Since 1966 the Compañía Española de Minas de Río Tinto has been working the 30 million tons of low-grade copper ores at Cerro Colorado in Huelva in collaboration with Rio Tinto-Zinc and the Patiño Mining Company of Canada. Of the joint venture, known as Río Tinto Patiño, SA, 45 per cent is owned by the Spanish company, 50 per cent by Patiño and 5 per cent by Rio Tinto-Zinc. The annual production target at Cerro Colorado is 20 000 tons of copper together with 30 tons of silver and 3 tons of gold, obtained from the oxidised Gossam surface ores. The copper ore concentrating plant at Cerro Colorado has a daily capacity of 10 000 tons and sends its output to the company's new copper smelter and electrolytic refinery at Huelva. In 1970 Río Tinto Patiño discovered a deposit of 11 million tons higher grade ore at Arinteiro in Santiago province and has invested £6 million to establish mines. This Galician field is expected to be in production by 1975 and to yield 7 000 tons of copper per year; after concentration on the site, the concentrates will be shipped to the smelter at Huelva. In 1968 a further source of copper was found by the Instituto Geológico y Minero de España at Cala in north-west Huelva, as part of a deposit of cupriferous magnetite, chiefly interesting as an iron ore. Production of copper concentrates from Cala climbed from 1 279 tons in 1970 to 6 000 tons in 1972. In 1973 the Cala magnetite body was extended by the discovery of a further 50 to 60 million tons, allowing expectations of annual copper concentrates production to rise to some 18 000 tons, with a metal content of 20 per cent. These will be treated partially in an integrated flotation plant at the site. Altogether, including imports, the Río Tinto Patiño refinery at Huelva is expected eventually to produce 75 000 tons of electrolytic copper per year.

The iron ore supply situation is rather more critical. In 1973 the Spanish iron and steel industry consumed 12.5 million tons, of which only 53 per cent came from home sources. By 1980 requirements probably will reach 18 to 20 million tons. The problem of increased home production of iron ore is one of quality and not quantity. Spain has reserves of over 2 000 million tons (equalling 676 million tons of metal), but most Spanish ore consists of 'fines' which need sintering or concentrating. Nevertheless, prospecting and assaying work by the Instituto Geológico y Minero and by private interests doubled Spain's assured iron ore reserves in the two years 1969 to 1970. In 1972 the Instituto began large-scale prospecting campaigns as part of the national mining research programme, which is included in the third economic and social development plan 1972–75. One of these campaigns, the so-called 'southwest iron' programme, has revealed particularly promising prospects for iron ore production in the provinces of Badajoz, Huelva and Seville. Between 1966 and 1972 the iron ore reserves of the southwest were increased by 74 million tons, and production of concentrates rose from 130 000 tons in 1966, with an average iron content of 50 per cent, to over 500 000 tons in 1972, with a 59 per cent metal content. The enlargement of the cupriferous magnetite deposit at Cala is expected to yield annually 700 000 tons of sinterised 'fines' in the near future, with an iron content of 60 per cent.

Elsewhere in Andalusia the Instituto Geológico y Minero, drilling in the Alquife-Marquesado district of Granada, has encountered a 27.5 m bed of iron ore, with a 58 per cent iron content, near to the existing mines. The United States Steel Corporation, with other Pittsburg-based partners, has set up the Compañía Minera de Sierra Nevada, SA, to prospect for and mine iron ore in the same zone. Drilling also is in progress not far away in Murcia, where an area of 452.5 km^2 near Cehegín has been surveyed for magnetites, which are present in sufficient quantity to justify mining. A prime factor behind these southeastern operations is the expansion of the steel works at Sagunto (Valencia) and the construction of the new Ford car factory nearby at Almusafes. In other parts of Spain, prospecting in the Sierra Menera (Teruel) has raised iron ore reserves from 20 to 110 million tons, with a 51 per cent iron content. The Instituto Geológico y Minero is active in Vizcaya and Santander, while the National Industrial Institute company Adaro, working under contract to the Instituto in Lugo, León and parts of Oviedo, has carried out an aerial magnetometry survey of 8 500 km^2 and is now drilling in the most promising places.

The national mining plan recognised the urgency of building more concentration plants to up-grade low quality Spanish iron ores to acceptable international standards. The commercialisation of the ores of northwest Spain, for example, is made difficult by their high phosphorus content. The government is thus helping to finance the construction of a sintering plant for the carbonate ores of Vizcaya, and pelletising plants for the ores of the northwest and of Sagunto. Only lack of capital should prevent Spain from increasing her iron ore production to meet at least national needs; output has grown to nearly 7 million tons in 1973. In the meantime, necessity is spurring activity abroad: in 1973 the National

Industrial Institute reached an agreement with the Companhia Minera Vale do Rio Doce to work iron ore in central Brazil, a source from which Spain hopes soon to obtain 2 million tons of ore per year to help cover the needs of the Institute's Ensidesa integrated iron and steel works at Avilés in Asturias.

The national mining plan is also concerned with the restructuring and expansion of Spain's other mineral sectors, in order to meet industrial demands, save currency and cope with the unparalleled upsurge in metal prices. The prices of tin, lead and zinc, especially, which are vital raw materials for many industries, have spiralled on the London Metal Exchange. Lead prices almost doubled between August 1972 and November 1973. In the latter month the exchange was forced to take the unprecedented step of restricting trading in zinc. There are now virtually no spare supplies of zinc for sale as the Communist bloc countries, particularly North Korea, which are normally the main suppliers, want zinc for their own industrial expansion. Another factor in the shortage has been the closure of smelters in the United States because of the age of plants, the low zinc prices of the 1960s, the freeze on American domestic producers' metal prices, and antipollution legislation. As regards tin, the price on the London Metal Exchange rose from £1 565 to £2 555 per ton over the year November 1972 to November 1973, due to strong world demand coinciding with a series of production setbacks such as smelter closures.

Production of each of these metals has been declining in Spain recently, but scarcities and systematic exploration by the Instituto Geológico y Minero have duplicated and even triplicated known and economically-viable reserves. Lead and zinc output should increase significantly with the discovery of new deposits and the re-evaluation of old fields at Rubiales (Galicia), Oso (Asturias), Monte de los Azules (near Cartagena, Murcia), Agueda (Soria) and Piedrafita de Cebrero. The last deposit alone should have an annual output of 40 000 tons of zinc and 14 000 tons of lead. In southern Murcia, near La Unión, the Instituto has expanded the known lead reserves by 31 per cent, equal to some 95 000 tons of metal content, while near Mazarrón, in the same province, zinc, iron and magnesium complexes have been discovered which, though difficult to treat metallurgically, are mined easily and contain some 165 000 tons of zinc metal. Spain now appears to have assured lead reserves for another twenty years and enough zinc for twenty-five years. The restructuring of the country's tin mines at Silleda, Beariz and Verín (Orense), Vitigudino (Salamanca) and in Extremadura is putting this sector on a firmer footing. Prospecting and research at a cost of £1.8 million is expected to lead to an output of 700 tons a year from these districts, not to mention production from other provinces.

Mineral deposits once regarded as too poor for consideration are beginning to arouse interest. An example is nickel, in which Spain has a complete trade deficit but possesses abundant deposits in Málaga of 0.16 per cent metal content, a level below the 0.20 per cent at present considered to be the viable minimum. Several mining companies are investigating the wolfram mines of Orense and Salamanca, the output of which has declined steadily since they supplied the belligerents in the Second World War. Inactive concessions are now being reworked and a pilot washing plant to handle 34 tons a day has been built in Extremadura. While uneconomic mercury mines, such as the Astur-Belga and Soterrana, have been partially closed, and those at Torno shut down completely, the modernisation of the Almadén mines, begun in 1972, has kept up national production, and in 1973 Russian technicians were called in to advise on the fuller extraction of mercury from the spoil at Almadén. In 1972 the prospecting teams of the Instituto Geológico y Minero discovered deposits of fluorite at El Chaparral in Córdoba. Yielding fluorspar, used as a flux in the steel industry, these reserves of 600 million tons could supply Spain's needs for many years. In 1973 the Instituto found alunite deposits at Riaza (Segovia), yielding aluminium of indeterminate extent but of high quality and a pilot plant is being built at Vitoria for their technical and commercial analysis. Other sectors in which the Instituto is fostering research and modernising production include bauxite at Fuentespalda (Teruel) and Llacuna (Tarragona), manganese, titanium, diamonds at Carratraca (Málaga), silver at Guadalajara and Alcudia, kaolin in Guadalajara, the clay mineral bentonite in Almería and potash in Catalonia. The Instituto's prospecting activities are paying dividends in the discovery of other resources. In 1973, for instance, the Underground Waters Division encountered an aquifer in the Júcar basin with a potential yield of over 1 000 million m^3 per year – a volume equal to that to be supplied by the great Tagus-Segura water transfer scheme and enough to supply the foreseeable future demands of the irrigation and tourist zone south of Valencia.

By 1980 the national mining plan hopes to have transformed Spain's foreign trade deficit in ores and metals, amounting to £80 million in 1970, into an export surplus of £30 million, in addition to a saving of £200 million in imports. By the same date, Spain may have recovered her historic position as the leading ore producer in western Europe.

References

ANON (1971) 'El Instituto Geológico y Minero de España', *España Hoy*, 13, 61–4.

ANON (1972–73) *Desarrollo*, 28 May, 24 December 1972; 15 May, 3 June, 2 September, 21 October 1973.

ANON (1973) *Información Española*, 11 August, 3 December.

COMISARÍA DEL PLAN DE DESARROLLO ECONÓMICO Y SOCIAL (1972) *III Plan de desarrollo económico y social 1972–1975*, Madrid.

INSTITUTO GEOLOGICO Y MINERO DE ESPAÑA (1961) *Mapa minera de España, excluidos hierro y carbón; Mapa de los criaderos de hierro de España*, Madrid.

INSTITUTO GEOLÓGICO Y MINERO DE ESPAÑA (1973) *Memoria 1972*, Madrid.

PRADO CALZADO, J. (1973) *La Minería en España (Temas Españoles Núm. 529)*, Madrid.

G. R. P. Lawrence

21 Pumped storage hydroelectricity

The generation of electricity by water turbines is not new; neither is the construction of an artificial reservoir to maintain a regular water supply for such generators. Many of the world's major rivers have been harnessed for this purpose but despite regulation of their flow there is always an ultimate dependence on a natural climatic variable, the precipitation received by the river basin. The uniqueness of pumped storage generation is that the water drawn from the supply reservoir is replaced, by pumping, from a lower level basin containing water that has already been used for power production. It is, in effect, a water recycling process. Electricity for the pumps is supplied from other sources, so these schemes cannot be used for the continuous supply of electricity, that is, for 'base load' needs, but can only be used intermittently, with 'off peak' electricity providing the power for pumping.

Pumped storage generation of electricity is particularly appropriate to the power requirements of industrialised countries, and many of these do not necessarily possess natural features suited to the establishment of conventional hydroelectric plant. In only a few cases (Alpine Europe) are major consuming areas of electricity adjacent to mountainous areas with high rainfall and a reliable runoff, provided by a single river or drainage basin, together with the possibility of a high head of water dropping to a satisfactory site for a power station. Pumped storage generation does not need a natural supply of water leading into its upper reservoir, and therefore this can be an entirely artificial construction. The principle was first applied at Schaffhausen, in Switzerland, over sixty years ago, but since then pumped storage installations have been commissioned in a number of other countries. The basic pattern of a pumped storage scheme comprises an upper reservoir, a power house and a lower reservoir, connected by tunnels, pipelines and shafts. A considerable diversity of layouts reflects the great variety of areas where such schemes have been developed. It is possible to distinguish between 'low head' and 'high head' schemes, between those where a large volume of water falling through a short height drives the turbines and those where a similar output can be effected by a smaller amount of water falling through a distance well in excess of 300 m. The relationship between water quantity (W, in cusecs) and head (H, in metres), ignoring losses, may be expressed as follows:

$$\text{kilowatts} = 0.276\ (WH).$$

The world distribution of schemes shows an overwhelming concentration in the industrialised countries of North America, Japan, western Europe and Australia (Table 21.1). Projects under consideration in other countries (Finland, Hungary) will not alter this distribution pattern. Plants have also been established in three South American countries and

Table 21.1 Total installed pumped storage generating capacity in the mid-1970s

Country	*Total capacity* (MW)	*Date of first scheme*
Chile	4	1923
Colombia	6.5	1947
Yugoslavia	33	1973
Portugal	36	1960
Sweden	48	1933
Norway	85	1955
Brazil	128.1	1939
Bulgaria	137.5*	1947
Canada	198	1957
Irish Republic	280	1973
Poland	424	1968
Belgium	432	1971
Rumania	500	†
East Germany	611	1932
Switzerland	713.8	1909
United Kingdom	1 060	1963
Luxembourg	1 096	1964
Austria	1 121.2	1928
France	1 181	1938
Australia	1 500	1974
Spain	1 902.1	1965
Italy	2 999	1940
West Germany	3 415.5	1930
Japan	7 049.8	1931
USA	10 043.5	1956

* MVA.
† No data.
Source: Anon, 1970.

in fact some of these were undertaken at an early stage in the development of this form of power production. They are all of low capacity, but Brazil has projected schemes of a substantial size for several locations, although as yet there is no date for their implementation.

The largest schemes are those currently under construction or very recently commissioned, including plants with more than 1 200 megawatts (MW) installed generating capacity at Castaic in the USA, Shin Takase in Japan and Tumut 3 in Australia (Table 21.2). Some interesting variations can be found between these large stations, and the smaller schemes of western Europe also show special features.

Table 21.2 Pumped storage schemes of the 1970s

Location	*Generating head* (m)	*Installed capacity* (MW)	*Date*
Australia			
Tumut 3	151	1 500	1974
Austria			
Zemm	672	230	1971
Belgium			
Coo	269	432	1971
Bulgaria			
Belmeken	700	83*	1972
Antonivanovci	122	50*	1973
France			
Vouglans	92	53	1973
Emosson	750	128	1972
Revin	245	660	1974–75
West Germany			
Hornberg	625	970	1976
Waldeck II	†	440	1975
Irish Republic			
Turlough Hill	282	280	1973
Italy			
Brassimone-Suviana	378	338	1973
Lago Delio	732	1 016	1971–72
St Fiorano II	1 403	280	1972
Fadalto	107	240	1971–72
Chiotras-Piastra	990	540	1974
Rovina-Piastra	540	105	1974

Table 21.2 – *continued*

Location	*Generating head* (m)	*Installed capacity* (MW)	*Date*
Japan			
Numappara	478	732	1973
Shin Takase	230	1 280	1975
Masegawa I	105	296	1972
Kisenyama	220	480	1970
Shin Toyone	230	1 150	1972
Atashika I	382	496	1978
Luxembourg			
Vianden 10	278	196	1971
Spain			
Guillena	230	216	1974
Switzerland			
Ova Spin	200	50	†
Hongrin	878	240	1976
Sarganserland	483	262	†
United Kingdom			
Foyers	173	300	1974
United States of America			
Kinzua	250	396	1970
Grand Coulee I	111	97	1971
Mormon Flat	44	50	1971
Northfield Mt	250	1 000	1972
Horse Mesa	79	100	1972
Ludington	110	1 872	1973
Blenheim-Gilboa	348	1 000	1973
Carters	132	250	1974
Jocassee	95	610	1974
Elbert I	135	100	1975
Kaysinger Bluff	24	160	1977
Castaic	324	1 200	1977
Yugoslavia			
Lisina	370	33	1973

* MVA.
† No data.
Source: Anon., 1970.

Pumped storage schemes

United States of America

The Castaic scheme forms part of the California State water project and is sited in Los Angeles County some 40 km north of Los Angeles (Fig. 21.1). It is part of a comprehensive development of water resources in the area for agricultural, industrial, domestic and recreational purposes, and the power production aspects include a conventional hydroplant as well as the pumped storage scheme. The National Environmental Policy Act of 1969 makes it necessary to design these schemes to harmonise with the surroundings, and this has prolonged the planning and early legal stages of such projects. The engineering design of the Castaic scheme is of particular interest since the site lies less than 16 km from the San Andreas fault and special precautions were taken to allow for surges of water in the tunnel and shafts which might result from seismic disturbances.

The major feature of the California State water project is the state aqueduct which brings water from the north of the state to the arid south. This divides into two at Tehachapi and along the western branch are two regulating reservoirs, Quail Lake and Pyramid Reservoir. The latter has been formed by flooding a barren basin in this semi-arid desert area and now provides recreation facilities of many kinds, including fishing, swimming and boating. From this waterbody, which forms the upper reservoir of the project, a 12 km tunnel (the Angeles Tunnel) leads to a 324 m drop to the power plant at the head of Castaic Lake, the

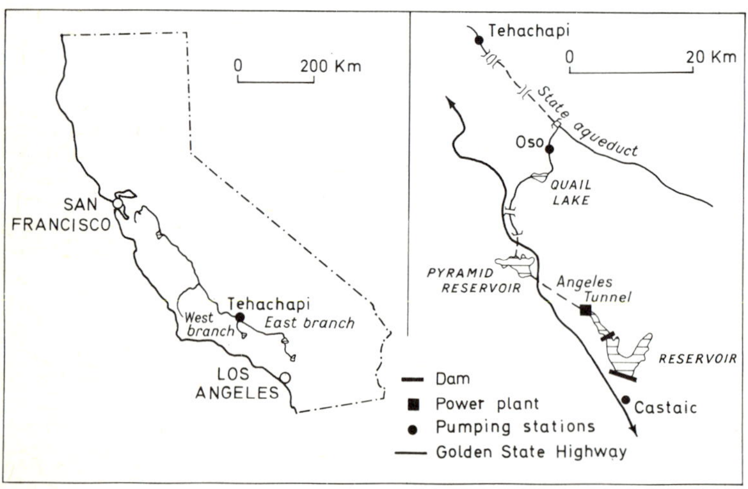

Fig. 21.1 The Castaic pumped storage scheme, California. Based on Haase, 1971

lower reservoir of the scheme. The plant contains six 200 MW reversible pump turbines and a separate 50 MW conventional unit giving a total installed generating capacity of 1 250 MW. The Castaic reservoir has a pumping forebay separated from the main section, so that the water level fluctuations normally associated with pumped storage schemes do not markedly affect this waterbody, which also provides extensive recreational facilities.

Japan

Japan has the world's second highest total of installed generating capacity by pumped storage methods, as befits a country with a high annual production of electricity (third after the USA and the USSR). The major industrial belt traverses Honshu to terminate in the Tokyo area and the Takase River project is part of the Tokyo Electric Power Company's development for the capital area. Scheduled for completion by 1977, it will generate peak supplies to supplement baseload electricity provided by thermal and nuclear power plants (Yoshida, 1972).

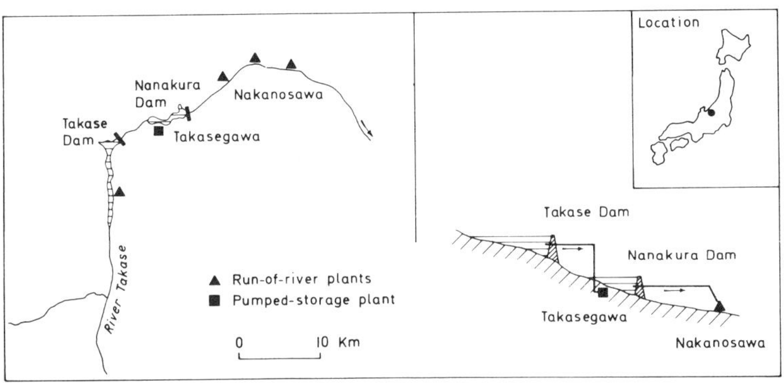

Fig. 21.2 The Takase River pumped storage scheme, Japan. Based on Mizukoshi, 1972

The upper and lower reservoirs of the project will be formed by damming the valley of the Takase River at two points so that a working head of some 229 m will be provided for the major part of the scheme (Fig. 21.2). At the power house at Takasegawa, the installed generating capacity will be 1 280 MW. The lower reservoir, dammed at Nakanosawa, will provide an operating head of a further 177 m and a regulated flow of water continuing along the valley to power a conventional project of up to 38 MW installed capacity. This will more than compensate the loss of two smaller 'run-of-river' hydropower plants on the Takase River which will be displaced by the new works. River flow below the schemes will be improved by regulating the discharge from the lower reservoir.

In earthquake-prone Japan, as in the USA, the planning of pumped storage schemes poses particular problems of dam construction. The basement rocks of the Takase region are mainly hard granite and the dams will be of triangular cross-section resting on a broad base. The power stations are being built in underground caverns and these too require careful excavation and construction procedures.

Australia

In Australia pumped storage power production is associated with major multipurpose water regulation schemes. Thus, the Snowy Mountains scheme in its northern section involves water diversion into the Tumut River and incorporates a number of power stations, the most recent of which is a pumped storage plant, Tumut 3, 1973, with an installed capacity of 1 500 MW.

The Tumut is a tributary of the Murrumbidgee and receives water from a series of lakes and reservoirs in the Snowy Mountains, in particular from the extensive Lake Eucumbene, which in turn is fed by piped water from the Tantangara Reservoir on the upper Murrumbidgee (Fig. 21.3). These waterbodies are at or just below the 1 200 m level on the Great Dividing Range and the various descents in the Tumut River are made first by pipeline to power the Tumut 1 and Tumut 2 conventional hydroschemes. Below these, with a surface level of about 500 m, is the Talbingo Reservoir and the Blowering Reservoir, both formed by damming the Tumut valley. Between these is the powerhouse of the Tumut 3 scheme, so that Talbingo and Blowering are respectively the upper and lower reservoirs of the project. In addition, the upper section of the Blowering is itself dammed to form the Jounama pondage, so that the variable water level for the pumped storage scheme is confined to this area. The flow of water required by the irrigation areas of the lower Murrumbidgee is not affected by the water needs of the power scheme. The various power schemes of the Snowy Mountains contribute peak load electricity to the connected state grid systems of Victoria and New South Wales and the addition of the Tumut 3 project will increase the absolute peak potential from this area (Johnson *et al.*, 1967).

Western Europe

While recent developments of major pumped storage schemes are particularly characteristic of the USA and Japan, the world's major concentration of such schemes is in Europe (Table 21.1). In addition to those in Alpine countries, which resemble conventional hydroelectric plants with the upper reservoir being formed by damming a high-level valley, pumped storage schemes have also been established in areas nearer the major regions of electricity demand, thus minimising costs and problems

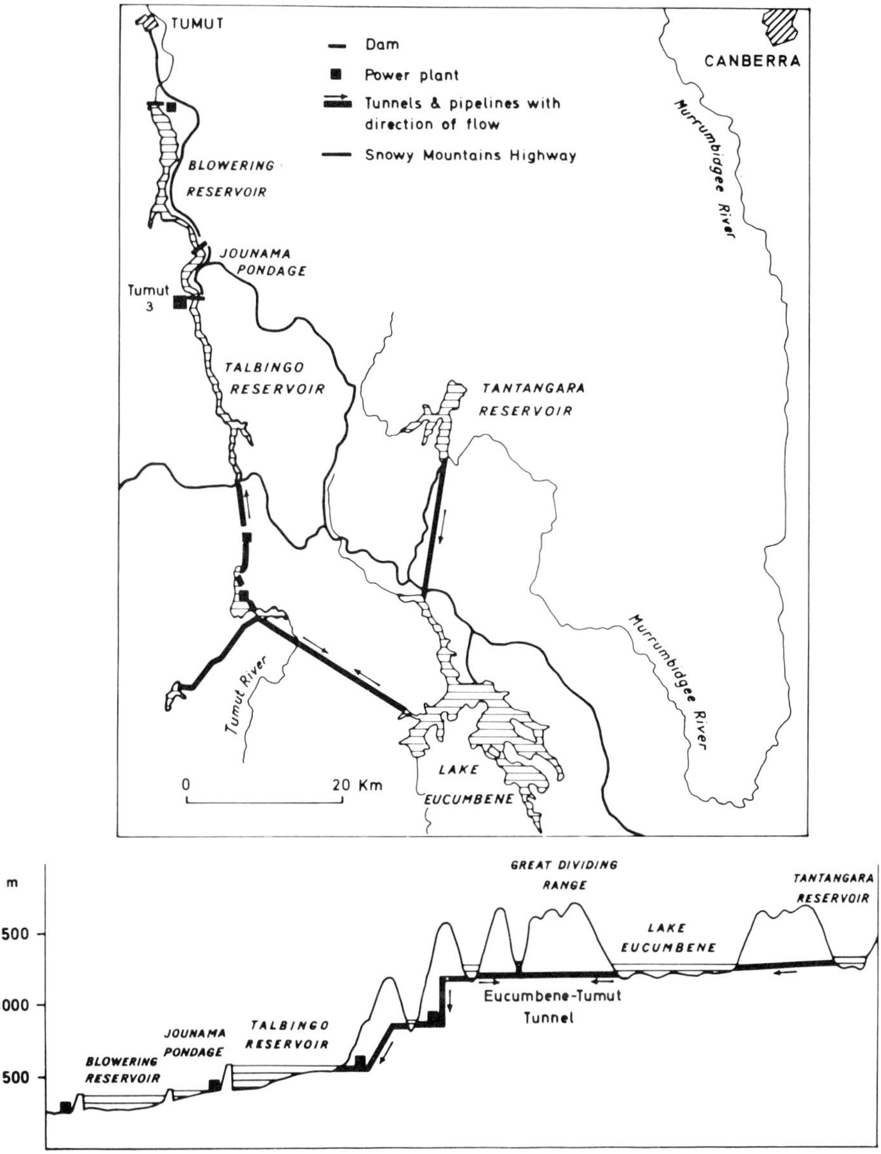

Fig. 21.3 The Snowy Mountains (Tumut 3) pumped storage scheme, Australia. Based on Snowy Mountains Authority data

of power transmission. This development illustrates one of the special distinguishing features of this form of generation – its lack of dependence on a continuing water supply except to replace losses due to seepage or evaporation. Schemes in West Germany, Belgium and Luxembourg have made use of the combination of high-level plateaus and

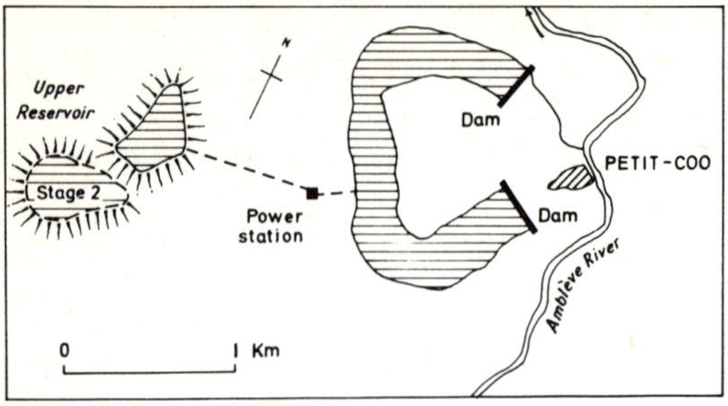

Fig. 21.4 The Amblève River pumped storage scheme, Belgium. Based on Lottes, 1972

incised river valleys by building artificial upper reservoirs on the plateau surfaces and lower reservoirs in the river valleys. This is well seen in the case of the Coo-Trois Points scheme on the Amblève in the Belgian Ardenne, and also in the Vianden scheme of Luxembourg.

The Amblève valley near Coo is distinguished by an abandoned meander loop, which has been dammed at both ends to provide the lower reservoir. The upper reservoir, on the plateau surface, is nearly 270 m higher, and contained by a circular earthwork dam (Fig. 21.4). This form of upper reservoir poses construction problems, particularly relating to water seepage, and a waterproof lining is generally necessary, using asphalt or some other mastic material. Only mechanised methods of earth removal, transport and grading of slopes have made the construction of large scale earthworks of this kind technically and economically feasible. The Coo power house has been built in an underground cavern and a second stage development is proposed which will double the installed capacity of the scheme, at present 432 MW.

Similarly, at Vianden, a pumped storage scheme makes use of the deeply incised valley of the Our for the location of the lower reservoir, contained by a concrete dam. An artificial earthwork surrounds the upper reservoir on the summit level surface of the Ardenne plateau and an underground power house with an installed generating capacity of 900 MW has been built between the two reservoirs. In addition to the main power scheme at Vianden, a small run-of-river hydroplant has been installed at the foot of the river dam to supply auxiliary and standby power, estimated at 12 million kilowatt-hours (kWh) annually. Although this plant is located in Luxembourg, the main output is fed eastwards into the Nordrhein–Westphalen grid, for it is here that the major peak demands occur which need to be met by this form of scheme.

In Britain, pumped storage schemes are comparatively recent, the first to be put into operation being that at Blaenau Ffestiniog which, with an installed capacity of 360 MW, was at the time of its opening in 1963 one of the world's biggest schemes. An upper reservoir (Llyn Stwlan) was formed by damming a high-level valley of characteristic cirquelike form, with the creation of a lower reservoir (Tan-y-Grisiau) in the main valley. Slate workings complicated the scheme and disused mines required sealing before the Llyn Stwlan Reservoir could be completed. Situated close to this plant is a contrasting base load power station, the nuclear power plant at Trawsfynydd. This provides the off peak power for the pumps at Blaenau Ffestiniog so that between them they partly meet the variable electricity demands of the industrial Midlands. In this way a high load factor is maintained for the nuclear power station.

The Foyers and Cruachan schemes in Scotland, using Loch Ness and Loch Awe respectively as the lower reservoirs, provide more recent examples of pumped storage generation and a further development is planned in North Wales at Dinorwic (Lovins, 1973). Another large scheme will be considered in the Scottish Highlands for commission in the 1980s. Nevertheless, compared with some countries, the British total installed generating capacity of pumped storage turbines (1 060 MW) is minimal.

Tidal schemes

A complete contrast with the more conventional schemes is provided by the tidal generating project at St Malo in France (Fig. 21.5). This makes use of the low head (11 m) but large volume of water at the mouth of the estuary of the River Rance and involves some complex operations. Here pumping is adopted to ensure that there is a sufficient water flow either from the sea into the empty basin behind the barrage or from a very full basin into the sea to produce electricity by a flow in both directions. The twenty-four reversible pump turbines in this scheme are each capable of producing 10 MW. The water flow is supplemented by the natural ebb and flow of tides so that pumping is only necessary to produce a head of water in the operating cycle when tidal movements are not favourable.

The use of sea water introduces particular problems in the construction and maintenance of a hydroscheme, in view of its corrosive action on metal. As a general guide sea water is more than ten times as corrosive as river water (Yoshimoto, 1973). Although moving saline solutions cause less trouble than stationary ones, unfortunately in a pumped storage scheme there are periods when the water will be at rest. Another problem is that of marine organisms which adhere to structures, destroying paint and obstructing the water flow.

Nevertheless, a further development of the use of sea water in hydroelectricity is currently under study in Japan. This is the Atashika sea

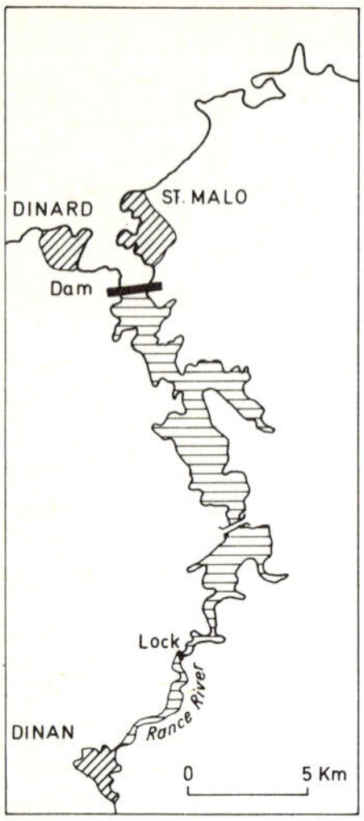

Fig. 21.5 The Rance estuary tidal scheme, France. Based on Jones, 1968

water pumped storage project which is sited on the Pacific coast of Honshu, some 200 km southeast of Osaka (Fig. 21.6). The upper reservoir, due for completion in 1978, is on a tributary of the Omata River. It will be formed by damming the valley and will receive some natural inflow from existing streams as well as the pumped salt water of the scheme. The lower 'reservoir' will be the sea at Atashika Bay. Environmental problems are particularly severe and include those of salt water seepage into the surrounding countryside from the upper reservoir, and the effects of moist salt air on farmlands and agricultural production. A number of countermeasures are proposed, including a secondary dam and special drainage areas to collect saline seepage water, the purchase of affected land areas, and the control of water and flood discharge from the upper reservoir by means of a special tunnel in sea waterproof concrete. The upper reservoir will have a high water level at 465 m above sea level and the effective generating head will be 382 m. With an installed generating capacity of 496 MW the entire project is in many ways comparable with western European schemes such as those at Coo and Vianden, rather than the Rance tidal plant.

The pattern of generation

The pumped storage method of generating electricity is of considerable value in meeting rapid peak demands, but as power is needed to operate the pumps it cannot be regarded as a net means of producing electricity. Pumped storage in fact is a device by which energy can be transferred from one portion of the load curve to another and many writers have

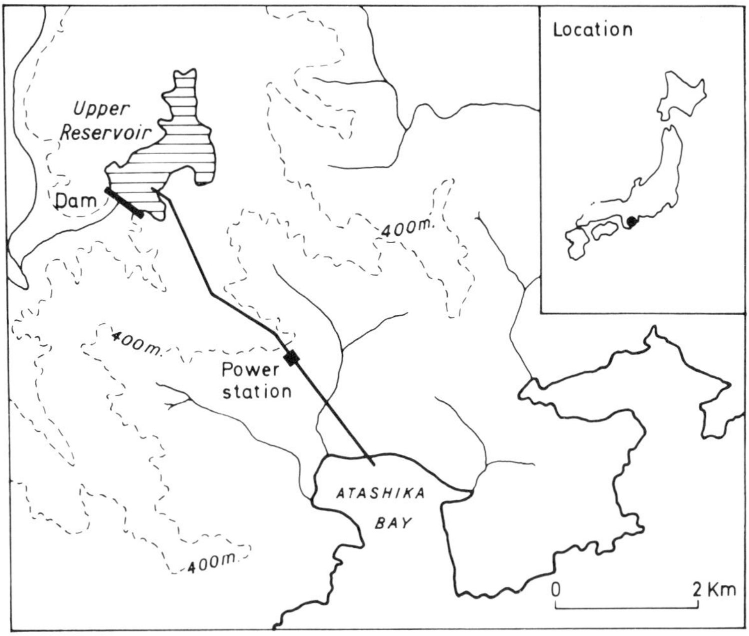

Fig. 21.6 The Atashika pumped storage project, Japan. Based on Yoshimoto, 1972

described it as a means of 'storing' electricity, the water retained in the storage reservoir being considered as 'potential' electricity.

The load curve of electricity production generally shows a diurnal wave, but in fact the pattern of operation of a 'peaking' pumped storage scheme may be more complex, as in the case of the fluctuating pattern of actual generation and pumping cycles for the Vianden plant (Fig. 21.7). On the other hand, pumped storage schemes may be linked with a seasonal regime, so that in the case of some Alpine stations glacier meltwater is conserved in the upper reservoir for use during periods of heavy demand in winter. Between these two examples, there is the case of the network which has a heavy series of working week peak demands compensated by a weekend lull, during which the upper reservoir can be replenished.

However, within most electricity supply systems the diurnal load pattern normally exhibits the expected daytime peak and nighttime trough (Fig. 21.8 A). Hydroelectricity fills at least two roles in the production of energy in this load pattern; it may be a base load producer or it may be used purely for peak load production, as in pumped storage schemes. Viewed in the context of likely future patterns of power production, pumped storage will not only help to meet peak demands but also will increase the demand for power during the offpeak times (Fig. 21.8 B).

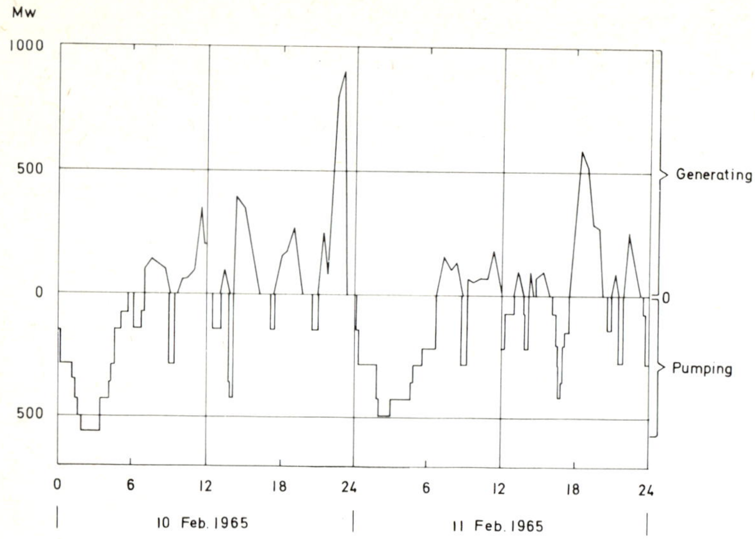

Fig. 21.7 Schedule curve for the Vianden pumped storage plant, Luxembourg. Based on Boehler, 1960 and 1961

Confirmation of the importance of conventional hydroelectricity and pumped storage power generation as satisfiers of peak demands is provided by a consideration of the response times of various forms of production (Fig. 21.9). Although initial peak demands may be met by complex switching procedures within an interconnected grid, it is usually necessary to bring plants on 'spinning reserve' into full production. Finally, idle plants will be started and it is here that water-driven turbines are clearly advantageous. It has been argued that, unless there are great economic advantages in coal, oil or gas-fired power stations in a given region, probably the most satisfactory future arrangement for electricity production is a combination of nuclear and hydro base load stations with pumped storage peak load schemes (Lottes, 1972).

Problems

The problems associated with the design and construction of pumped storage hydroelectricity undertakings are related to geological, technological and economic factors. Of the first group, mention has been made of the particular hazards of pumped storage schemes in areas liable to earthquakes, such as Japan and the western USA. In addition, water-retaining properties are important requirements in the construction of the two reservoirs and unless the subsoil is impermeable expensive lining procedures are necessary. Constant fluctuations of the water level increase the need for careful construction of reservoir banks and even special surfacing, at least over the 'tidal' range. Internal pressures within

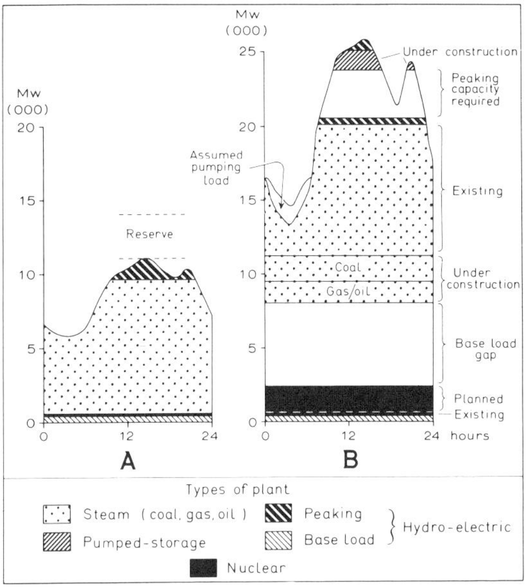

Fig. 21.8 The load curve for a typical large grid sypply system

A. Diurnal load pattern in August 1967;

B. Projected diurnal load pattern for electricity supply plants in the 1980s.

Based on Thon, 1970

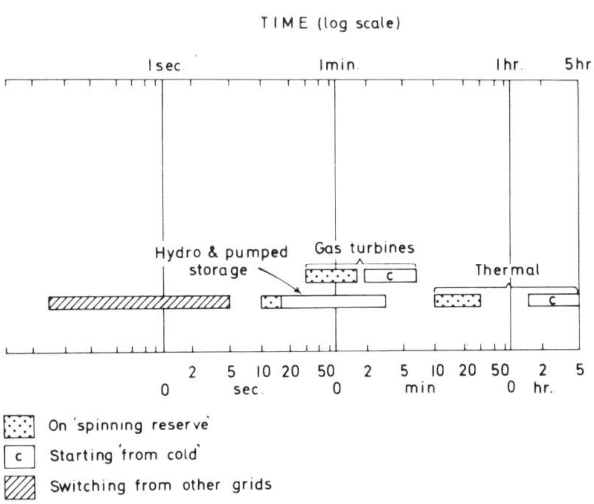

Fig. 21.9 Response times for an instantaneous demand of 200 MW by various means of power generation. Based on Thon, 1970

the rock mass are greatly affected by the establishment of pumped storage schemes and pose civil engineering problems. Stresses will be produced by a fluctuating weight of water in the upper reservoir and some of these will operate on the interconnecting pipes and shafts, as well as on the power house, if this is in an underground cavern. Technological problems connected with the turbines driving the generators and with the pumps required for replenishing the upper reservoirs have been met in various ways. Some pumped storage schemes employ separate turbines and pumps but increasing use is made of a combined turbine pump or 'reversible turbine'. The first of these was commissioned in the 1939 Pedreira project in Brazil and the great majority of schemes now under construction are designed to use reversible turbines.

Economic factors relating to a pumped storage project may be divided into those of the overall electricity production pattern, and those specifically associated with this form of generation. In the first case the competing claims of thermal, nuclear and hydroelectric schemes must be weighed before basic decisions are made as to the form of electricity production. In a given pumped storage situation, particular costs resulting from design and construction aspects must be examined. A major cost of any scheme, usually about 30 per cent, is incurred in the construction of waterways, especially by tunnelling (Thon, 1970). Additionally, the cost of providing pumping energy must be evaluated as this may be supplied by old and comparatively inefficient thermal plants. During the expected life of a pumped storage plant, however, the cost of supplying pumping power may vary. One factor which can be argued in favour of these schemes is the virtual doubling of available output when a pumped storage plant changes from pumping to generation. At this point the grid ceases to supply power for pumping and so electricity becomes available for other consumers, thus reducing the load by an amount which may approximately equal the output capacity of the plant. When the plant starts, a similar amount of power is fed into the grid and so the net gain in these two operations is likely to be double the plant's normal capacity.

The high cost of installing hydroelectric plants has been given as a reason for favouring other means of generation. In addition, as the most satisfactory sites for such plants are a considerable distance from areas of high electricity demand, transmission costs are increased. Although capital costs are high, it has been suggested that effective savings can be produced by using pumped storage, but that the optimal capacity of such schemes should lie between 15 and 35 per cent of the maximum load in the grid system concerned (Lottes, 1972).

In examining the various arguments relating to costs, it is generally agreed that the low head scheme is nearly twice as expensive per unit of production as the high head installation. Technical limitations of the high head schemes have been reduced successively and, in order to utilise their economic advantages, at least one ingenious proposal has been made to introduce an artificial head of water independent of the topography. This proposal involves extensive underground storage areas to replace the lower reservoir. A further refinement is the possibility of dividing the working head into two parts with an intermediate reservoir and power house. The envisaged total head between the surface reservoir and the main underground storage reservoirs in such a scheme might well be of the order of 1 200 m (Warnock and Willett, 1973).

The construction of installations of this type creates problems, a major one being the design of the reversible pump turbine to work in these conditions. This device would represent but a further stage in the recent rapid development of equipment for pumped storage generation. Although most major schemes at present have working heads only a little in excess of 400 m, where reversible pump turbines are installed, schemes are already in existence in Austria with working heads of over 1 000 m but with separate pumps.

Conclusions

The construction of any power station usually provokes some form of criticism. This ranges from the aesthetics of the appearance of a conventional thermal plant with cooling towers, to the possible dangers thought to be associated with nuclear power schemes. Atmospheric pollution stemming from thermal plants in or near large cities has also been a source of controversy. In the case of hydroelectricity installations the form of generation can be considered 'clean', but arguments centre on conservation and aesthetic grounds. Regulation of river flow and the construction of reservoirs will change the appearance of what is, in many cases, an area of some landscape value. If the project involves dam construction and valley flooding in a region of particular scenic importance (for example, a designated national park) the economic arguments in its favour must be convincing. The construction of the power station itself need pose only a few problems as it is likely that this will be underground. Normally, surface works and buildings can be sited carefully and landscaped to cause a minimum of intrusion (Galze, 1972, 1973).

Arguments for and against the construction of pumped storage schemes accept that additional peak load generating capacity is required within any electricity supply system. Technical improvements have made thermal plants more efficient but they can only meet peak demands if they are already producing some power on 'spinning reserve'. Modern gas turbine plants are another means of meeting unexpected demands as these can be brought quickly into use and their capital costs are likely to

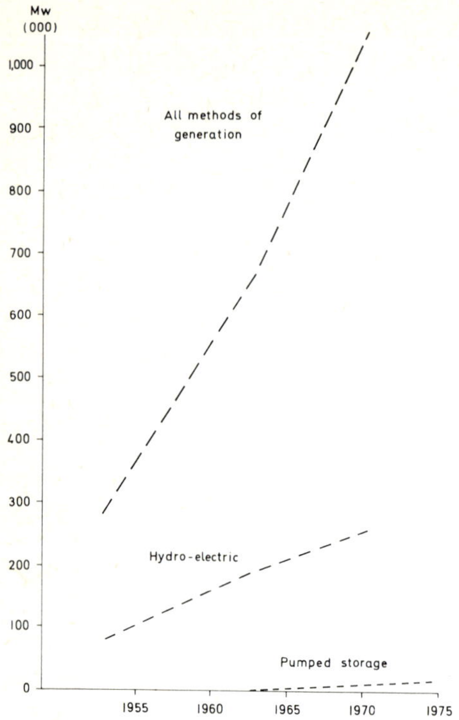

Fig. 21.10 World installed electricity generating capacity, 1955–75. Based on United Nations, 1973

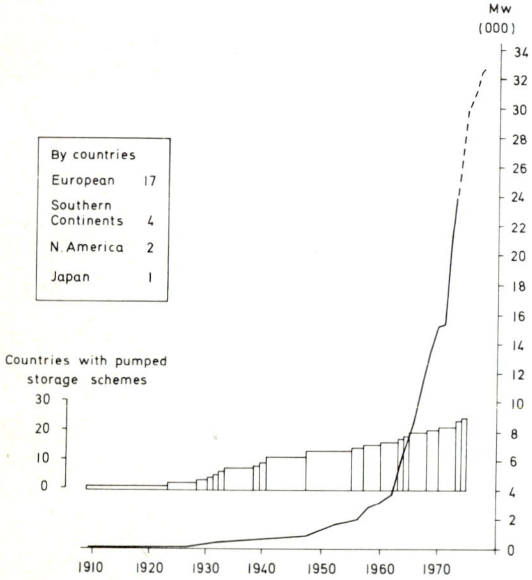

Fig. 21.11 Installed pumped storage generating capacity and schemes under construction. Based on Anon, 1970

be less than those of a major pumped storage scheme. Although advocates of hydroschemes may point to their lack of atmospheric pollution compared with almost every other form of power generation, one argument advanced against pumped storage is that the power used for pumping must be generated elsewhere, probably in a conventional plant. It is thus argued that tomorrow's potential peak 'clean' hydroelectricity is being generated by tonight's pollution-producing thermal station!

Pumped storage schemes are capable of development in different topographic situations and the latest proposal for underground lower reservoirs could make them even more adaptable. Schemes of this kind may then be located relatively near the regions requiring peak electricity supplies and the clean air advantages of hydroelectricity could be of significance. The most telling arguments in favour of any hydroelectricity scheme would seem to be those where the project is of a multipurpose nature. In such a scheme, power production (by conventional and/or pumped storage hydroelectricity stations) forms only part of the overall development. Water supply for agricultural, industrial and domestic needs is likely to form a major part of the undertaking, as in the Australian and Californian examples described. Additionally, existing recreational facilities may be augmented by the creation of a new water body, as shown at Castaic, although as European examples have indicated, a major power station scheme is likely to have a tourist potential in its own right.

At present the role of pumped storage electricity generation in the overall picture of power production is small. Less than 1 per cent of the world's total installed generating capacity uses this method and it represents only 5 per cent of total installed hydroelectricity power (Fig. 21.10). These figures belie the fact that there was a rapid growth in the installation of large pumped storage schemes during the 1960s (Fig. 21.11). This was also a time when different means of power generation were being evaluated to meet a rapidly expanding demand and marked a stage when nuclear energy was in its infancy while coal-burning plants were being converted to oil or gas. A number of large pumped storage schemes are now under consideration and, while some may be little more than projected planning exercises, it is noteworthy that a recent survey (Anon, 1970) showed that some fifteen countries were actively investigating this form of power production.

References

ANON (1970) 'Survey of the world's pumped-storage stations', *Wat. Power*, 22, 96–101.

BOEHLER, K. (1960 and 1961) 'The Vianden pumped-storage plant', *Energiewirtschaftliche Tagesfragen*, 85 and 94, special reprint.

GALZE, A. R. (1972, 1973) 'The environmental impact of a multi-purpose pumped-storage project', *Wat. Power*, **24**, 459–63 (Part 1) and **25**, 25–6 (Part 2).

HAASE, C. D. (1971) 'Castaic pumped-storage project in California', *Wat. Power*, **23**, 333–41.

JOHNSON, K. E., AUBREY, K. G. and MACPHERSON, P. M. (1967) 'Tumut 3 project – concept and investigations', *J. Instn Eng. Aust.*, **39**, 123–37.

JONES, I. E. (1968) 'The Rance tidal power station', *Geography*, **53**, 412–15.

LOTTES, G. (1972) 'The development of European pumped-storage plants', *Wat. Power*, **24**, 22–33.

LOVINS, A. (1973) 'Things that go pump in the night', *New Scient.*, **58**, 564–66.

MIZUKOSHI, T. (1972) 'The Takase River pumped-storage project', *Wat. Power*, **24**, 64–70.

SNOWY MOUNTAINS AUTHORITY (n.d.) *Water and power from the Snowy Mountains scheme*, Cooma, NSW.

THON, J. G. (1970) 'General aspects of pumped-storage hydro-power', *Wat. Power*, **22**, 89–95.

UNITED NATIONS (1973) *World energy supplies, 1968–71*, Statistical Papers, J.16, New York.

WARNOCK, J. G. and WILLETT, D. C. (1973) 'Underground reservoirs for highhead pumped-storage stations', *Wat. Power*, **25**, 81–87.

YOSHIDA, M. (1972) 'Development of pumped-storage projects, situation and trends', *J. Jap. Soc. Civ. Engrs*, **57**, 81–95.

YOSHIMOTO, T. (1972) 'The Atashika seawater pumped-storage project', *Wat. Power*, **24**, 57–63.

P. T. Kivell

22 Postwar urban residential growth in north Staffordshire

North Staffordshire[1] developed in the nineteenth and early twentieth centuries as a number of separate and largely independent settlements, until, in 1910 the major 'pottery towns' were federated into the borough, and later the city of Stoke-on-Trent, with the adjacent town of Newcastle-under-Lyme remaining administratively separate. In the postwar period, the processes of urban growth together with increasing personal mobility and a growing *de facto* social and economic unification have resulted in the emergence of a north Staffordshire metropolitan area of half a million people focused on the two boroughs of Stoke and Newcastle.

This essay attempts to account for the extensive postwar urban expansion and trace its spatial pattern. It might be assumed that urban growth is a direct consequence of population growth but in fact the overall population increase has been very modest, certainly not sufficient in itself to create substantial pressures for development. The 1951–71 period saw an addition of 25 000 people to the subregion's population, an increment which in round figures would have generated a demand for approximately 8 000 more houses. In fact, 69 000 houses have been built in north Staffordshire since the war and its population growth, although an important contributory factor, cannot account entirely for this scale of housebuilding and urban expansion.

Population change

Since the war the population of north Staffordshire has shown a net increase, which between 1951 and 1971 has been in the order of 5 per cent (Table 22.1). The corresponding figures for England and Wales and the west Midlands were 11.1 and 15.4 per cent respectively; thus in the national and regional contexts north Staffordshire's growth has been

Table 22.1 Population changes 1951–71

Local authority area	*Total population* (000s)			*Percentage change 1951–71*
	1951	*1961*	*1971*	
Stoke CB	279.9	277.2	265.2	–5.2
Newcastle MB	71.1	76.7	77.0	8.2
Biddulph UD	10.9	13.8	17.4	59.4
Kidsgrove UD	16.3	19.8	22.0	34.8
Leek UD	19.4	19.2	19.4	0.0
Stone UD	8.4	8.8	11.0	30.8
Cheadle RD	31.2	32.4	40.1	28.5
Leek RD	12.7	12.6	13.2	4.1
Newcastle RD	17.1	18.2	20.6	20.5
Stone RD	16.7	18.5	22.1	32.8
Total	483.7	497.2	508.0	5.0
Total excluding Stoke	203.8	220.0	242.8	19.2

Source: Census 1971, *Preliminary Report*.

very slow. The 1961–71 growth rate of Stoke Metropolitan Area[2] is ranked thirty-seventh among the fifty-two metropolitan areas of England and Wales (Kivell, 1972). This net figure of 5 per cent growth conceals two basic trends: the population of the core area, that is Stoke, has fallen sharply, while, with only two exceptions, the surrounding areas have made substantial gains.

Population loss

Although the area's population grew by 25 000 during the period 1951 to 1971, that of Stoke-on-Trent declined by 15 000, and its share of the subregional total fell from 57.9 to 52.2 per cent. An analysis of the percentage population loss at ward and parish level for the decade 1961 to 1971 indicates two basic patterns (Fig. 22.1). The major population declines have occurred in (1) inner city areas, especially the eastern parts of Stoke CB which are at the very heart of the conurbation, and central Newcastle; and (2) remote rural parishes.

The former losses, which are common in many British cities, represent a complex movement of people from central Stoke and Newcastle. In part this is due to a rundown in basic industries, and in part to the redevelopment of older areas and the resulting processes of decentralisation and peripheral growth. The urban fabric of these districts is for the most part decayed or decaying, and bears many reminders of earlier, more active phases of manufacturing and extractive industries. In these areas

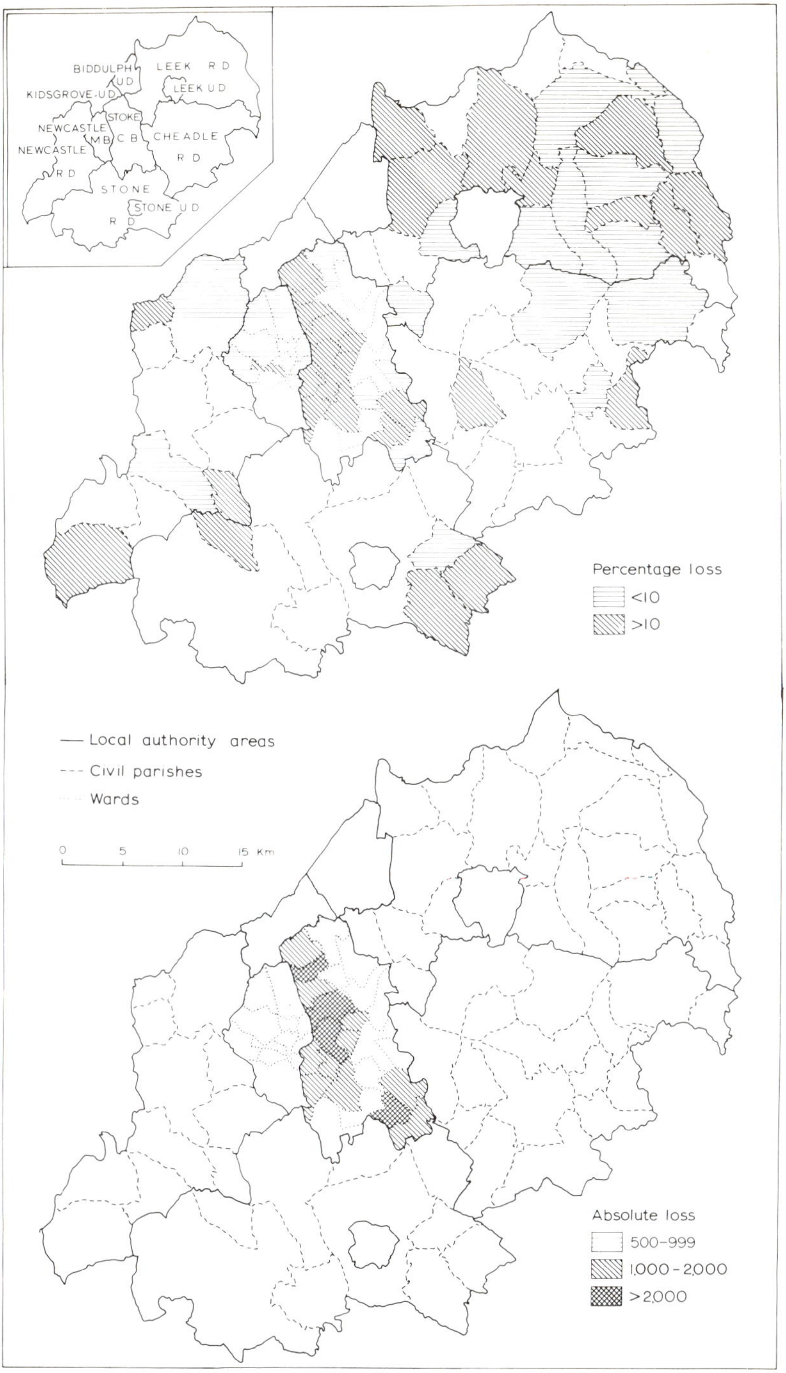

Fig. 22.1 Population losses in north Staffordshire, 1961–71. Based on Census, 1961 and 1971

today the intermingling of nineteenth-century housing, industry and service functions creates an unsatisfactory residential environment.

The substantial population losses in rural areas are due to a declining labour demand in basic activities, particularly in the agricultural sector, where increased mechanisation and improved techniques are reducing the number employed on the land and the dependent secondary rural population.

The map of percentage losses overemphasises the position of many rural parishes with small population totals. This is remedied by examining absolute loss which is concentrated in Newcastle and Stoke (Fig. 22.1). The distribution of absolute population loss reveals that instead of the relatively simple pattern of inner area loss which might be expected in a city with a single nucleus, the polycentric nature of the north Staffordshire conurbation produces a pattern based on several 'central areas'. Thus the locations of the centres of Tunstall and Longton, close to the borough boundary, have resulted in substantial losses from peripheral areas which, in a more conventional city, could well be thriving outer suburbs. This existence of at least six centres in place of one dominant urban focus is responsible for many apparent anomalies in the pattern of population change and residential development of north Staffordshire.

Population gain

Although the area as a whole has lacked a substantial population increase in postwar years, certain districts have made marked gains (Fig. 22.2). These points of exceptional growth may be attributed initially to two specific factors: (1) the location of employment opportunities, and (2) growth in commuting.

Some towns in the area have been more fortunate, or more active in the attraction of new employment than others. Among the north Staffordshire employment exchange areas, only Newcastle and Kidsgrove have been showing net gains, with increases of 3 966 and 4 934 jobs respectively between 1951 and 1971. Within Stoke the new jobs created have been offset by continuing declines in coal mining and the pottery industries. Specific employment sources account in some cases for unusually high population gains in certain rural parishes. Thus the 1961–71 gains of 57.1 per cent for Keele and 37.3 per cent for Madeley can be attributed largely to the growth of the university and the college of education respectively.

Newcastle and Stoke are virtually encircled by parishes with high postwar population gains. This growth in ostensibly rural districts can be explained largely in terms of the urban area extending beyond its administrative boundaries, a result not of a too restrictive boundary but

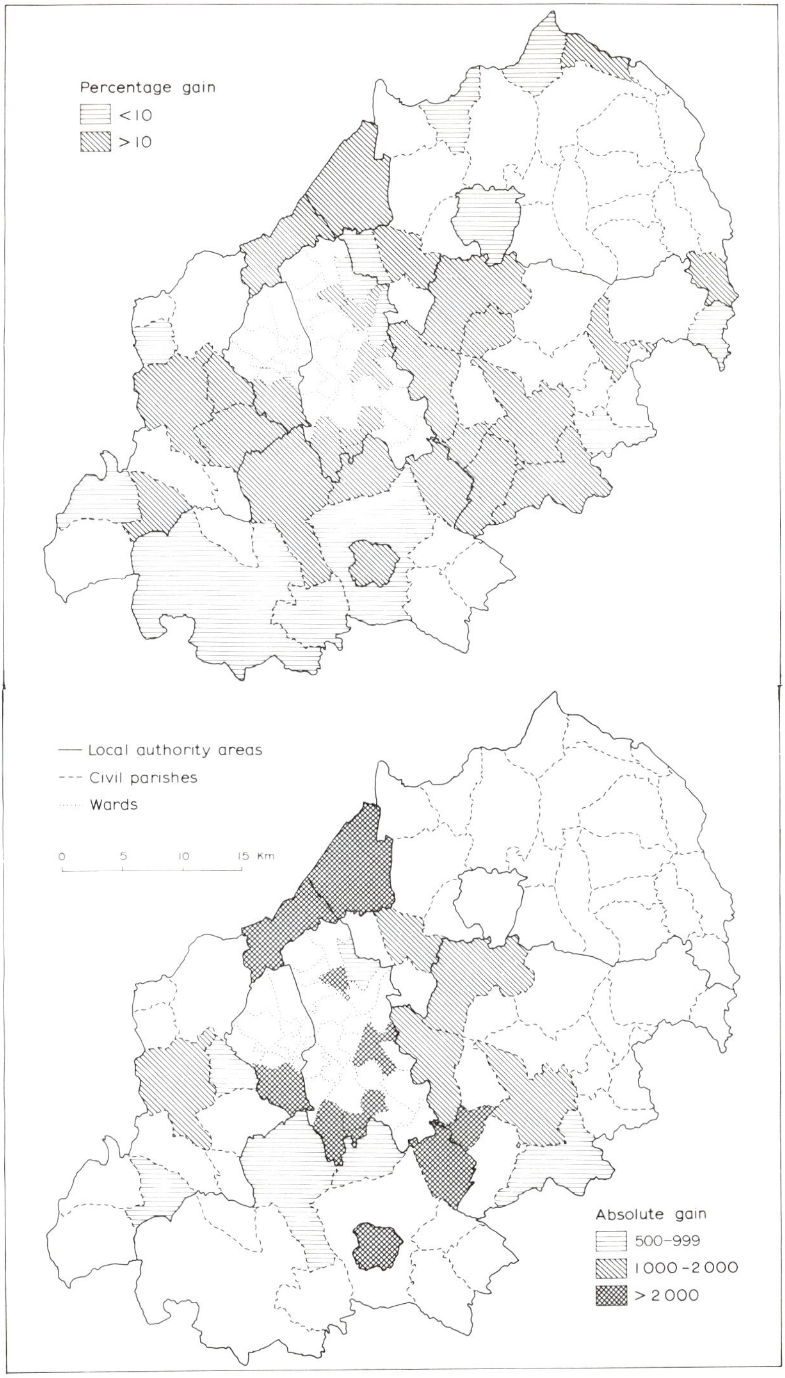

Fig. 22.2 Population gains in north Staffordshire, 1961–71. Based on Census, 1961 and 1971

of the way in which the car has enabled place of work and place of residence to become further apart. A wealthy commuter fringe is now a feature common to most British cities, but in north Staffordshire its growth has received additional stimulus from the unattractive inner urban environment and the comparative lack of middle-class suburban areas.

With few exceptions the pattern of parish gains (Fig. 22.2) can be explained by this recent upsurge of commuting and the consequent development of a 'dormitory' role by certain villages surrounding the urban area. This is the case in such parishes as Ashley, Swynnerton, Madeley and Cheddleton. Other parishes, closer to the Newcastle and Stoke boundaries, have experienced population gains for the same reason, but are now virtually contiguous parts of the built-up area. This situation has been acknowledged at Endon, Stanley, Fulford and Caverswall, which have been incorporated as part of the city (1965 extensions).

Although the population of the urban area has declined, Stoke remains dominant as an employment centre. These two factors taken together underlie the importance of commuting. In 1966 Stoke provided over 34 000 jobs for people living outside its boundary and, in addition, Newcastle had a daily inflow of 8 860 workers. The residences of these commuters into Newcastle and Stoke are shown in Table 22.2, which also confirms Newcastle's role as a dormitory in relation to Stoke. It also shows that with the exception of Leek RD, the towns and rural districts surrounding Stoke and Newcastle contributed more commuters in 1966 than in 1951.

The significance of commuting can be illustrated by the rural district of Newcastle, where boundary changes between 1951 and 1966 were of

Table 22.2 Number of residents in different local authority areas working in Newcastle and Stoke (000s)

Local authority area	*1951*		*1961*		*1966*	
	Stoke	*Newcastle*	*Stoke*	*Newcastle*	*Stoke*	*Newcastle*
Newcastle MB	14.0	–	14.8	–	14.7	–
Biddulph	0.9	0.0	1.7	0.0	1.9	0.2
Kidsgrove UD	2.9	0.6	3.5	0.7	3.5	0.8
Leek UD	0.3	0.0	0.3	0.0	0.3	0.0
Stone UD	0.2	0.0	0.3	0.0	0.4	0.0
Cheadle RD	2.7	0.0	4.6	0.2	3.6	0.2
Leek RD	2.7	0.1	3.9	0.1	2.3	0.1
Newcastle RD	0.8	1.8	1.3	2.4	1.6	2.0
Stone RD	1.3	0.1	1.8	0.1	2.7	0.1

Source: Census 1951, Workplace Tables; Sample Census 1966.

little importance. In this period there was a 41 per cent increase in the number of commuters to Newcastle and Stoke. As the 1966 Census gives the total economically active work force of Newcastle RD as 8 290, within the limitations of the sample accuracy it may be suggested that 43 per cent of the active residents were employed in Newcastle and Stoke.

Residential expansion: alternative explanations

The need for new housing and urban expansion is related to factors in addition to population change. Perhaps the most important are those which affect the deterioration and replacement of the existing housing stock and in this respect Stoke's nineteenth-century heritage is of great significance. A 1968 study indicated that 37 000 dwellings in the city (41 per cent of the total) were built before 1914. Not all of these were unfit by existing standards, but the situation had serious implications for both current and future housing policies. In 1955 the city's initial slum clearance programme identified 12 000 unfit dwellings, but continuing deterioration, together with improved minimum acceptable standards, have meant that almost 17 000 houses will have been dealt with when the programme is concluded in 1974 (Stoke City Architect, 1972).

Even with a static population important changes constantly occur in the number of households and in the degree of shared accommodation or of overcrowding. The population of the city remained virtually unchanged between 1961 and 1971[3] but an increase in the number of separate households from 84 664 to 90 685 and a fall in occupancy rates from 0.71 to 0.60 persons per room added to the demand for housing.

Boundary changes during the 1951–71 period make the comparisons of population totals and household numbers in individual areas difficult, but it is clear that even in this short period the average size of households decreased. In 1951 a subregional population of 476 461 occupied 142 546 households, an average of 3.34 persons per household. By 1971 the corresponding figures were 498 235, 171 635, and 2.90. Assuming a static population from 1951 to 1971 and the operation of the trend to smaller households, 164 296 households would have been formed by 1971. This, changing headship rates and the decrease in average household size accounted for an additional 21 750 households, independent of overall population increases.

The overall trend towards owner-occupation further influences the pattern of residential expansion in terms of the type and location of housing required. Despite short-term fluctuations produced by the housing market itself and by mortgage availability, a distinct movement towards owner-occupancy occurred between 1961 and 1971

Table 22.3 Tenure of housing

	1961		*1971*	
Local authority area	*Percentage owner-occupied*	*Percentage local authority*	*Percentage owner-occupied*	*Percentage local authority*
Stoke CB	41.8	31.0	50.8	33.7
Newcastle MB	47.7	33.7	52.6	35.4
Biddulph UD	49.7	21.8	62.4	25.8
Kidsgrove UD	46.5	28.1	57.0	29.5
Leek UD	51.0	26.3	59.5	27.8
Stone UD	31.3	28.7	52.5	24.6
Cheadle RD	57.2	13.3	73.4	12.4
Leek RD	63.8	5.3	73.3	5.9
Newcastle RD	50.3	17.3	60.2	19.5
Stone RD	43.3	18.4	57.2	22.4

Source: Census 1961, 1971.

(Table 22.3). As the proportion of houses rented from local authorities has increased in all but two cases, major decreases must have taken place in the privately rented sector.

The housing response

The natural and manmade constraints acting on the residential building programme in north Staffordshire have been already identified (Moisley, 1951; Staffordshire County Council, 1973). Derelict land is a fundamental problem in Stoke. Despite the reclamation of 225 hectares between 1969 and 1971, the city possessed in 1972 570 hectares of derelict land, in addition to 405 hectares potentially derelict, more than any other borough in the country. These figures understate the scale of the problem, for actual derelict sites effectively sterilise a far greater area for development. This dereliction, together with subsidence produced by past and current extractive industries, bars residential development from a large part of Stoke and from an area around the anticlinal limb of the coalfield to the north and west of Newcastle.

To the east and north of Stoke dereliction is replaced by altitude and steeply sloping land as the major development constraint. Much of the urban area is embraced by the 152 m contour and the Pennine fringe, represented arbitrarily by the 230 m contour, comes close to Stoke's eastern boundary. An additional hazard is posed by possible flooding in the Trent and Foxley valleys. The southern part of the urban area is free of such constraints, but expansion here has resulted in the loss of much good quality agricultural land. The agricultural and recreational potential of land in this sector does itself act as an additional constraint.

Despite these problems, neither the city nor the subregion is short of building land. Stoke has a reserve of approximately 640 hectares and there is sufficient building land for many years in the rest of the subregion. Quality of potential sites, however, is as important as quantity and in Stoke especially the widespread reminders of the city's nineteenth-century industrial activity are not conducive to residential development.

A city can cater for its housing requirements in two ways; (1) the renovation and replacement of existing stock, and (2) the development of virgin land, either within the city or, with the cooperation of an adjacent authority, outside its boundary. These are not exclusive and the region has pursued both courses. The former is a crucial element in the north Staffordshire context, because the subregion's high proportion of houses built before 1914 will continue to be the most important single factor in determining housing demand in the public sector.

Postwar building

An active programme of housebuilding was carried out in the interwar period in both the public and private sectors, the latter being concentrated in Newcastle. These developments were criticised in the consultative report on north Staffordshire submitted to the Minister of Town and Country Planning (Abercrombie and Jackson, 1949). Attention was drawn to the lack of services and facilities, the poor attempts to integrate large new estates with existing urban areas and the emergence of private and local authority estates with monotonous similarities of house-type and social class. These criticisms, together with the advice contained in the 1944 *Housing Manual* (Ministries of Health and Works, 1944) on the development of neighbourhood units, suggested a reappraisal of goals within the public housing sector. The neighbourhood concept was adopted in various forms by Stoke but many of the 1949 criticisms still apply today in both private and public sectors.

North Staffordshire in common with many urban areas experienced a postwar building boom which in Stoke reached a peak in the mid-1950s, to be followed by a slump and a lesser peak in the mid-1960s (Fig. 22.3). Stoke and Newcastle have dominated the postwar situation in terms of buildings constructed (Table 22.4) but relative to their total stock most other districts have far higher proportions of postwar houses. Staffordshire has a high proportion (53 per cent) of its total housing stock in the form of postwar houses, but this is largely accounted for by the areas of rapid expansion in the south of the county. The figure for north Staffordshire at 40 per cent is slightly below the west Midlands regional average of 42 per cent, but above the national average of 37 per cent.

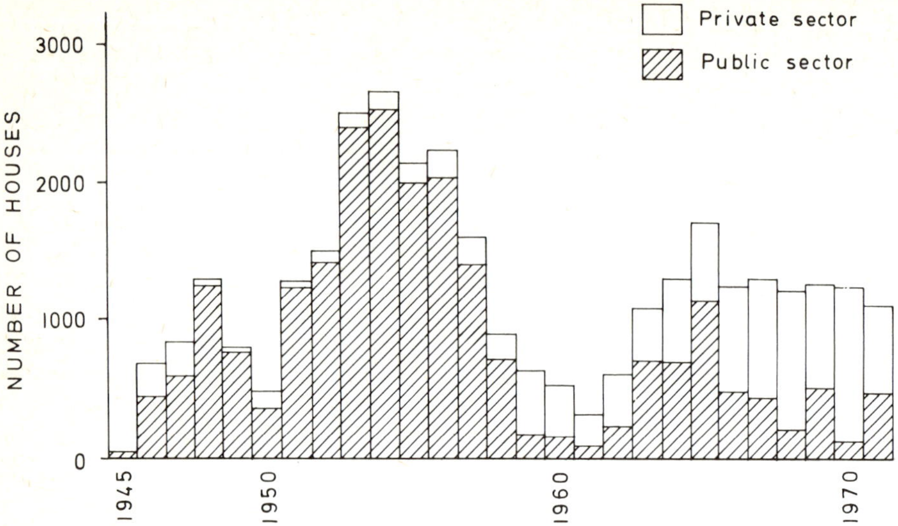

Fig. 22.3 Stoke-on-Trent: Postwar house building. Based on Plant, 1957, 1960

Table 22.4 Distribution of dwellings 1971, and number of houses built since 1945

Local authority area	*1971 total* (000s)	*Total built since 1945* (000s)	*Percentage built since 1945*
Stoke CB*	92.5	32.4	35
Newcastle MB	26.6	10.6	38
Biddulph UD	5.6	3.1	55
Kidsgrove UD	7.3	4.0	55
Leek UD	7.2	2.2	31
Stone UD	3.8	1.8	46
Cheadle RD	13.0	7.5	54
Leek RD	4.6	2.5	54
Newcastle RD	6.7	2.8	42
Stone RD	6.6	2.5	38
Total	173.9	69.4	40

* Data for Stoke refer to the end of 1971, for other areas to 1 April 1971.
Source: Staffordshire County Council, 1973; City of Stoke-on-Trent Housing Department.

Public and private sector housing

Postwar housing can be divided fairly evenly between the private and public sectors. Of a total of 67 637 houses constructed in north Staffordshire between 1945 and the end of 1972, 46 per cent were built for the private sector, a marginally smaller proportion than the national average

Table 22.5 Houses built in north Staffordshire between 1945 and 1972*

Local authority area	*Public sector*		*Private sector*	
	number (000s)	*percentage*	*number* (000s)	*percentage*
Stoke CB	22.6	69.3	9.9	30.7
Newcastle MB	6.3	62.3	3.9	37.7
Biddulph UD	1.0	32.5	2.2	67.5
Kidsgrove UD	1.6	42.7	2.2	57.3
Leek UD	1.2	65.8	1.0	44.2
Stone UD	0.6	38.8	1.1	61.2
Cheadle RD	1.5	19.6	6.3	80.4
Leek RD	0.2	13.1	1.3	86.9
Newcastle RD	0.2	11.2	1.9	88.8
Stone RD	1.3	51.4	1.3	48.6
Total	36.5	54.0	31.1	46.0

* Stoke 1945–71.
Source: Staffordshire County Planning Department; Stoke Planning Department.

of 49.4 per cent. The vast majority of the remainder were local authority houses although this total includes a small number built by other 'public' bodies such as the National Coal Board and local housing associations.

There are, however, significant spatial and temporal variations within this housing total. In Stoke and Newcastle local authority housing has formed approximately two-thirds of the total, whereas in most of the other areas the emphasis has been on the private sector (Table 22.5). After the war there was an emphasis on construction of housing in the public sector and few private dwellings were built. In more recent years this balance has been reversed. In Stoke, for example, which possesses almost half the total number of postwar houses in the subregion, the public sector accounted for over 90 per cent of total houses built in the 1950s. By 1959 the private sector became dominant and, except for the 1963–65 period, it has maintained this position. Present trends suggest that an increasing proportion of building will take place in the private sector.

Public sector housing

Local authority housebuilding played a very large role in the extensive suburban growth of the interwar period. The particular nature of this expansion varied but many large peripheral estates were 'tacked on . . . to the edge of the existing built up area, so as to avoid unnecessary expenditure on sewage disposal works and long extensions to sewers, water, gas and electricity supplies' (Kirby, 1971, p. 250). In addition to these peripheral estates, where low density development prevailed, there was also a certain amount of higher density building, usually in the form of terraced houses or maisonettes, on a few virgin or redeveloped sites nearer town centres.

Although the polynuclear nature of the north Staffordshire urban area causes certain deviations, this general pattern of development has been followed. Stoke and Newcastle carried out widespread extensions to their built-up areas during the interwar years. In the public sector Newcastle pioneered a modest scheme of twenty-nine corporation cottages as early as 1915; subsequent developments were on larger estates at Poolfields, Knutton and Cross Heath–Hempstalls north and west of the town. In the case of Stoke large peripheral municipal estates were built after 1919 at Trent Valley and during the 1920s and 1930s increasingly large scale developments were initiated at Abbey Hulton and Meir.

By 1945 peripheral local authority housing estates were concentrated in three main areas in Stoke: (1) east of the city centre at Abbey Hulton, Carmountside and Bucknall; (2) on the southeastern fringe at Meir; and (3) in the north around Burslem and Goldenhill (Fig. 22.4). The first two groups consist of five large estates which together accounted in 1945 for 40 per cent of the city's total local authority housing. In Newcastle the concentration was more marked with Poolfields, Knutton and Cross Heath–Hempstalls comprising nearly 57 per cent of the total.

After the war the two boroughs were faced with severe reconstruction problems, a product not of wartime damage but of the need to replace outworn housing and renew the nineteenth-century infrastructure (Burnett and Scott, 1962). However, the severity of the problem was mitigated by first, the subregion's slow growth in population, easing the demand for new housing and second, the existence within Stoke's boundary of a plentiful supply of land suitable for residential development.

In the immediate postwar years a qualitative element was added to residential development: 'In the past it was considered that a housing programme had been satisfactorily accomplished merely by the provision of a number of houses . . . clearly something more must be added to the old ideas on housing' (Plant, 1960, p. 32). In consequence the neighbourhood unit, an aggregation of approximately 10 000 people complete with centralised services and facilities, was adopted as the basic module for residential planning. The concept of the neighbourhood unit was pursued in two senses. In its development plan, Stoke was divided into thirty-eight neighbourhood units (Stoke CB, 1957). Most of these encompassed existing residential areas where only modest changes such as a reduction in population density were foreseen. However, some were areas on the fringe of the existing city where major expansion and the formation of neighbourhood units *sensu stricto* were planned.

The best examples of this latter category are the Ubberley–Bentilee and adjoining Berry Hill neighbourhood units. The Ubberley–Bentilee

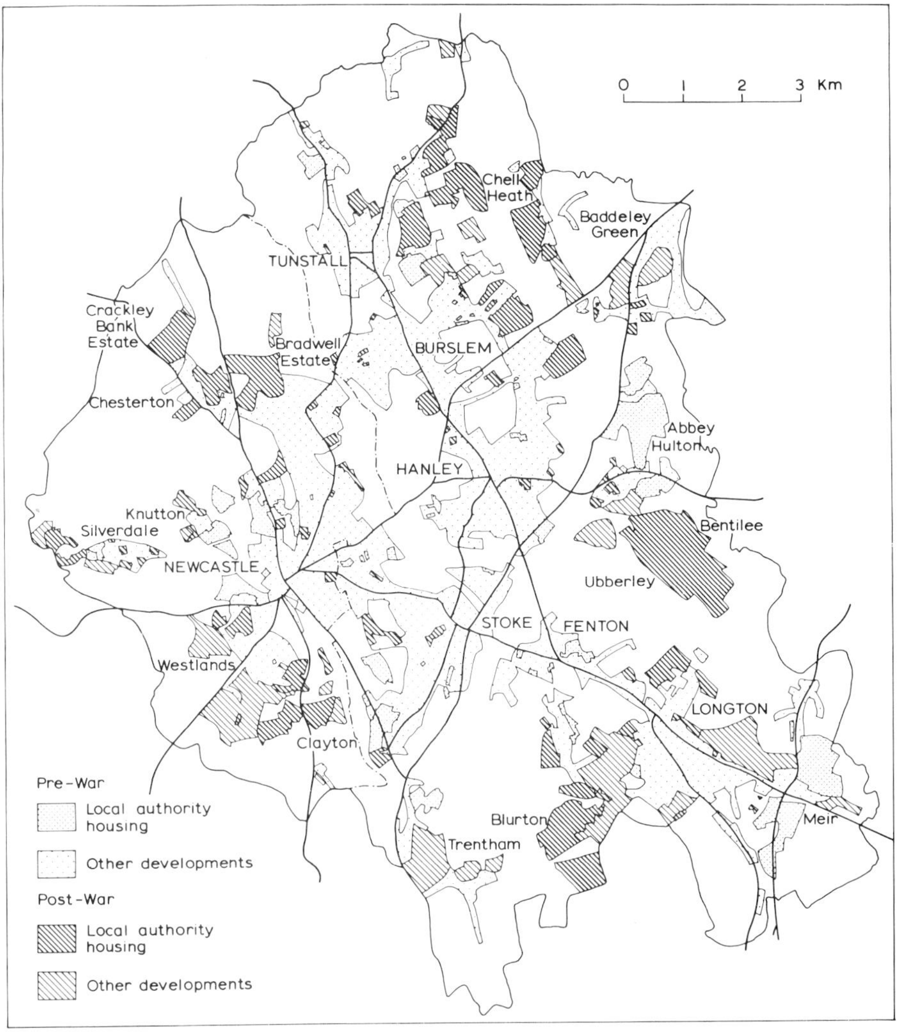

Fig. 22.4 Stoke-on-Trent and Newcastle: Pre- and post-Second World War developments

unit was constructed in the late-1950s and by 1965 it contained 3 500 homes and a population of 14 000. From the point of view of housing standards, environmental quality and the provision of local facilities, this kind of estate offers many advantages over both the unplanned growth of an earlier era and the prewar estates. It has been suggested however that its size and elongation inhibit its success as a neighbourhood or community unit (Herbert and Rodgers, 1967) and it perpetuates on a large scale the monotonous similarity of house-type and social class described in the 1949 report.

The development of local authority housing estates in Newcastle and

Stoke between 1945 and 1972 occurred at both large and small scales according to location and land availability (Fig. 22.4). There is a basic division between large new estates on the fringes of the built-up area and smaller developments on a few remaining sites nearer the town centres. The largest estates, including the massive Ubberley–Bentilee development, are located in the eastern part of the borough of Stoke, continuing the eastward expansion of the city begun before the war. With this expansion the physical constraints have become more apparent and limit future developments in this direction. Other large estates have been developed to the north at Chell Heath (993 houses) and in the south at Blurton (1 476 houses). In this way the main urban area has been encircled by major peripheral estates. The complexity of the overall urban form, however, is accentuated by the way in which some of the constituent towns, especially Longton, have experienced peripheral development of a more localised nature, thus promoting the fusion of the individual townships into a single complex urban agglomeration.

In Newcastle there is a clearcut division between two major development locations, one in the Crackley Bank and Bradwell districts north of the town and the other in the Clayton and Seabridge districts to the south. Both locations are peripheral and contain large estates – the Bradwell estate with 1 125 houses being the largest in the borough. The development of local authority estates south of Newcastle town centre in the 1960s is significant for previously this area had been exclusively one of private housing.

Throughout Newcastle and Stoke a number of smaller estates (25–250 houses) have been established on cleared sites or by infilling pockets of undeveloped land. The large estate is however most characteristic of both areas; in Stoke approximately 70 per cent of postwar public sector housing has been provided on estates of more than 250 houses and in Newcastle the percentage is only slightly lower at 65.

Although the postwar building programme has attempted to cater for the needs of specific sectors of the community by mixing its house types, and constructing flats for small family units and bungalows for the elderly, conventional two-storey houses still predominate. In Newcastle, 65 per cent of postwar housing stock is in two-storey form (compared with over 80 per cent of the prewar stock), and although figures are not available for Stoke field evidence confirms the same basic pattern. On the peripheral estates, such as Ubberley–Bentilee, where gross densities may be as low as 21.7 houses per hectare the two-storey semidetached house is the norm, whereas on smaller developments, closer to the town centres, short terraces of six or eight houses at higher densities are more common. Although both authorities make provision for accommodation in the form of flats, the area is notable for its almost complete absence of high-rise buildings.

Private residential development

The pattern of private house building in the postwar years has been very different from that in the public sector, the greatest contrast being the weak development of private housing in Stoke compared with the rest of the area. Although the absolute number of private houses constructed in the city was greater than in any other local authority area (Table 22.5), Stoke accounted for less than one-third of the total, in spite of its half-share of the subregion's population. A mere 10.8 per cent of Stoke's 1971 housing stock consisted of postwar private dwellings compared with percentages of 14.7 for Newcastle MB, 29.9 for Kidsgrove, 38.6 for Biddulph and 48 for Cheadle RD. This relatively low level of private development in Stoke can be explained largely by the city's abnormal economic and social structure, together with the poor quality of its environment. None of these characteristics is conducive to private housing development, but too much emphasis should not be placed on separating Stoke from the rest of the area because to a large extent each of the local authority areas forms part of a large housing market which functions irrespective of administrative boundaries.

The pattern of private housing development is linked closely with the growth of commuting. For this reason much of the private housing demand generated by Newcastle and Stoke has been satisfied in adjacent areas. Some of this housing is contiguous with the main urban area as at Endon in the northeast and Blythe Bridge in the southeast, but large scale private developments have also taken place in more distant villages such as Swynnerton, Madeley and Ashley.

With the exception of limited developments around Baddeley Green–Milton and Bucknall–Berry Hill postwar housing within Stoke and Newcastle is concentrated in the southern part of the built-up area (Fig. 22.4). This pattern is a direct result of land availability and land quality, but also a reflection of social criteria. The urban area varies in social status, and the aspirations and values of people who comprise the house-buying element of the subregion's population must be considered in any attempt to explain the development of different residential areas. In Newcastle, for example, when the Westlands estate was developed in the 1930s in the Garden City mould, it became a desirable residential area. This desirability, a product of its exclusive nature, ensured that property prices in the area would rise. In postwar years this district has been one of the major development areas for private housing and it has emerged as one of the few distinct middle-upper income suburbs in the conurbation. Social values, coupled with pleasant surroundings and adequate transport links with the various town centres, also contributed to private residential development in the Trentham area south of Stoke. Further east a third major area of private housing has grown up around Longton, as allowed for in the 1957 development plan.

Ex-urban growth

More than one-half of the subregion's postwar private building has been located outside the main urban area of Stoke and Newcastle. In view of the limited environmental attraction of much of Stoke, and constantly improving personal mobility, private housing will probably continue to be concentrated in the peripheral dormitory villages and commuter settlements as typified by Madeley, 8 km west of Newcastle.

Several reasons may be advanced to explain the rapid development of Madeley in the past decade. By the end of the war it had acquired a semi-industrial character with local employment in various extractive industries and in the remnants of wartime installations. Although the 1949 report suggested that residential development should be discouraged because of the value of the surrounding agricultural land it was nevertheless designated in 1951 a Group I village (Staffordshire County Council, 1951). Although this by no means guaranteed expansion it was a relevant feature in the village's subsequent growth. A further stimulus was provided by the location of a large college of education within the village and of the nearby University of Keele. Although both establishments provided most of their own residential requirements they stimulated the local economy and the demand for housing. In addition to these specific factors there is the more general consideration of Madeley's situation as a large village with a wide range of basic services and facilities and within easy commuting distance of Newcastle and Stoke.

The scale of expansion at Madeley can be seen from its population changes: between 1931 and 1951 the population was static at almost 3 000; by 1961 it had grown to 3 444 and by 1971 to 4 730, a 65.4 per cent increase in twenty years. Approximately 800 new houses have been constructed in the village since 1945, exclusive of the building directly connected with the college of education, and slightly more than half of this total are private developments.

Five major postwar estates have been added to the village:

1. The Moss – 307 houses constructed by Newcastle RDC and the National Coal Board between 1946 and 1965;
2. Brown's Farm estate – a mixed private development of 171 bungalows and two-storey houses built in the 1960s;
3. Greenmeadows – a private estate of 93 detached and semidetached houses built in the 1960s;
4. Bryn Wood – a private development of 65 architect-designed houses in wooded surroundings, still under construction;
5. Hillwood – a private estate of 94 bungalows and detached houses completed in 1973.

Despite earlier objections, much of Madeley's expansion has been achieved without substantial loss of valuable agricultural land. The Moss can be considered a logical development in terms of consolidating and 'rounding-off' the main village. Building at Bryn Wood and Hillwood has taken place on steeply sloping wooded sites of only minor agricultural value. Madeley is not unique in developing as a commuter village for Newcastle and Stoke and many other examples can be found in the rural districts of Newcastle, Stoke and Cheadle.

Conclusions

Population growth has played a minor part in the process of postwar residential growth in north Staffordshire. The need to renew older dwellings, the trend towards smaller household size and a reduction in overcrowding have been the major stimuli. In spatial terms there has been a noticeable dichotomy between the emphasis on the public building sector within the urban core of the region and private development around the fringe. This is a pattern common to many British cities, but here it has been accentuated by the environmental conditions within Stoke and Newcastle. The development of local authority estates and a small amount of private building within the urban area has taken the form mainly of extensions on the fringe of the existing built-up area in a fashion which can be described as controlled sprawl. Attempts to give this sprawl some recognisable form or community structure by the use of neighbourhood units has met with only limited success. The detailed pattern of both public and private housing within Newcastle and Stoke has been determined in part by land availability and suitability, but in general public building has been concentrated in the east and private building in the south of the urban area, a continuation of prewar trends. Much of the pressure for private building has been satisfied by ex-urban growth in surrounding commuter villages.

One notable trend in this pattern of growth is the development of a dichotomous city in social terms, with poor quality prewar housing and local authority estates comprising the main urban area (Newcastle–Stoke), and privately owned housing concentrated in a few suburban localities on the edge of the commuter belt. This is by no means unique to north Staffordshire but it is a situation which is exacerbated by the area's nineteenth-century industrial and economic legacies. Stoke and Newcastle have only 7.6 and 13.2 per cent of their population respectively within the census category of employers, managers and professional workers, whereas in the surrounding rural districts this percentage ranges from 15.0 (Cheadle) to 24.8 (Stone). This polarisation is a natural process of social sorting and segregation. If the north Staffordshire conurbation is to avoid a split between an inner zone, almost exclusively of public housing and low socio-economic groups, and an outer one of private housing and high socio-economic status, the attention of planners must be directed increasingly to this issue.

Notes

1. Defined as the local authority areas of Stoke-on-Trent CB, Newcastle-under-Lyme MB, Biddulph, Kidsgrove, Leek and Stone UDs, and Cheadle, Leek, Newcastle and Stone RDs.
2. The International Urban Research definition of the Stoke Metropolitan Area is identical to that in note 1 with the addition of Alsager UD (Cheshire).
3. In fact there was a decline, as shown in Table 22.1, but a boundary change in 1965 added an extra 15 000 so that the population of the administrative city as defined in 1961 is almost identical to that as defined in 1971.

References

ABERCROMBIE, P. and JACKSON, H. (1949) *North Staffordshire Plan*, Ministry of Town and Country Planning, London.

BURNETT, F. T. and SCOTT, F. S. (1962) 'A survey of the housing conditions in the urban areas of England and Wales', *Sociol. Rev.*, **10**, 64.

HERBERT, D. T. and RODGERS, H. B. (1967) 'Space relationships in neighbourhood planning', *Tn Ctry Plann.*, **35**, 196–98.

KIRBY, D. (1971) 'The interwar council dwelling', *Tn Plann. Rev.*, **42**, 250–68.

KIVELL, P. T. (1972) 'A note on metropolitan areas, 1961–71', *Area*, **4**, 179–84.

MINISTRY OF HEALTH AND MINISTRY OF WORKS (1944) *Housing Manual*, London.

MOISLEY, H. A. (1951) 'The industrial and urban development of the north Staffordshire conurbation', *Trans. Inst. Br. Geogr.*, **17**, 149–65.

PLANT, J. W. (1957) *The Modern Conception of Housing Development: City of Stoke-on-Trent Housing, 1919–57*, Stoke.

PLANT, J. W. (1960) *City of Stoke-on-Trent Housing, 1919–60*, Stoke.

STAFFORDSHIRE COUNTY COUNCIL (1951) *County Development Plan 1951. Survey report and analysis*, Stafford.

STAFFORDSHIRE COUNTY COUNCIL (1973) *County Structure Plan. Policies and proposals*, Stafford.

STOKE-ON-TRENT CB (1957) *Development Plan 1957. Programme map*, Stoke.

STOKE-ON-TRENT CITY ARCHITECT (1972) *Report to the City Reconstruction Committee*, Stoke.

D. T. Herbert

23 Urban neighbourhoods and social geographical research

The concept of neighbourhood is of considerable importance to the social geographer and appears frequently in urban literature as a multi-dimensional and polymorphic idea. Despite these complexities, the neighbourhood concept can still be used in measurable and relevant ways and its particular interest to geographers stems from the assumption that it refers to a territorially based unit and possesses a range of significant spatial qualities. Neighbourhood in these terms can be seen both as a distinctive segment of the city (which is distinguished from other segments by its particular features in terms of the urban social environments) and as a location within the city which has qualities derived from its geographical position. In the former, neighbourhood can be viewed in ecological terms, though not with the symbiotic connotations of early ecology, and in the latter as a location in the wider system of spatial linkages and interactions.

One of the best known definitions of neighbourhood is that by Queen and Carpenter (1953) who described it as an area in which there is the habit of visiting one another, or exchanging articles and services and, in general, of doing things together. Glass produced a similar definition though this had the important qualification of shifting the emphasis from area to group: 'a territorial group the members of which meet on common ground within their own area for primary school activities and for spontaneous and organised social contacts' (Glass, 1948, p. 124). This distinction is important though it is by no means as crucial as some sociologists believe. Geographers and ecologists using the term 'area' regard it as a surrogate for the territorial group, since any discussion of neighbourhood must be concerned with people.

Much of the difficulty surrounding the use of the neighbourhood concept is derived from the assumption that it is a single-stranded entity. It is usually viewed as a socially interactive community as both the definitions quoted suggest. However, it must be acknowledged that there are several types of neighbourhood, which may possess different kinds of qualities and can be studied in varying ways and one recent researcher envisaged a continuum along which various types of neighbourhood could be placed (Blowers, 1973). The extremes of the continuum were determined by the extent of social interaction and common ties; the range was therefore from neighbourhoods of which the only distinguishing feature was location to neighbourhoods distinguished by high levels of social interaction and community identity. Most single definitions have been aimed at this latter kind of neighbourhood, which is the most demanding in terms of qualities. The five elements of Blowers's typology of forms of neighbourhood were described as *arbitrary*, general localities with definite names but imprecise limits; *physical*, distinct environments with clear boundaries; *homogeneous*, areas of uniform physical environments and population characteristics; *functional*, areas united by typical activity patterns; and *community*, close-knit groups engaging in primary contacts and social interaction. All types of neighbourhood were regarded as spatial entities; they varied according to the extent to which spatial qualities were merely coincidental or formed a major contribution towards the understanding of social patterns.

Geographers can use each of these meanings of neighbourhood as a basis for valid and useful research. Sociologists, however, have generally only used the neighbourhood concept in a socially interactive form and have found grounds for criticism in the fact that it does not always exist in that form. There are strong links between this view of neighbourhood and the sociological concept of community. Dennis (1968) for example, criticised the neighbourhood community idea, using the concept in its socially interactive sense. However, community itself is not universally regarded and Bell and Newby (1971) in their analysis of community studies, stated that a satisfactory definition is remote. They reported the findings of Hillery (1955) that there were ninety-four definitions of community, and no fewer than seventy of these included an area term of reference, only marginally less than those which included common ties and social interaction as essential qualities. A tentative definition of community by Parsons was 'that aspect of the structure of social systems which is referable to the territorial location of persons and their activities'. The important qualification to this is that by saying 'referable to' Parsons did not mean determined exclusively or predominantly by, but rather observable and analysable with reference to, location as a focus of attention and a partial determinant (Bell and Newby, 1971, p. 32). This type of sociological definition of community, which is closely analogous to one type of neighbourhood, still retains a central territorial frame of reference.

More recent geographical research into neighbourhoods has adopted wider terms of reference than the sociological concept of community neighbourhoods. Identification of neighbourhoods with uniform structural characteristics has received a considerable amount of attention in the research field of residential differentiation; interest in perception studies and spatial images has found several applications in geographical research into urban neighbourhoods. Analyses of spatial interaction at the locality level have helped test the reality both of functional and of community neighbourhoods. Geographical research which is more problem orientated has sought to define problem areas within the city, which might rank as priority areas for social action, and has also been concerned with levels of satisfaction in various kinds of neighbourhood, including planned municipal estates, new suburbs and older parts of the central city. Examples of some of these research procedures can be drawn from studies in South Wales.

Residential differentiation and homogeneous neighbourhoods

Residential differentiation is now a well established research field in social geography and a wide range of techniques and methodologies has been developed, with an increasing tendency to adopt more quantitative forms of analysis. In this field of study, in which the most familiar methodologies are social area analysis and factorial ecology, the term neighbourhood has been interchanged often with that of subarea or social area, but the whole procedure leads to the identification of homogeneous neighbourhoods. Input for these studies is obtained from published sources, principally the national censuses, and the objectives are to identify 'social dimensions' and to use these as a basis for urban residential differentiation. Both social area analysis and factorial ecology proceed by initially identifying divisions in social space and subsequently translating these into geographical space. This procedure is in contrast with the older natural area approach which began by identifying a territory and then studied its social organisation.

Geographical research into residential differentiation can be illustrated by reference to work in the city of Cardiff (1966 population 253 000) using data from the 1966 Sample Census. A principal components procedure was employed, with a varimax rotation, using an input of forty census variables recorded for 119 enumeration districts (Evans, 1973). Scores on the five leading components were used as the input to a cluster analysis procedure which produced a classification of enumeration districts within the city. The dendrogram (Fig. 23.1) shows in graphical form the steps involved in the cluster analysis while Fig. 23.2 illustrates the spatial groups which emerged at two selected stages. As the input to the cluster analysis was formed by scores of the five leading components, enumeration districts were classified according to their similarities in this

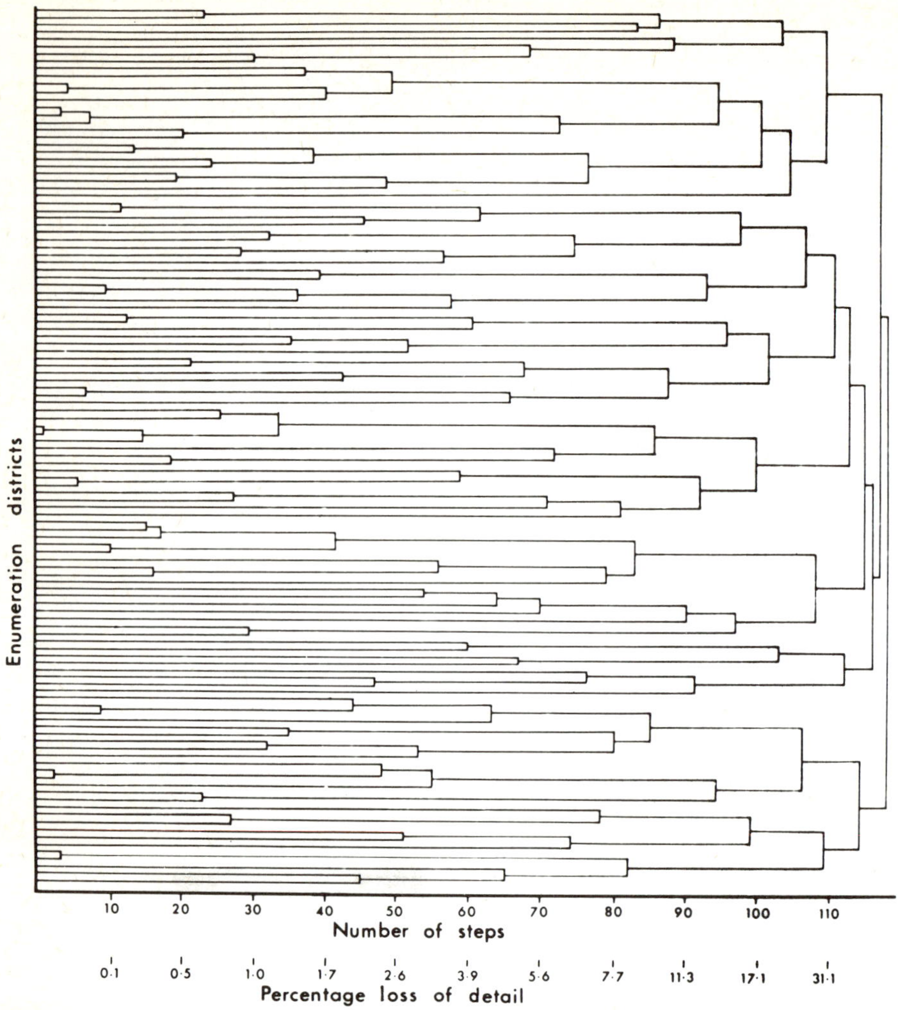

Fig. 23.1 Cluster analysis for Cardiff (1966): dendrogram

five-dimensional space. The character of the eventual areas was determined by the nature of the components as social dimensions; in the principal components analysis, for example, the three leading components were described as measures of socio-economic status, tenure and life cycle stage, and residential quality respectively. Neighbourhood homogeneity was primarily based on these characteristics. There is no formal procedure for identifying optimal solutions in the cluster analysis procedure, normal practice being to study the dendrogram and identify 'natural breaks'. From the Cardiff analysis, the nine-group classification was judged to be the best solution (Fig. 23.2 A), and the characteristics of these groups may be summarised.

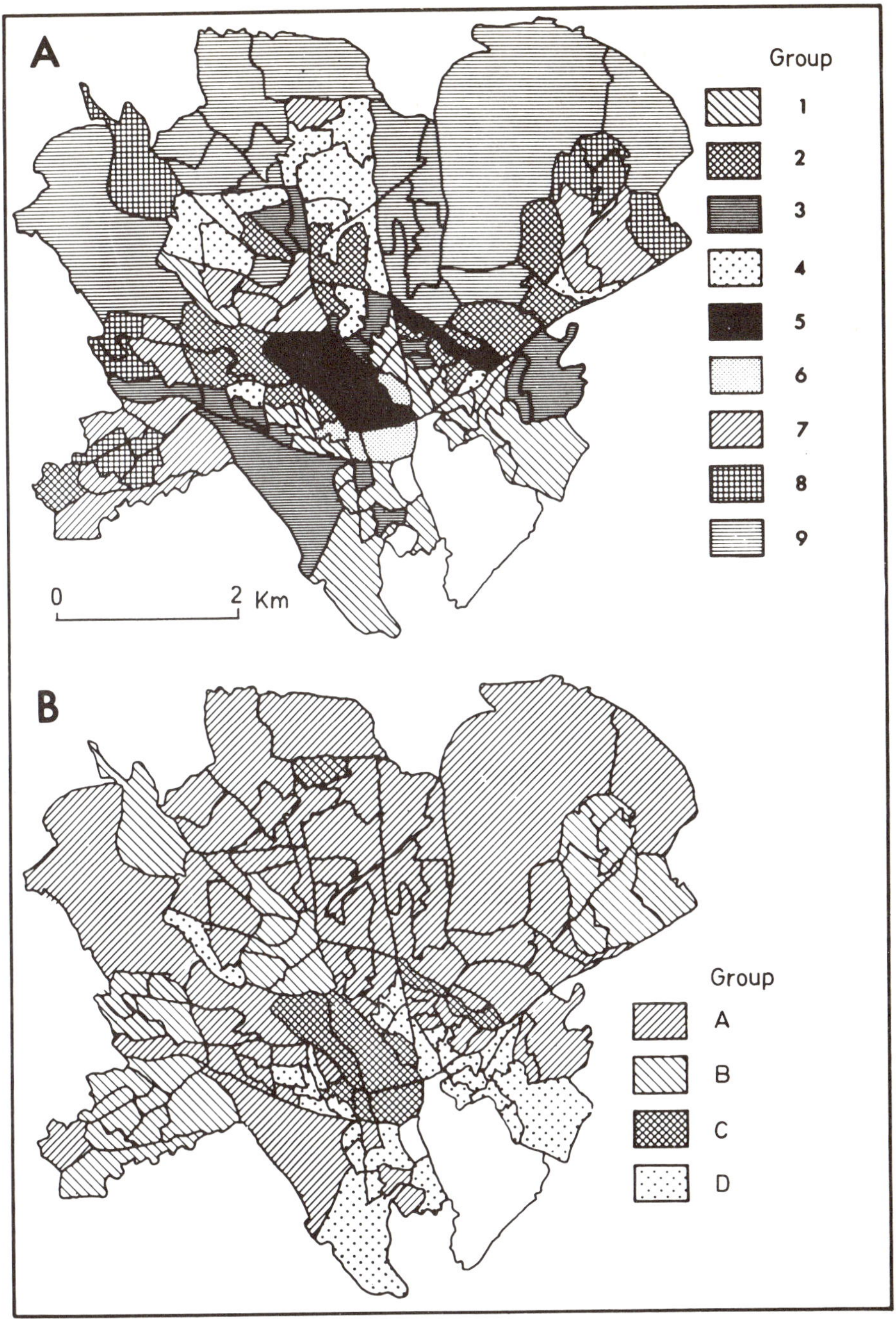

Fig. 23.2 Cluster analysis for Cardiff (1966): spatial groupings with nine groups (A) and four groups (B)

Group 1

Almost all these areas are located in central parts of the city and are terraced rows dating from the mid-nineteenth century. In terms of housing quality these are the most substandard element in the city and are the locations of contemporary urban redevelopment projects. The houses are often structurally unsound and lacking in modern facilities while the occupants are lower social class and rather overrepresented in terms of elderly people. Some of the areas (particularly Butetown) have substantial numbers of foreign-born population; other areas in this group include Splottlands, Adamsdown, Cathays and Grangetown.

Group 2

Residential areas in this category are mainly better quality terraced rows in Penyreol, Fairwater, Llandaff and Heath in what might be termed the intermediate ring of urban growth. They are above average in social class terms, as is evident from main occupational groups and levels of car ownership. Levels both of fertility ratio and of old people are below average for the city and density of dwelling occupance is low. Owner-occupance is the dominant form of tenure.

Group 3

This is a more mixed group in terms of housing type, including both private ownership areas and parts of municipal estates. Generally low in social class terms it is variable in fertility ratios but high in proportions of elderly people. Density of occupance is low and areas are in both the central city and the intermediate ring, including parts of Canton, Fairwater and Birchgrove, together with the predominantly local authority housing at Tremorfa.

Group 4

Dominantly located in Llanishen and Rhiwbina, these are modern suburbs of the city. Social class is high and housing is of good quality. Fertility ratios and proportions of old people are both low, and density of occupance is marginally high.

Group 5

This comprises terraced rows of slightly better quality, located close to the central city in the Pontcanna district. Social class is variable, reflecting the adoption of some parts of the areas by younger professional families. Fertility is low and the proportion of elderly people is high. There is a low density of occupance but a feature is the comparatively high incidence of private rented accommodation.

Group 6

There are a limited number of enumeration districts in this group and these are largely located close to the city centre in the Riverside district. Social class is low, and fertility and proportions of old people are average. Morphologically these are districts of large terraced houses which have been converted to flats and there is a high incidence of shared dwellings.

Group 7

This is a large group which contains predominantly local authority estates at Rumney, Mynachdy and Caerau. Social class is low, and fertility ratios are below average as are proportions of old people. Density of occupance is well above average in these more established council housing areas.

Group 8

This is the second group of local authority estates including Llanrumney, Ely and parts of Fairwater. These are generally more recent council houses, and fertility ratios and densities of occupance achieve the highest averages for all the enumeration districts in the city. Social class is low and there are very few elderly people.

Group 9

This group comprises the prestigious residential districts of Cardiff at Cyncoed, Radyr, Rhiwbina, Whitchurch and St Mellons. Social class is very high and fertility well below average. There is a high proportion of old people and a low density of occupance. Most of these districts contain modern and comparatively expensive houses.

These nine groups are aggregated to form larger clusters at successive stages of the analysis procedure. At Step 112, seven groups emerge by the amalgamation of former groups 2 and 4 and of groups 5 and 6. The former combination is of two areas in the intermediate to suburban ring of relatively high status housing; the latter includes the more transient, shared dwelling districts of Riverside and Pontcanna. At Step 115

(Fig. 23.2 B), four groups remain which comprise the following combinations of groups from the former nine group solution:

A	7 with 8
B	9 with 3 with 2 with 4
C	1
D	5 with 6

This macroscale of grouping produces the local authority housing areas (A), the most substandard districts (C), the transient, shared dwelling area (D), and a large group (B) which includes all more suburban areas in private tenure, with above average social class composition.

Residential differentiation and the identification of homogeneous neighbourhoods may be ends in themselves and studies have produced dimensions of residential structure and spatial generalisations or models (Rees, 1969). These procedures may also be used to identify area sampling frameworks from which particular types of residential areas may be selected for detailed investigation. Social area analysts, for example, have studied social interaction and participation in contrasted neighbourhoods (Bell, 1969); in the Cardiff study, Fig. 23.2 A shows part of an area sampling framework from which six enumeration districts were selected (Herbert and Evans, 1974). This set (Fig. 23.3) was selected to cover a range of socio-economic characteristics and included old terraced rows, local authority estates and a high status suburb. Information on rates of delinquency residence was also incorporated into the selection procedure for this particular study. Enumeration districts were chosen as the sample units because they formed the basic unit for census statistics; they were also selected as representatives of the larger homogeneous neighbourhoods of which they were part. Area sampling procedures of this kind place studies of individual neighbourhoods in a holistic context and also allow qualities of neighbourhood other than structural homogeneity to be tested.

Information in the Cardiff social survey can be used for examining some of these other qualities in two of the selected study areas. The examples chosen are patterns of social interaction, spatial images and mental maps, and assessment of neighbourhood qualities. The two study areas provide a sharp contrast as types of residential district. Adamsdown is a low-income terraced row area near the city centre and adjacent to heavy industries which provide a high proportion of local employment. The nearby district of Splott has experienced considerable housing clearance in recent years and parts of Adamsdown will be similarly affected. Rhiwbina, the other study area, is a prestigious suburban district north of the city centre which is part of the modern peripheral expansion of Cardiff. Using a standard occupational classification, 75 per cent of the householders of Rhiwbina could be categorised as middle-

Fig. 23.3 Survey areas in Cardiff

class compared with 15 per cent of the Adamsdown householders. Both areas were predominantly owner-occupied, though Adamsdown had a significant minority – 25 per cent – in private tenancies. In terms of age structure and household size, Adamsdown had an older population with more small households. It also had a much more established population, 51 per cent having lived in the district for more than twenty years, whereas in Rhiwbina, 54 per cent had moved into the area within the previous five years.

Patterns of social interaction

An essential of both functional and community neighbourhoods is the quality of linkages and of levels of locality based social interaction. This has been a dominant theme in geographical analysis of neighbourhood. Buttimer (1969) has elaborated the French concept of social space, based not on principles of social stratification but on the hypothesis of a hierarchy of spaces within which groups live, move and interact. One level within this hierarchy was described as neighbourhood space, or the network which encompassed daily and local movement. Boal (1969) studied four types of movement in central Belfast: movement to a bus stop for a city centre bus; movement to buy groceries; movements associated with visits to or from relatives or friends; and linkages suggested by pre-marriage addresses of interviewee and spouse. A standard working-class pattern characterised by an overlapping series of close-knit networks was identified. A special feature of these activity systems in the Belfast context was the rigidity of the divide between Roman Catholics and Protestants. Other geographical studies have investigated activity patterns within planned neighbourhoods; a study by Herbert and Rodgers (1967) of a large local authority estate in Stoke-on-Trent discovered significant variations from one segment of the neighbourhood to another. The central portion was closely integrated with the neighbourhood but the eastern and western peripheries had more linkages to centres outside the neighbourhood.

Results from the Cardiff study are less concerned with the spatial nature of linkages than with the general variation in levels of different types of social interaction in the two districts. Table 23.1 contains data on the social visiting patterns of respondents and their husbands or wives; the question required information on the number of people in the neighbourhood who were known well enough to visit.

There are marked differences between the two neighbourhoods, which may be summarised by the statement that levels of social interaction, expressed by these categories of visits, are much higher in the middle-class area. Adamsdown has much higher proportions of people who never interact with neighbours in each of the four categories. Its proportions with no visits range from 36 to 69 per cent in contrast with the Rhiwbina range of from 8 to 36 per cent. Among those respondents who did socially interact with their neighbours in these ways, there were still significant differences between the two areas. In Rhiwbina the proportions who nominated six or more neighbours ranged from 40 per cent, for people they would go out with in the company of husband or wife, to 72 per cent for people they would visit on their own. The range of scores in Adamsdown on the same measure was from 14 to 29 per cent. On this evidence levels of locality based social interactions are markedly higher in the middle-class area, despite the fact that Adamsdown portrays many of the qualities of the close-knit working class community. In order to examine more closely the bases of these apparent area differences the whole sample (for the six survey areas) was analysed in terms of several other parameters which included age, sex, length of residence and social class. There were some significant variations, the over sixty-five age group, for example, being consistently low in terms of visits of

Table 23.1 Types of social unit to households in Adamsdown and Rhiwbina

	Adamsdown				*Rhiwbina*			
Number of households	*Visit by husband or wife*	*Visit by husband and wife*	*Evening out for husband or wife with neighbour*	*Evening out for husband and wife with neighbours*	*Visit by husband or wife*	*Visit by husband and wife*	*Evening out for husband or wife with neighbour*	*Evening out for husband and wife with neighbours*
	(percentage of respondents)				(percentage of respondents)			
0	36	50	66	69	8	22	22	36
1–2	14	9	7	6	8	8	12	8
3–5	12	10	12	10	12	14	16	16
6–10	10	11	9	8	26	20	22	24
11–20	6	6	3	8	28	26	26	14
Over 20	11	12	2	1	18	10	2	2

Source: Field data.

all kinds. By far the most striking variation, however, was by social class. Proportions recording no interaction at all in the various categories were persistently highest in the low social class groups; proportions with no visits in the company of spouse, for example, ranged from 24 per cent in class AB (middle and upper-middle class) to 72 per cent in class E (lowest working class). The clear and progressive social class differences serve to validate the area differences, as social class is the main contrast between Adamsdown and Rhiwbina.

As a further test regarding patterns of interaction, information was collected on the extent to which friendships were maintained with people outside the immediate neighbourhood but within the city. There were in fact no striking differences between the two areas in this kind of interaction. Of the Rhiwbina sample 32 per cent recorded no friends in other parts of the city with whom they could exchange visits, compared with 35 per cent in Adamsdown. Proportions nominating at least two such friends elsewhere in Cardiff were 32 per cent in Rhiwbina and 23 per cent in Adamsdown. Geographical distributions of such friends in relation to each of the two areas were different in terms of types of districts – friends were found in similar status neighbourhoods – but were not dissimilar in distance and directional terms. Friends of the Adamsdown population were typically from nearby terraced row districts and in a minority of cases (17 per cent) from peripheral local authority estates. Linkages with friends from Rhiwbina were longer and rather more dispersed but the nearby suburbs of similar status figured prominently in the pattern (Fig. 23.4).

This information on levels of interaction within neighbourhoods does not confirm the high levels of local interaction usually associated with older working-class areas, nor does it demonstrate a substantially higher non-locality based social interaction for the middle-class area. The average number of neighbourhood friends per household was calculated at 1.5 in Adamsdown and 3.8 in Rhiwbina; average numbers of friends in other parts of the city was 1.0 in Adamsdown and 1.3 in Rhiwbina. The implications of these findings for Adamsdown are that the traditional highly interactive working-class community may have diminished qualities. There may be local bases of explanation, such as an ageing population as younger families move to new estates and an allied reduction in levels of occupance and population density. Alternatively more general aspects of societal change, such as new work-residence relationships or the reduction in community-binding forces such as common poverty or deprivation, may have led to an increase in home-centred activities. From the Rhiwbina evidence, some important social geographical forces may be discerned. Despite current theories on an emerging 'cosmopolite' middle class and the increasing significance of non-place communities (Webber, 1963), there is no support here for these

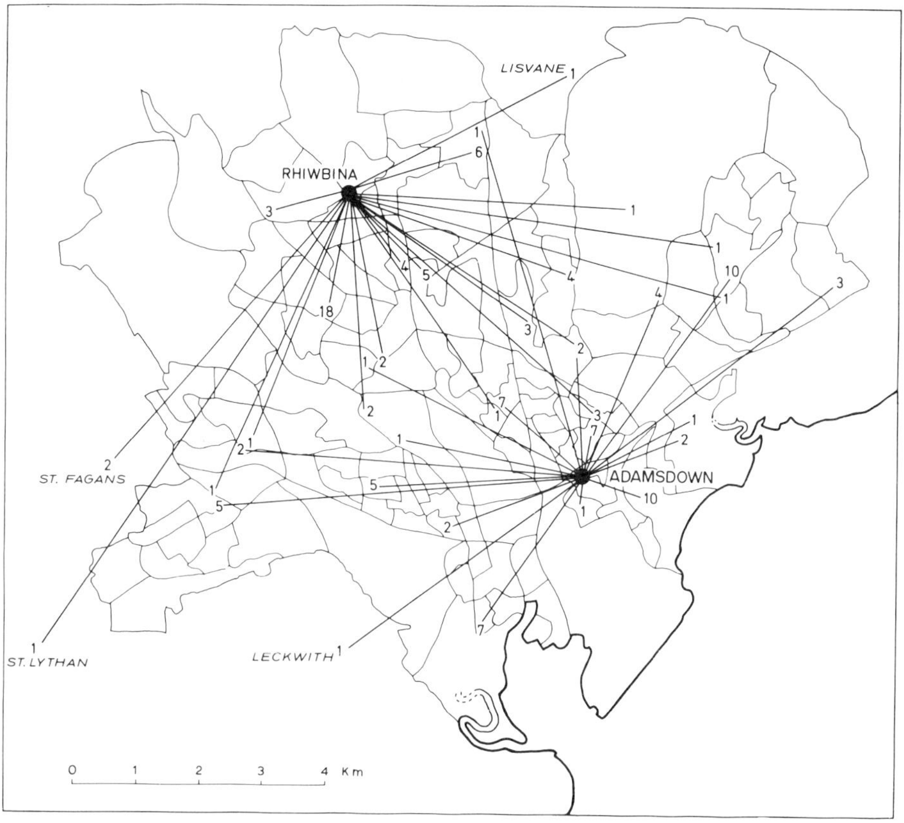

Fig. 23.4 Spatial linkages for friendships for two survey areas

contentions. The Rhiwbina population may have dispersed friendship patterns and linkages determined by specialist interests, but a striking proportion of their social interaction is in fact locality based and urban place or neighbourhood is a significant reference in comprehending this type of behaviour.

Spatial images

The idea of a perceived or symbolic neighbourhood, which may be investigated in terms of spatial images or mental maps, has emerged as a strong field of interest in geographical research. Eyles (1968), for example, studied residents' perceptions in spatial terms of Highgate Village. A sample of the local population was asked to mark the boundaries of the 'village' on a map and the results were examined for order and consensus comparing subgroups with different ages and lengths of residence in the area. As part of his more general study of neighbourhoods in Cambridge, Lee (1968) asked housewives to define on maps the limits of the neighbourhoods within which they lived. Analysing these results in conjunction with information on patterns of activity, Lee found a high degree of individual variation in terms of definition of neighbourhood, of levels of involvement in locality and of attitudes. He suggested that although there was no close agreement on the meaning of neighbourhood, there was a 'consentaneity' of views, implying agreement and interdependence but not necessarily in reciprocal systems. Identification of the spatial image of neighbourhood as perceived by its inhabitants is important both in academic and practical terms; the bases for variation may be examined and the reality of planned neighbourhoods can be assessed.

Methodology in the measurement of mental maps has been variable and a rigorous and consistent set of techniques has yet to be established. Researchers have asked respondents to draw boundaries on maps, to name boundaries or limits, or to list the streets which comprise the neighbourhood. Some of these approaches are liable to distortions, use of maps, for example, making the unwarranted assumption that all

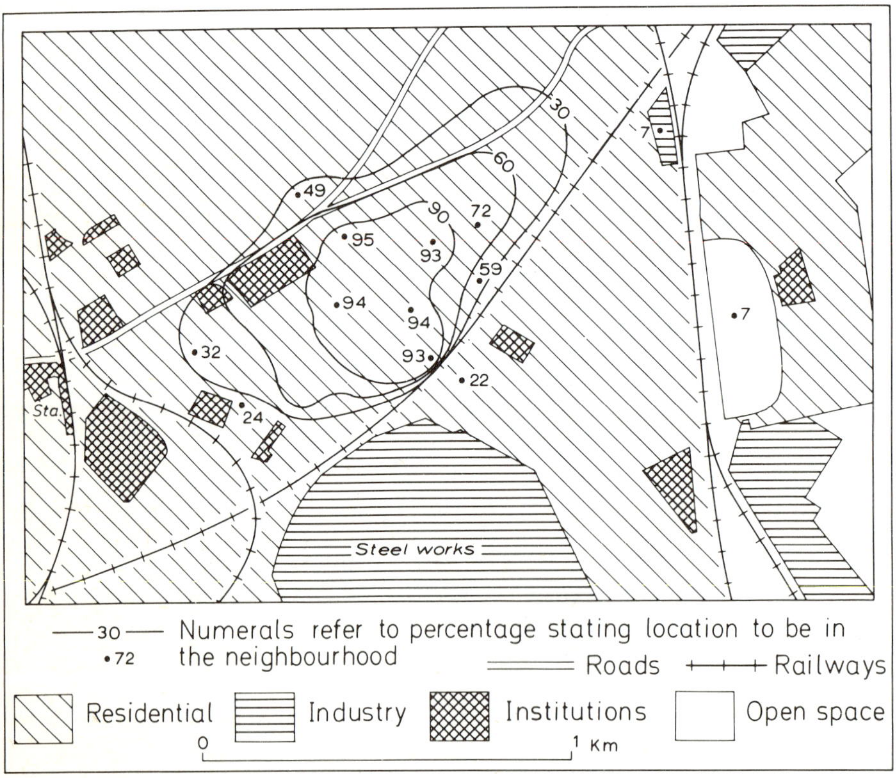

Fig. 23.5 Spatial images of Adamsdown

people can understand spatial patterns. One technique employed at a regional scale has involved presenting the respondent with a list of locations (some of which are streets and others landmarks) and requesting a rating of each according to whether it is definitely in, possibly in, or definitely not in the neighbourhood (Herbert and Edwards, 1972). This is a compromise in that it scores a set of points and the lines linking these have to be speculative, but it does seem to operate at a level of detail which people can readily understand.

This approach has been applied to the two Cardiff neighbourhoods (Figs 23.5 and 23.6). One basis of differences between Adamsdown and Rhiwbina was clearly related to urban morphology and densities; the

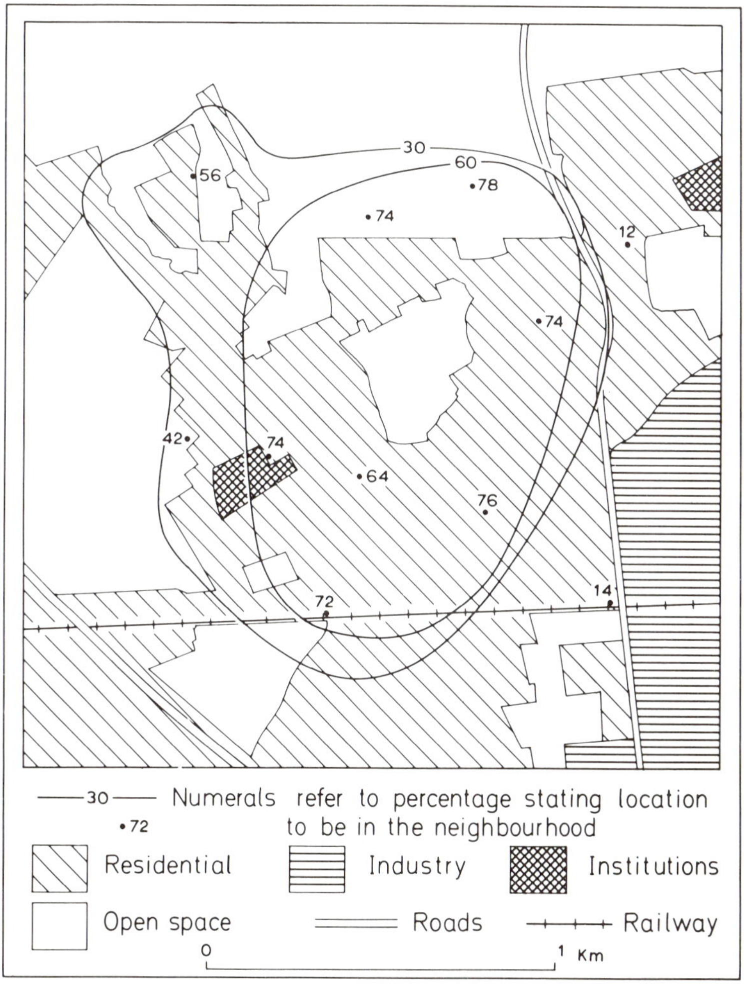

Fig. 23.6 Spatial images of Rhiwbina

perceived neighbourhood in Adamsdown was much more compact and spatially constrained than that in Rhiwbina. For the Adamsdown sample there was a high level of consensus as to what area comprised the neighbourhood. The locations scoring over 90 per cent, that is the proportion regarding the location as definitely in the neighbourhood, define a distinctive and meaningful area. Scores decreased gradually towards the eastern periphery, but breaks were much sharper on all other sides. These breaks could be related to natural or ecological barriers, particularly in the south where the railway line divided Adamsdown from the Splottlands. Although Newport Road on the north side is a major traffic artery, almost half the respondents regarded Cardiff High School as being definitely in the neighbourhood. There are two possible explanations for this: first, the fact that some of the children in these households attend the school; and second, that areas north of the road gradually become of higher status. Interviewers in the field noticed these distinctions being made; residents of Adamsdown were willing to regard themselves as part of Roath (essentially north of Newport Road) but were adamant that they did not live in Splott (south of the railway). To the west, Howard Gardens was only regarded as definitely in the neighbourhood by one-third of the respondents and scores declined sharply towards the industrial area. This fall in scores to the southwest was marked despite the absence of any strong ecological barriers.

A feature of the perceived neighbourhood in Rhiwbina was the lower level of consensus even in the core of the district. Highest scores were typically 70 to 75 per cent rather than those of over 90 per cent which characterised Adamsdown. This level of high scores, however, was found over a much more extensive area than was the case in Adamsdown, in part a reflection of morphology and density but also suggesting a less spatially constrained image. The only sharp declines in scores occurred to the east and southeast where Thornhill Road and Caerphilly Road provide marked barriers and residential areas begin to be typically lower status.

Applied aspects of neighbourhood research

Besides studies of neighbourhood which are primarily academic but may have applied value, geographers have begun to analyse neighbourhood with a more specific orientation towards practical ends. Buttimer (1972) has applied concepts of social space to planned neighbourhoods in Glasgow with a primary aim of assessing the conventional standards used in residential area design by examining residents' attitudes towards, and evaluations of, the design and service provision of selected housing estates. Rejecting the argument that individuals are so variable in terms of values and perceptions of the ideal residential environment as to make generalisations for planning impossible, she stated the case for matching environmental design to the needs of its residents. The more general and persistent proclamations of the need for relevance and practical applicability in geographical research have emerged from American studies of inner city areas. Peet (1972) offered a general statement that geographers have had very little to say about social problems despite the fact that such problems are more characteristic of some places than of others and are amenable to geographic analysis. It is the problem areas which have provided the obvious arenas for involvement and action; neighbourhoods typified by deprivation have been analysed as communities and principles of community development have been applied.

One approach has attempted to measure levels of satisfaction with local urban environments and to obtain assessments of individual aspects of neighbourhood quality. This was the procedure adopted by Buttimer (1972) and some similar results are reported for the two Cardiff neighbourhoods, though full analyses are not yet available. On the question of satisfaction with the neighbourhood as a place in which to live, majorities in both areas were quite satisfied, though there was a higher proportion in Rhiwbina, 96 per cent compared with 67 per cent in Adamsdown. A fuller examination of attitudes towards neighbourhood facilities was obtained by recording responses to a list of qualities on a three point scale.

There was a keen awareness of the poor quality of their urban environments by the inhabitants of Adamsdown. Only small minorities, for example, regarded it as a good area in terms of qualities such as cleanliness, noise, general appearance, and availability of open space; Rhiwbina,

Table 23.2 Assessment of neighbourhood qualities

	Adamsdown			*Rhiwbina*		
	good	*fair*	*poor*	*good*	*fair*	*poor*
Quality	(percentages)			(percentages)		
Cleanliness	16	48	36	74	24	2
Convenient shops	97	3	0	94	4	2
Neighbourly contact	63	26	11	58	32	10
Safety	37	43	20	54	38	8
Privacy	72	20	8	78	20	2
Play facilities	0	16	84	16	52	32
Noise	11	39	50	76	24	0
General appearance	6	47	47	82	18	0
Open space	0	9	91	50	38	12
Social facilities	16	62	22	42	50	8
Accessibility	61	39	0	52	30	18

Source: Field data.

by contrast, had high positive scores on all of these qualities. People saw Adamsdown as having most advantages in terms of convenience to shops, neighbourly contact and accessibility; in each of these qualities it scored higher than Rhiwbina but only marginally. For social facilities, Rhiwbina scored higher, but both areas had very low positive scores on the question of play facilities for children. To a large extent assessments of this kind are based on the particular set of values of the individual, but there were high levels of consensus within each neighbourhood on several of the nominated qualities. Neighbourhood studies of this kind, if taken further, may demonstrate the extent to which apparent deprivation is real in the perceptions of the inhabitants and may help planning decisions.

Conclusions

This discussion has illustrated some aspects of geographical analyses of neighbourhood but many avenues of further research remain. It is likely that most of these will rest on fuller comprehension of social values and on more detailed analyses of attitudes, behaviour and perceptions in a spatial context. Besides the methodology which will develop from within the subject, there are still derivative concepts from the sociological sciences which have yet to be used to their full potential. Reference group theory, for example, offers a relevant set of concepts and the Sherifs (1964, p. 272) argued forcibly that society should be studied in group terms: 'Groups are facts of human existence. There are groups which engage in socially laudable and socially obnoxious activities.' Reference groups are defined as those to which the individual identifies or aspires to belong; understanding the norms of the group may be regarded as one key towards the understanding of individual behaviour. A question for geographers is the extent to which neighbourhood or territorial group constitutes a significant reference group. Although reference group theory has stimulated a considerable literature, little of this has possessed spatial implications. Recently, Eyles (1973) considered the relevance of reference group theory to geographical research, though his particular emphasis was upon the relationship between reference groups and feelings of deprivation. Lee (1968, p. 261), who developed a sociospatial schema concept for neighbourhood, had doubts on the general applicability of reference group theory to the collective neighbourhood, but suggested: 'It could be argued that the neighbourhood schema itself, in its social aspects, has reference group properties.'

Another approach which may be relevant to the geographical study of neighbourhood is social network analysis, developed principally by social anthropologists. Connell (1973), writing on social network analysis in a geographical context, suggested that it was not a systematic theory but a series of techniques used to arrive at an explanation for urban social relations. As yet most social network analyses have not used a spatial parameter, but there seems clear potential to do so. Irving (1973), in an exploratory study, suggested that there were problems in applying the techniques but that suitable indices might be constructed. Lee (1968), saw social network analysis as a potential way of studying neighbourhood relationships if it could begin to take account of the physical spatial referent for social relationships.

These concepts and others promise interesting possibilities for social geographers. Some geographical approaches, such as the methodology for classifying homogeneous neighbourhoods, now seem to have reached logical conclusions; others, such as the investigation of symbolic neighbourhood as a spatial image, are at early stages of analyses. Perhaps the most challenging prospect of all for social geographers lies in the study of functional and community neighbourhoods. Here, besides identifying the patterns, the problem is to examine the extent to which neighbourhood is a referent in understanding attitudes and behaviour. Although the evidence for true cause and effect must be handled with care, there are already good grounds for asserting the relevance of a geographical perspective which emphasises the importance of spatial qualities and the reality in behavioural terms of territorially located groups.

References

BELL, C and NEWBY, H. (1971) *Community Studies*, Allen & Unwin.

BELL, W. (1969) 'Urban neighbourhoods and individual behaviour', in P. Meadows and E. H. Mizruchi, eds, *Urbanism, Urbanisation and Change: comparative perspectives*, Addison-Wesley, pp. 120–46.

BLOWERS, A. (1973) 'The neighbourhood: exploration of a concept', in *The City as a Social System*, Open University, Urban Development Unit 7, pp. 49–90.

BOAL, F. W. (1969) 'Territoriality on the Shankill–Falls divide, Belfast', *Ir. Geogr.*, 6, 30–50.

BUTTIMER, A. (1969) 'Social space in interdisciplinary perspective', *Geogr. Rev.*, 54, 417–26.

BUTTIMER, A. (1972) 'Social space and the planning of residential areas', *Environ. & Behav.*, 4, 279–318.

CONNELL, J. H. (1973) 'Social networks in urban society', *Inst. Br. Geogr. Spec. Pubn.*, 5, 41–52.

DENNIS, N. (1968) 'The popularity of the neighbourhood community idea', in R. E. Pahl, ed., *Readings in Urban Sociology*, Pergamon Press, pp. 74–92.

EVANS, D. J. (1973) 'A comparative study of urban social structures in South Wales', *Inst. Br. Geogr. Spec. Pubn.*, 5, 87–101.

EYLES, J. D. (1968) *The Inhabitants' Images of Highgate Village, London*, LSE Department of Geography, Graduate Discussion Paper 15.

EYLES, J. D. (1973) 'Spatial opportunity and the concept of the reference group', *Prof. Geogr.*, **25**, 121–3.

GLASS, R. (1948) *The Social Background of a Plan: a study of Middlesbrough*, Kegan Paul.

HERBERT, D. T. and EDWARDS, J. A. (1972) 'Images of Randstad', *Tijdschr. Econ. Soc. Geogr.*, **63**, 239–43.

HERBERT, D. T. and EVANS, D. J. (1974) 'Urban sub-areas as sampling frameworks for social surveys', *Tn Plann. Rev.*, **45**, 171–88.

HERBERT, D. T. and RODGERS, H. B. (1967) 'Space relationship in neighbourhood planning', *Tn Ctry Plann.*, **35**, 196–8.

HILLERY, G. (1955) 'Definitions of community: areas of agreement', *Rur. Soc.*, **20**, 111--23.

IRVING, H. (1973) *Report on the Pilot Investigation for the Neighbourhood Study of Hull and Haltemprice, 1971*, Social Sciences Research Council.

LEE, T. R. (1968) 'Urban neighbourhood as a socio-spatial schema', *Hum. Relat.*, **21**, 241–67.

PEET, R. (1922) 'Some issues in the social geography of American poverty', in *Geographical Perspectives on American Poverty*, Antipode Monograph 1, Worcester, Mass., pp. 1–16.

QUEEN, S. A. and CARPENTER, D. B. (1953) *The American City*, New York.

REES, P. H. (1969) 'Concepts of social space: toward an urban social geography', in B. J. L. Berry and F. E. Horton, eds, *Geographic Perspectives on Urban Systems*, Prentice-Hall, pp. 306–94.

SHERIF, M. and SHERIF, C. W. (1964) *Reference Groups*, New York, Harper & Row.

WEBBER, M. (1963) 'Order in diversity: community without propinquity', in L. Wingo, ed., *Cities and Space*, Johns Hopkins Press, pp. 23–54.

H. B. Rodgers

24 The demand for outdoor and active recreation: a decade of research in retrospect

Almost a decade has passed since the first contacts were made between the University of Keele and the British Travel Association (now the British Tourist Authority) from which the Pilot National Recreation Survey was developed in the Department of Geography at Keele. The time was then opportune for the launching of a first attempt to measure the national pattern of demand for outdoor and active recreation, which, in the future, would make a greater call on national land and investment resources. Real incomes had risen considerably: the greater material prosperity of the average family was expressed most dramatically in the rapid spread of car-ownership. Although actual weekly hours of work had not diminished much (House of Lords, 1973), the five-day week had become well established and the majority of the population had at least two weeks' paid holiday. The recreational needs of the affluent society were beginning to be perceived, although initially they had been recognised more clearly outside Britain, notably in the long series of reports of the Outdoor Recreation Resources Review Commission (1962) in the USA. Recreation research in Britain itself was still in a pioneer stage: Burton and Wibberley (1968) had begun work on aspects of rural recreation and Dower's *Challenge of Leisure* (1965) was influential in stimulating greater public interest and concern. At Westminster, policies were being framed to bring government more explicitly into recreational planning. The advisory Sports Council began its work in 1965 and in 1968 the Countryside Act replaced the National Parks Commission by a new Countryside Commission with a broader remit (though still with too little real power).

The early 'demand' surveys in retrospect

There was plainly a need for a data base for recreational planning and for some statement and analysis of the recreational use of leisure to serve as a framework of reference for both the shaping of policy and decision-taking. By the quirks of both fate and politics, not one but two surveys were to be initiated to fill this void. As the joint Keele and British Travel Association (BTA) project began to evolve, so a number of government departments commissioned from the Government Social Survey a recreational survey, eventually published as *Planning for Leisure* (Sillitoe, 1969). To have had two repetitive inquiries where one would have done may seem wasteful mismanagement. It may be argued that a single collaborative survey, larger in sample size and stronger in resources, would have been of greater service to the national need, but doubts exist. A doubling of sample size would not have improved radically the statistical reliability of the results, while the elimination of one or other survey would have reduced their joint informational content. In fact the two exercises, beginning from much the same brief, developed very differently.

To summarise, the Keele study was more general, sampling both the urban and rural population of the whole of Britain. The official inquiry was limited to the urban population of England and Wales. Designed to meet the needs of several government departments, it concentrated on the recreational habits of people in new towns and inner London, incorporating separate subsamples. The Keele inquiry, in contrast, examined in some depth the characteristics of recreational trips (BTA, 1967) and, in a second report, it alone took the considerable statistical risks of analysing regional differences (BTA, 1969). As a result, more is known probably from the two separate surveys than from a joint one, which would have pleased only the statisticians, but still left them critical.

Reviewing the reports of these two studies, and writing self-critically as the author of one, it is clear that too much was expected of them at the time, especially by their professional users – the planners, civil servants and both public and private entrepreneurs. They were reconnaissance surveys into a methodological wilderness as there was little theory building or conceptual work to guide them. Neither could attempt much more than simple description based on elementary analysis and it would have been premature for either study to attempt to develop any theory of recreation behaviour framed within the concept of a 'demand model'. Moreover, given the time-scale and available resources, it was impossible to develop predictive analysis beyond an attempt to discern recreational change at a speculative and inferential level. Basically, both were datum studies.

The accent in subsequent research has shifted towards correcting these omissions and shortcomings, reflecting both official research policy and spontaneous academic interest. From its inception the old advisory (now an executive) Sports Council has sought to stimulate and guide recreation research, often in cooperation with the agencies represented in the Countryside Recreation Research Advisory Group. Both the Sports Council and Countryside Commission have identified priorities for research into areas of ignorance in the recreation demand field (Sports Council, 1971; Countryside Commission, 1971). Among these were inquiries into regional and subregional variations in recreational demand. Little of substance is known about spatial variance in recreation, save that it exists and at different scales. Contrasts appear not only at the major regional scale, but also at the subregional scale. Almost certainly, each major local authority within the new administrative structure is recreationally idiosyncratic to some degree, generating a distinctive system of demand. However, virtually nothing is known about demand at this 'operational' spatial scale, at which major planning decisions are made. At a still smaller scale, separate communities at the neighbourhood or district level must generate distinctive recreational needs (if only because of their pronounced social contrasts): this is the scale of recreational activity spaces and movement networks, in which trip lengths to participant recreations are very short. At the larger regional and even subregional scales, surveys have begun to add detail to the pattern (National Opinion Polls, 1969: North West Sports Council, 1972), but at the smaller scale of spatial variance little guidance is available.

The recreational future: the prediction of demand

The central problem in the demand side of recreation research is that of forecasting future growth and change and is certainly incapable of exact solution. Whether the planner has, today, an acceptable datum statement of the size and structure of demand for recreation is arguable, but there is no doubt that he has no forecasts better than half-informed guesses about even the short-term planning time-scale of the next ten or fifteen years. Coupled with this problem of temporal prediction, there is the question of 'latent' demand, that is the increased participation a population is capable of generating, given its present socio-economic attributes but improved access to facilities. Each new pool, sports centre or squash court translates a latent demand in its catchment population into a high level of additional participation. As yet, this increment cannot be quantified, though several impact studies of the effect of new facilities on the catchment population are in progress. Clearly supply leads demand, and to this extent large investments in new facilities may be merely remedial. They correct present underprovision but may not be able to absorb the future growth of demand. In order to gain a sharper understanding of some of these severely practical problems, the Sports

Council has given some degree of priority to fundamental work of a conceptual and methodological type. This work might lead to operationally acceptable demand models in which, ideally, every influence on participation would be simulated and all factors synthesised, in what could prove a very powerful aid to policy and decisions at any scale of operation. This ideal is remote and it is questionable whether so mercurial an aspect of human behaviour as recreation can be treated within the framework of normative response to given stimuli (Burton, 1971).

Prediction, the influence of supply on demand and attempts to build demand models have all prompted a growing volume of research interest, although it would be misleading to suggest that this has conveyed a clear message to the planner. In the pioneer national demand surveys, prediction took the form of inference, based on the socio-economic profiles of participants in particular activities. Golf was shown to be a sport of the prosperous, car-owning professional or executive man (and his wife), with a steeply rising level of participation across the middle to higher ranges of the income scale, and high levels of participation persisting late in middle age. Affluence will increase the incidence of some of these attributes in the total population: the number of potential golfers will rise, *pro rata*; and so golf is a 'growth sport'. Team games have stronger low income associations; they are not related closely to car ownership or to other indicators of affluence, and there is a pronounced age-related decay of participation through the twenties age group (with a sharp downturn after marriage). Thus team sports have little potential for relative growth. From such arguments (backed by the results of direct questioning about activities that people wanted, or firmly planned, to take up) it was possible to distinguish between recreations with great growth potential, and those with relative stability or even actual decline. Sailing and boating, golf and camping, informal countryside and coastal recreation and perhaps fishing belong to the former category. This kind of inference would be criticised as being little better than guesswork, saved from complete incredibility by frequent statements of the obvious.

Since these early days much more advanced mathematical procedures have been brought to bear on the problem of prediction. Whatever their level of sophistication these all suffer from precisely the same logical flaws as the process of simple inference without benefit of formulae. All depend on statements of a relationship between participation and sociodemographic attributes, and all assume that this relationship will prove stable over time. All these assumptions are almost certainly wrong. Empirical work in the USA has suggested that the attributes of, for example, the typical golfer are by no means unchanging: participation spreads to new social segments, or is abandoned by old ones. Nevertheless, planners and recreational providers demand forecasts based on the best available evidence, and it is right that attempts should be made to supply their needs, provided that the results are used cautiously.

Perhaps the simplest of the analytical devices employed in attempts at prediction (and the most closely related to simple inference) is the 'discriminant' procedure adopted in the northern region leisure survey (National Opinion Polls, 1969). In effect, this isolates from the mass of social variance among the following of a particular recreation those few critical characteristics that best 'explain' the incidence of participation in the total population. For example, sex, social class and car ownership are the best 'predictors' of participation in golf: men with cars in social class AB are six times more at risk of succumbing to golf than the population at large. If the number of such men is known in the postulated regional population of 1980, a 'yield' of golfers can be calculated. In short the method divides a future population into a number of categories (based on the predictive indicators) and calculates yields of activity in particular recreations. For many activities only one or two attributes appear efficiently discriminatory (for example, only sex for bowls) and most of the social variance in participation profiles is lost and 'wasted' in this approach to the problem.

To correct this weakness, a different approach to the prediction problem was developed in the northwest recreational survey (North West Sports Council, 1972). For eleven recreational activities, participation levels were tested against fifteen 'profile' factors thought to have a potential explanatory significance. Among these factors were subregion of residence, age, sex, socio-economic group, a number of measures of family composition and access to a car. Four measures of the balance of work and leisure time were included, to test directly the extent to which participation in some (especially time-consuming) recreations responded to variations in the amount of leisure available. Multiple regression equations were developed for each recreation, to relate participation rates to these fifteen variables, or to as many as revealed a substantial influence on participation. In brief, all possible influences on participation were tested and a complex multidimensional relationship built up to describe and 'explain' participation in terms of the attributes of participants. These regression equations can be used directly for prediction. In effect they reveal the degree of likelihood that an individual with particular attributes will swim, or fish or take day trips in either the present or a future population. The fishing equation shows that a thirty-year-old male who drives has a 12 per cent chance of being an angler: every 100 such men – in today's population or that of 1981 – will produce twelve anglers. Given that these relationships remain stable over time (and they will not, perfectly) and that the distribution of the relevant attributes can be predicted in the 1981 population (which can only be approximated) a predictive tool is available. This multiple

Table 24.1 Predicted increase in the number of participants, 1967–80, northern region

Activity	*Percentage participation rate, 1967*	*Percentage change in number of participants, 1967–80*	*Number of additional participants, 1980* (000s)
Golf	3.02	74	58
Country trips	35.37	42	389
Camping–caravanning	4.17	64	69
Coastal trips	38.87	23	233
Water activities	7.20	20	37
Walking	44.77	2	22
'Large pitch' games	12.92	2	7
Swimming	17.77	0	0
Cycling	6.55	3	5
Tennis	5.75	–3	–5
Cricket	5.07	–1	–1

Data refer to 'normal' participation levels (percentage participating over three months before interview).

Source: National Opinion Polls, 1969.

regression approach to prediction is still the subject of continuing research, sponsored by the Sports Council, and a report, as a supplement to *Leisure in the North West*, is forthcoming.

In general the results of the attempts at mathematical prediction have confirmed, but expressed more precisely, the overall conclusions based on inference. Results of the northern region survey demonstrated the intense pressure likely to fall on the coast, the countryside, water spaces and the rural fringes of urban areas (Table 24.1). They indicate also the weakening relative position of many mass activities that absorbed large quantities of the total recreational demand into relatively compact, but intensively used urban spaces. The implications for land use policy are evident. However it must be stressed that many well-known and established trends in recreational change do not emerge from these data – the rise of squash and badminton, the immense growth of indoor five-a-side football and of other activities provided in multisports complexes. The vigorous development of the newer indoor sports is reducing the demand for recreational land, but at the same time generating enormous capital needs on a scale that few local authorities have as yet attempted to meet.

It is questionable whether any analysis of the relationships between social attributes and recreational participation can provide acceptable predictions of future trends. Work of this kind can never reveal causal processes: people do not play golf or camp as an inevitable response to possessing a car. The influence of car ownership is merely permissive, not causal. The conceptual basis of all analytical work of this nature is that a system of constraints operates on the individual to limit the possible range of his recreational behaviour. There is an underlying assumption that everyone would be infinitely active were their choices of recreation not constrained by age, income, mobility, household composition and other allied attributes. This is a negative approach, which does nothing to reveal the direct stimuli that make one man a golfer and another (socially identical) a sailing enthusiast. Little is known about motivation at this *personal* level and no amount of household survey data subjected to refined analysis will remedy this situation. Depth interviewing techniques are now showing how specific recreations meet personal needs (Willmott and Young, 1973) and a small number of sociologists and psychologists have begun to examine the links between work, recreation and personality (Parker, 1972). If predictions are to be made about the recreational future based on a real understanding of stimulus and response, the work of the statistician on social survey data must be underpinned more firmly than hitherto by a conceptual framework derived from sociological and psychological research.

The modelling of recreational demand in a spatial context

One of the areas of recreation research attracting the greatest interest concerns attempts to construct models – either mathematical or graphical in expression – that would relate and integrate the social and the spatial factors that interact to create local patterns of recreational demand, focused on particular centres of provision. It would be beyond the scope or purpose of this review to take up this topic, apart from a brief summary of the elements and objectives of the model-building approach to recreation demand. Indeed, a more elaborate treatment would be premature, for few would claim that this approach has progressed beyond a theoretical and experimental stage.

Such models have taken several forms, simple and complex. As an example of the latter, an attempt has been made to develop a model to explain the use made of the Swiss Cottage swimming pool by its catchment area population (Maw, 1969). The entire zone within which the catchment falls is divided into 100 m grid squares. From a complex of postulates, assumptions and estimates, the number of swimmer users of the centre is calculated for each grid square, and then compared with empirical survey data for a particular week's use of the pool. The logic of the model is explained more clearly than its operation. Grid square populations are estimated, as are their relevant socio-economic characteristics; some broad assumptions about the quantities of available leisure are adopted; and from these estimates an attempt is made to derive a total potential demand for swimming. Whether this total demand will express itself as swimming trips to the Swiss Cottage centre depends on other modifying factors, in effect sieves through which total demand is passed to transform it into expressed demand. The extent to which potential swimmers are aware of the pool raises the complex problem of the perception of the recreational environment. Accessibility, too, modifies the demand situation; as does the competition of rival pools, which will share the total demand in a complex way in which awareness, accessibility, the quality of the pools and peripheral facilities play their roles.

The chief value of generalised models of this type is that they point to gaps in understanding the way in which recreational demand is generated, and satisfied in patterns of movement to specific facilities. For example, it can be postulated that a 'demand surface' exists for any specific recreation. Participation in most activities is so strongly linked to socio-economic attributes that different types of community generate very different levels of demand. The distribution of golfers in the city, for example, as a proportion of local populations, is far from uniform. Given clear statements of the associations between participation levels and social characteristics (for example, the type of regression equations developed from the northwest survey data) it becomes possible to attempt to develop a demand surface, which is essentially a map of demand contours. The 1971 Census data, particularly in their geocoded, grid square presentation, will provide a reasonably adequate range of base data for this sort of work, though they are clearly imperfect, since they contain no information about income. This form of spatial analysis of recreational demand would not only create a more secure platform for model-building, but also provide substantively useful data for the planner and developer.

One of the problems in developing a theoretical framework that would convert recreation research from a conglomerate of *ad hoc* inquiries into a disciplined area of social research is that so little leisure behaviour is normative in the sense that it follows rational laws to produce stereotyped reactions. For example, there is no general application of the principle of distance-minimisation: people do not, in many recreations, use the nearest facility, but perversely seek out others. The northwest leisure study attempted to discover what proportions of participants in a number of sports used the nearest available facility. The results varied (and some are not acceptable for technical reasons), but in no case in the twelve sports analysed was the great bulk of participation at the nearest point of supply: 60 per cent of golfers, and 80 per cent of bowlers, for example, last used facilities that were not the nearest to their homes. Similarly there is little or no tendency towards distance-minimisation (though perhaps a tendency to optimise distance) in the taking of day trips for informal countryside or coastal recreation. Indeed many of these are to multiple destinations, so that the notion of seeking a desirable commodity at the nearest point of supply does not operate. For reasons like these, the recreational trip is much less simple than the journey-to-shop. Creation of a model of recreational behaviour as definitive as the central place model of shopping behaviour cannot be expected. The networks of movement set up by the interaction of spatially varying patterns of demand with a system of supply – the location of accessible recreational facilities and resources – do not, on present evidence, exhibit strongly normative or regular characteristics.

For this reason the application of rather simple economic analogies to the recreational situation would seem to have limited scope. Clawson in the USA, followed by Smith and Kavanagh (1969, see also Smith, 1971) in Britain, has applied the concept of the demand schedule, which relates consumption to price, to the use of recreational facilities. The argument is that the real cost of a recreational experience is a complex sum composed of several elements – the costs of entry, the journey, food and other items. Demand is assumed to be constrained by total cost, although in this total cost travel and entry charges are taken to be critical. The outward decrease with distance in the attraction of a facility is seen therefore as a negative response of demand to a total cost increasing with distance: in short, distance-decay is interpreted in economic terms.

A relationship between total cost (including entry charge) and

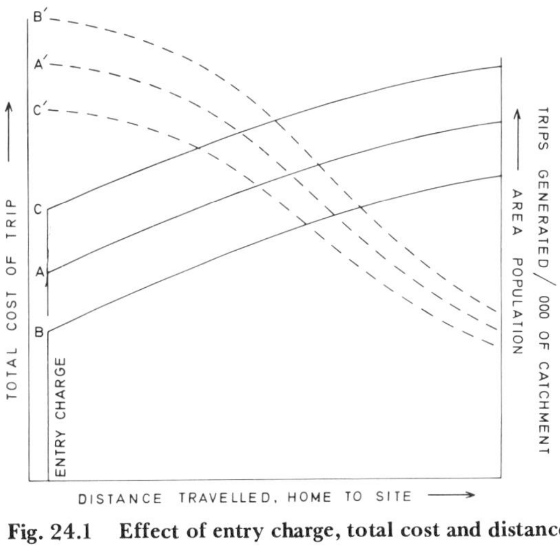

Fig. 24.1 Effect of entry charge, total cost and distance travelled on trip generation rates.

An entry charge at level A produces the total cost curve, rising with distance, as shown: the associated distance-decay effect on trip-generation rates follows the dotted line from A^1. The form shown is typical, with rapid decay over the middle distance ranges. If entry charges are reduced to B or raised to C trip-generation rates are increased or decreased as lines B^1 and C^1 indicate. The proportionality effect of altering entry charges is such as to produce a convergence of the trip-generation rates with distance. In a quantified example the size of the vertical increments in trip-generation rates (A^1–B^1: A^1–C^1) could be read by regarding any change in entry charge as a cost-distance variation on curve A, with its consequent and measurable effect on trip generation along curve A^1, referred to the right-hand scale

distance travelled is illustrated in Fig. 24.1, showing that the trip-generation rate per 1 000 of the population decreases with distance-cost. If total cost constrains demand then the effect on participation rates of varying the entry charge can be read easily from the diagram. The trip-generation curve is lifted or depressed by an increase or decrease, for people are 'brought nearer', or 'moved further' from the facility by the price change. Here is a device for investigating the effect of pricing policy on the volume of use of a centre. A research study commissioned by the Sports Council has used this approach to examine management policies for southern swimming pools. Its success depends on whether journey cost is seen by users as a proxy for admission charge. This may be the case in relation to facilities like pools, but it cannot be so where there is no distance minimisation principle in the generation of trips, for example those for informal recreation in the countryside. Another application of the 'Clawson curve' is to cost-benefit studies. It is argued that if, for example, 20 per cent of the users of a country park are prepared to travel perhaps 50 km at a travel cost of £1 per head a notional price is put on a free recreational experience. Those who travel a shorter distance at lower cost therefore enjoy a consumer surplus, calculable in total, which can be regarded as a cash benefit that flows from the creation of the facility, and as an offset to capital invested. Whatever its hazards, this is an interesting approach to social accounting in assessing the value of recreational investment. All such arguments, however, are based on the concept of normative response, and people are at their least rational in their recreational behaviour.

The journey-to-recreation and the nature of catchment areas

Studies of recreational trips and the catchments of sites and facilities suggest that there is much irrationality and irregularity in the movements of people at leisure, though the larger the scale of aggregation the more 'normal' the tendencies. The work of Duffield and Owen (1971) is particularly valuable in relating a demand system to an environment of supply through a study of trip origins and destinations. What has emerged from this and other work is that there are two types of trip-systems in recreation: those minimising and those that tend to optimise journey length. In general, the latter are the rural and the former the urban recreations: equally the latter concern informal and the former more specific, purposive recreations like active sports.

Thus catchments vary in nature. Some show a regular distance-decay effect, with diminishing proportions of users from increasing distance ranges, often in steeply exponential form. These are chiefly the specific, developed facilities: the aggregate catchments of five indoor sports

Table 24.2 Distance travelled to five indoor sports centres

	Distance (km)						
	<1.6	*1.6–3.1*	*3.2–4.7*	*4.8–6.3*	*6.4–7.9*	*8.0–16.0*	*>16.0*
Percentage of total trips	35	24	11	0	3	8	10

Source: Birch, 1971.

centres (Table 24.2) illustrate the point (Birch, 1971). Other catchments of some countryside locations tend towards a peaked distribution of trip lengths somewhere in the middle of the distance range and not immediately explicable in terms of the distribution of potential visitor population. The data for Cannock Chase are not untypical (Burton, 1974), but not all countryside 'honeypots' follow the same pattern. Some reveal strong and simple distance-decay characteristics. The difference is probably hierarchical: the Cannock Chase pattern of an 'optimal' trip length is that of a regional facility, but rural sites with merely local significance generate a distance-minimising pattern of use.

Table 24.3 Journey times to Cannock Chase

	Time (minutes)				
	<15	*16–30*	*31–45*	*46–60*	*>60*
Percentage of total trips	14	31	22	12	21

Source: Burton, 1974.

Given these contrasts, attempts to develop demand-supply-trip-generation models on simple principles (for example, the gravity model) are not likely to reflect reality. Approaches that embody an hierarchical principle offer better prospects: certainly the use pattern of open space in London is hierarchical in nature. The little that is known about recreational catchments suggests that they are oddly irregular and distorted, much more so than their shopping centre equivalents. The catchments of Liverpool parks are intensely eccentric: some draw high proportions of visitors from limited sectors of their potential catchments, with other sectors – of 180° in extent in extreme cases – quite negative (Balmer, 1972). Birch found the same was true of indoor sports centres: of the five studied two (Bracknell and Harlow) had regular catchments in which areas contributed to the use of the centre roughly in proportion to distance and population. Two others were most eccentric (Stockton and Newcastle), with large negative sectors in the catchments. This was explicable at Newcastle, where the Tyne intervenes to deprive the centre of half its potential service area, but not at Stockton whose catchment has little penetration to the east and so fails to serve the population of some of the fastest developing parts of Teesside. The fifth centre, at Port Talbot, had an almost axial east–west catchment, which failed to extend into even the lower sections of the mining valleys to the north. The Rickmansworth Aquadrome, one of the major water recreation facilities of the London region, has a similarly distorted catchment which illustrates a principle that probably has a wide application in the case of major conurbation-fringe facilities. Its catchment is essentially sectorform: the great bulk of its usage is from a narrow sector reaching from inner London through the northwest suburbs, broadening to include towns like Watford and St Albans on the northwest margin of the city region. The principle of competition also applies: there is virtually no penetration of the catchment area to the south, where the Thames offers similar facilities.

Conclusions

A pessimistic appraisal of the last decade's work on the nature of recreational demand would be that it had mapped out more precisely the areas of ignorance but had achieved little more. It has defined problems but discovered few solutions. It has neither built an acceptable theory of recreational behaviour nor yielded empirical conclusions that the planner and the recreation industry can accept with confidence. At best, a few broad approximations to the major trends in demand are available, backed by some crude generalisations about the influences that shape them. More success has been achieved in describing recreations themselves – there is an enormous body of data on the nature of activities – than in understanding the motives of the people who practise them. For this latter task the household survey, the principal tool of inquiry to date, is a very blunt instrument, and the more refined techniques of the sociologist and psychologist are required for illumination and guidance.

Some of the major research initiatives of the immediate future therefore lie outside the field of the geographer yet; there is still much for him to do, even in demand studies, quite apart from his central role in the evaluation of recreational resources and the elucidation of problems of capacity. The mapping of demand surfaces, the exploration of ways in which the recreational environment is perceived, and the analysis of networks of movement by which the demand surface is related to both the real and perceived systems of supply are among the most promising areas of inquiry. In the broadest terms, recreation research neatly illustrates the constant dilemma of the social sciences, that it is impossible to build behaviour theories until observational methods have assembled initial bodies of empirical data. Until a body of theory begins to emerge, observation, measurement and analysis lack direction and intellectual purpose. In recreation research this dilemma is only now being resolved, in such a fashion that more purposeful empirical work and theory building can reinforce each other.

References

BALMER, K. (1972) 'The use of open space in Liverpool', unpublished PhD thesis, University of Liverpool.

BIRCH, J. G. (1971) *Indoor Sports Centres*, Sports Council, London.

BRITISH TRAVEL ASSOCIATION and UNIVERSITY OF KEELE (1967, 1969) *Pilot National Recreation Survey: reports*, Nos 1 and 2, London.

BURTON, R. C. J. (1974) *The Recreational Carrying Capacity of the Countryside*, Keele University Library.

BURTON, T. L. ed. (1971) *Experiments in Recreation Research*, Allen & Unwin.

BURTON, T. L. and WIBBERLEY, G. P. (1968) *Outdoor Recreation in the British Countryside*, Wye College.

COUNTRYSIDE COMMISSION (1971) *Research Priorities*, London.

DOWER, M. (1965) *Fourth Wave: the challenge of leisure*, Civic Trust, London.

DUFFIELD, B. S. and OWEN, M. L. (1971) *Leisure plus Countryside Equals: a geographical appraisal of countryside recreation*, 2 vols, University of Edinburgh, Department of Geography.

HOUSE OF LORDS (1973) *Second Report from the Select Committee on Sport and Leisure*, London.

MAW, R. (1969) 'Construction of a leisure model', *Off. Archit. Plann.*, **42**, 35–57.

NATIONAL OPINION POLLS (1969) *Outdoor Leisure Activities in the Northern Region*, North Regional Planning Committee, Newcastle upon Tyne.

NORTH WEST SPORTS COUNCIL (1972) *Leisure in the North West*, Manchester.

OUTDOOR RECREATION RESOURCES REVIEW COMMISSION (1962) *Outdoor Recreation for America*, Washington.

PARKER, S. R. (1972) *The Future of Work and Leisure*, Paladin.

SILLITOE, K. K. (1969) *Planning for Leisure*, London.

SMITH, R. J. (1971) 'The evaluation of recreational benefits: the Clawson method in practice', *Urban Stud.*, **8**, 89–102.

SMITH, R. J. and KAVANAGH, N. J. (1969) 'The measurement of benefits of trout fishing', *J. Leis. Res.*, **1**, 316–32.

SPORTS COUNCIL (1971) *Research Priorities for Sports Provision*, London.

WILLMOTT, P. and YOUNG, M. (1973) *The Symmetrical Family*, Routledge & Kegan Paul.

A. D. Walton

25 Air photograph interpretation and the Keele University collection

The arrival at a disused brickworks at Madeley near Keele of consignments of air photographs and associated storage equipment from RAF Medmenham, Berkshire, in the summer and autumn of 1962 was not S. H. Beaver's first encounter with such items. It represented, however, the culmination of many years of endeavour on his part to ensure the safekeeping of the unique aerial record of the landscapes of wartime Europe. As early as 1944 he had initiated the acquisition of prints by the Ministry of Town and Country Planning, a collection which was to become the extensive cover of England and Wales constituting the National Air Photograph Registry of the Department of the Environment. Beaver thus had a working knowledge of the kind of operation which was then being set in motion.

In 1946, D. L. Linton, following a long war experience in the interpretation of air photographs, had begun to explore the possibilities of using photographs from Air Ministry sources in geographical studies and training. Early in 1947 he had reported his findings to a conference of heads of university departments of geography, and during the summer of that year a short course in air photograph interpretation was organised by S. H. Beaver and R. R. Rawson in the Joint School of Geography of King's College and the London School of Economics. This was presented for the first time in Spring Term 1949 to students graduating in the summer of 1950.

Meanwhile a House of Commons question by Sir Edward Keeling elicited from Mr Ivor Thomas the answer that the Royal Air Force wartime photographs were being preserved while the best arrangements for their future use were under review with all concerned. A further consideration at this time was a request to the Air Ministry, concerning the establishment of a National Air Photograph Library. This was not to be resolved for many years.

A. D. Walton

Origins of the photographs at Keele

The collection of exclusively vertical line-overlap photographs at Keele is a product of the 'fly higher – fly faster' philosophy of wartime British aerial reconnaissance. This was initiated and developed between September 1939 and June 1940 by the late S. H. Cotton, an Australian pilot, and his colleagues in what was known variously as the Photographic Development Unit, No. 2 Camouflage Unit and the Heston High Level Flight.

The Royal Air Force entered the war with a concept of reconnaissance involving the use of the Westland Lysander for short range work of a tactical nature and the Bristol Blenheim for longer range strategic flights at heights of 3 250 m to 4 900 m. Neither aircraft was suited to its task, especially in the face of the excellent fighters which had been developed and were being delivered to the Luftwaffe. Even when the cameras functioned efficiently, which, because of condensation problems, rarely occurred at the Blenheim's operating height, the casualty rates on reconnaissance flights were prohibitive and the results unsatisfactory. During this early stage the Heston Flight had two Spitfire aircraft, specially adapted for reconnaissance work. They proved to be appreciably faster than the standard version and, flying at altitudes of 9 800 m and above, the duck-egg green aircraft were virtually invisible against the background of sky and cloud. They were able to obtain essential clandestine cover of politically sensitive areas and to penetrate German air space to photograph well-defended targets such as Wilhelmshaven and the enemy arsenal in the Ruhr. Good cover of the German positions in the Siegfried Line was also gained. Operations were conducted in the winter of 1939–40 from Heston, Seclin airfield near Lille, Chalerange near the Luxembourg border and Coulommiers near Paris.

Originally the Spitfires were fitted with the F.24 camera carrying a lens of 5 inch focal length and a magazine sufficient for 250 exposures on a 5 x 5 inch format. The negative and contact print scales were very small, of the order of 1 : 79 200 from 10 000 m or 1 : 72 000 from 9 800 m. They were, however, capable of interpretation by skilled personnel. In an attempt to improve the scale a civil aerial survey camera was borrowed, and the immediate results were significantly better, with a negative scale of 1 : 18 000 from a height of 9 800 m. During 1940 the F.52 camera was developed and brought into service to provide a 9 x 7 inch format with 500 exposures in the magazine. The range of lenses available eventually included focal lengths of 14, 20 and 36 inches. The last, which was to become the standard reconnaissance lens by 1941, was capable of producing, from 9 800 m, negatives at scales of 1 : 10 000, slightly larger than the OS 6 inch map. On 26 June 1940 the Heston Flight returned from their French bases, having obtained good quality cover, restricted only by weather conditions, of German and Italian targets. Much cover from these early days of the war has survived at Keele.

The unit was taken over by the Royal Air Force, renamed No. 1 Photographic Reconnaissance Unit, and moved from Heston to Benson in Oxfordshire in late 1940. In September 1941 the Spitfire was joined by the de Havilland Mosquito, which increased the range and flexibility of the reconnaissance force. Its use brought more of Norway, the German shore of the Baltic as far east as East Prussia, Berlin, the important sites at Peenemunde and Zinnowitz, and, indeed, the whole of north Germany, within comfortable range.

By April 1944 the Photographic Reconnaissance Unit at Benson had grown to comprise five squadrons, with detachments at Wick, Leuchars and St Eval. Missions covered the coasts of Europe from Norway to the Spanish frontier and extended deep inland and into the Baltic. The work of this unit (now termed 106 Group, Coastal Command) was augmented by reconnaissance carried out by squadrons of the Second Tactical Air Force, providing large-scale cover over areas rather than point targets. Cover of the Mediterranean lands of Europe is extensive and often of very high quality. There is rather more cover at 1:15 000 than at 1:10 000 scale, being flown with cameras of 20 inch rather than 36 inch focal length. Many sorties consist of a series of point targets such as ports, airfields, marshalling yards, junctions and bridges. There is much coastal and island cover especially in the Adriatic and the Aegean Seas, but only scattered inland cover of Greece and Yugoslavia. Linear searches along railways are common and occasionally total cover of a limited area exists.

The aircraft carried usually a split vertical pair of cameras, F.21 on the port side of the flight line and F.22 on the starboard. Sometimes, as in the case of the Benson reconnaissance on the morning after the attack of the Mohne and Eder Dams, a single vertical camera was mounted. On many other occasions an additional third vertical camera of short (6 inch) focal length was carried in addition to the main cameras. This produced simultaneous small-scale cover which has often proved to have qualities not possessed by the more detailed large-scale cover from the main cameras. The small-scale cover shows the broader landscape features, while the large-scale specific details. The ultimate fate of all the wartime oblique photography is not known. Prints of certain spectacular and significant events are not in the Keele collection, possibly due to unauthorised 'souvenir hunting', or to a policy of dispersal to other archives, such as the Imperial War Museum.

Custody of the library

No further action regarding the establishment of a library for these wartime photographs was taken until 1957. In September of that year, on the initiative of the Geography Department at Keele, the matter was raised again with the Government and S. H. Beaver took the lead in the ensuing long and complicated discussions between the Air Ministry and the universities as a whole, and between individual universities interested in acquiring the collection. In June 1958 an Air Ministry invitation was issued to senior university representatives to inspect the former Allied Central Interpretation Unit print library at RAF Medmenham and the nearby negative store at RAF Benson. At that time the collection was estimated to comprise 18 million prints. Five topics of discussion emerged which involved problems of security, international diplomacy, sorting the prints, the condition of the film backing and whether the collection should be dispersed among universities or retained as a unit in one university. The chief contributors to this discussion were Professors Beaver, Linton, Odell and Peel and Dr J. K. St Joseph. A decision was taken to retain the prints in one unit and to house them at Keele in temporary accommodation rented by the university at Madeley. The negatives, whose storage would have been costly and abortive, were eliminated from further consideration.

Another period of inactivity followed until July 1960 when the Air Ministry, acknowledging the value of the collection, stated that a decision on its disposal would soon be taken. The Air Council, meeting in September 1960, rejected the proposal for the establishment of a National Air Photograph Library. As an immediate result of this decision the photographs of England and Wales were to be transferred to the Ministry of Housing and Local Government, cover of Scotland to the Department of Health for Scotland and prints of Northern Ireland to the Stormont Department of Finance. With these prints went the onus of providing library facilities not less extensive than those hitherto provided by the Air Ministry. There remained for disposal at RAF Medmenham the prints of overseas territories, three-quarters of which were wartime cover of enemy occupied Europe. The Air Ministry view was that they had no requirement for these photographs and that no further expenditure on them was justified.

S. H. Beaver had begun negotiations for financial aid from the University Grants Committee with a view to retaining the accommodation at Madeley and meeting the Air Ministry's proposals and conditions for storage. By August 1961 staff of the Central Reconnaissance Establishment had estimated that the transfer of the Medmenham Library to Keele would take a year to complete. By July 1962 the transfer had begun, being completed in August 1963. It was then clear that the total delivered would fall short of the original estimate of 60 000 sorties of overseas cover. The shortfall was due to the claims on prints of overseas countries, such as independent members of the Commonwealth, to the requirements of national security and to miscalculations from incomplete records at Medmenham. The total reached 5.5 million prints in 40 000 boxes. After a stay of three years in the temporary accommodation at Madeley, the collection was moved between March and May 1966 to the permanent library accommodation provided in the Geography Building at Keele.

The security and diplomatic problems of access to the collection had still to be resolved. Because of the United Kingdom's relationship with some former enemy occupied countries, of which photographs were taken, the vetting process carried out at Medmenham had reduced the geographical extent of the collection so that the eastern limit was just short of the 'Communist bloc' countries. No cover of eastern Europe was released to Keele's custody. The cover now held is limited to countries of western Europe, formerly occupied by Germany, extending from the North Cape to Cape Matapan, including Crete, and from the 'Communist bloc' to Brest. Thanks to the endeavours of S. H. Beaver, the Air Ministry and the Foreign and Commonwealth Office, access to prints over specified, limited areas of West Germany was granted by the Federal German Government in February 1967. This was followed in the same year by similar clearances from the governments of Luxembourg, Belgium, Italy, Denmark and the Netherlands. Access to cover of sites in France, Greece, Norway and Yugoslavia is available only after representation to the respective government. The contents of the library remain the property of the Ministry of Defence and the Public Record Office, being on permanent loan. The university is custodian of the library and the Geography Building has been recognised by the Lord Chancellor's Office as a proper repository for this kind of national archive. The contents of the library are subject to the provisions of the Official Secrets Act and the Public Record Acts.

Organisation of the library

The material transferred to Keele consists of boxes of prints grouped by sorties and numbered, together with their associated sortie plots. Unfortunately the wartime cover traces were not delivered to Keele from RAF Brampton; without these transparent overlays showing the outlines and reference numbers of sorties and used with small scale base maps, the precise location of cover is made difficult and time-consuming. The sorties are arranged in strict numerical sequence, corresponding closely to date sequence in the case of the northwest European theatre of operations, the exception being cover of the Low Countries from the last quarter of 1944. The cover of the north coasts of the Mediterranean is separate, housed in number sequence but not in chronological order.

To convert the collection into a unit more readily suited to geographical requirements, the entire cover has been indexed by a system of coordinates. A modified version of that used by the Royal Air Force, the Keele system is based on ten-minute squares of latitude and longitude. Each square is titled uniquely by global quadrants and by the six-figure coordinates of the corner nearest to the origin of the graticule at the intersection of the Equator and the Greenwich Meridian. For each square in which cover is found a card is filled in, recording the one degree square reference of the area, the name of a key point in the square and the full two-letter and six-figure reference of the ten-minute square. Entered on the card in numerical sequence are the library reference numbers of all the sorties which contain cover of any part of the square. The possible existence of cover over a site can be rapidly established from latitude and longitude data to the nearest minute.

Backing this completed card index are the transfer sheets, which show in number sequence for each sortie its library reference number (as indicated on the boxes and the sortie plots), the sortie number (as shown on the plots and print titling), the date of the sortie, its scale group and its location by one degree square of latitude and longitude. The sortie plots themselves contain the same data, together with the mean time of photography, the flying height, the focal lengths of the cameras used, notional scales from flight data and the positions on the base map of flight lines of photography and selected prints. Thus photography referring to particular events can be located easily in time as well as space. The original master sortie plots are usually on colour-printed map extracts but in some cases they are on duplicated flimsy prints referred by grid line intersections and by orientation to specified sheets of the Geographical Section, General Staff wartime maps.

Use of the library

The library facilities first became available with the move into the new Geography Building at Keele in 1966 (Walton, 1969). During the following eight years the chief users of this photographic 'Domesday of Europe 1939–45' have been archaeologists. The collection has been used to almost the same extent by modern historians, whose interest has mounted as the memory of the war has receded and as the respectability of studies of wartime activity has increased. City archivists, state agencies, air attachés at European embassies, legal advisers, authors, publishers and explosive ordnance disposal units have all consulted cover. Geographers have not been well represented either in the number of requests or in the number of successful searches, as in many cases their areas of research lie beyond the scope of the library's cover. For example, parts of Norway which are now of interest to geomorphologists and glaciologists were never of military concern during the war and consequently were not photographed. The use of the library as a source of precise evidence of the state of specific sites thirty or more years ago to provide additional datum points in the detection of change and in monitoring rates of change has been too infrequent. Indeed, little interest has yet been shown in the cover of landscapes of western and Mediterranean Europe as they existed before the large scale modification that occurred between 1943 and the end of the war. The existence of cover of these areas at scales of between 1:10 000 and 1:15 000 provides a unique source for the reconstruction of such landscapes.

No coast in the world has been so closely or so frequently photographed as that of occupied Europe between 1940 and 1945, yet the study of the varied coastal landforms recorded at that time has attracted little attention. Great changes have taken place in the cultural landscape of this part of Europe since the war, as for example in the programme of *remembrement* in France. Extensive cover of northern France, Belgium, Holland and Germany illustrates these landscapes as they were in the 1940s. In both France and Germany little aerial cover is available before 1950. At Keele the limit can be pushed back seven or eight years in many cases. Similarly, little interest has been shown in the cover of village morphology and rural settlement in general. There is a wealth of detail (at scales of 1:10 000 and larger) in the cover for transport studies, ranging from civil engineering structures on railways to now defunct equipment, such as locomotive depots, associated with steam traction. In addition the pioneer motorway networks built in Germany and Italy in the 1930s are well represented.

Conclusions

The two major tasks of plotting cover by ground position and by date have now been accomplished and the services of the library are now fully in operation. Cover accurate to within eight hours of a particular event and accurate to the location of a single house in a Rhineland town can be produced. What has not yet been attempted is a detailed search for fine quality cover of examples of particular features of physical and human geography. So far, identification of such features has only been incidental to other searches, but it is intended eventually to compile an index of cover of such topical features. Although this collection of wartime prints forms the core of the air photograph library at Keele, cover of other regions continues to be added. In all, the collection at Keele constitutes an invaluable source of visual information on western and Mediterranean Europe of the 1940s.

Reference

WALTON, A. D. (1969) 'Air photograph library', *Area*, 1, 31.

Notes on contributors

R. P. BRADSHAW	Undergraduate and postgraduate student, Keele University,* 1962–69; temporary lecturer in geography, Keele University, 1970–71; now lecturer in geography, Nottingham University.
B. T. BUNTING	Demonstrator in geography, Keele University, 1954–55; now associate professor of geography, McMaster University, Ontario, Canada.
ROSEMARY BURTON	Demonstrator and research fellow in geography, Keele University, 1968–72.
R. A. BUTLIN	Demonstrator in geography, Keele University, 1961–62; now senior lecturer in geography, Queen Mary College London.
L. M. CANTOR	Lecturer and senior lecturer in education, Keele University, 1957–69; now professor of education, Loughborough University of Technology.
MONICA M. COLE	Lecturer and senior lecturer in geography, Keele University, 1951–64; now professor of geography, Bedford College London.
G. COLEMAN	Demonstrator in geography, Keele University, 1968–70; now lecturer in development studies, University of East Anglia.
E. DERBYSHIRE	Undergraduate, Keele University 1950–54; postgraduate student and demonstrator in geography, Keele University, 1956–60; lecturer, senior lecturer and now reader in geography, Keele University, 1966–
A. M. D. GEMMELL	Demonstrator in geography, Keele University, 1971–73; now lecturer in geography, Aberdeen University.

* University College of North Staffordshire until 1962; from then Keele University.

D. T. HERBERT	Postgraduate student, demonstrator and lecturer in geography, Keele University, 1960–65; now senior lecturer in geography, University College of Wales, Swansea.
P. T. KIVELL	Postgraduate student, Keele University, 1967–70; now lecturer in geography, Keele University, 1972– .
G. R. P. LAWRENCE	Demonstrator in geography, Keele University, 1956–58; now lecturer in geography, King's College London.
A. MOYES	Demonstrator in geography, Keele University, 1965–68; now lecturer in geography, University College of Wales, Aberystwyth.
J. NAYLON	Lecturer and now senior lecturer in geography, Keele University, 1957– .
L. W. F. PAGE	Lecturer in geography, Keele University, 1965– .
A. D. M. PHILLIPS	Lecturer in geography, Keele University, 1968– .
M. J. PULLEN	Undergraduate, Keele University, 1956–60; now senior lecturer in economics, Keele University.
H. B. RODGERS	Lecturer, senior lecturer, reader and professor of social geography, Keele University, 1964–71; now professor of geography, Manchester University.
M. E. C. SANT	Undergraduate, Keele University, 1960–64; now director of East Anglian studies, University of East Anglia.
ELIZABETH M. SHAW	Research assistant in meteorology, Keele University, 1960–62; now lecturer in civil engineering, Imperial College London.
HEATHER M. TINSLEY	Demonstrator in geography, Keele University, 1972–74.
B. J. TURTON	Demonstrator, lecturer and now senior lecturer in geography, Keele University, 1960– .
A. D. WALTON	Lecturer in geography, Keele University, 1954– .
J. R. WALTON	Demonstrator in geography, Keele University, 1972–73; now lecturer in geography, University College of Wales, Aberystwyth.
P. D. WILDE	Undergraduate, Keele University, 1960–64; postgraduate student, Keele University, 1967–70; now lecturer in geography, University of Tasmania.

M. J. WISE	Professor of geography, London School of Economics; succeeded S. H. Beaver to the Sir Ernest Cassel Readership in Economic Geography, London School of Economics, 1950.
P. WORSLEY	Undergraduate, Keele University, 1958–62; now lecturer in geography, Reading University.
E. M. YATES	Lecturer in geography, Keele University, 1951–54; now reader in geography, King's College London.